Lecture Notes in Computer Science 11204

Commenced Publication in 1973
Founding and Former Series Editors:
Gerhard Goos, Juris Hartmanis, and Jan van Leeuwen

More information about this series at http://www.springer.com/series/7407

Shijian Li (Ed.)

Green, Pervasive, and Cloud Computing

13th International Conference, GPC 2018
Hangzhou, China, May 11–13, 2018
Revised Selected Papers

 Springer

Editor
Shijian Li
Zhejiang University
Hangzhou, China

ISSN 0302-9743   ISSN 1611-3349 (electronic)
Lecture Notes in Computer Science
ISBN 978-3-030-15092-1   ISBN 978-3-030-15093-8 (eBook)
https://doi.org/10.1007/978-3-030-15093-8

Library of Congress Control Number: 2019934095

LNCS Sublibrary: SL1 – Theoretical Computer Science and General Issues

This Springer imprint is published by the registered company Springer Nature Switzerland AG
The registered company address is: Gewerbestrasse 11, 6330 Cham, Switzerland

Preface

On behalf of the Organizing Committee, it is our pleasure to welcome you to the proceedings of the 13th International Conference on Green, Pervasive, and Cloud Computing (GPC 2018), held in Hang'zhou, China, during May 11–13, 2018. The conference was organized by Zhejiang University City College, China. GPC aims at bringing together international researchers and practitioners from both academia and industry who are working in the areas of green computing, pervasive computing, and cloud computing. GPC 2018 was the next event in a series of highly successful events focusing on pervasive and environmentally sustainable computing. In the past 10 years, the GPC conference has been successfully held and organized all over the world: Taichung, Taiwan (2006), Paris, France (2007), Kunming, China (2008), Geneva, Switzerland (2009), Hualien, Taiwan (2010), Oulu, Finland (2011), Hong Kong (2012), Seoul, Korea (2013), Wuhan, China (2014), Plantation Island, Fiji (2015), Xian, China (2016), and Cetara, Italy (2017).

This year the value, breadth, and depth of the GPC conference continued to strengthen and grow in importance for both the academic and industrial communities. The strength was evidenced this year by having a number of high-quality submissions resulting in a highly selective program. GPC 2018 received 101 submissions on various aspects including green computing, cloud computing, pervasive sensing, data and knowledge mining, and network security. All submissions received at least three reviews via a high-quality review process involving 48 Program Committee members and a number of additional reviewers. The review and discussion were held electronically using EasyChair. On the basis of the review results, 35 papers were selected for presentation at the conference, giving an acceptance rate lower than 35%. In addition, the conference also featured seven invited talks given by Sajal K. Das, Hai Jin, Jiannong Cao, Minyi Guo, Daqing Zhang, Yong Lian, and Jukka Riekki. We sincerely thank all the chairs, Steering Committee members, and Program Committee members. Without their hard work, the success of GPC 2018 would not have been possible. Last but certainly not least, our thanks go to all authors who submitted papers and to all the attendees. We hope you enjoy the conference proceedings!

May 2018

Ling Chen
Qing Lv
Shijian Li

Organization

General Chairs

Anind Dey   Carnegie Mellon University, USA
Gang Pan   Zhejiang University, China
Marco Conti   National Research Council of Italy (CNR), Pisa, Italy

Technical Program Chairs

Ling Chen   Zhejiang University, China
Qin Lv   University of Colorado Boulder, USA

Steering Committee

Hai Jin (Chair)   Huazhou University of Science and Technology, China
Nabil Abdennadher University of Applied Science and Arts
                     Western Switzerland, Switzerland
Christophe Cerin   University of Paris XIII, France
Sajal K. Das   Missouri University of Science and Technology, USA
Jean-Luc   University of California – Irvine, USA
Kuan-Ching Li   Providence University, Taiwan, China
Cho-Li Wang   University of Hong Kong, SAR, China
Chao-Tung Yang   Tunghai University, Taiwan, China
Laurence T. Yang   St. Francis Xavier University, Canada

Special Session Committee

Bin Guo   Northwestern Polytechnical University, China
Zhiyong Yu   Fuzhou University, China

Local Organization Chairs

Hui Yan   Zhejiang University City College, China
Guanlin Chen   Zhejiang University City College, China

Publication Chairs

Shijian Li   Zhejiang University, China
Lin Sun   Zhejiang University City College, China

Technical Program Committee

Saadatm Alhashmi	University of Sharjah, United Arab Emirates
Zeyar Aung	Masdar Institute of Science and Technology, United Arab Emirates
Jorge G. Barbosa	University of Porto, Portugal
Ling Chen	Zhejiang University, China
Xiaofeng Chen	Xidian University, China
Xiaowen Chu	Hong Kong Baptist University, SAR China
Raphaël Couturier	University of Burgundy – Franche - Comté, France
Changyu Dong	Newcastle University, UK
Jean-Philippe Georges	University of Lorraine, France
Dan Grigoras	University College Cork, Ireland
Jinguang Han	University of Surrey, UK
Fu-Han Hsu	National Central University, Taiwan
Kuo-Chan Huang	National Taichung University of Education, Taiwan
Dan Johnsson	Umeå University, Sweden
Erisa Karafili	Imperial College London, UK
Helen Karatza	Aristotle University of Thessaloniki, Greece
Dong Seong Kim	University of Canterbury, New Zealand
Ah Lian Kor	Leeds Beckett University, UK
Cheng-Chi Lee	Fu Jen Catholic University, Taiwan
Chen Liu	Clarkson University, USA
Jaime Lloret	Polytechnic University of Valencia, Spain
Jiqiang Lu	Institute for Infocomm Research, Singapore
Rongxing Lu	University of New Brunswick, Canada
Xiapu Luo	The Hong Kong Polytechnic University, SAR China
Mingqi Lv	Zhejiang University of Technology, China
Victor Malyshkin	Russian Academy of Sciences, Russia
Daniele Manini	University of Turin, Italy
Mario Donato Marino	Leeds Beckett University, UK
Fabio Mercorio	University of Milano Bicocca, Italy
Alessio Merlo	University of Genoa, Italy
Alfredo Navarra	University of Perugia, Italy
Marek Ogiela	AGH University of Science and Technology, Poland
Ronald H. Perrott	Oxford e-Research Centre, UK
Dana Petcu	West University of Timisoara, Romania
Florin Pop	University Politehnica of Bucharest, Romania
Kewei Sha	University of Houston—Clear Lake, USA
Roopak Sinha	Auckland University of Technology, New Zealand
Fei Song	Beijing Jiaotong University, China
Pradip Srimani	Clemson University, USA
Ruppa Thulasiram	University of Manitoba, Canada
Simon Tjoa	St. Poelten University of Applied Sciences, Austria
Marcello Trovati	Edge Hill University, UK
Ding Wang	Peking University, China

Lizhe Wang   Chinese Academy of Sciences, China
Wei Wang   The University of Texas at San Antonio, USA
Yu Wang   Deakin University, Australia
Meng Yu   The University of Texas at San Antonio, USA
Yanmin Zhu   Shanghai Jiao Tong University, China

Contents

Pervasive Application

Data Mining and Knowledge Mining

Posters

Network, Security
and Privacy-Preserving

A Complex Attacks Recognition Method in Wireless Intrusion Detection System

Guanlin Chen[1,2], Ying Wu[1,2], Kunlong Zhou[1], and Yong Zhang[1(✉)]

[1] School of Computer and Computing Science, Zhejiang University City College, Hangzhou 310015, People's Republic of China
zhangyong@zucc.edu.cn
[2] College of Computer Science and Technology, Zhejiang University, Hangzhou 310027, People's Republic of China

Abstract. During recent years, the challenge faced by wireless network security is getting severe with the rapid development of internet. However, due to the defects of wireless communication protocol and difference among wired networks, the existing intrusion prevention systems are seldom involved. This paper proposed a method of identifying complicated multistep attacks orienting to wireless intrusion detection system, which includes the submodules of alarm simplification, VTG generator, LAG generator, attack signature database, attack path resolver and complex attack evaluation. By means of introducing logic attack diagram and virtual topological graph, the attach path was excavated. The experimental result showed that this identification method is applicable to the real scene of wireless intrusion detection, which plays certain significance to predict attackers' ultimate attack intention.

Keywords: Mobile Internet · Wireless intrusion · Multi-step attack

1 Introduction

In recent years, wireless networks are becoming more and more popular. Wireless Local Area Networks are deployed both in the company and in public or home. A vibrant access point AP is also hugely convenient for people, especially those who use mobile terminals. The number of mobile users has been increasing these years, and the field is continually covering social, games, video, news, finance and so on. More and more users tend to use the mobile to interact, which is the natural advantage of the mobile terminal, but also make it a target of public criticism.

Therefore, how to make up for the loopholes in the mobile Internet as much as possible, and how to detect and prevent a variety of known and unknown intrusion is a critical thing.

This paper proposed a method of identifying complicated multistep attacks orienting to wireless intrusion detection system (MSWIDS), which includes the submodules of alarm simplification, virtual topology graph (VTG) generator, logic attack graph (LAG) generator, attack signature database, attack path resolver and complex attack evaluation. Using introducing logic attack diagram and virtual topological graph, the

S. Li (Ed.): GPC 2018, LNCS 11204, pp. 3–17, 2019.
https://doi.org/10.1007/978-3-030-15093-8_1

attach path was excavated. The experimental result showed that this identification method applies to the real scene of wireless intrusion detection, which plays certain significance to predict attackers' ultimate attack intention.

1.1 Related Works

Intrusion detection as a vital network security technique, which has an extraordinary ability to detect illegal attacks and intrusion and has been widely concerned and applied. After more than 30 years of development, the research of intrusion detection system in three aspects, such as distributed, intelligent and mobility made significant progress. It has become a content-rich, wide area involved synthetically subject.

In 2013, Aparicio-Navarro, Kyriakopoulos and so on proposed a multi-layer WiFi-based attack detection method based on D-S evidence theory [1]. In 2016, Afzal, Rossebø et al. Proposed metrics [2] that can accurately identify de-authentication attacks and evil twin attacks. In 2016, Victor and Carles et al. Studied the anomaly detection method for wireless sensor network (WSN) [3].

Liang and Nannan proposed a multistep attack scenario identification algorithm MARP [4] based on intelligent programming.

In 2015, Shameli-Sendi and Louafi et al. Proposed a defence-centric model based on Attack-Defense Tree (ADT) to assess the damage of complex attacks [5]. There are two types of calculation for the value of a complex attack. One is attack-centric, the other is defence-centric. The former is only for the final target. If the attacker does not reach, the damage value is 0, and the latter considers each At-tack steps. There are two types of ADT node, attack node and defence node, if the parent node is the same type is the refinement relationship, if not the same is the countermeasure relation-ship. Each ADT has a root node as the primary target and extends downward from the root node. If the root node is an attack node, an attacker starts from a leaf node that is the attack node in the tree and arrives at the root node as an attack path. The refinement relationship under a node can be disjunctive or conjunctive, i.e., to satisfy at least one of the true or true at the same time.

In 2016, Bi, Han and Wang proposed a probabilistic algorithm for dynamically generating a top K attack path for each node [6]. An attack path can be a single-step attack or a complex attack. The time complexity of calculating top K for a complete attack graph will be exponential, and the algorithm dynamically generates these paths and eventually forms an incomplete attack graph.

In 2016, Wang and Yuan et al. Proposed MaxASP, a maximal sequential pattern mining algorithm without candidate sequences, which can efficiently obtain the maximum attack sequence [7].

However, with the diversification of attacks on the mobile Internet, existing intrusion detection [8] technologies often fail to detect the attack accurately. Therefore, we also study the wireless intrusion alarm [9] aggregation method for mobile Internet [10] and apply its research in wired networks [11] to the new environment of the mobile Internet.

1.2   Organization of the Paper

The WIACM method this paper presents includes three steps: alert formatting, alert reduction and alert classification. The rest of this paper is organized as follows. Section 2 introduces the general framework of the WIACM method. Section 3 gives this problem decomposition. Section 4 describes the experimental environment and gives the experiment results. We conclude with a summary and directions for future work in Sect. 5.

2   MSWIDS General Designs

The architecture figure of the MSWIDS designed in this paper is shown in Fig. 1 below. The system includes four modules: data acquisition, single-step attack recognition, complex attack recognition and information display interface.

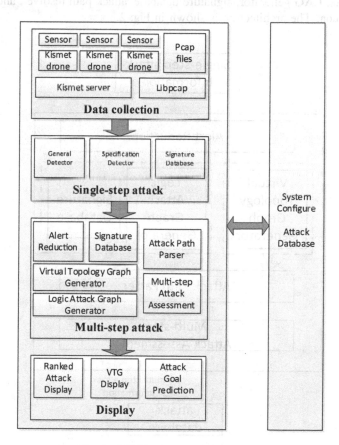

Fig. 1. MSWIDS general framework

Data Collection. Distributed acquisition environment flow, for the subsequent analysis of data sources.

Single-step Attack. Identify a single-step attack, used for subsequent complex attack identification.

Complex Attack. Complex Attack Identification Module: Identify complex attacks and the final intention of the attacker.

Display. Real-time display system identifies the attack process.

3   MSWIDS Detailed Design

The single-step attack identification module generates a series of SAIs that are stored in the Attack Database for analysis by the complex attack recognition module. The complex attack recognition module includes sub-modules such as alarm reduction, VTG generator, LAG generator, signature database, attack path resolver, and complex attack evaluation. The architecture is shown in Fig. 2.

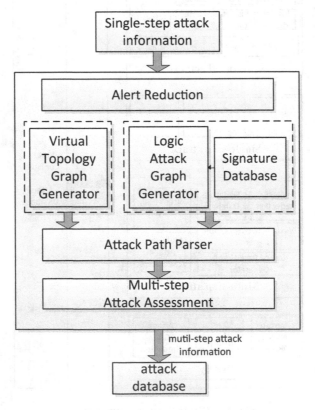

Fig. 2. Complex attack recognition module

The Complex Attack Identification Module corresponds to the alarm association phase of an IDS, while the attack graph used for alarm correlation. In this paper, a complex attack identification module uses a LAG and a VTG to mine the attack path.

There are two main steps in this module. First of all need to pretreatment, remove irrelevant alarms and repeating alarms, streamline the number of alerts. The second step is to sort through the hyper-alert to identify the attacker's easy-to-take attack sequence and predict the attackers' follow-up and final intentions.

When identifying attack sequences, logically there are three categories of alerts. One is Alert Detected (ADE), which belong to some single-step attack; the other is Alert Undetected (AUD), which may be a missing alarm; one is Alert Predicted (APR), the result of the forecast. Also, considering the attacker may make some meaningless attacks during the attack to confuse the real purpose, the first type of alert will be divided into two kinds: Alert Real (AR) and Alert Disturbed (AD).

3.1 LAG Generator

The LAG generation process is shown in Fig. 3 below.

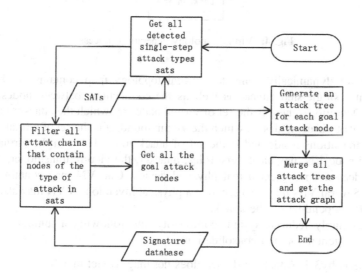

Fig. 3. Logic attack graph generation process

The above dynamically generated attack graph contains all attack chains that have detected single-step attacks and which they may exist. Select the relevant attack chain from the signature database and then merge and generate the attack tree. The reason for attacking the graph is that the full attack graph made by the entire signature database becomes more complicated, which is not conducive to subsequent mining because the size of the signature database increases. A simple attack scene will generate a simple attack map, and a complex attack scene will produce a complex attack map, so more matches.

3.2 VTG Generator

The VTG generation process is shown in Fig. 4 below.

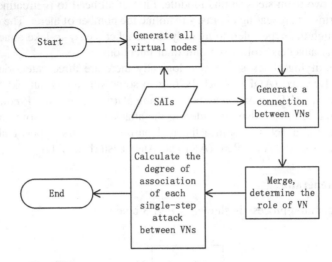

Fig. 4. Virtual topology generation process

The above dynamically generated virtual topology map contains all the nodes involved in a single-step attack, as well as the connections between nodes. Virtual Topology Definition, where is the set of virtual nodes on which the attacker is located, and is the set of virtual nodes on which the victim found. It is the set of virtual nodes on which the transitions reside and is the set of attack traffic between the virtual nodes. The virtual node VN may have three roles: attacker, the victim, transitioner. A virtual node VN does not represent a real physical device. One VN may contain different addresses. Several VNs may also separate a physical invention. A VN may also assume various roles depending on the attack.

For reasonably reduced system complexity, the following assumptions need to make for VN generation and consolidation:

Hypothesis hy3.2: Attacking device does not map as victim VN.
Hypothesis hy3.3: One VN corresponds to a single role associated with an attack.
Hypothesis hy3.4: If a VN has more than one address, it indicates that address spoofing is being used and is considered an attacker.

The address of VN is addr_vn (mac_addr, ip_addr, ssid, vendor). If the mac_addr and ip_addr of the two addresses are at least the same, they are considered as belonging to the same VN. If mac_addr and ip_addr are the same but have different values ssid or vendor, and they also belong to the equal VN.

To determine the role of the VN method:

(1) The address conflict between VNs;
(2) src of VN address that contains SAI belongs to VN_s, including dst belongs to VN_t, including trans belongs to VN_m.

Calculate the corresponding single-step attack correlation between VNs:

Formula (1) For attack A, the degree of association between VN_1 and VN_2 is as follow:

$$assd_A(VN_1, VN_2) = \frac{\sum_{i=1}^{k} SAI_{VN_1->VN_2}(i).AP}{k} \tag{1}$$

Where k is the alert number of the single-step attack alarm set $SAI_{VN_1->VN_2}$ from VN_1 to VN_2.

3.3 Attack Path Analysis

Combined with VTG and LAG attack path analysis, the flow chart shown in Fig. 5 below.

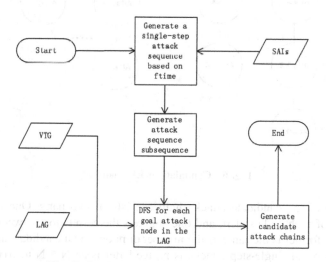

Fig. 5. Logic attack graph generation process

Each path represents an attack chain. In the DFS process, every attack node A passes through an attack path in the VTG. If the attack path from the VTG attacker node, Edge A can form an attack path to the victim node B, then put it corresponding to The SAI information is added to the LAG attacking node to determine whether the condition from the attacking node to the next attacking node meets the state. If satisfied, the recursion is performed and the impact at the LAG side takes effect.

Formula (2) VTG to find a path to attack, calculate the degree of relevance as follows:

$$assd_A(VN_1, VN_2, \ldots, VN_k) = \prod_{i=i}^{k} assd_A(VN_k) \qquad (2)$$

And map it to the corresponding node of the candidate attack chain.

3.4   Complex Attack Assessment

In the previous step, for each target attack node, a set of candidate attack chains will be generated. Suppose one of these attack chains is the attacker's actual attack chain and attack intention. As shown in Fig. 6 below:

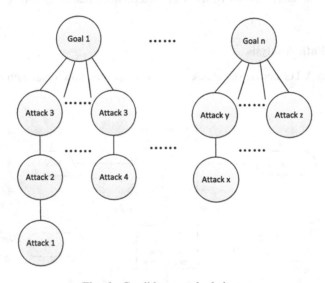

Fig. 6. Candidate attack chain set

The score of the candidate attack chain consists of two parts. One indicates the effectiveness of each node's alert, and one suggests the correlation between alerts. The number of the former is the same as the number of nodes in all candidate attack chains. If the number of all single-step attacks is N, the latter is an N * N matrix.

For each target attack node, calculate the scoring element of each candidate attack chain assc1.

$$\begin{aligned} assc1(Attack\_chain) = \prod_{i=i}^{k} \big(& assd(Attack\_chain(i)) \\ & - averge\_assd(Attack\_chain(i)) + 1 \big) \end{aligned} \qquad (3)$$

averge_assd representing a candidate node is a global attack such attacks node assd average.

The correlation between attack A1 and attack A2 is calculated as follows:

Find the edge sets E1 and E2 of all attack types A1 and A2 in the VTG.

$$asse(e1, e2) = \begin{cases} 0, e1 \text{ and } e2 \text{ have no common endpoint or } e1.ftime > e2.ftime \\ 0.5, e1 \text{ and } e2 \text{ starting with same point but end with different point} \\ 0.5, e1 \text{ and } e2 \text{ starting with same point but end with different point} \\ 1, e1 \text{ and } e2 \text{ have the same endpoint} \end{cases}$$

(4)

$$assa(Attack\_chain(i), Attack\_chain(i+1)) = \frac{\sum\limits_{i=1}^{n}\sum\limits_{j=1}^{m} asse(E2(i), E2(i))}{n * m}$$ (5)

$$assc2(Attack\_chain) = \prod_{i=i}^{k-1} (assa(Attack\_chain(i), Attack\_chain(i+1)))$$ (6)

assc1 is a positive number, assc2 is a number in the range [0, 1]. The total score of the candidate attack chain is:

$$assc(Attack\_chain) = \alpha \ln(assc1(Attack\_chain)) + \beta assc2(Attack\_chain)$$ (7)

This article parameters α, β are 1.

After getting each attack chain score, you sort it by rating for each target attack node.

Then discard the following attack chains and target attack nodes:

(1) target attack node without attack chain;
(2) attack chain score below the threshold of 0.5;
(3) If all candidate attack chains of the target attack node are at a low level, the entire attack chain set is deleted.

Sort all the remaining attack chains, find out the top k target attack nodes and their attack chains with the highest scores, and output them as multi-step attack information (MAI).

The MAI contains the VTG visualization map, sorting attack chain, and attack target prediction, as shown in Fig. 7.

The MSWIDS system displays the current VTG in real time. Through the VTG, users can observe whether the existing network is attacking or not and highlight the position of essential nodes such as the attacker and victim through statistics on traffic between virtual nodes. The sorted attack chain displays all currently possible attack chains in real time, and the attack intent corresponding to one or several attack chains in the front is the current intention of the attacker.

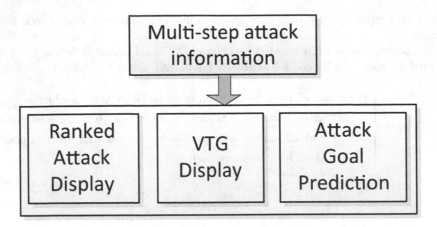

Fig. 7. MAI

4 Experimental Environment and Result Analysis

This experiment uses a win7 desktop, two win8 notebooks and three kali Linux virtual machines on it. Also, three USB cards are required for the attacker and the MSWIDS collector, and the laboratory wireless network environment AP. The experimental equipment and environmental conditions are shown in Tables 1 and 2.

Table 1. Experimental equipment configuration

No.	Equipment	OS	Role
1	Desktop	Win7	victim
2	VM on Desktop	Kali linux	MSWIDS server
3	VM on notebook A	Kali linux	attacker
4	Notebook A	Win8	victim or transition
5	VM on notebook B	Kali linux	attacker or drone
6	Notebook B	Win8	victim or transition
7	Related APs	Linux	victim or transition

Table 2. Lab environment

AP Type	Number
Public certification	6
WEP encryption	2
WPA encryption	7
Hidden SSID	1

The roles of devices 1 to 3 are fixed, and other device roles use different roles according to various attack scenarios.

WEP encryption and hidden SSID are experimental preparation, another AP belongs to the daily environment.

The experimental scenario deployment platform is shown in Fig. 8 below.

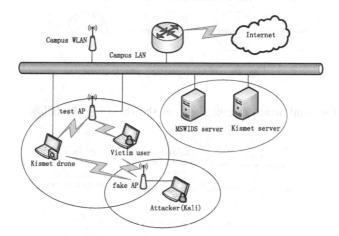

Fig. 8. Experimental scene deployment platform

4.1 De-authentication Attack Scenario

There are four types of attacks involving De-authentication attacks, as follows:

Chain 1: attack intention is the denial of service attack, the attack chain as shown in Fig. 9 below.

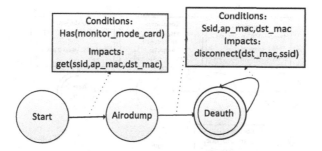

Fig. 9. Dos attack chain

Chain 2: The final attack is MAC Spoofing, as shown in Fig. 10 below.

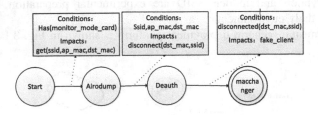

Fig. 10. MAC Spoofing attack chain

Chain 3: The final attack is Evil Twin, as shown in Fig. 11 below.

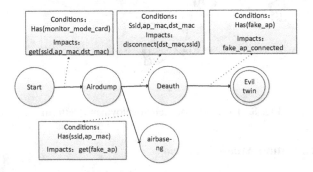

Fig. 11. Evil twin attack chain

Chain 4: The attack intention is Key Crack, as shown in Fig. 12 below.

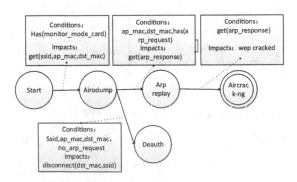

Fig. 12. Key Crack attack chain

4.2 Recognition Result

The attacks on the four kinds of attack intents described in 4.1 are tested respectively. Experiments to identify the attack behavior as valid data, each attack intention of 50 times. The experimental results show that the recognition rates of Chain 1 and Chain 4 are 0.68 and 0.78 respectively, while the recognition rates of Chain 2 and Chain 3 are low, which are 0.42 and 0.36, respectively. Chain 1 flooding attacks have the highest degree of discrimination with other attack chains due to non-stop implementation of authentication-off attacks. Chain 4 generates a large number of ARP packets during the attack, and the Arp replay attacking node is easy to identify. Chain 2 and Chain 3 because of the dominant attack only cancel the certification attack, so although the final intention is different, but not enough distinction between each other. The experimental results are shown in Table 3 below.

Table 3. De-authentication attack scene recognition result

No.	Chain 1	Chain 2	Chain 3	Chain 4	Total attack	Rate
1	34	5	4	7	50	68%
2	3	21	17	9	50	42%
3	5	22	18	5	50	36%
7	2	6	3	39	50	78%

Figure 13 shows the bar distribution of the authentication result of the un-authentication attack scene:

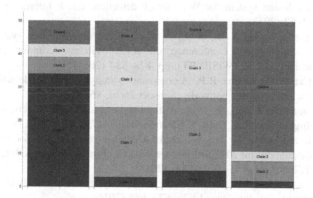

Fig. 13. De-authentication attack scene recognition result

5 Conclusions

Experiments show that using the method proposed in this paper to determine the attackers' intention to attack, one of the critical factors is the discrimination between the possible attack chains. The method will yield better predictions if there is a clear characteristic of an attack committed by the real attack intent. However, if both attacks attempt to use the same or similar attack steps, they will have similar evaluation scores in the final predicted candidate attack chain, and the prediction result may not be suitable for using only one attack intention corresponding to the attack chain. Besides, the MSWIDS system proposed in this paper, from data acquisition, single-step identification to complex attack recognition, the validity of the information generated by each module will affect the processing of subsequent modules. The validity of the system also depends on Algorithm and should include consideration of the possibility of misjudgment in a similar single-step attack identification. In this paper, the intermediate results in each step are described by using the evaluation value rather than the simple boolean measure. The experiment proves that it is of certain significance to predict the attacker's ultimate attack intention.

Acknowledgements. This work was partially supported by Zhejiang Provincial Natural Science Foundation of China (No. LY16F020010), Hangzhou Science & Technology Development Project of China (No. 20150533B16, No. 20162013A08) and the 2016 National Undergraduate Training Programs for Innovation and Entrepreneurship, China (No. 201613021004).

References

1. Aparicio-Navarro, F.J., Kyriakopoulos, K.G., Parish, D.J.: An automatic and self-adaptive multi-layer data fusion system for WiFi attack detection. Int. J. Internet Technol. Secur. Trans. **5**(1), 42–62 (2013)
2. Afzal, Z., Rossebø, J., Talha, B., et al.: A wireless intrusion detection system for 802.11 networks. In: International Conference on IEEE Wireless Communications, Signal Processing and Networking (WiSPNET), pp. 828–834 (2016)
3. Victor, G.F., Carles, G., Helena, R.P.: A comparative study of anomaly detection techniques for smart city wireless sensor networks. Sensors **16**(6), 868 (2016)
4. Liang, H., Nannan, X., Erbuli, N., et al.: A multi-stage attack scenario recognition algorithm based on intelligent planning. Chin. J. Electron. **41**(9), 1753–1759 (2013)
5. Shameli-Sendi, A., Louafi, H., He, W., et al.: A defense-centric model for multi-step attack damage cost evaluation. In: 2015 3rd International Conference on Future Internet of Things and Cloud (FiCloud), pp. 145–149. IEEE (2015)
6. Bi, K., Han, D., Wang, J.: K maximum probability attack paths dynamic generation algorithm. Comput. Sci. Inf. Syst. **13**(2), 677–689 (2016)
7. Wang, Z., Yuan, P., Huang, X., et al.: Research of a novel attack scenario constructing method. J. Southwest Univ. Sci. Technol. **31**(1), 55–60 (2016)
8. Pan, S., Morris, T., Adhikari, U.: Developing a hybrid intrusion detection system using data mining for power systems. IEEE Trans. Smart Grid **6**(6), 3104–3113 (2015)
9. Julisch, K.: Mining alarm clusters to improve alarm handling efficiency. In: Proceedings of the 17th Annual Computer Security Applications Conference (ACSAC 2001), New Orleans, USA, pp. 12–21. IEEE Press (2001)

10. Jiang, Z., Zhao, J., Li, X.-Y., et al.: Rejecting the attack: source authentication for Wi-Fi management frames using CSI information. In: Proceedings of the 32nd IEEE Conference on Computer Communications (INFOCOM 2013), Turin, Italy, pp. 2544–2552. IEEE Press (2013)
11. Thangavel, M., Thangaraj, P.: Efficient hybrid network (wired and wireless) intrusion detection using statistical data streams and detection of clustered alerts. J. Comput. Sci. **7**(9), 1318–1324 (2011)

Malware Detection Using Logic Signature of Basic Block Sequence

Dawei Shi and Qiang Xu[✉]

Jiangnan Institute of Computing Technology, Wuxi, Jiangsu 214083, China
sdave@126.com, yytxql1990@126.com

Abstract. Malware detection is an important method for maintaining the security and privacy in cyberspace. As the most mainstream method currently, signature-based detecting is confronted with many obfuscation methods which can hide the true signature of malware. In our research, we propose a logic signature-based malware detecting method to overcome the shortcoming of being susceptible to disturbance in data signature-based method. Firstly, we achieve the logic of basic block based on Symbolic execution and Static Single Assignment, and then use a set of expression trees to represent the basic block logic, the trees set will be filtered to pick out the remarkable items. Depending on basic block logic trees set, we use n-gram method to select features for the discrimination of malicious and benign software. Every feature of program is a sequence of basic block logic and the feature matching is based on edit distance calculating. We design and implement a detector and evaluate its effectiveness by comparing with data signature-based detector. The experimental results indicate that the proposed malware detector using logic signature of basic block sequence has a higher performance than data signature-based detectors.

Keywords: Logic signature · Basic block logic · Expression tree · Basic block sequence

1 Introduction

Malware is used for illegal actions like information stealing, system breakdown, hardware damage, and other malicious purposes. In the current, malware has had a tremendous impact on the security of Internet, and is become more complicated and more diverse. As a technique aimed at countering malware, malware detector [1] ensure the security of system by detecting the behavior of malware, its effectiveness depend on the empirical means of detecting methods. The purpose of malware detection is maintaining a green and security environment of Internet, and promoting the development of society.

Signature-based detection still plays a dominant role in malware detecting at present, it generates signatures from known malware and use them to determine the maliciousness of a program. We can conclude that the signature of malicious behavior is the core for detection, while the extracting and selecting of signature will have a great impact on the effectiveness of detection. Many signatures can be selected to identify malware, and according to the granularity, these signatures can divide as string

S. Li (Ed.): GPC 2018, LNCS 11204, pp. 18–32, 2019.
https://doi.org/10.1007/978-3-030-15093-8_2

signatures [2, 3]; opcode signatures [4, 5]; basic block signatures [6, 7]; system call signatures [8–10], program signatures [11] and other more. The detecting method based on different granularity of signature is unlike, but they all focus on generating remarkable signature that could distinguish malware from benign software accurately and rapidly.

Based on various kinds of signatures, a number of methods have been proposed to detect malware. Griffin [2] features a scalable model that estimates the occurrence probability of arbitrary byte sequences to generate string signature and detect malware. Martin [3] generates a frequency representation which is related to the low level opcodes operations and is able to discriminate malware and benign-ware with no disassembly process. Santos [4] and Ding [5] both present a control flow-based method to extract executable opcode behaviors, their feature generating are based on opcode sequence. Vinod [6] compares instructions at basic block of original malware with that of the variants using longest common subsequence (LCS) to extract signature for malware detection. Adkins [7] applies heuristic detection via threshold similarity matching between basic blocks, and the matching is actually calculating the similar between two binaries. Mehdi [8] use a variable-length system call representation scheme and present an in-execution malware detector. Elhadi [9] proposes a framework combines signature-based with behavior-based using API graph system.

But on the other hand, malware use lots of anti-analysis methods for evading detection. These methods contain encryption, packer, polymorphism, metamorphism and obfuscation [12], they transform the data-flow and control-flow of malware for hiding their true intentions and true signatures. And unfortunately, most of previous detecting methods are based on signatures of data, such like the sequence [4, 5, 7], the frequency [3, 10], the possibility [2, 6, 8], while these signature-based detecting principles could easily to evade. For instance, encryption and packer could change string signatures; polymorphism and rootkit could hide system call signatures; code reordering could disturb opcode signatures and so on. Once the malware developer gets clear on the signatures selected for detecting, he could use multiple anti-analysis methods to avoid being detected. So our signature-based detection should not only concentrate on the data signatures, but also take logic signature into consideration. Because the logic reflects the function of code, it would not be easy to change during execution. In analysis, logic could be extracted from instructions, basic blocks, control flow, system call APIs. By comparison, instruction logic is too simple and cannot identify malware significantly, control flow logic and system call API logic are too complex which could not distinguish malicious and benign software conveniently, thus we choose basic block logic as the most appropriate signature for malware detection.

The difficulty in logic-based signature extracted contain three aspects, the first is how to extract basic block logic; the second is how to generate feature vector of logic; the third is how to calculate the similarity between logic. We design our malware detecting method to overcome the above problems, and the contributions of this paper are listed as follows:

(1) We propose a basic block logic extracting method based on Symbolic Execution and Static Single Assignment, and generate formulas of variables calculation expression and conditional jump expression to represent the logic of basic blocks.

(2) We design a signature generating method by transform expressions into a set of trees and choose n-gram method to select features for classification. Each feature is a sequence of basic block logic, and we screen the set of trees for reduce the size of basic block logic.

(3) We develop an edit-distance based method to calculate the similarity between two logics of basic blocks. This method not only calculates the edit-distance between two trees, but also judge similarity through features of the set of trees.

(4) We employ supervised classifiers to train and classify executable, namely, the K-Nearest Neighbor, Bayesian Networks, Decision Tree, Support Vector Machine. The result shows that our approach is efficient and effective while the evaluation indicates a high detection precision of 96.2%.

2 Structure of Malware Detector Based on Logic Signature

Malware detector is essentially a classifier with transformation algorithm **D** which is used to identify the property of the program under inspection. The input of classifier is the program sample **p**, the output **D(p)** is the result of classifier, while one value means the program is malicious, and the opposite value means the program is benign.

In general, the transformation algorithm in classifier contains several steps for transforming program sample into identification result, as is shown in Fig. 1. First, the sample will be analyzed and extract file information or runtime information, the information extracting can be static or dynamic. Second, some information of sample will be selected and generate feature vector, the feature vector should significant reflect the characteristic of program. Third, according to the machine learning algorithm, feature score is calculated, such as Bayesian Networks algorithm [13], the score is possibility of malicious or benign code. Forth, classification result is achieved by comparing the feature score with the threshold.

Our detecting method is a supervised method contains two phases: training and testing [1]. In the training phase, the detector will summary and select the remarkable signature in distinguish malware from benign software. On the other hand, during the testing phase, the detector will classify the sample by comparing its signature with those which are selected from training phase. Corresponding to the four steps in transformation algorithm, the training and testing phases also consist of four procedures.

In the training phase, the detector first extracts the information of samples, and then generates feature vector, after that, the detector will choose classifier and calculate the score of feature, and next set the threshold and finish classification. The result of classification is sometimes not perfect, so we should adjust the scoring formula according to the classification result and the true property of sample, what is more, the feature can be reselected based on the score. The training may be executed iteratively for optimizing the parameters of detector.

The testing phase is relatively simple compared to training phase. The detector still extracts information firstly, then selects features and calculates their score, finally it will compare the score with threshold to achieve the classification result.

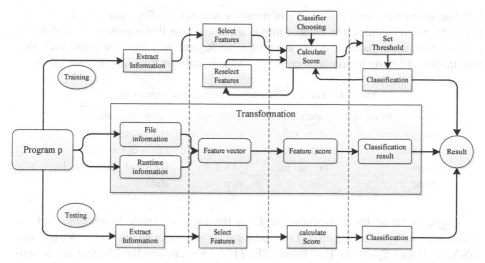

Fig. 1. A structure of malware detector.

In this paper, we use static analysis method to extract the information of program and achieve the control flow which consists of basic blocks. As explained earlier, the selected features will greatly affect the accuracy of detection, while our method chooses basic block logic sequence to extract features. During the scoring and classification, we select 4 kinds of popular classifier as follows: the K-Nearest Neighbor [14], Bayesian Networks [13], Decision Tree [15], Support Vector Machine [16].

3 Extracting Logical Behavior of Basic Block

A basic block is a sequence of instructions that contains no branches, once the first instruction of basic block is executed, all the instructions will be executed sequentially. A program is composed of basic blocks, and there exist many relations among these basic blocks, the graph which expresses the basic blocks and their relations is control flow graph. Normally, a control flow graph (CFG) is a directed graph G = (N, E, entry, exit), where N is the set of basic blocks, E is the set of jump between basic blocks, entry is the entrance node and exit is the exit node. Since the basic block sequence is a sub-graph of control flow graph, we could easily extract it from control flow graph. In this section, we will introduce the method of extracting the logic in basic block and generating features which represent the logic of basic block sequence.

3.1 Basic Block Logic Extracting

The data signature of basic block could be concealed by instruction reordering and instruction replacing, but these obfuscation would not disturb the logic signature. In basic block logic extracting, the instruction is the elementary item for analysis, so we firstly concentrate on the logic of assembly instructions. According to the logic of every assembly instruction, we summary the operation of instructions and classify them into 4 classes: assignment, calculation, jump, none (show in Table 1). Include them, assignment is the instruction that assign the value of memory, register, or eflags, and calculation is the instruction that calculate the value, jump is the instruction that represent the transfer of control-flow, none means the instruction has no logical meaning. We could infer that calculation and jump instructions could remarkably indicate the logic of basic block.

Table 1. The classification of instruction operations.

Instructions category	Opcodes	Examples	Logic of examples
Assignment	mov, xchg...	mov eax, ebx	ebx → eax
Calculation	add, and, not...	add eax, ebx	eax + ebx → eax
Jump	jo, jnc, ja...	jo loc_1	if OF==1, jump loc_1
None	nop...	nop	none

Logic of basic block is collected from the logic of every instruction, and for analyzing the logic among related instructions, we introduce Static Single Assignment (SSA) [17] and Symbolic Execution (SE) [18]. SSA enables the efficient implementation of many important decompiler components, while each variable is assigned exactly once, and every variable is defined before it is used. SE is used to analyze the relationship between inputs and program execution, the core of symbolic execution is using symbolic values as input and then tracking the propagation of variables. Through the method of SSA and SE, we could extract the logic of data-flow and control-flow. The major two steps in logic extracting are listed as below:

(1) We list all the instructions in basic block, and then with the help of SSA format, we perform slicing to backtrack a chain of instructions with data and control dependencies;

(2) With the use of symbolic execution, we analysis the dependency chain and extract the logic of variable calculations and jump conditions.

Here we will use an example to explain our method, show as Fig. 2, and the object in example is a typical basic block. Figure 2(a) shows a source-level view of basic block, Fig. 2(b) shows an assembly-level in accordance with Fig. 2(a), and Fig. 2(c) is SSA format of basic block logic, the V here represents new variable. Depend on the sequence of instructions in Fig. 2(b), we use symbolic execution to backtrack the dependencies. The dependency of variable calculation $[ebp + y]$ is shown by arrows on the left of Fig. 2(b), and the dependency of jump condition jle is shown on the right. Use the dependency and SSA format logic, we can conclude the formula expression of $[ebp + y]$: $V15 = ((V1 + 2) + V5 * V5) * (V5 * V5)$, and the formula expression of jump condition is $((V1 + 2) + V5 * V5) \leq 2$.

During logic extracting, we can find that in one basic block, there exist more than one variable calculation (like $[ebp + x]$, $[ebp + y]$......), by only contains one jump condition, so the basic block logic can be expressed as a combination of a set of variable calculations $\mathcal{C}$ and a jump condition $\mathcal{J}$.

$$\mathcal{L} = \{\mathcal{C}, \mathcal{J}\} = \{(c1, c2, c3...), \mathcal{J}\} \qquad (3-1)$$

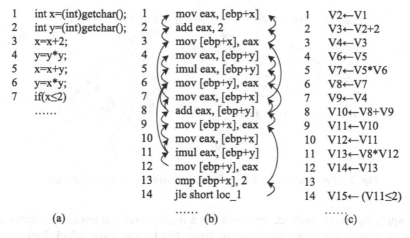

	(a)		(b)		(c)
1	int x=(int)getchar();	1	mov eax, [ebp+x]	1	V2←V1
2	int y=(int)getchar();	2	add eax, 2	2	V3←V2+2
3	x=x+2;	3	mov [ebp+x], eax	3	V4←V3
4	y=y*y;	4	mov eax, [ebp+y]	4	V6←V5
5	x=x+y;	5	imul eax, [ebp+y]	5	V7←V5*V6
6	y=x*y;	6	mov [ebp+y], eax	6	V8←V7
7	if(x≤2)	7	mov eax, [ebp+x]	7	V9←V4
	……	8	add eax, [ebp+y]	8	V10←V8+V9
		9	mov [ebp+x], eax	9	V11←V10
		10	mov eax, [ebp+x]	10	V12←V11
		11	imul eax, [ebp+y]	11	V13←V8*V12
		12	mov [ebp+y], eax	12	V14←V13
		13	cmp [ebp+x], 2	13	
		14	jle short loc_1	14	V15← (V11≤2)
			……		……

Fig. 2. An example of basic block and its logic extracting of SSA format.

3.2 The Expression of Basic Block Logic

The logic of basic block is a set of formula expressions which consist of variables and operators. Although these formula expressions are intuitive, their format is not suit for comparison and analysis, thus this paper overcome this shortcoming by converting formula expression into expression tree.

In expression tree, there exist three kinds of nodes: immediate operands, symbolic variables and operators. Normally, the leaf nodes must be immediate operands or symbolic variables, and operators must be non-leaf nodes. The generating method of expression tree is similar to abstract syntax tree (AST) generating. In our method, the generated expression tree has two properties:

(1) The nodes of symbolic variables can be regarded as undifferentiated.
(2) The child nodes of some operators do not need to stipulate their positions, and the other should make restriction, the detail is shown in Table 2.

We still take the example of Fig. 2, and consider the expression of $[ebp + x]$ and $[ebp + y]$, their expression tree is generated as Fig. 3.

Compared with the formula expression, expression tree is suit for analysis, but in malware detection, the size of expression tree is variable-length, while it is not fast in feature matching, so we choose another form to express the tree. The core idea of our expression form is using the features of tree to replace the expression tree, and the features contain:

(1) The number of nodes in expression tree;
(2) The number of operators in expression tree;
(3) The number of variables in expression tree;
(4) The operators which hold the highest frequency in expression tree, and the count of this operators;
(5) The count of variable which hold the highest frequency in expression tree;
(6) ……

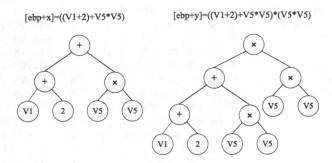

Fig. 3. Two expression trees of $[ebp+x]$ and $[ebp+y]$ respectively.

By applying above method, we could use a fixed expression contain the features of expression tree to represent the logic of basic block. The basic block logic can be expressed as:

$$\mathcal{L} = \{(c1, c2, c3\ldots), \mathcal{J}\} \Rightarrow \{(t_c 1, t_c 2, t_c 3 \ldots), t_\mathcal{J}\} \Rightarrow \{(f_t 1, f_t 2, f_t 3 \ldots), f_{t\mathcal{J}}\} \quad (3-2)$$

3.3 Feature Generating and Selecting Based on the Basic Block Logic

Feature generating is carried out to extract features from the logic of basic block, and feature selecting is applied to reduce the dimensionality of the features set to remove redundant and noisy feature which reduce the performance of classification. Here, we use N-Gram [19] method to generate features of the program under inspection, and select features based on the Information Gain.

N-gram method selects a sequence of information as features and n represents the length of sequence. The n-gram feature generating method is widely used in the representation of system call sequence and opcode sequence, thus we could employ it in our detection. In feature generating, we first choose an execution path $\mathcal{P}$ based on the control-flow graph, this path is composed of many sequential basic blocks $\mathcal{P} \triangleq (\mathcal{B}1, \mathcal{B}2, \mathcal{B}3, \mathcal{B}4\ldots)$. Based on the execution path, we use a sliding window to extract basic block n-grams, the length of sliding window is n and the moving step is one. After that, we will extract n sequential basic blocks each time, such like '$\mathcal{B}1, \mathcal{B}2, \mathcal{B}3$', '$\mathcal{B}2, \mathcal{B}3, \mathcal{B}4$' …. Then we will extract the logic of basic block and form n sequential basic block logic, as '$\mathcal{L}1, \mathcal{L}2, \mathcal{L}3$', '$\mathcal{L}2, \mathcal{L}3, \mathcal{L}4$' … , and each logic $\mathcal{L}$ could expressed as trees or tree features. After the traversal of all execution paths in control-flow graph, we will generate all the features which represent as n sequential basic block logic.

After feature generating, we achieve so many features, we first delete the features whose expression is too short, and next calculate the information gain for filtering out the features which hold better performance in classification. Information gain represent the decrease of entropy cause by remove a feature from the feature set. It is given by equation:

$$IG(Ng) = \sum_{V_{Ng} \in \{0,1\}} \sum_{C \in C_i} P(V_{Ng}, C) log \frac{P(V_{Ng}, C)}{P(V_{Ng})P(C)} \qquad (3-3)$$

Ng is the n-gram feature. V_{Ng} is the value of Ng, $V_{Ng} = 1$ if Ng appear in the program samples, it is 0 otherwise. C is the category of sample, in our problem, it is malware or benign software. $P(V_{Ng}, C)$ is the proportion of Ng appear in category C. $P(V_{Ng})$ is the proportion of Ng appear in all samples. $P(C)$ is the proportion of the category C in all samples. The features which get the higher IG will be selected for malware detecting.

3.4 The Comparison of Features

Unlike the features generated from system call sequence or opcode sequence, the feature of basic block logic sequence is composed of a set of trees, thus it is difficult to compare. The comparison of features is used in both training phase and testing phase. During training phase, we use feature comparison to merge the similar features, and in testing phase, the comparison is to determine whether this feature belongs to malware features or begin software features. Because the features we generated are sequences of logic expression, and they have two formats, the one is a sequence of tree sets, and another is a sequence of tree features.

In the comparison of tree sets sequence, we use the graph matching algorithm based on edit distance as our basic method. The graph matching could identify if the two trees are similar, then conclude if the logic of two basic blocks are similar, and only all the basic block logics are similar in sequence can we determine the matching of the two features. This method can describe as:

$$\forall t_c i \in \mathcal{L}_B, t'_c i \in \mathcal{L}_{B'} \quad t_c i \approx t'_c i \Rightarrow \mathcal{L}_B = \mathcal{L}_{B'} \qquad (3-4)$$

$$\forall i \in [1, n] \ \mathcal{L}_{B_i} = \mathcal{L}_{B'_i} \Rightarrow \mathcal{F} = \mathcal{F}' \qquad (3-5)$$

Before the tree matching, we firstly convert the generated tree from binary tree to multi branched tree. This conversion is duo to the characteristic of operators, such like the operators of addition, the formula expression $a + b + c$ can express as three kinds of binary tree, show as Fig. 4, but they are same in logic, so we make conversion to overcome this drawback, and the example of $[ebp + x]$ is as Fig. 4. And the operators which should make adjustment are listed as Table 2.

Furthermore, according to the condition show as formula 3-4, we can achieve that the similar of two logics of basic block requires all the trees to be alike, this is time-consuming in analysis. So we would reduce the size of the set of trees. In our optimization, we only take the most complicated tree of calculation and the tree of jump condition into consideration when apply tree matching.

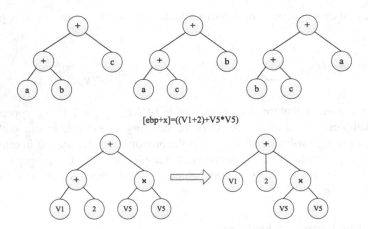

Fig. 4. Examples of adjustment of expression tree.

The tree matching is use the method of edit distance [20], an edit distance is defined as the minimum cost edit path between two graphs, it meaning how many distortion operations should execute to transform one graph to another. A standard set of distortion operations is given by insertions, deletions, and substitutions. When the edit distance is less than the threshold value, the two graphs consider being similar.

Table 2. The adjustment of operators.

Operators	Add	Subtract	Multiply	Divide	Modulo	And/or/xor	Not
Child nodes number limitation	No	2	No	2	2	No	1
Position restriction of child nodes	No	Yes	No	Yes	Yes	No	No
Need adjustment	Yes	No	Yes	No	No	Yes	No

In the comparison of tree features sets sequence, we will compare the features of logic tree, such as the number and the frequency. The feature of logic tree is a set of several IDs which represent the operators and the counts, and we will use a formula to calculate the difference between two logic trees. In this formula, f_t is the element of tree feature set, n is the number of features to describe a logic tree, $f_t[i]$ is the ith element in f_t, w_i is the weight of ith element in calculation.

$$D = \sum_{i=1}^{n} w_i(f_t[i] - f_t'[i])^2 \qquad (3-6)$$

In our detection method, we could also apply the comparison of tree features to reduce the time-cost of tree matching, and the skill is to screen candidates through tree features before tree matching.

4 Experimentation

In order to test the effectiveness and efficiency of our method, we evaluate our prototype on several experiments. All our experiments were performed on a 3.1 Ghz Pentium quad-core processor with 8 GB of RAM, the operation system is Windows 7 SP1.

4.1 Data Collecting

For the experimentation, we employ a library consists of 352 malicious software and 156 benign software. Parts of the malicious software are chosen from VXheawens and the others are captured by our own honeypot. These malicious samples contain almost all categories of malware, such like virus, worm, Trojan, backdoor and so on. Including the malicious samples, many are selected from different malware families, as Bagle, Klez, Nctsky, Mydoom, Minmail and others. Two datasets are created based on the library, the first dataset is used for training, and another is for testing. The ratio of training dataset size to testing dataset size is 4:1, and during analysis, the elements of the two dataset will be changed and then apply detection repeatedly.

4.2 Training and Testing

Multiple classifiers are used in our experimentation, and these classifiers are based on supervised machine-learning. Several of them are listed as follows:

K-Nearest Neighbor (KNN): The K-Nearest Neighbor is one of the simplest supervised machine-learning algorithms for classifying samples. This method classifies each new pattern using the evidence of nearby sample observation. In this paper, K is set to 10 according to knowledge of previous experiments.

Bayesian Networks (BN): A Bayesian Network is an annotated directed graph of joint multi-variate probability distributions. As a probabilistic model for multivariate analysis, Bayesian Network classifies the new pattern depending on the probability distribution function. In this paper, we use Gaussian distribution.

Decision Tree (DT): A decision tree uses a tree-like model of multistage decisions for approximating discrete-valued functions. The basic idea of decision tree is to divide a complex decision into a union of several simpler decisions. The C4.5 algorithm is used in this paper.

Support Vector Machine (SVM): Support Vector Machine generates a hyperplane to classify the samples which are represented as n dimensional format. The kernel function in SVM is used for transforming the original data to higher-dimensional space. We use polynomial kernel functions in our method.

In our experimentation, we use machine learning tool WEKA to implement the classification algorithms mentioned above.

4.3   Evaluation Criteria

The performance of the classifier is evaluated using three criteria: accuracy, the false negatives rate (FNR), the false positive rate (FPR). They are calculated using equations:

$$Accuracy = \frac{TP + TN}{TP + TN + FP + FN}$$
$$FPR = \frac{FP}{TN + FP} \qquad FNR = \frac{FN}{TP + FN}$$

Where TP is the number of malware correctly classified as malware, TN is the number of benign software correctly classified as benign software, FP is the number of benign software incorrectly classified as malware, FN is the number of malware incorrectly classified as benign software.

4.4   Experimental Results

Our malware detection method is explained in Sect. 3, and the prototype is implemented in python language. The control flow graph is generated with Angr [21].

The logic extracting is based on basic block analysis, and during experimentation, we extract almost 12400 distinct basic block logic, after remove the logic whose expression is too short, the number of effective logic is about 5700.

The features we selected for representing the signature of logic have two formats: the sequence of basic block logic tree set, and the sequence of basic block logic tree features set. We experiment our detector with these two formats and their combination, while the combination method is use the sequence of basic block logic tree features for matching firstly, and then verify the matched basic blocks by using basic block logic tree sets. The result is shown in Table 3. We can find that the accuracy of detection based on logic tree set is the best, but its time-cost in testing is the worst, this is because the time in graph matching is time-consuming. While the detection based on logic tree features set hold the least time-cost, however its accuracy is the worst. The good news is that the combination of these two format behave well both in accuracy and time.

Table 3. Experiment results of two signature representation formats.

Classifier	Logic tree set		Logic tree features set		Combination	
	Accuracy (%)	Time (s)	Accuracy (%)	Time (s)	Accuracy (%)	Time (s)
KNN	94.0	85.5	92.6	3.4	93.8	4.2
BN	94.5	84.9	92.3	3.4	94.1	4.2
DT	96.8	83.2	93.8	3.3	96.2	4.1
SVM	90.4	83.1	87.5	3.3	89.6	4.2

The choosing of n in n-gram method will significantly affect the results of detector, so we use 2-grams, 3-grams, and 4-grams to generate features, and apply classifier to make comparison. The result in Table 4 can clearly indicate that although 2-grams hold the minimum time of about 2.3 s, its accuracy is not perfect, especially in classifier BN and SVM, the accuracy is much less than that of 3-grams. What is more, the result of

Table 4. The results of classifier using n-gram method with different length n.

Classifier	2-grams		3-grams		4-grams	
	Accuracy (%)	Time (s)	Accuracy (%)	Time (s)	Accuracy (%)	Time (s)
KNN	91.6	2.3	93.8	4.2	94.5	7.5
BN	88.5	2.3	94.1	4.2	90.7	7.4
DT	92.8	2.3	96.2	4.1	91.7	7.5
SVM	79.8	2.2	89.6	4.2	86.3	7.4

4-grams is still worse than that of 3-grams. The best result is classifier DT using 3-grams method, its accuracy is the highest of 96.2 and time-cost is 4.1 s.

According to the previous 2 experiments, we select the combination method to represent their signatures and then select features, and 3-grams method is used for extracting features. As is mentioned in paper [1], the number of features is an important parameter in classifier. We selected feature numbers of 100, 200, 400, 600, 800, and test their effect toward classification. The results of accuracy, FNR, FPR are shown in Table 5.

Table 5. The detecting results using different number of features.

Classifier	100		200		400		600		800	
	Accuracy	FNR/FPR	Accuracy	FNR/FPR	Accuracy	FNR/FPR	Accuracy	FNR/FPR	Accuracy	FNR/FPR
KNN	91.0	9.9	92.8	5.9	93.5	7.3	**93.8**	**4.9**	92.4	6.9
		8.5		7.8		8.9		6.7		7.9
BN	88.3	7.8	91.4	6.6	92.9	8.0	**94.1**	**4.6**	92.6	9.3
		13.4		9.4		6.6		6.4		6.6
DT	93.2	13.2	92.5	12.0	96.0	4.9	**96.2**	5.7	95.8	6.7
		3.9		5.5		4.5		2.9		3.1
SVM	83.9	17.0	85.0	9.2	85.4	8.8	89.6	5.9	**90.1**	9.4
		15.6		17.5		17.1		12.3		**8.7**

The accuracy of detection is better with the increasing of feature number and will reach its maximum when the number is set as 600 in KNN, BN, DT classifiers, as 93.8%, 94.1% and 96.2%. The top value of accuracy in SVM classifier is 90.1% while number of features is 800, however, the accuracy in SVM with 600 features is already not bad when compare with others, and it could be better than that of 800 features in time-cost.

Furthermore, it can be seen from Table 5, that the classifiers which set the feature number with 600 would hold an optimal or suboptimal result in FPR and FNR. In KNN, BN, 600 features selected method is the best both in FPR and FNR, and in DT, it is best in FNR and hold the minimum value in the sum of FPR and FNR. In SVM, the 800-features perform best in FNR, and the 600-features perform best in FPR, and their summary of FNR and FPR is almost the same value. So we could conclude that the best selection of the feature number is 600 in our experiments.

Compared with other classifiers, the Decision Tree method gets the best preference, it has the highest accuracy and the lowest FNR, and its FPR is not bad.

We make a comparison between logical signature-based detector and data signature-based detector for proving the effectiveness of our method. The data signature-based detecting method is chosen from paper [4], and its features are generated from opcode sequences. As is shown in Table 6, we use 4 classifiers to test the accuracy, FPR and FNR of both detectors, while we also adjust the feature number for achieving their best performance.

Figure 5 clearly show the accuracy of four classifiers based on logic features and data features, and the result indicate that the logic-based detector is better than data-based detector. The "-L" represent that the classifier uses logic features, and "-D" means it uses data features. The average accuracies of KNN-L, BN-L, DT-L, SVM-L are 92.75%, 92.08%, 94.58% and 86.88%. The average accuracies of KNN-D, BN-D, DT-D, SVM-D are 91.75%, 91.25%, 93.02% and 85.48%. We can infer the accuracy of logic-based detector is approximately 1.2% higher than the accuracy of data-based detector. We select the best performance of these classifiers respectively, and record their accuracy, FPR, FNR and the best selected feature numbers in Table 6. We can calculate that the logic-based detector is better than data-based detector toward accuracy with approximately 2.1% higher in average. The highest accuracy value is in DT-L, it gets the accuracy of 96.2%, and its FPR is 5.7, FNR is 2.9, which is the lower when compared with others.

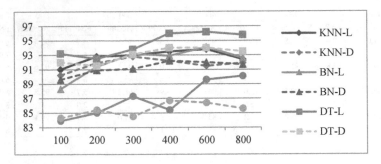

Fig. 5. The comparison of four classifiers based on logic features and data features.

Table 6. The best performance of classifiers based on logic and data features.

Classifier	Accuracy	FPR	FNR	Best number of features
KNN-L	93.8	4.9	6.7	600
BN-L	94.1	**4.6**	6.4	600
DT-L	**96.2**	5.7	**2.9**	600
SVM-L	90.1	9.4	8.7	800
KNN-D	92.8	13.1	5.3	300
BN-D	92.2	7.2	9.5	400
DT-D	94	8.5	4.8	400
SVM-D	86.7	15.8	12.1	400

5 Conclusions and Future Work

As more and more anti-analysis techniques are applied in malicious software, the detection of malware based on data signature will be disturbed. In this paper, we describe an approach of using logic signature to classify malicious and benign software. Symbolic execution and Static Single Assignment method is employed to extract the logic of basic block, and then generate the basic block logic tree for features generating and selecting. The results of detection accuracy, false negatives rate and false positive rate indicate that our method is better than data signature-based detector according to our experimentation. The advantage over data signature-based detector is duo to the elimination of many disturbances introduced by obfuscation. We also test our prototype to achieve the best parameters in detecting, and the result show that under the condition of 3-gram features generating, and select the number of features as 600, and used Decision Tree classifier, we achieve the highest accuracy of 96.2%, which is 2.2% higher than the best performance of data signature-based method.

In our detection method, software behavior is extracted from static approach. Since static approach could fast extract information of control flow graph, it still holds the shortcoming at the accuracy of CFG, this is because of the difficulties introduced by encryption and packer. Therefore, in our future work, we will choose dynamic analysis for extracting software behavior, while this dynamic method can naturally eliminate interruptions. In order to overcome the low speed in traversal of all possible paths of program, we use forced execution to force program execute along different paths as quickly as possible. What is more, we observe that the feature generating is time-consuming, and the main cost in feature generating is the generating of logic tree, so we will continuously search for other filters to reduce the frequency in tree generating.

Acknowledgment. This research was supported by the National Natural Science Foundation of China (91318301), and the National High Technology Research and Development Program ("863" Program) of China (2012AA7111043).

References

1. Idika, N., Mathur, A.P.: A survey of malware detection techniques. Purdue University (2007)
2. Griffin, K., Schneider, S., Hu, X., Chiueh, T.: Automatic generation of string signatures for malware detection. In: Kirda, E., Jha, S., Balzarotti, D. (eds.) RAID 2009. LNCS, vol. 5758, pp. 101–120. Springer, Heidelberg (2009). https://doi.org/10.1007/978-3-642-04342-0_6
3. Martín, A., Menéndez, Héctor D., Camacho, D.: String-based malware detection for android environments. Intelligent Distributed Computing X. SCI, vol. 678, pp. 99–108. Springer, Cham (2017). https://doi.org/10.1007/978-3-319-48829-5_10
4. Santos, I., et al.: Opcode sequences as representation of executables for data-mining-based unknown malware detection. Inf. Sci. **231**(9), 64–82 (2013)
5. Ding, Y., et al.: Control flow-based opcode behavior analysis for Malware detection. Comput. Secur. **44**(2), 65–74 (2014)

6. Vinod, P., et al.: Static CFG analyzer for metamorphic Malware code. In: International Conference on Security of Information and Networks, Sin 2009, Gazimagusa, North Cyprus, October, pp. 225–228. DBLP (2009)
7. Adkins, F., et al.: Heuristic malware detection via basic block comparison. In: International Conference on Malicious and Unwanted Software, pp. 11–18. The Americas IEEE (2014)
8. Mehdi, B., et al.: Towards a theory of generalizing system call representation for in-execution malware detection. In: IEEE International Conference on Communications, pp. 1–5. IEEE (2010)
9. Elhadi, A.A.E., Maarof, M.A., Osman, A.H.: Malware detection based on hybrid signature behaviour application programming interface call graph. Am. J. Appl. Sci. 9(3), 283–288 (2012)
10. Natani, P., Vidyarthi, D.: Malware detection using API function frequency with ensemble based classifier. In: Thampi, S.M., Atrey, P.K., Fan, C.-I., Perez, G.M. (eds.) SSCC 2013. CCIS, vol. 377, pp. 378–388. Springer, Heidelberg (2013). https://doi.org/10.1007/978-3-642-40576-1_37
11. Chandramohan, M., Tan, H.B.K., Shar, L.K.: Scalable malware clustering through coarse-grained behavior modeling. In: ACM SIGSOFT, International Symposium on the Foundations of Software Engineering, p. 27. ACM (2012)
12. You, I., Yim, K.: Malware obfuscation techniques: a brief survey. In: International Conference on Broadband, Wireless Computing, Communication and Applications, pp. 297–300. IEEE (2010)
13. Jensen, F.V., Nielsen, T.D.: Bayesian networks and decision graphs. Technometrics 50(1), 97 (2007)
14. Denœux, T.: A k-nearest neighbor classification rule based on dempster-shafer theory. In: Yager, R.R., Liu, L. (eds.) Classic Works of the Dempster-Shafer Theory of Belief Functions. STUDFUZZ, vol. 219, pp. 737–760. Springer, Heidelberg (2008). https://doi.org/10.1007/978-3-540-44792-4_29
15. Landgrebe, D.: A survey of decision tree classifier methodology. IEEE Trans. Syst. Man Cybern. 21(3), 660–674 (2002)
16. Suykens, J.A.K., Vandewalle, J.: least squares support vector machine classifiers. Neural Process. Lett. 9(3), 293–300 (1999)
17. Van Emmerik, M.: Static single assignment for decompilation. UQ Theses (RHD) - UQ staff and students only (2007)
18. Khurshid, S., PĂsĂreanu, C.S., Visser, W.: Generalized symbolic execution for model checking and testing. In: Garavel, H., Hatcliff, J. (eds.) TACAS 2003. LNCS, vol. 2619, pp. 553–568. Springer, Heidelberg (2003). https://doi.org/10.1007/3-540-36577-X_40
19. Mira, F., Huang, W., Brown, A.: Improving malware detection time by using RLE and N-gram. In: International Conference on Automation and Computing, pp. 1–5 (2017)
20. Bille, P.: A survey on tree edit distance and related problems. Theor. Comput. Sci. 337(1), 217–239 (2005)
21. Shoshitaishvili, Y., et al.: SOK: (state of) the art of war: offensive techniques in binary analysis. In: Security and Privacy, pp. 138–157. IEEE (2016)

Verifiable and Privacy-Preserving Association Rule Mining in Hybrid Cloud Environment

Hong Rong[✉], Huimei Wang, Jian Liu, Fengyi Tang, and Ming Xian

School of Electronic Science, National University of Defense Technology,
Changsha, China
r.hong_nudt@hotmail.com, freshcdwhm@163.com, ljabc730@gmail.com,
tangfengyi91@163.com, qwertmingx@sina.com

Abstract. Nowadays it is becoming a trend for data owners to outsource data storage together with their mining tasks to cloud service providers, which however brings about security concerns on the loss of data integrity and confidentiality. Existing solutions seldom protect data privacy whilst ensuring result integrity. To address these issues, this paper proposes a series of privacy-preserving building blocks. Then an efficient verifiable association rule mining protocol is designed under hybrid cloud environment, in which the public unreliable cloud and semi-honest cloud collaborate to mine frequent patterns over the encrypted database. Our scheme not only protects the privacy of datasets from frequency analysis attacks, but also verifies the correctness of mining results. Theoretical analysis demonstrates that the scheme is semantically secure under the threat model. Experimental evaluations show that our approach can effectively detect faulty servers while achieving good privacy protection.

Keywords: Association rule mining · Privacy preservation · Integrity verification · Hybrid cloud

1 Introduction

With the advent of big data era, analyzing the huge volume and variety of data in-house is beyond most data owners' capabilities, especially for those lack of computational resources and expertise. Therefore, outsourcing the data storage with its mining tasks to cloud service providers (CSPs) becomes an optimal solution for the resource-constrained clients. This paradigm [1] has already come true in real-life, e.g., Amazon Machine Learning [2], Cloud Machine Learning Engine [3], Azure Machine Learning services [4], etc. These services aim at facilitating clients to build their prediction models in a cost-efficient way.

Nevertheless, this new paradigm brings about serious security issues. One of them is privacy leakage, that is, the cloud server may illegally access the outsourced database to learn sensitive information. Another important issue is that the integrity of data or mining results may be incorrect due to server hardware

© Springer Nature Switzerland AG 2019
S. Li (Ed.): GPC 2018, LNCS 11204, pp. 33–48, 2019.
https://doi.org/10.1007/978-3-030-15093-8_3

or software failures, etc. In this paper, we focus on solving these problems of association rule mining outsourcing in hybrid cloud environments to guarantee privacy, integrity, as well as efficiency.

In order to preserve data privacy in outsourcing scenario, data owners usually encrypt their databases locally before uploading them to the cloud [5]. Substitution ciphers can be used to hide the meaning of items, whereas they expose the supports of items, and thus are vulnerable to frequency analysis attack. Wong et al. [1] first addressed this by proposing a one-to-n item mapping to enhance protection. It was improved through k-support anonymity technique which requires each authentic itemsets must have the same support as at least $k - 1$ other itemsets [6,7]. These works require inserting artificial itemsets or transactions, whereas the cost to create fake dataset is considerably high for data owners [8].

Another line of research is to employ homomorphic encryption (HE) techniques to improve security. Yi et al. [8] proposed three solutions based on distributed ElGamal cryptosystem to achieve different privacy needs. FHE (fully homomorphic encryption) was used in [9] that allows mining on the encrypted data. Li et al. [10] proposed a new symmetric HE to detect fictitious transactions inserted according to [7]. Generally, HE alike schemes are computationally expensive because of exponential modular operation [11], key switching [12], etc.

To verify the correctness and completeness of mining results, the previous works [13] utilized a set of fake (in)frequent itemsets based on the assumption that the probability for the server to cheat on true and fake itemsets are equal. A deterministic approach was proposed by constructing cryptographic proofs, which achieves verification with 100% guarantee but higher overhead [14]. Additionally, none of these works consider privacy preservation as an issue.

In this paper, we propose a hybrid-cloud-assisted privacy-preserving association rule mining solution over the joint datasets. To the best of our knowledge, this is the first work on outsourced association rule mining that not only protects the privacy of the itemsets and frequent patterns, but also verifies the correctness of results. The contributions of this paper are three-fold:

- First, by introducing the hybrid cloud system model, we propose a series of privacy-preserving building blocks based on Shamir's secret sharing scheme [15]. These building blocks allow the cloud servers to perform addition, multiplication, comparison operations over split data shares. By leveraging the redundancy of cloud servers, we provide an efficient scheme that can verify the integrity of outsourced computation without constructing fake items or cryptographic proofs beforehand.
- Second, we propose a verifiable and secure outsourced frequent pattern mining protocol based on aforementioned building blocks. Theoretical analysis demonstrates that the protocol is resistant against frequency-based attacks under semi-honest threat model. Nothing regarding input or output privacy is revealed to cloud servers.

– Last but not least, we conduct evaluations over the real datasets. The experimental results show that our solution not only mines frequent itemsets correctly and efficiently, but also identifies computation errors effectively.

Table 1 summarizes qualitative comparisons of existing privacy-preserving outsourcing protocols from different aspects. FAA, CPA, and ITS denote Frequency Analysis Attack, Chosen-Plaintext Attack, and Information-Theoretic Security, respectively. From Table 1, it can be observed that none of the existing works except ours, achieve both privacy protection and integrity verification at the same time. Besides, our solution does not need the data owners to participate in the outsourcing process, thus incurring relatively smaller costs on them.

Table 1. Comparative summary of existing privacy-preserving outsourced association rule mining protocols

Protocol	Encryption scheme	Security model	Integrity verification	No owner participation
Wong et al.'s [1]	1-to-n Map	FAA	×	√
Tai et al.'s [6]	k-support	FAA	×	√
Giannotti et al.'s [7]	k-privacy	FAA	×	√
Liu et al.'s [9]	FHE	FAA & CPA	×	×
Yi et al.'s [8]	HE	FAA & CPA	×	√
Li et al.'s [10]	IHE	FAA & CPA	×	×
Ours	Secret sharing	FAA & TIS	√	√

The rest of the paper is organized as follows. In Sect. 2, we briefly introduce association rule mining. The system model, threat model, and design goals are presented in Sect. 3. The design details of our proposed privacy-preserving building blocks are described in Sect. 4. We present the complete outsourcing protocol in Sect. 5. We also analyze the security of our scheme in Sect. 6. Section 7 shows the theoretical and experimental evaluations. Finally, we conclude the paper and outline future work in Sect. 8.

2   Preliminaries

In this section, we briefly introduce the definition of association rule mining, which is the fundamental outsourcing problem for our solution.

Let $I = \{i_1, i_2, ..., i_m\}$ be a set of items, and a database D be a set of transactions, where each transaction t_i is a subset of I, i.e., $D = \{t_1, t_2, ..., t_w\}$, and $t_i \subseteq I$ for $i \in [1, w]$. Given an itemset $A \subseteq I$, a transaction t_i contains A if and only if $A \subseteq t_i$. The support of A is the number of transactions in D containing A, denoted by $Supp(A)$. Given support threshold T_s, A is considered to be frequent itemset if and only if $Supp(A) \geq T_s$. An association rule is expressed as $A \Rightarrow B$, where $A \subseteq I, B \subseteq I$, and $A \cap B = \oslash$. Let $Conf(A \Rightarrow B)$ denote the confidence of $A \Rightarrow B$, which means that $Conf(A \Rightarrow B) = Supp(A \cup B)/Supp(A)$. Given the confidence threshold T_c specified by the data miner, $A \Rightarrow B$ is regarded as an association rule if and only if $Supp(A \cup B) \geq T_s$ and $Conf(A \Rightarrow B) \geq T_c$. As mining frequent itemsets account for the major workload of association rule mining, we focus on its representative Apriori algorithm in this paper.

3   Problem Statement

In this section, we formally describe our system model, threat model and design objectives.

3.1   System Model

In our system model, we mainly focus on how hybrid cloud servers tackle frequent pattern mining request in a privacy-preserving manner. There are two types of entities, i.e., Data Owners and Hybrid Cloud Environment, as shown in Fig. 1. The details of the parties are as follows.

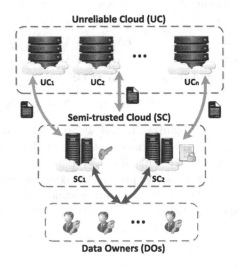

Fig. 1. System model

- Data Owner (DO): DO holds a vertically or horizontally partitioned dataset. DO who lacks data storage, computational resources, and expertise is willing to outsource their data and mining tasks to the cloud service providers. Multiple DOs request the cloud to perform collaborative association rule mining over their aggregated database.

- Semi-trusted Cloud (SC): SC comprises two semi-honest servers, which may come from private cloud or commercial cloud with high availability guarantee. SC possesses higher computing power than a single DO, but it's still insufficient to support large-scale data storage and computation due to limited server quantity.
- Unreliable Cloud (UC): UC is composed of a large number of unreliable servers, which provides public storage and computation services. In our context, unreliable servers means they may not only breach data privacy, but also return incorrect results accidentally due to hardware or software failures. Compared to SC, the price of renting UC is much cheaper at the expense of downgrade service security, availability, and integrity.

In Fig. 1, SC_1 and SC_2 act as the proxy for DO, which mainly performs auxiliary calculations, such as verifying intermediate results from UC, generating random numbers, etc. We assume there are n servers in UC, denoted by $UC_1, ..., UC_n$, where $n > 2$. UC servers store the outsourced datasets, all the intermediate results and final outputs in encrypted form, while they execute mining protocol with assistance of SC.

The system model with hierarchical trust is reasonable and practical in real life, because (1) it's commonsense that the cloud service price accords with its quality; (2) it's an economic and feasible option for clients to combine different kinds of public cloud services to balance security, efficiency, as well as cost; (3) it's possible to design appropriate schemes to guarantee data privacy and result integrity in the meantime, which will be illustrated in the following sections.

3.2   Threat Model

In the threat model, SC servers are assumed to be semi-honest, which means that they follow the protocol but try to infer private information about owners' data and final results by using the messages they record during execution.

UC servers are also interested to learn data belonging to other parties. Most UC servers behave like semi-honest servers in most time, but some of them may accidentally collapse or fail to work correctly. There are many causes in real-life, such as hardware or software failure, saving computation power for profits, or being compromised by malicious attackers, etc.

More specifically, we assume: there are t UC servers that may produce errors, including broken data blocks, inaccurate computation results, etc, where $0 \leq t \leq n$. The failing servers are randomly distributed of UC, i.e., the probability of getting false result from any server in UC is equal. At most $k - 1$ UC servers may collude together to learn data privacy.

Furthermore, both SC and UC have some knowledge of some private items and their corresponding frequencies, thus they can launch frequency analysis attack. We assume there is no collusion between two different clouds and SC_1, SC_2. In addition, all the interactive data are transmitted through secure communication channels, so eavesdropping attack is not considered.

3.3 Design Objectives

The objectives of our proposed privacy-preserving association rule mining protocol for outsourced database are as follows:

- **Privacy.** Neither itemsets in the aggregated database, nor the itemset supports and support threshold should be revealed to SC and UC. Moreover, the mining result such as the frequent patterns should be protected from CSP.
- **Integrity Verification.** The integrity verification means verifying integrity of the outsourced data as well as the mining results. If one or more cloud servers produce false computation results or corrupted data, the proposed solutions should be able to detect the errors precisely and timely.
- **Efficiency.** The most computations should be processed by clouds in an efficient way while DOs are not required to involve in outsourcing.

4 Proposed Privacy-Preserving Building Blocks

In this section, we first present the basic scheme based on Shamir's secret sharing algorithm. Then we show how to perform addition, multiplication, comparison, and integrity verification. They serve as the privacy-preserving building blocks of our outsourced mining solution.

4.1 Secure Data Outsourcing Scheme

Shamir's secret sharing aims at sharing a secret value s among n servers by dividing s into n shares through a random polynomial. To recover the secret, at least k shares are needed to reconstruct the polynomial of degree $k - 1$. It has been proven that Shamir's scheme provides perfect information-theoretic security against recovering s from less than k shares [15]. We combine secret sharing into the secure data outsourcing scheme, consisting of the following protocols:

- **Key generation protocol** ($\mathsf{KeyGen}(\kappa) \rightarrow X$)
  $\mathsf{KeyGen}(\cdot)$ takes a security parameter κ as input, and outputs a secret key X, which is a vector of random numbers, where $X = \langle x_1, ..., x_n \rangle$. Generally, x_i can be derived from any field, for $1 \leq i \leq n$. In this paper, we assume x_i randomly chosen from the finite field $\mathbb{Z}_p$, where $x_i \neq 0$ and p is a large prime with κ bit size. SC runs $\mathsf{KeyGen}()$ and broadcasts X to DO.
- **Secret dividing protocol** ($\mathsf{SD}(m) \rightarrow S$)
  $\mathsf{SD}(\cdot)$ takes the secret key X and a secret value $m \in \mathbb{Z}_p$ as inputs. The secret holder generates such a random polynomial $q(x)$ of degree $k - 1$ that $q(x) = \sum_{i=1}^{k-1} a_i x^i + m$, where the coefficient $a_i \in_R \mathbb{Z}_p$, for $i \in [1, k - 1]$. We use $\in_R \mathbb{Z}_p$ to denote selecting a random number in $\mathbb{Z}_p$. Then secret holder computes $s_i = q(x_i)$ for $i = 1, ..., n$. The output of $\mathsf{SD}(m)$ is $S = \langle s_1, ..., s_n \rangle$, where s_i is uploaded to UC_i for $i \in [1, n]$.

- **Secret recovering protocol** ($\mathsf{SR}(S) \to m$)

  $\mathsf{SR}(\cdot)$ takes S and X as inputs, and outputs the secret m. Since there are k unknown coefficients in the random polynomial $q'(x) = \sum_{i=0}^{k-1} a_i' x_i^i$, where $a_i' \in_R \mathbb{Z}_p$ for $i \in [1, k-1]$ and $a_0' = m$, at least k shares from S are required to determine $q'(x)$ of degree $k-1$. Given any k tuples like $(x_1, s_1), ..., (x_k, s_k)$, m can be recovered by Lagrange interpolation or by solving k linear equations.

- **Data outsourcing protocol**

  Given that m is the data to be outsourced by DO, SC_2 first generates a random value $r \in_R \mathbb{Z}_p$ and computes $S_r \leftarrow \mathsf{SD}(-r)$. r is sent to DO, who uses it to randomize m by computing $m' \leftarrow m+r$ and sends m' to SC_1. After that, SC_1 splits m' into $s_{m',i}$ via running $S_{m'} \leftarrow \mathsf{SD}(m')$ and distributes ith share to UC_i, for $i = 1, ..., n$. With $S_{m'}$, UC_i removes the randomness of r by $s_{m',i} + s_{r,i}$ based on additive homomorphism. In this way, the majority burden of DO is shifted to cloud servers. As SC_1 and SC_2 do not collude, m is not revealed to SC_1.

4.2 Secure Addition Protocol (SA)

Design Details. Let $C_1 = \langle \alpha_1, ..., \alpha_n \rangle$ denote the split shares of m_1, and $C_2 = \langle \beta_1, ..., \beta_n \rangle$ denote the shares of m_2, respectively. Given that α_i, β_i are held by UC_i for $i = 1, ..., n$, the goal of **SA** is to compute the shares of addition of m_1 and m_2, without revealing anything about m_1, m_2, or $m_1 + m_2$.

Let $C_{Add} = \langle \gamma_1, ..., \gamma_n \rangle$ denote the final shares. $\forall i = 1, ..., n$, UC_i obtains γ_i through calculating $\gamma_i \leftarrow \alpha_i + \beta_i$. The correctness of **SA** is straightforward. Given that the random polynomials for m_1, m_2 generated by DO are $q_1(x) = \sum_{i=1}^{k-1} a_i x^i + m_1$ and $q_2(x) = \sum_{i=1}^{k-1} b_i x^i + m_2$ respectively, where a_i, b_i are random coefficients for $1 \le i \le k-1$. Since X is the same for both shares, it's apparent to get $\gamma_i = \sum_{j=1}^{k-1} w_j x_i^j + m_1 + m_2$, where $i \in [1, n]$, $w_j = a_j + b_j$, for $1 \le j \le k-1$. Due to only k unknowns, any k computed shares can be used to recover $m_1 + m_2$. In other words, **SA** is achieved by this additive homomorphic property.

4.3 Secure Multiplication Protocol (SM)

Design Details. Given that divided shares of m_1, m_2 are held by $UC_1, ..., UC_n$ respectively, the goal of **SM** is to compute the shares of multiplication of m_1 and m_2. During execution of this protocol, nothing regarding $m_1, m_2, m_1 \times m_2$ should be known to UC and SC. The notations of C_1 and C_2 are the same as in **SA**. The details of **SM** are as follows:

Step-1 (@SC_2): SC_2 generates two random numbers $r_1, r_2 \in_R \mathbb{Z}_p$. With X, SC_2 computes $C_{r_1} \leftarrow \mathsf{SD}(r_1)$, $C_{r_2} \leftarrow \mathsf{SD}(r_2)$, and $C_{r_3} \leftarrow \mathsf{SD}(r_1 r_2)$ respectively, in which $C_{r_1} = \langle f_1, ..., f_n \rangle$, $C_{r_2} = \langle g_1, ..., g_n \rangle$, $C_{r_3} = \langle h_1, ..., h_n \rangle$. After that, the split shares f_i, g_i, h_i are sent to UC_i together with r_1 and r_2, for $i \in [1, n]$.

Step-2 (@UC): $\forall i = 1, ..., n$, UC_i randomizes α_i by computing $\alpha_i' \leftarrow \alpha_i + f_i$; likewise, $\beta_i' \leftarrow \beta_i + g_i$. After that, α_i', β_i' are transmitted to SC_1.

Step-3 (@SC_1): With $\alpha'_1, ..., \alpha'_n$ received from UC, SC_1 selects any k shares to reconstruct the randomized messages of C_1, C_2 (denoted by m'_1, m'_2, respectively). Afterwards, SC_1 computes their multiplication by $\varepsilon \leftarrow m'_1 \times m'_2$. SC_1 then divides ε by running $\langle \rho_1, ..., \rho_n \rangle \leftarrow \mathsf{SD}(\varepsilon)$. ρ_i is sent to UC_i for $i = 1, .., n$.

Step-4 (@UC): Let $C_{mul} = \langle \eta_1, ..., \eta_n \rangle$ denote the desired output. UC_i first computes $\alpha''_i \leftarrow \alpha_i \times r_2$ and $\beta''_i \leftarrow \beta_i \times r_1$, for $i \in [1, n]$. To remove the blinding factors in ρ_i, UC_i calculates $\eta_i \leftarrow \rho_i - \alpha''_i - \beta''_i - h_i$, which is the share of $m_1 m_2$.

Correctness of SM. If UC and SC execute Step-1 and Step-2 strictly, SC_1 can setup a set of equations to solve m'_1 such that $\alpha'_i = \sum_{j=1}^{k-1} w_j x_i^j + m'_1$, where $w_j \in \mathbb{Z}_p$, $i \in [1, n], j \in [1, k-1]$. To obtain m'_1, SC_1 computes $m'_1 \leftarrow \mathsf{SR}(\alpha'_1, ..., \alpha'_n)$. Likewise, m'_2 can be recovered. Based on **SA**, it's easy to verify that $m'_1 = m_1 + r_1$ and $m'_2 = m_2 + r_2$. Hence, we have $\varepsilon \leftarrow m_1 m_2 + r_2 m_1 + r_1 m_2 + r_1 r_2$. In the end, UC_i perform three subtractions to get η_i, in which $\eta_i = \sum_{j=1}^{k-1} y_j x_i^j + m'_1 m'_2 - r_1 m_2 - r_2 m_1 - r_1 r_2$. Let $\sum_{i=1}^{k-1} s_i x^i$ denote the polynomial used to split $m'_1 m'_2$, and let $\sum_{i=1}^{k-1} t_i x^i$ denote the polynomial to split $r_1 r_2$, in which $s_i, t_i \in_R \mathbb{Z}_p$ for $1 \le i \le k-1$. It can be easily observed that $y_i = s_i - r_1 b_i - r_2 a_i - t_i$ and the constant coefficient y_0 is $m_1 m_2$ due to removing randomness of r_1 and r_2. Therefore, the correctness of the **SM** protocol holds.

4.4 Secure Comparison Protocol (SC)

Design Details. Given that divided shares of m_1, m_2 are held by $UC_1, ..., UC_n$ respectively, the goal of **SC** is to compare the relationship between m_1 and m_2, i.e., $m_1 \le m_2$, or $m_1 > m_2$. During the execution of **SC**, neither m_1, m_2 nor $m_1 - m_2$ should be revealed to cloud servers.

Note that the secrets involved in association rule mining (including m_1, m_2) are represented by integers, the range of which is restricted in $[-N, N]$, where $N > 0$ and $\mathcal{L}(N) < \mathcal{L}(p)/8$. Hereafter, $\mathcal{L}(\cdot)$ denotes the length of value in bits. Recall that p is such a large prime that is used as finite field modulus. The details of the **SC** protocol are in the following:

Step-1 (@SC_2): SC_2 generates two random numbers $r \in_R \mathbb{Z}_p$ and $s \in_R \{0, 1\}$, s.t., $\mathcal{L}(r) < \mathcal{L}(p)/4$. SC_2 also computes shares of 1 by $S_1 \leftarrow \mathsf{SD}(1)$. r, s and S_1 are sent to UC.

Step-2 (@UC): $\forall i = 1, ..., n$, if $s = 1$, UC_i calculates the following by using its received shares $\theta_i \leftarrow 2 \times (\alpha_i - \beta_i)$. If $s = 0$, UC_i calculates $\theta_i \leftarrow 2 \times (\beta_i - \alpha_i)$. Supposing that $S_1 = \langle \mu_1, ..., \mu_n \rangle$, UC_i then randomizes the difference with r by $d_i \leftarrow r \times (\theta_i + \mu_i)$. After that, d_i is transmitted to SC_1.

Step-3 (@SC_1): Upon receiving $d_1, ..., d_n$ from UC, SC_1 selects any k of them to reconstruct the blinded difference by $\delta' \leftarrow \mathsf{SR}(d_1, ..., d_n)$. If $\mathcal{L}(\delta') > \mathcal{L}(p)/2$, SC_1 denotes $\theta = 1$ and $\theta = 0$ otherwise. Then θ is transmitted to UC servers through secure channel.

Step-4 (@UC): Once θ is received, UC_i determines the relationship of m_1 and m_2. Let θ^* be the minimum indicator. If $s = 1$, $\theta^* \leftarrow \theta$; otherwise, $\theta^* \leftarrow 1 - \theta$. It can be inferred that $\theta^* = 0$ means that $m_1 > m_2$ while $\theta^* = 1$ means that $m_1 \le m_2$.

Correctness of SC. Recall that split shares of $SD(m_1)$ are $\alpha_i = \sum_{j=1}^{k-1} a_j x_i^j + m_1$ while shares of $SD(m_2)$ are $\beta_i = \sum_{j=1}^{k-1} b_i x^j + m_2$, for $i \in [1, n]$. The polynomial for $SD(1)$ is $\mu_i = \sum_{j=1}^{k-1} \nu_j x^j + 1$, where $\nu_j \in_R \mathbb{Z}_p$ for $1 \le j \le k - 1$. During Step-3, given that $s = 1$, SC_1 can recover δ' by any k from $d_1, ..., d_n$. Supposing that $\delta = 2(m_1 - m_2) + 1$, it's easy to derive that $\delta' = r\delta$.

Since $-N \le m_1, m_2 \le N$, we have $1 \le \delta \le 4N + 1$ if $m_1 \ge m_2$; otherwise, $-4N + 1 \le \delta \le -1$. By multiplying r, we have $r \le \delta' \le r(4N + 1)$ for $m_1 > m_2$ and $r(-4N + 1) \le \delta' \le -r$ for $m_1 \le m_2$. However, as these are modular operations, $r(-4N+1) \le \delta' \le -r \Rightarrow p+r(-4N+1) \le \delta' \le p-r$. Because $\mathcal{L}(r) < \mathcal{L}(p)/4$ and $\mathcal{L}(N) < \mathcal{L}(p)/8$, it can be verified that $\mathcal{L}(p)/8 < \mathcal{L}(\delta') < 3/8 \cdot \mathcal{L}(p)$ if $m_1 > m_2$, whereas if $m_1 \le m_2$, we have $\mathcal{L}(\delta') > \log(p - 2^{3/8 \cdot \mathcal{L}(p)+1} + 2^{\mathcal{L}(p)/4}) > \mathcal{L}(p)/2$ and $\mathcal{L}(\delta') < \log(p - 2^{\mathcal{L}(p)/4}) < \mathcal{L}(p)$. Hence, $\mathcal{L}(p)/2$ can be regarded as judgement threshold and the correctness of the **SC** protocol holds.

Furthermore, the reason why we use the form $2(m_1 - m_2) + 1$ instead of $m_1 - m_2$ is to avoid private information leakage. If $\delta = m_1 - m_2$, SC_1 definitely knows $m_1 = m_2$ when $\delta' = 0$. Thereby, without knowledge of r, s, nothing regarding actual values of m_1, m_2, or $m_1 - m_2$ are revealed to SC_1.

4.5 Integrity Verification Scheme (IV)

Design Details. Given that $\tau_1, ..., \tau_n$ are the split shares of $m \in \mathbb{Z}_p$ from UC, the basic idea of **IV** is to perform verification at the moment of reconstructing m. Recall that there are t failing UC servers, so the number of correct results is $n - t$. Thus, solving $\binom{n-t}{k}$ different linear systems of k equations should give the same recovered message. Emekci et al. [16] proved that any linear system each with at least one malicious party would give a different value. We extend this to Theorem 1 as follows:

Theorem 1. *Given two different sets of linear equations that produce the same results, the probability that the shares from both sets are correct is very large.*

Proof. The two sets of equations forms two different linear systems, denoted by $A_i c_i = s_i$ for $i \in \{1, 2\}$, where c_i is the polynomial coefficients to solve, s_i is a vector of divided shares, and A_i is the matrix composed of exponents of secret key, i.e., x_j^d, $j \in [1, n], d \in [0, k - 1]$. If $c_1 = c_2$, it's easy to infer that $s_2 = A_2 A_1^{-1} s_1$. However, it's almost impossible to compute A_1 and A_2 in advance, because neither the secret key X, nor which shares are chosen are revealed to UC. Without loss of generality, we assume s_1 contains t_1 false shares while s_2 contains t_2 false shares, where $t_1, t_2 \ge 1$. The probability of $c_1 = c_2$ is equal to the probability of selecting $t_1 + t_2$ random numbers from $\mathbb{Z}_p$, which is $1/p^{t_1+t_2}$. Contrarily, the probability of getting the correct shares if the solutions are equal is $1 - 1/p^{t_1+t_2}$, which can be infinitely large as long as p is large enough.

Based on Theorem 1, our verification strategy is straightforward: KH (key holder) randomly selects two sets of shares from $\tau_1, ..., \tau_n$ to recover secret messages, by which it judges whether the results from two different linear systems are equal. If they are equal, the result is considered correct; otherwise, at least one of UC servers make mistakes. Therefore, it is possible to verify the correctness of results from UC. The major steps are shown in Algorithm 1.

Algorithm 1. $IV(\tau_1, ..., \tau_n) \rightarrow true/false$

Require: KH has $\tau_1, ..., \tau_n$ and X.
 1: KH randomly chooses k out of $\tau_1, ..., \tau_n$ as set Γ_1 to compute the polynomial coefficients $\bar{c}_1$;
 2: KH randomly chooses k shares as set Γ_2 to compute the polynomial coefficients $\bar{c}_2$, s.t., $\Gamma_1 \neq \Gamma_2$;
 3: **if** $\bar{c}_1 == \bar{c}_2$ **then**
 4:     KH considers Γ_1 and Γ_2 are correct;
 5:     **return** $false$;
 6: **else**
 7:     KH considers Γ_1 or Γ_2 contains errors;
 8:     **return** $true$;
 9: **end if**

In terms of verifying the above building blocks, both DO and SC_1 can play the role of KH. For instance, to verify if the split shares are compromised by UC for secure data outsourcing scheme, DO executes **IV** based on its received shares. As for **SA**, DO may validate its calculated result by downloading the shares and recovering them. For **SM** and **SC**, SC_1 takes part in verification in that it reconstructs intermediate results by shares from UC at Step-3. The integrity of m'_1, m'_2 in **SM** and δ' in **SC** are validated by SC_1 while the output of Step-4 should be checked by DO. We stress that to verify the final result of **SC**, DO can simply check whether θ^* from all servers are equal rather than invoking **IV** because they should be the same if $t = 0$.

5   Proposed Outsourced Mining Protocol

In this section, we present the privacy-preserving outsourced association rule mining protocol by utilizing the building blocks proposed in Sect. 4.

5.1   Design Details

The entire outsourcing process can be divided into three stages, namely, Data Preprocessing Stage, Outsourced Mining Stage, as well as Result Retrieval Stage. The details of each stage is presented as follows:

Data Preprocessing Stage. The outsourced database can be vertically or horizontally partitioned. Let's take horizontally partitioned database as an example, i.e., each DO's dataset contains only a subset of total transactions whereas the item set is complete. This stage includes three steps:

Step-1 DO converts D into a matrix D_M, in which each row is a bit vector indicating the existence of certain items just like [9]. Specifically, element value "1" means the item exists while "0" means absence. Each column stands for a kind of item, so the count of columns is the total number of items.

Step-2 DO then randomly permutates the rows of D_M to obtain a transformed dataset, denoted by D_T. We use π_r to represent the permutation function. Note that π_c should be stored by DO in order to recover the true itemsets.

Step-3 Given that κ is security parameter, SC runs KeyGen(κ) to generate X. After that, DOs exploit secure data outsourcing in Sect. 4.1 to divide each element in each transaction of D_T. The result is denoted as D'_T. Since the underlying scheme is probabilistic, shares of the same secret are entirely different. Finally, the random shares are uploaded and converged at corresponding UC_i.

Outsourced Mining Stage. During this stage, cloud servers run Binary Apriori algorithm, which is a variant Apriori for mining data in bit vector representation [9]. We call the outsourced Apriori protocol as **OA**. Given that UC servers hold shares of the joint datasets and the support threshold, they aim at finding all the frequent itemsets, as shown in Algorithm 2.

Algorithm 2. OA(D'_T, T'_s) → L

Require: UC stores D'_T and T'_s, while SC holds X, where D'_T denotes split shares of the joint dataset and T'_s is the split shares of support threshold.
1: Initialize 1-item candidate set $C_1 = \{attr_i | i = 1, ..., m\}$, where $m = |I|$;
2: **for each** candidate $c \in C_1$ **do**
3:     $c.supp' \leftarrow$ **ComputeSupport**(c, D'_T);
4:     **if** $true ==$ **IsFrequentItemset**$(c.supp', T'_s)$ **then**
5:        $L_1 \leftarrow L_1 \cup c$; // Initial $L_1 \leftarrow \emptyset$
6:     **end if**
7: **end for**
8: **for** $(l = 1; L_l \neq \emptyset; l + +)$ **do**
9:     $C_{l+1} \leftarrow$ **GenerateCandidateSet**(L_l);
10:     **for each** candidate $c \in C_{l+1}$ **do**
11:        $c.supp' \leftarrow$ **ComputeSupport**(c, D'_T);
12:        **if** $true ==$ **IsFrequentItemset**$(c.supp', T'_s)$ **then**
13:           $L_{l+1} \leftarrow L_{l+1} \cup c$; // Initial $L_{l+1} \leftarrow \emptyset$
14:        **end if**
15:     **end for**
16: **end for**
17: **return** $L \leftarrow \{L_i \neq \emptyset | i = 1, ..., l + 1\}$;

To start with, UC servers concurrently generate candidate set C_1 with 1-item by selecting the individual attribute id $attr_i$ (at line 1). Then UC and SC jointly compute the frequent 1-item set L_1 by calling the **ComputeSupport** and **IsFrequentItemset** functions (at lines 2 to 6). After that, frequent l−itemsets for $l \geq 1$ are computed in an iterative fashion (at lines 8 to 16). Firstly, UC servers invoke **GenerateCandidateSet** to generate candidate set C_{l+1} of length $l + 1$. Secondly, similar with the steps of computing L_1, cloud servers calculate divided support shares for $\forall c \in C_{l+1}$. The frequent itemset L_{l+1} is obtained by identifying candidates in C_{l+1} whose supports are no less than predefined threshold T_s. The iteration terminates until all frequent itemsets are found.

There are three functions used in **OA**, namely, **ComputeSupport**, **IsFrequentItemset**, and **GenerateCandidateSet**, which are described as follows:

- **ComputeSupport**: it takes a candidate itemset c and D'_T as inputs, and outputs the shares of c's support, where $c \in C_l$. According to our data preprocessing schemes, we use attribute numbers to represent items. To compute the support of $c = \langle attr_1, ..., attr_l \rangle$, UC and SC compute Ω_i (i.e., the split indicator of c in transaction t'_i for $i \in [1, w]$) by performing bitwise AND-operation on columns $attr_1, ..., attr_l$ at i-th transaction. This bitwise operation $\prod_{j=1}^{l} t'_i(attr_j)$ is achieved by calling **SM**. With $\Omega_1, ..., \Omega_l$, UC executes **SA** to add them together to get the desired support shares, denoted as $c.supp'$.
- **IsFrequentItemset**: it takes $c.supp'$ and support threshold T'_s as inputs. The function is realized by UC and SC via executing **SC**$(T'_s, c.supp')$. If the output of **SC** is 0, we have $c.supp < T_s$; otherwise, $c.supp \geq T_s$ that implies that c is a frequent itemset, in which $c.supp$ and T_s are the real value of $c.supp'$ and T'_s, respectively.
- **GenerateCandidateSet**: it takes frequent itemsets L_l of length l ($l \geq 1$) as inputs, based on which UC computes a candidate set C_{l+1} of length $l + 1$. Since UC_i holds L_l for $i \in [1, n]$, the candidate itemset $c \in C_{l+1}$ is created by $\zeta \cup \epsilon$, where $\zeta \in L_l$, $\epsilon \in I \land \epsilon \notin \zeta$. Recall that I is the set containing all items. This function utilizes the monotone property that any subset of a frequent itemset must be freuqent.

As **IV** scheme has been integrated into the proposed privacy-preserving building blocks, the protocol can verify integrity of intermediate results if some UC servers are breakdown. Note that DO does not participate in mining stage directly, so only **SM** and **SC** can be verified by SC_1. Nevertheless, it is enough to verify the intermediate results of **ComputeSupport** and **IsFrequentItemset**. Even though SC_1 has no knowledge about when and which UC server may fail, a simple strategy is carried out that it aborts the execution once an error is detected during the outsourcing stage. Besides, none of the above functions require participations from DO.

Result Retrieval Stage. During this stage, DO retrieves all the frequent itemsets L and the divided shares of corresponding supports from UC. As the itemsets are attribute sequence numbers that are permuted at preprocessing, the real itemsets should be recovered by π_c^{-1}, while the real supports can be reconstructed with X. The correctness of the results can also be verified via **IV**.

6 Security Analysis

In the secure data outsourcing stage of our scheme, UC learns nothing regarding the outsourced datasets as long as the underlying secret sharing is semantically secure [15]. Since the support threshold are also encrypted, UC knows neither exact support of itemsets and threshold. So it is very hard to launch frequency analysis attack based on its background knowledge.

UC cannot learn anything regarding inputs or outputs during **SA** based on additive homomorphism. **SM** is achieved by cooperation between UC and SC. No secret is revealed to UC due to no collusion between UC and SC. Though SC_1 is able to obtain m_1', m_2' which are randomized by r_1, r_2, it does not learn the original messages and the multiplication result. Likewise, UC cannot solve the polynomial to know the inputs and δ in **SC**. SC_1 only gets randomized δ' so that the different δ is protected. Nothing is revealed to SC_2 for it does not hold any data. Furthermore, the real relationship between inputs is also protected from SC_1 because of random s. **IV** scheme is invoked during the execution of **SM** and **SC** at Step-3. Since all input data are blinded, no privacy is known to SC. Hence the proposed building blocks is secure under our threat model.

During Outsourced Mining Stage, **OA** algorithm invokes three functions iteratively, which are composed of the building blocks and set operations. Most of inputs, outputs, and intermediate results within these functions are protected from clouds. The only information revealed to UC is the output of **IsFrequentItemset**, which indicates whether the target itemset is frequent or not. But it is only used to update candidate itemsets. By random permutation over dataset columns, UC cannot deduce real frequent itemsets from D_T' and L. Therefore, our scheme protects data privacy based on Composition Theorem [17].

Combining above illustrations and Theorem 1, the outsourced mining protocol achieves both security and integrity objectives.

7 Performance Analysis

In this section, we analyze the performance of our solution from both theoretical and experimental perspectives.

7.1 Theoretical Analysis

Let `Mul` denote the modular multiplication operation. Given that Lagrange interpolation is used to recover the polynomial coefficients, it takes DO $n(k-1)$`Mul` to divide a secret, while it takes $(k-1)$Mul for recovery. **SM** takes $(2nk + 2k - n + 1)$Mul with $6n\mathcal{L}(p)$ communication traffic. The computational cost of **SC** is $(2n + k)$Mul and its communication cost is $2n\mathcal{L}(p)$. **IV** only needs two times of SR$(\cdot)$ operations. Note that the cost of outsourcing protocol depends on the frequent candidate itemset count in each iteration. To compute the support for l-itemset, the cloud servers call $w(l-1)$ times of **SM** and $w-1$ times of **SA**. **SC** is called once in comparison with threshold. For simplicity, we evaluate the overhead in one iteration. Thus, **OA** repeats $|C_l|$ rounds of **ComputeSupport**, **IsFrequentItemset**. The computation and communication complexity of our scheme are $O(|C_l| \times w \times l \times n \times k)$ and $O(|C_l| \times w \times l \times n)$, respectively.

7.2   Experimental Analysis

To the best of our knowledge, there is no prior works that consider data privacy and integrity verification simultaneously. So we only evaluate our protocols under different settings. We carry out tests using the real datasets from FIMI[1], in which the Chess datasest contains 3196 transactions with different 75 items. The protocols are implemented upon Intel Xeon E5-2620 CPU with 8GBytes of memory using Crypto++ 5.6.3 library. We choose $\mathcal{L}(p) = 512$ throughout our tests and transform the raw dataset into its binary representation like [9].

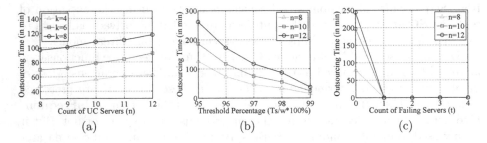

Fig. 2. Cloud computational cost with varying parameters on the Chess dataset.

First, we assess the overhead on cloud servers with varying UC server count (n) and selected shares (k) when the support threshold $T_s = 97\% w$. The results are given in Fig. 2(a). The cloud running time includes computation time for UC and SC. It can be seen that the time grows gradually with the increase of n and k. Since k defines the degree of polynomial coefficients while n determines the number of equations, increasing them definitely incurs more costs during calling **SM** and **SC** while improving the system redundancy.

We then evaluate the outsourced mining protocol with varying T_s and n. Figure 2(b) depicts that the cloud running time declines with the growth of T_s, because less frequent itemsets suffice the higher threshold condition. However,

---

[1] http://fimi.ua.ac.be/data/.

it is noteworthy that the impact of support threshold depends on the density of dataset, thus the same threshold ratio may result in different amounts of frequent itemsets for different datasets.

Lastly, we evaluate the integrity verification performance of our solution. In the setup, we choose $T_s = 97\%w$ and $k = n/2$. Figure 2(c) shows the cloud running time sharply drops to near zero with the increasing count of failing UC servers (t) irrespective of n. This is because our scheme identifies the appearance of servers' errors and aborts the execution. Throughout the tests, **IV** achieves 100% accuracy of integrity verification. Furthermore, we find that the communication overhead of **OA** presents similar tendency as computation cost, whereas this part of analysis is omitted due to space limitation.

8 Conclusion

In this paper, we proposed a verifiable and privacy-preserving solution for outsourced association rule mining. On the basis of secret sharing scheme, we proposed a series of efficient privacy-preserving building blocks to evaluate addition, multiplication, comparison over encrypted data and to transfer majority workload to clouds. The verification scheme checks the validity of the mining results by using the redundancy of servers. Our solutions protect privacy of the outsourced database and itemset frequencies under the hybrid-cloud threat model. Experiments on the real dataset show that our protocol performs efficiently and identifies errors effectively in case of failing servers. As future work, we will focus on further improving the efficiency and security of the techniques for privacy protection and integrity verification to make it more practical in real-life.

References

1. Wong, W. K., Cheung, D.W., Hung, E., Kao, B., Mamoulis, N.: Security in outsourcing of association rule mining. In: Proceedings of the VLDB, pp. 111–122 (2007)
2. Amazon Machine Learning. https://aws.amazon.com/machine-learning/
3. Cloud Machine Learning Engine. https://cloud.google.com/ml-engine/
4. Azure Machine Learning. https://azure.microsoft.com/en-us/services/machine-learning-services/
5. Mehmood, A., Natgunanathan, I., Xiang, Y., Hua, G., Guo, S.: Protection of Big Data Privacy. IEEE Access **4**, 1821–1834 (2016)
6. Tai, C., Yu, P. S., Chen, M.: k-support anonymity based on pseudo taxonomy for outsourcing of frequent itemset mining. In: Proceedings of the 16th International Conference on Knowledge Discovery and Data Mining, pp. 473–482 (2010)
7. Giannotti, F., Lakshmanan, L.V.S., Monreale, A., Pedreschi, D., Wang, H.: Privacy-preserving mining of association rules from outsourced transaction databases. IEEE Syst. J. **7**(3), 385–395 (2013)
8. Yi, X., Rao, F., Bertino, E., Bouguettaya, A.: Privacy-preserving association rule mining in cloud computing. In: Proceedings of the ASIA CCS, pp. 439–450 (2015)

9. Liu, J., Li, J., Xu, S., Fung, B.C.M.: Secure outsourced frequent pattern mining by fully homomorphic encryption. In: Madria, S., Hara, T. (eds.) DaWaK 2015. LNCS, vol. 9263, pp. 70–81. Springer, Cham (2015). https://doi.org/10.1007/978-3-319-22729-0_6

10. Li, L., Lu, R., Choo, K.K.R., Datta, A., Shao, J.: Privacy-preserving-outsourced association rule mining on vertically partitioned databases. IEEE Trans. Inf. Forensics Secur. **11**(8), 1847–1861 (2016)

11. Gennaro, R., Jarecki, S., Krawczyk, H., Rabin, T.: Secure distributed key generation for discrete-log based cryptosystems. J. Cryptol. **20**(1), 51–83 (2007)

12. Gentry, C., Halevi, S.: Implementing gentry's fully-homomorphic encryption scheme. In: Paterson, K.G. (ed.) EUROCRYPT 2011. LNCS, vol. 6632, pp. 129–148. Springer, Heidelberg (2011). https://doi.org/10.1007/978-3-642-20465-4_9

13. Wong, W.K., Cheung, D.W., Hung, E., Kao, B., Mamoulis, N.: An audit environment for outsourcing of frequent itemset mining. In: Proceedings of the VLDB, pp. 1162–1173 (2009)

14. Dong, B., Liu, R., Wang, W.H.: Integrity verification of outsourced frequent itemset mining with deterministic guarantee. In: ICDM, pp. 1025–1030 (2013)

15. Shamir, A.: How to share a secret. Commun. ACM **22**(11), 612–613 (1979)

16. Emekci, F., Methwally, A., Agrawal, D., Abbadi, A.E.: Dividing secrects to secure data outsourcing. Inf. Sci. **263**, 198–210 (2014)

17. Goldreich, O.: The Foundations of Cryptography - Volume 2, Basic Applications. Cambridge University Press, Cambridge (2004)

Resource and Attribute Based Access Control Model for System with Huge Amounts of Resources

Gang Liu$^{(\boxtimes)}$, Lu Fang, Quan Wang, Xiaoqian Qi, Juan Cui,
and Jiayu Liu

School of Computer Science and Technology,
Xidian University, Xian 710071, China
gliu_xd@163.com, l_silence@164.com, qwang_xd@163.com

Abstract. In information systems where there are a large number of different resources and the resource attributes change frequently, the security, reliability and dynamics of access permissions should be guaranteed. The changing raises security concerns related to authorization, and access control, but existing access control models are difficult to meet practical requirements. In this paper, a resource and attribute based access control model named RA-BAC was proposed. The model bases on attribute-based access control (ABAC) and links access control policy with resource, and redefines the access control rules. Besides, we compare RA-BAC and ABAC from the perspective of theory and simulation experiment respectively to show the advantage of RA-BAC model. We give a detailed analysis combining with instances to show the practicability of the RA-BAC model. RA-BAC solves the problems of policy conflict and policy library expansion in the ABAC model when there are too many resources and the attributes of resources are changed frequently in the system. Using RA-BAC model in system can makes permission query efficient and reduce workload of the system administrator of managing the policy library.

Keywords: Access control · Resource · Attribute ·
Attribute-based access control · Policy conflict

1 Introduction

With the advances in information technology and the information explosion, there are more and more systems that maintain huge amounts of information and the access to such systems should be secure, reliable and dynamic. Access control is an essential mechanism that controls what operations the user may or may not be able to do [9] and it's one of the key technology to solve these problems.

But in terms of traditional access control models, it is hard to apply them to the distributed environment of large information systems [8], so many researchers study on the attribute based access control model (ABAC) [6] at present. However, ABAC model is not fully apply in the environment where quantity of subject and resource is very large. For example, in an auto parts production workshop, there are hundreds of different kinds of parts and each of these parts need different operations. A strong

© Springer Nature Switzerland AG 2019
S. Li (Ed.): GPC 2018, LNCS 11204, pp. 49–63, 2019.
https://doi.org/10.1007/978-3-030-15093-8_4

collaboration and connection exist between these operations, and different people perform different operations. If we deploy ABAC model in this system, the workers are the subjects and the parts are the resources, the environment condition is used for restraining access control permissions. However, due to the specification and maintenance of the policies, ABAC model has the great complexity and the heterogeneity of user information even increases the complexity [1]. Thus, millions of access control rules should be predefined if using ABAC model in this system, and when the number of subject attributes, resource attributes and environment attributes increase, there are significantly increased in the numbers of access control rules. Besides, ABAC policy rules are often complex making them prone to conflicts [10] and too many rules may lead to policy library expansion.

Thus, this paper proposes a resource and attribute based access control model named RA-BAC. Unlike ABAC model, the access control policy doesn't store in the policy library in the RA-BAC model, there is an access control list (ACL) linked with each resource and the permissions which the subject has to access the resource are stored in the ACL. And at the same time, a new access control rule is developed according to the permissions in the RA-BAC model. In the RA-BAC model, system decide if the subject has permission to access the resource through the permissions stored in the ACL which is linked with the resource. Because there are only permission sets in the access control policy, so there is no policy conflict. And in the RA-BAC model, there is no policy library and the access control policy is stored separately in the ACL linked with the resource, so there are no policy library expansion when the number of the resources is large. And the way that access control policy stored in the ACL separately can increase the efficiency of the permission query and make the administrator add or delete the policy more efficient, which reduces the heavy task of the administrator in the ABAC model. Compared with the ABAC model, RA-BAC model is more suitable for the system which has a large number of resources and the attributes of the resources change quickly, such as the internet system, in which each node is a resource and the access control policy can be stored in it, and the auto parts production workshop which is proposed in this paper. In conclusion, RA-BAC model not only solve the problems of policy conflict and policy library expansion in the ABAC model, but also make the permission manage more efficient, flexible and universality.

This paper is organized as follows: Sect. 2 introduces related works, which contain the definition of the ABAC model and the deficiencies of the ABAC model. Section 3 gives a formal definition of RA-BAC and defines its access control rules firstly, and next interprets its operational framework, finally gives a evaluation of the RA-BAC and the ABAC model from the angles of theory and simulation experiments respectively. Section 4 describes a scenario for illustrating the operation mode of the RA-BAC model and the model definition and operation process are described further, which proves the availability of the model. Section 5 provides the summary and provides the future development tendency.

2 Related Works

Attribute-based access control (ABAC) is a model which is raised for fine-grained access control, and ABAC is used for the problems of massive user extension dynamically in complex information system, and the basic architecture of ABAC are shown in Fig. 1 [5]. The ABAC can provide fine-grained access control, and the attributes of ABAC include subject, resource, operation, environment, whose attributes can be described as subject attribute (SA), resource attribute (RA), operation attribute (OA) and environment attribute (EA), that's A = {SA, RA, OA, EA}. The access control rules which made up of subject attributes, resource attributes, operation attributes and environment attributes are stored in the Access Control Mechanism (ACM), When a subject access to a resource, the ACM receives the subject's access request and gets the attribute values of the subject, resource and operation, then detect the environment attribute values, and examines these attributes against a specific policy stored in the ACM, the ACM determines if the subject can perform upon the object. If the request is allowed, the subject will be granted permission to access resources, otherwise rejected.

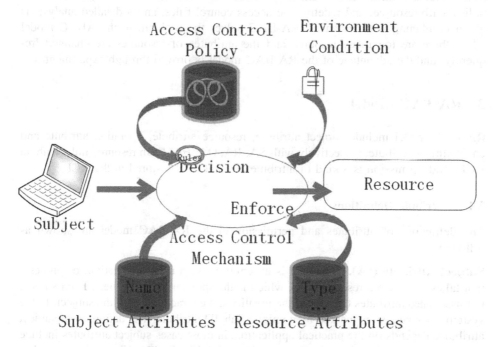

Fig. 1. Overview of the ABAC model.

The main characteristic of ABAC model is that the authorization in the ABAC model is based on the attributes, and the authority judgement is according to the policy in the policy library, and ABAC can apply to fine-grained access control well. But ABAC has some deficiencies in terms of practical application: if there are lots of subjects and resources in a system, there will be too many rules, and when the number of subject attributes, resource attributes and environment attributes increase, the numbers of access control rules will increase significantly, too many rules may lead to policy conflict and policy library expansion, the authority judgement efficiency will go low and the tasks of the system administrator will be very heavy.

In [1], Aftab et al. analyze the drawbacks of the ABAC model in the case of too many attributes in detail, and they raised a new model which combines ABAC model and RBAC model and the authority judgement is using attributes through the role, the access control model they developed that accumulates the strengths of RBAC and ABAC and also remove some of the deficiencies of above stated models. [7] takes a similar approach to solve the problem of the policy library expansion in ABAC. In [4], Hein et al. raise a conflict detection and resolution algorithm, and manage the life cycle of the access control rules, so the effectiveness of conflict detection and digestion algorithm can be improve. In [3], Fatema et al. improve the rules and forward the rules of conflict resolution formally, which can resolve policy conflict issues.

But these papers all study on one aspect of the ABAC model, and there is no practical scenario for the research. Thus, this paper proposes a resource and attribute based access control model named RA-BAC, this new model links access control policy with resource, and redefine the access control rules, and a detailed analysis is given combining with instances. RA-BAC solves the problem in the ABAC model when there are too many resource and the attributes of resources are changed frequently, and the advantage of the RA-BAC model is proved through experiment.

3   RA-BAC Model

RA-BAC model include subject attribute, resource attribute, operation attribute and environment attribute, respectively with SA, RA, OA, EA. Each resource links with an ACL, and permission is a kind of attribute element of RA stored in the ACL.

3.1   Attribute Definitions

The definitions of attributes and permission in the RA-BAC model are shown as follows:

Subject Attribute (SA). A subject is an entity (e.g., a user, application, or process) that takes action on a resource [12], which is the sponsor of the request. Each subject has associated attributes that define the identity and characteristics of the subject. In the system, each subject is marked by a unique Sub_ID attribute. The definition of subject attributes depends on the practical application, in most cases subject attributes include the subject's ID, subject's name, among others, and the subject attribute is expressed as SA = {Sub_ID, Sub_name, ...}.

Operation Attribute (OA). An operation is an action which is executed by a subject when it has access to an object (such as read and write), and it is expressed as "O" We can describe operation attribute as OA = {read, write, ...}.

Environment Attribute (EA). The environment attribute describes the system state when a subject has access to an object. These attributes describe the operational, technical, and even situational environment or context in which the information access occurs [12]. Example attributes include current date and time, the current environment temperature when access activities occur, and the system security level when hackers attack system. An environment attribute is expressed as EA = {date, time, ...}. Similar to the environment attributes in ABAC, attribute authentication is based on conditions in the environment when a request is made in ABAC [2]. In RA-BAC, an environment attribute is applied to the access control policy but it does not depend on a particular user and resources.

Resource Attribute (RA). Resource is accessed by a subject in RA-BAC, a resource attribute has a different definition in different application scenarios, such as network service, data structure or system components [13]; each resource is marked by a unique Res_ID attribute, and access control list, which consists of permissions; and what operation is allowed to be performed by a subject in a particular environment is linked with every resource. Resource attribute includes Res_ID attribute, resource name attribute, resource size attribute, etc., and it varies with different environments, but permission is indispensable, so we defined permission as attribute element in the resource attribute sets in RA-BAC, so RA = {Res_ID, permission, ...}.

Permission. Permissions store numerous permission data, and attribute expression in RA-BAC is defined by the system administrator in advance. Every permission includes a group of subject attribute and operation attributes, and environment attribute. Because SA and OA can be obtained from the process that subject send an access request to resource, but system should detect the EA from sensor equipment etc. So EA is different from SA and OA, and system will not detect EA in some special cases, so we make a formal definition for permission in RA-BAC and describe it by a binary group:

$$Permission = (PA, EA), PA = (SA, OA)$$

PA is a binary group (SA, OA), which consists of SA and OA, and SA and OA are the subject attribute and operation attribute, respectively, where a subject is allowed to operate on an object. EA is a set of environment attributes where permission is allowed. The environment attribute set is also divided into two cases:

(1) When EA is empty, permission is not constrained by environmental attribute, and access request is allowed only if $(s, o) \in PA$ is true; and s and o express the subject attributes of the request subject and the operation attributes of accessed resource, respectively.
(2) When EA is not empty, authorization is allowed under certain environmental conditions. Thus, even though $(s, o) \in PA$ is true, environmental attribute still needs to be true so the access request can be allowed, and e expresses the current environment attributes when the access occurs.

3.2   Access Control Policy

In the previous section, we defined attribute, and in this section, we introduce access control policy in the RA-BAC model, and the access control policy is as follows:

1. S, R, O, and E are subjects, resources, operations and environments, respectively;
2. $SA_k(1 \leq k \leq K)$, $OA_p(1 \leq p \leq P)$, and $EA_n(1 \leq n \leq N)$ are the predefined attributes for subjects, operations and environments, respectively;
3. ATTR(s), ATTR(o), and ATTR(e) are attribute assignment relations for subject s, resource r, operation o and environment e, respectively:

$$ATTR(s) \subseteq SA_1 \times SA_2 \times \ldots \times SA_K$$
$$ATTR(o) \subseteq OA_1 \times OA_2 \times \ldots \times OA_p$$
$$ATTR(e) \subseteq EA_1 \times EA_2 \times \ldots \times EA_n$$

In the RA-BAC model, a access control rule is a Boolean function of the subject attribute, resource attribute, operation attribute and environment attribute, and it decides on whether a subject S can perform an operation O on resource R in a particular environment E. The access control rule is defined as follows:

Step1: $ATTR(s) \times ATTR(o), e$
Step2: If $EA \neq \emptyset$, $can\_access(s, r, o, e) \leftarrow (ATTR(s) \times ATTR(o)) \in PA \wedge (e \in EA)$
Else $can\_access(s, r, o, e) \leftarrow (ATTR(s))$

The RA-BAC model use this rule to determine if the authorization is legal, and the evaluation process is shown in Fig. 2. The first step is to calculate the value of the subject attributes and operation attributes $ATTR(s) \times ATTR(o)$ and collect the value of the environment attribute e. The second step is to estimate whether EA is empty firstly, if EA is not empty, then the system will estimate whether Cartesian product's any subset of s and o exist in the PA of the permission, and at the same time, the current environment attributes e should exist in EA of the permission; If EA is empty, then the system only need to estimate whether Cartesian product's any subset of s and o exist in the PA of the permission, and there is no need to estimate whether. If the condition is satisfied, then the value of the Boolean function is T, thus, the subject is allowed to operate on the resource, return Permit; otherwise, the value of the Boolean function is F and the access request is refused, return Deny. Using this access control policy in RA-BAC can estimate the access request effectively. For example, there is a system where only permitting the subject whose ID is more than 10 to access the resource whose ID is less than 50 and read it under the circumstance that the system security level is less than 3, thus, $permission = (P = (Sub\_Id10) \times (O = read), E = (systemsecurelevel3))$ exists in the permission of the resource whose ID is less than 50. According to the access control rule above, when a subject A whose ID is 15, request to read the resource whose ID is 30 in the system under

IF $EA \neq \phi$

    IF $(ATTR\,(s) \times ATTR\,(o)) \in PA$ and $e \in EA$

        return *Permit* ;

    ELSE

        return *Deny* ;

    END IF

ELSE

    IF $(ATTR\,(s) \times ATTR\,(o)) \in PA$

        return *Permit* ;

    ELSE

        return *Deny* ;

END IF

Fig. 2. The process of policy evaluation

the circumstance that the system security level is 2, such calculation will be made as follows:

Step1: $ATTR(s) \times ATTR(o) = (Sub\_ID = 15) \times (O = read)$ and $e = $ (system secure level = 2)

Step2: $can\_access(s, r, o, e) \leftarrow (ATTR(s) \times ATTR(o)) \in PA) \wedge (e \in EA)$

According to the system rules, the result of the Boolean function is true, and the access request is allowed.

3.3   Structure of RA-BAC Model

This RA-BAC architecture integrates XACML in actual implementation. eXtensible access control markup language (XACML) is an OASIS standard that describes both a policy language that is implemented in XML and an access control decision request/response language that is implemented in XML [11]. A typical authorization architecture of the RA-BAC model is illustrated in Fig. 3.

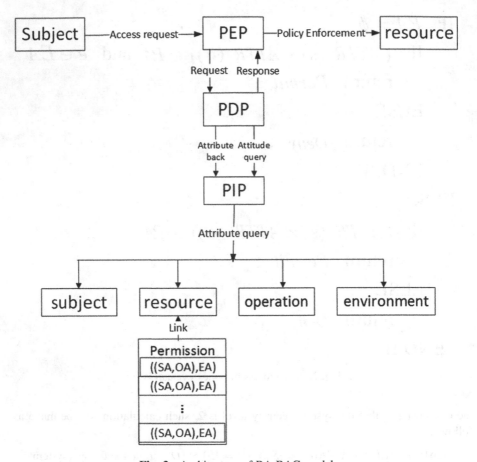

Fig. 3. Architecture of RA-BAC model.

- Policy enforcement point (PEP). The subject in the system sends an access request to PEP, and PEP performs access control according to the authorizations that were returned from PDP. In the RA-BAC model, PEP is a connection point between the subject and the resource. PEP receives the access requests and transfers them to the policy decision point (PDP). After receiving an access control decision from PDP, if the decision is Permit, then PEP permits the subject to access the resource; if the decision is Deny, then the subject cannot access the resource. If the decision is not applicable, meaning there is no matching access control rule, this request is considered illegal by default, and PEP denies the subject access to the resource.
- Policy information point (PIP). PIP is responsible for collecting the subject attributes, resource attributes, operation attributes and environment attributes and transforms the attributes to PDP so that PDP can make an access control decision through these attributes. In the RA-BAC model, because a resource links with ACL directly and there are permissions in ACL, PIP can collect permission attributes through the resources.

- Policy decision point (PDP). PDP is responsible for applying the applicable policies to evaluate subject attributes, resource attributes, operation attributes and environment attributes and making the authorization decision (permit or deny). PDP is a very important policy execution engine. PDP calculates subject attributes and operation attributes $ATTR(S) \times ATTR(O)$ and estimates whether the value of the subject attribute and operation attributes exists in the $PA = (SA, OA)$ of the permission attribute. At the same time, if the environment attribute is not empty, whether the current environment attributes exist in the environment attribute E of the permission attribute should still be considered. If the environment attribute is empty, then this step can be omitted. If $ATTR(S) \times ATTR(O)$ and e both exist in the permission, then the access request is permitted; otherwise, it is not. PDP returns the access control decision to PEP and PEP will perform appropriate operation.

The information flow of the RA-BAC model is described as above, and compared with ABAC, the RA-BAC model has more advantage in the illustrative scenario where there is a large number and many different kinds of resources. The authorization in the RA-BAC model is based on attributes, so it has all the advantages of the ABAC model. Besides, there are only permission sets in the access control policy, so there are no policy conflict. And in the RA-BAC model, there is no policy library and the access control policy is stored separately in the ACL linked with the resource, so there is no policy library expansion when the number of the resources is large. And the way that access control policy is stored in the ACL separately can increase the efficiency of the permission query and make the administrator add or delete the policy more efficient, which reduces the heavy task of the administrator in the ABAC model.

3.4   Evaluation of RA-BAC Model

In order to show the advantage of RA-BAC further, we compare RA-BAC and ABAC from the perspective of theory and simulation experiment respectively in this section.

Firstly, ABAC uses the access control policy which is stored in the policy library as the basis for authorizations, When there are too many policies, it is more likely to lead to policy conflict and policy library expansion. But in RA-BAC, there is no policy library and access control policy is stored separately in the ACL linked with the resource, and there are only permission sets in the access control policy, so there is no defect of policy conflict and library expansion.

Secondly, all the access control policies are stored in the policy library, policy management tasks are overweight for system administrator. For instance, if there are n resources need to be deleted, in the ABAC model, system administrator needs to find the policy that corresponds with the original n resources in the policy library and then delete them. To facilitate the calculation, suppose that each resource needs m operations and the m kinds of operations can be performed by k subjects. If using ABAC in this situation, system administrator want to modify the worker's permissions, they will need to find $n \times m \times k$ kinds of access control policy that corresponds with the original resources in the policy library, and then delete them without considering environmental factors. However, if using RA-BAC, the system administrator only needs to delete original n resources and then ACL linked with them will be deleted as well, the tasks of system administrator can be reduced greatly. And the bigger the value of m and k, the advantage of RA-BAC model will be more obvious. Besides, if there are correlations between the operations, modifying the permission is quite complex. In the RA-BAC model, there is only the need to find the resource and modify the linked permission, which greatly reduces the complexity of modifying permissions. Thus applying RA-BAC in the system that has enormous resources and the attributes of resources changed frequent, the tasks of system administrator can be reduced greatly.

Using RA-BAC model in a system that has a large number of access control policies, the efficiency to estimate whether a subject can operate on a resource will be improved effectively. Analyzing from the perspective of time complexity, supposing there are m resources and n policies, and to facilitate the calculation, suppose that each resource has k related policies. In the ABAC model, the system should compare each subject attribute, resource attribute, operation attribute and environment with each rule and find the rules in the policy library when the system estimate the policy, and the time complexity of sort search is $O(n)$. In the RA-BAC model, when the system estimate the policy, system should find the source firstly and find rules in the ACL linked with resource, and the time complexity of sort search is $O(m + k)$. In the ABAC model, the access control policy consist of $SA \times RA \times OA \times EA$. And in the RA-BAC model, the access control policy consist of $((SA, OA), EA)$, there are only permission sets in the access control policy. In the system where there are a large number and many different kinds of resources, each resource has many resource attributes, so $n \gg m$ and $n \gg k$. Thus, the time complexity of RA-BAC is lower than ABAC, that's to say, RA-BAC model can increase the efficiency of the permission query.

Finally, a simulation experiment was given in this paper. Firstly, a database named RA-BAC was created in MYSQL, next we create subject table, resource table and policy table in the database; Then we create the permission table named with resource id, and the permission data in the permission table is consistent with the policy in the policy table. In this simulation experiment, in order to keep the consistency of policy in ABAC and RA-BAC, the policy stored in the policy table are permissible policies generated by rules, so the policy search time in the ABAC model in practice will be longer. That's say, the compare result of the simulation experiment is close to reality but the difference value is lower, the advantage of RA-BAC will be more obvious in practice. We store the permission table of the RA-BAC model and the policy table of the ABAC model in the database respectively.

The experimental process is: First of all, database generate 1000 subjects and 1000 resources randomly, and then generate access control policy according to the relationship between the subject attribute, resource attribute, operation attribute and environment attribute and store them in the policy table, there are 144739 policies in the policy table. Then, we create 1000 resource tables in the RA-BAC database named with the resource id, and add policies in the permission table according to the policy table so that the experimental are controlled. Because the advantages of the RA-BAC model is more obvious in the system where there is a large number and many different kinds of resources, so we change the number of the resources in the experimental and have a further test. To prevent the chance of the experiment, we do a lot of experiments on each data and take the average as the result, and the result are shown in Table 1. As shown in the table, the time of the permission inquiries, the time of the system administrator delete one resource, the time of the system administrator delete one permission of a resource and the time of the system administrator add one permission of a resource in the RA-BAC model all shorter than in the ABAC model. And when the numbers of the resources increase, permission inquiries, delete one resource and delete one permission of a resource will take more time in the ABAC model, and the time of these operations won't change in the RA-BAC model, so the more resources in the system, the more efficient of the RA-BAC model. The time of add one resource is related with the number of policies about the resource, and in our experiment, we suppose there are 500 policies related with the resource. In this condition, the time of add one resource in the RA-BAC model is the same as the ABAC model, and when we change the numbers of policies, the result remains the same.

In conclusion, the RA-BAC model is an improved ABAC model, which links the permission with resource and resolves the problems of policy conflict and policy library expansion in the ABAC model, and the RA-BAC model can be well applied in the system that has enormous resources and the attributes of resources changed frequent. The correctness of the theory mentioned above can be proven through the experiment, that's using RA-BAC model in the system where there are many different kind of resources and too many policies can increase efficiency and reduce the tasks of the system administrator to manage the policy library, and at the same time there is no policy conflict and policy library expansion in the RA-BAC model, so there are more advantages of the RA-BAC model in practical application.

Table 1. The return time comparison results of ABAC and RA-BAC

Number of resources		1000	3000	5000	10000
Permission queries	ABAC model	0.130 s	1.456 s	3.710 s	7.523 s
	RA-BAC model	0.003 s	0.003 s	0.0005 s	0.007 s
Delete one resource	ABAC model	0.2830 s	1.4809 s	3.8120 s	7.1601 s
	RA-BAC model	0.1320 s	0.1300 s	0.1290 s	0.1320 s
Add one resource	ABAC model	14.014 s	14.000 s	13.950 s	14.010 s
	RA-BAC model	13.944 s	13.945 s	13.951 s	14.001 s
Delete one permission of a resource	ABAC model	0.1320 s	1.4790 s	3.6950 s	7.0110 s
	RA-BAC model	0.0260 s	0.0580 s	0.0370 s	0.0350 s
Add one permission of a resource	ABAC model	0.0290 s	0.0280 s	0.0270 s	0.0290 s
	RA-BAC model	0.0280 s	0.0260 s	0.0270 s	0.028 s

4 Illustrative Scenario

This section illustrates the run mode of RA-BAC using the permission management of an auto parts production workshop as an instance, and shows the practicability of RA-BAC in a practical application.

There are many kinds of parts in an auto parts production workshop, and workers must operate different kinds of parts in different ways, and a strong collaboration and connection exist between these operations. Different tasks, such as casting, heat treatment, forging, cold stamping, welding, machining, and time quota and material quota also need to be accomplished by various workers. Likewise, a certain sequence needs to be followed in operations even on the same parts. Therefore, different tasks and positions have different operating permissions on the parts. Once a worker does an illegal operation on a particular part, its size and quality will be influenced and the enterprise will suffer great losses. However, control of the workers' operating permissions may be too complex. This is where our access control model in the system is applied, and then the rationality of the operation can be satisfied and the interests of the enterprise can be guaranteed effectively. We now show the architecture of a practical scenario applying RA-BAC, as follows.

Workers need to operate differently on a part and there is a strong connection between these operations. A part must be machined in this order: casting, forging, cold stamping, welding, machining, and heat treatment. The heat treatment must satisfy the condition that the quenching temperature be in 840 ± 10 °C and tempering temperature be in 525 ± 25 °C. In such a practical scenario described above, we make a definition of the attributes in RA-BAC, just as follows.

Subject Attribute (SA). The workers are the subjects, and the workers are marked by the unique number (SUBID), and the system estimate whether a worker has the permission to operate on a part based on the worker's ID, working time and the worker's position. Subject attributes: worker's ID, worker's name, working time and worker's position, which can be described as SA = {SUBID, SUBNAME, WORKTIME, position}.

Resource Attribute (RA). The parts are the objects, and the parts of the same attributes are divided into one group and marked by the same number (RESID), and different attributes of parts have different numbers. Resource attributes: part's ID, part's state, the material, part's size and linked geometry, which can be described as RA = {permission, RESID, State, Material, Size, Geometry}.

Environment Attribute (EA). In this case, this is the current temperature when the part is processed, and we describe it as EA = {Current Temperature (CT)}.

Operation Attribute (OA). The operation that workers can operate on a part, OA = {attributes (O)}. And the attribute values include casting, forging, cold stamping, welding, machining and heat treatment, which can be described as (casting, forging, cold stamping, welding, machining, heat treatment).

System Administrator. The system administrator is responsible for checking, approving and updating the permission lists.

In an auto parts production workshop, only the parts whose state is Machining can proceed to heat treatment by workers whose position task is heat treatment during normal working hours, and at the same time, the processing temperature must be between 820 °C and 850 °C. If any conditions above are not met, such as the workers not being at work or a worker is in the wrong position, if the latter, then the part is also in the wrong position or the processing temperature is wrong, as it was processed by a worker who does not have the permission to operate on the part.

If a heat treatment worker whose ID is 10010011 requests to have a heat treatment operation on the part whose ID is 1110002 during one's work time, with the right processing temperature, and the processing temperature is right. PDP needs to evaluate the policy using the access control policy to determine whether the workers have operation permissions after PIP has collected the relevant attributes which include the workers' number attribute, working time attribute, the parts' number attribute, the parts' state attribute, operation attribute and the environment temperature attribute, and so, the policy is shown as follows:

$$ATTR(s) \times ATTR(o) = (sub\_NAME = HT) \wedge (sub\_worktime(8:00, 17:$$
$$00)) \wedge (res\_ID = 1024) \wedge (res\_state = Machining) \times (o = HT)$$
$$e = CT \in (820\,°C, 850\,°C)$$

Then, PDP searches the value in the access control list associated with parts, which is aimed at estimating the value of $(ATTR(s) \times ATTR(o) \in PA) \wedge (e \in EA))$. Because the request is qualified, the evaluated value is true and the worker whose id is 10010011 is permitted to have a heat treatment operation on the part whose number is 1110002.

Next, if foundry workers request to have a heat treatment operation on the part whose number is 1110002 during their work time and the processing temperature is right. PIP will calculate the attributes $ATTR(s) \times ATTR(o)$ and evaluate the policy after PIP collects the attributes and transfers them to PDP. Then, we know that $(ATTR(s) \times ATTR(o) \in PA) \wedge (e \in EA)) = F$, so the foundry workers cannot have a heat treatment operation on the part whose number is 1110002.

In conclusion, deployment RA-BAC model in an auto parts production workshop to control the workers' operating permissions can prevent workers from illegal operation on parts effectively. Besides, using RA-BAC model to control the workers' operating permissions can reduce the heavy burdens of system administrator, and at the same time the worker's permission queries can increase effectively, so that the production efficiency of the enterprise can be improved.

5 Conclusion

We propose a resource and attribute based access control model named RA-BAC whose policy is based on attributes, it can be applied well in an environment that has a large number of resources and policies. There are only permissions in the access control policy in the RA-BAC model, so there is No disadvantaged of policy conflict. The way policy linked with the resource can increase the efficiency of the permission query and make the administrator add or delete the policy more efficient, we compare RA-BAC and ABAC from the perspective of theory and simulation respectively in this paper to prove the advantage of RA-BAC model. Besides, we defined the attributes of RA-BAC and access control policies, and applied the model in the permission management of an auto parts production workshop, which can prove the practicability of the RA-BAC model in actual scene.

Acknowledgments. This work is supported by the National Natural Science Foundation (NNSF) of China (Grant No. 61572385).

Conflict of Interest Statement. There is no conflict of interest regarding the publication of this paper.

References

1. Aftab, M.U., Habib, M.A., Mehmood, N., Aslam, M., Irfan, M.: Attributed role based access control model. In: Information Assurance and Cyber Security, pp. 83–89 (2016)
2. Covington, M.J., Sastry, M.R.: A contextual attribute-based access control model. In: Meersman, R., Tari, Z., Herrero, P. (eds.) OTM 2006. LNCS, vol. 4278, pp. 1996–2006. Springer, Heidelberg (2006). https://doi.org/10.1007/11915072_108
3. Fatema, K., Chadwick, D.W., Van Alsenoy, B.: Extracting access control and conflict resolution policies from European data protection law. In: Camenisch, J., Crispo, B., Fischer-Hübner, S., Leenes, R., Russello, G. (eds.) Privacy and Identity 2011. IAICT, vol. 375, pp. 59–72. Springer, Heidelberg (2012). https://doi.org/10.1007/978-3-642-31668-5_5
4. Hein, P., Biswas, D., Martucci, L.A., Muhlhauser, M.: Conflict detection and lifecycle management for access control in publish/subscribe systems. In: IEEE International Symposium on High-Assurance Systems Engineering, pp. 104–111 (2011)
5. Hu, V.C., et al.: Guide to attribute based access control (ABAC) definition and considerations. ITLB (2014)
6. Hu, V.C., Kuhn, D.R., Ferraiolo, D.F., Voas, J.: Attribute-based access control. Computer **48** (2), 85–88 (2015)
7. Kuhn, D.R., Coyne, E.J., Weil, T.R.: Adding attributes to role-based access control. Computer **43**(6), 79–81 (2010)
8. Ouldslimane, H., Bande, M., Boucheneb, H.: WiseShare: a collaborative environment for knowledge sharing governed by abac policies. In: International Conference on Collaborative Computing: Networking, Applications and Worksharing, pp. 21–29 (2012)
9. Riad, K., Yan, Z., Hu, H., Ahn, G.J.: AR-ABAC: a new attribute based access control model supporting attribute-rules for cloud computing. In: IEEE Conference on Collaboration and Internet Computing, pp. 28–35 (2015)
10. Shu, C., Yang, E.Y., Arenas, A.E.: Detecting conflicts in abac policies with rule reduction and binary-search techniques. In: IEEE International Symposium on Policies for Distributed Systems and Networks, pp. 182–185 (2009)

11. Singhal, A., Winograd, T., Scarfone, K.: Guide to secure web services. NIST Spec. Publ. **800**(95), 4 (2007)
12. Yuan, E., Tong, J.: Attributed based access control (ABAC) for web services. In: IEEE International Conference on Web Services, pp. 561–569 (2005)
13. Zhong, J., Hou, S.J.: Attribute-based universal access control framework in open network environment. J. Comput. Appl. **30**(10), 2632–2631 (2010)

Urban Data Acquisition Routing Approach
for Vehicular Sensor Networks

Leilei Meng[1], Ziyu Dong[2], Ziyu Wang[3], Zhen Cheng[1],
and Xin Su[1(✉)]

[1] College of IOT Engineering, Hohai University, Changzhou 213022, China
2831114887@qq.com, lte_5g@yeah.net, leosu8622@163.com
[2] 1600 Amphitheatre Pkwy, Mountain View, CA 94043, USA
dongziyu6@gmail.com
[3] Nanjing Ivtime, Town of Future Network, Nanjing 210000, China
wangziyu@ivtime.com

Abstract. Vehicular sensor networks have emerged as a new wireless sensor network paradigm that is envisioned to revolutionize the human driving experiences and traffic control systems. Like conventional sensor networks, they can sense events and process sensed data. Differently, vehicular sensors are equipped on vehicles such as taxies and buses. Thus, data acquisition is a hot issue which needs more attention when the routing protocols developed for conventional wireless sensor networks become unfeasible. In this paper, we propose a robust urban data acquisition routing approach, named Multi-hop Urban data Requesting and Dissemination (MURD) scheme. It consists of a base station, several static roadside replication nodes and many moving vehicles. They work together to realize the real-time data communications. The simulation results show that the proposed MURD is a flexible data acquisition routing approach which outperforms conventional approaches in terms of packet delivery ratio and data communication delay.

Keyword: VSN · Hybrid architecture · Data acquisition ·
Real-time communication

1 Introduction

The vehicular ad hoc networks (VANETs) have been obtaining commercial relevance because of recent advances in inter-vehicle communications and decreasing costs of related equipment. This situation stimulates a brand new family of visionary services for vehicles, i.e., from entertainment applications to advertising information, and from driver safety to opportunistic transient connectivity to the fixed Internet infrastructure [1–4]. Particularly, vehicular sensor networks (VSNs) [5] is emerging as new infrastructure for monitoring the physical world, especially in urban areas where a high concentration of vehicles equipped with onboard sensors is expected. VSNs inherently provide a perfect way to collect dynamic interest of information, and sense various physical quantities with very low cost and high accuracy. VSNs represent a significantly novel and challenging deployment scenario, which is considerably different

© Springer Nature Switzerland AG 2019
S. Li (Ed.): GPC 2018, LNCS 11204, pp. 64–76, 2019.
https://doi.org/10.1007/978-3-030-15093-8_5

from conventional wireless sensor networks (WSNs) [6, 7], thus requiring innovative solutions on data routing. In fact, vehicular sensors are not affected by strict energy constrains and storage capabilities because they can be equipped with powerful processing units and wireless transmitters. Consequently, energy dissipation and data storage space are not considered as routing designing issues in VSNs. The rapidly changing topology substitutes them becoming the major designing issues in recently research.

In this paper, we aim to design a data acquisition routing protocol for VSNs with the hybrid architecture [8], where the vehicular sensors are moving fast and keeping transmission in an ad hoc way. At the same time, they can exchange the information with roadside gateways keeping the network connectivity. The proposed routing approach focuses on some real-time applications, in which they are highly delay constraints. For example, after fearful nature disasters, such as Haiti earthquake and Morak typhoon, the VNSs are needed to build a real-time wireless communication system when the main urban wired networks are destroyed by the nature disasters. In this case, the sensed attributes of vehicular sensor could be "the detected life information by using the infrared ray", "the information of toxic gas leak" and so on. Therefore, the major designing issues for this paper are: how to design a feasible, efficient and robust VSN framework to monitor urban stimuli and provide desired and reliable information for users within a certain period of time from the moment data is sensed. In order to achieve this purpose, we set up a vehicular wireless network that consists of three kinds of system units including a base station, several static roadside replication nodes [9], and many vehicular sensors. Likely in [10], the roads are separated into small segments, each of which has a certain amount of fixed roadside replication nodes. Moreover, we assume that each segment is labeled with a unique ID grouped together and worked in a zone manner. Because the article is addressed to design an optimal data communication scheme for VSNs, several assumptions have been made. They are follows:

- Vehicles participating in the VSN are equipped with GPS devices, so they can know their own position, speed, and moving direction correctly.
- All the system units are equipped with identical preloaded digital maps.
- All the system units have plentiful space for storage and power efficient.
- All units in network work normally. There is no abnormal one being generated. Sensors detect and recognize the information correctly.
- There are obstacles such as buildings and mountains have potentially influence on blocking the transmission among the system units.

We focus on the collaboration of information exchange among vehicles, roadside sensors, and base station. Some of questions summarized and will be explained in the paper as follows:

- Do we need clustering scheme? Is it more efficient than homogeneous hierarchical approaches?
- What is roadside replication nodes used for? Why they are necessary in the system?
- How to decrease the negative influence to connectivity caused by the mobility?

In order to improving routing delay constraints, routing reliability and scalability, what kind of routing method should we used (Unicasting, Geocasting, and Broadcasting)?

2   System Models

There are three system models in the article, including the network model, the distributed event occurrence model, and the vehicles mobility model.

2.1   Network Model

In this article, suppose that all the transmitting units communicate with each other by using IEEE 802.11 standard. It is difficult to disseminate packets to different road segments [10] shadowed by obstacles because the urban area is crowded with tall buildings around intersections. Routing protocols need to predict an optimal path between the vehicles and the central base station in terms of distance and current status by using GPS. We label each road segment with a unique ID called segment ID (SID). A network zone is composed by several segments which can form a loop, and then the whole experiment area can be divided into zones as illustrated in Fig. 1. There is one base station located in the center of network area. The black filled circles denote roadside replication nodes located in the corner of zones and the middle of each road segment. The road segments are divided by the intersections, and several segments looped in a circle can form a network zone.

The network communication system can be composed of three units, including base station, moving vehicular sensors and roadside replication nodes. The base station is a large information storage unit and has the longest transmission range that can easily cover the vicinal zones. It can perform as an emergent center after the nature disasters of the application in Sect. 1 and periodically receives fresh interesting data from vehicular sensors while stores and responses the historical information to them. The second unit is the rapidly moving vehicular sensors which events-drivingly request interests. The third unit is the immobile roadside replication nodes that can be deployed at roads intersections or middle of long road segments storing and replying the interests. We assume that roadside replication nodes are power rechargeable that have plentiful storage space and long transmission range. Each of them keeps an optimal routing table for transmitting the sensed interests to the base station with multi-hop method. There are three reasons of utilizing roadside replication nodes: the first one is the application requirements that we described in Sect. 1, where the roadside replication nodes are used to keep the network connectivity when only few vehicles are deployed in the urban area after nature disasters. Due to the vehicular sensors limited transmission range and the obstacle blocking effect, it is difficult to ensure that every vehicular sensor can directly send the messages to the base station. In order to keep the network connectivity, the roadside replication nodes can take responsibility of information exchange between vehicular sensors and base station. The second one is due to the data collision avoidance. In the case of base station sending data to vehicles, it is difficult to predict each vehicle's precise position when the networks is large-scale and has hundreds of vehicles. If the base station broadcasts the messages, the packets collision rate will be high. Particularly, what if the forwarded data is not interested for every vehicle, using broadcast method cannot realize the efficient usage of the system resources. To handle this problem, we use roadside replication nodes to realize the geocasting scheme. The base station can geocast messages to a group of replication nodes located in the same network zone, and then they can easily unicast the message to the corresponding vehicles with low packets collision probability and high accuracy.

In the system, the vehicular sensors can forward the sensed data to either neighboring vehicles or the nearest roadside replication nodes. At the same time, they can send information request packets to neighboring vehicles and the nearest replication nodes to get interesting data. If the neighboring vehicles and the nearest roadside replication nodes have the requested interests, they immediately respond the requests; otherwise, the nearest roadside replications nodes aggregates the request packets with its SID and forwards it to the base station in a multi-hop method.

2.2 Distributed Event Occurrence Model

In the proposed system, the roadside replication nodes and vehicular sensors involved in routing process are driven by the event stimuli. Considering the application example in Sect. 1, the stimuli can be the events of detected people life and toxic gas leak. The nearest moving vehicular sensor was triggered to capture the stimuli and push the information to the emergent center-the base station. We assume that the average probabilities of event-driven vehicular sensors (VSs) are followed Possion Distribution. By considering the entire simulation time as a unit time interval, we can abbreviate the Possion distribution function as:

$$P(X = x) = \frac{e^{-\lambda_{VS}}(\lambda_{VS})^x}{x!}, x = 1, 2 \ldots, etc. \tag{1}$$

The λ is a positive real number which equal to the expected number of occurrences that occur during the given interval. It is variable with number of the system units simulated in the different scenarios. For instance, a practical default value for λ_{VS} is set as 0.1, that is, the average percent of event-driven vehicular sensors of all is 10%.

2.3 Mobility Models

We describe the mobility model for vehicles in proposed the scheme by summarizing the factors which affect mobility, and then establish two mobility models for all vehicles.

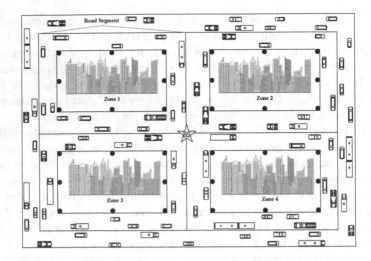

Fig. 1. An example network model in urban scenario.

2.3.1 Factors Affecting Vehicles Mobility

The mobility pattern of vehicles in a system influences the route discovery, maintenance, reconstruction, consistency and caching mechanisms. We illustrate several key factors that affect the mobility of vehicles in the system as follows.

Street layouts: streets force vehicles to confine their movement to well-define paths. It constrains movement pattern determines the spatial distribution of vehicles and their connectivity. Streets can have either single or multiple lanes and can allow either one-way or two-way traffic.

The obstacles: the obstacles, such as buildings determine the number of intersections in the area, as well as determine whether vehicles at the neighboring intersections can hear each other's radio transmission. Lager obstacles make the network more sensitive to clustering and degrade performance. In the paper, we assume that some virtualized obstacles located beside each road segment and influence the packets transmission among the system units.

Traffic control mechanisms: the most common traffic control mechanisms at intersections are stop signs and traffic lights. These mechanisms result in the formation of clusters and queues of vehicles at intersections and subsequent reduction of their average speed of movement. We set every street at an intersection has a virtual traffic lights. If vehicles follow each other move to an intersection with a red light, the vehicles form a street queue line at the intersection. Each vehicle waits for at least the required wait time once it gets to the head of the intersection after other vehicles ahead in the queue clear up. The traffic light gives the vehicles a probability, denoted as prob, to stop at the intersection when the vehicles reach an intersection with an empty queue. With the probability 1-prob, the vehicles can directly cross the intersection without stop. If it decides to wait, the amount of waiting time is randomly chosen between 0 and T seconds. Any vehicle that arrives later at a non-empty queue will have to wait for the remaining wait time of the previous vehicle plus one second which simulates the startup delay between queued vehicles. Whenever the signal turns green, the vehicles begin to cross the signal at intervals of one second, until the queue becomes empty. The next vehicle that arrives at the head of an empty queue again makes a decision on whether to stop with a probability prob and so on. Additionally, the vehicles are influenced by the movement pattern of their surrounding vehicles. The vehicle would try to maintain a minimum distance from the one in front of it by increase or decrease its speed, or may change to anther road lane.

2.3.2 The Mobility Models Implementation

In the system, there are two categories of the vehicles. One is event-driven vehicles which push or pull information from the base station, and we call them the Requesting Vehicles (RVs). The rest of vehicles are Common Vehicles (CVs). The Manhattan mobility is used for all the vehicles which move in a grid road topology mainly proposed for the movement in urban area, where the streets are in an organized manner. In this mobility model, the mobile vehicles move in horizontal or vertical direction on an urban map. The Manhattan model employs a probabilistic approach in the selection of vehicles movements. At each intersection, the probability of taking a left turn is 0.5 and that of right turn is 0.25 in each case. Thus the 0.25 is used for vehicles which move straightly. In order to highlight proposed MURD's real-time communication strongpoint and get a convincing result, we added the Dijkstra's algorithm mobility model for the RVs. At first, the vehicles which perform as RVs with the probability P (function 1) move in the Manhattan manner. After it receives the requested data from base station, it immediately moves in the Dijkstra's algorithm mobility manner. The destinations are chosen at the most distant road segment from the current one by using identical preloaded digital maps. Each RV follows the shortest path through the roads to its destinations. Then, it will changes back to the CV's status after the RV arrival at the destination. The reason for implement of Dijkstra's mobility will be explained in the Sect. 5.

3   The Proposed MURD

We present the proposed data acquisition scheme for vehicular sensor networks in this section, along with the urban scenario network models. The proposed MURD is composed of two system phases: the clustering and the data acquisition process.

3.1   Clustering

First, we explain the cluster formation of moving vehicles in proposed scheme. The sensors are located on the moving vehicles with GPS devices. Each vehicle knows its speed, moving direction, the precise location and the SID. The easy way for deciding to form a moving vehicular cluster is using SID. The setup of moving cluster occurs when some vehicles move into the same road segment and share a unique SID. The cluster release or reselection occurs when the moving cluster head vehicle moves out of the current segment. Figure 2 illustrates the proposed vehicular cluster formation method. When some moving vehicles, e.g. A, B, C, D, E move together with the equivalent speed and same direction. Each of them broadcasts their current speeds, and positions. The vehicle C which has speed close to the average speed and moves in the middle of this group of vehicles will automatically perform as a moving cluster head (MCH). Then, it sends clustering schedule information to form a moving cluster. When other moving vehicles, e.g. F, H move into the communication range of the MCH, they send the join-request with moving direction information to the MCH. If the moving direction is opposite to the MCH, e.g. H, the MCH will not allow it to join the moving cluster. In contrast the vehicle F which moving in a same direction with the MCH can receive the join-schedule packet from MCH and participate to the cluster. The reselection of the moving cluster occurs when the MCH moves out the current road segment. The reason for using clustering scheme is that the proposed MURD should decrease the probability of the data transmission between the vehicles and the roadside sensors abating the Doppler Effect. Using one MCH is more efficient than allowing all event-driven vehicles communicating with static roadside replication nodes at the same time, especially given a high default value of the λ_{VS}.

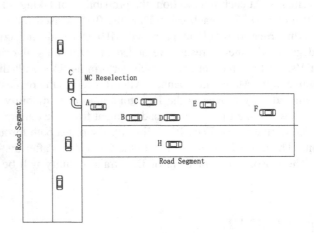

Fig. 2. The example of MC formation

3.2   Data Acquisition Process

In this section, we describe in detail that how data is acquired of the proposed scheme for successful transmission from the source to the destination. There are two data acquisition process, including data push process and data pull process. In the data push process, the sensed interests are forwarded from the vehicular sensors to the base station. On the other hand, the interests can be pulled from the base station via the data requesting and dissemination steps. Because the proposed routing mechanisms of these two processes are same, in the rest of paper, we only consider the data pull process. The entire process is divided into four steps: Data Requesting Phase where the data requesting packets are forwarded from vehicular sensors to roadside replication nodes; Data Forwarding Phase where the data requesting packets are aggregated with SID and routed from roadside replication nodes to the base station; Data Responding Phase where the requested interests are geocasted from base station to the roadside replication nodes; and Data Disseminating Phase where the interests are disseminated from the roadside replication nodes to the vehicular sensors.

3.2.1 Data Requesting Phase

An RV which requires interesting information sends a request control packet to its MCH vehicle via the unicasting method. The control packet includes the information of the vehicle, such as the vehicular ID, the current speed, the moving direction, and the information of destination position. The MCH which responds this packet will forward the packet to the nearest roadside replication node also via a unicasting method with help of city map information provided by the navigation system. Then the roadside replication node aggregates its own SID into this packet, and addresses to find an optimal way to forward it to the base station.

3.2.2 Data Forwarding Phase

By assuming that influences of the obstacles, there is not a directly way to forward packets to the base station. Thus, we need a multi-hop method to realize the data communication between the roadside replication nodes and the base station. Because the roadside replication nodes are assumed static with efficient power supply and have enough storage space, each of them can keep the optimal routing table realizing the proactive routing. There are three conditions for them choosing the routes with unicast method. They are illustrated in Fig. 3. The first one is aware of MCH's moving direction in a road segment. That is, the aggregated requesting packets of the nearest replication node must be hopped along with the MCH's moving direction in the current road segment. For example, the requested nearest replication node A receives the requesting packet from the MCH and aggregates its SID to this packet. The node A will ignores the neighbor node K as a hop route because it is located in opposite direction of MCH's moving direction. The second condition is the base station position-aware condition. Suppose the replication node B, C, D located in the signal range of node A, only node B and C can change the message forwarding direction that closes to the base station. Thus, node A will choose node B or C as the next hop node. The third condition is Most Forward with Radius scheme [11], in which makes node A send packet to node C because it is located farther than node B. Consequently, node C has

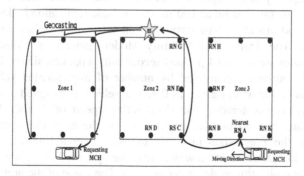

Fig. 3. An example of data forwarding phase for urban scenario

high priority to reach the base station. For the same reason, after node C sends the packet to node G, and then the packets can be directly send to the base station. The example routing path of proposed MURD is MCH-RNA-RNC-RNG-BS.

3.2.3 Data Responding Phase

After base station receives the requesting control packet of the vehicles, it can use this packet's information to determine the RV's current position, of which zone it is currently located in. Then, the base station uses geocasting method to send interesting data to the corresponding zone's roadside replication nodes. The replication nodes which can directly receive data from the base station will send the interesting data to the neighbor nodes in the same zone as illustrated in Fig. 3 in order to realize the data harvest.

3.2.4 Data Dissemination Phase

When all of the roadside replication nodes in the same zone receive the interesting data and anyone of them hears beacon messages that are periodically exchanged by RVs, the nearest replication node unicasts the interesting data to the RV's MCH immediately. The MCH then broadcasts the data until it receives the ACK message from the RV.

4   Simulation and Result

4.1   Simulation Environment

The primary goal of our proposed scheme is to establish a real-time vehicular communication system, which can maximize the delivery rate of offered packets and minimize the latency between source and destination. We implemented proposed scheme by using the ns-2 simulator, which is a discrete event simulator developed by the University of California at Berkeley and VINT project. The version of ns-2 is ns-2.34 [12] that implemented and based on the Monarch extensions to ns-2. The simulator models node mobility, allowing for experimentation with ad hoc routing protocols that must cope with frequently changing network topology. It implements the IEEE 802.11 Medium Access Control (MAC) protocol. The IEEE 802.11DCF is used with a channel capacity of 2 Mbps for vehicles, roadside sensors and base station. The vehicles transmission range is set as 100 m, the roadside replication nodes transmission range is fixed as 200 m and the base station has the longest transmission range which is given as 300 m. The Traffic and Mobility Models generation Constant Bit Rate (CBR) traffic sources are used of 4 packets/second with a packets size of 128 bytes. The number of CBR source is determined by number of implemented vehicles in each simulation and the value of λ_{VS} given as 0.1. The vehicle is assigned a maximal speed of 60 km/h with the accelerating and decelerating speed of 3 m/s2. Vehicles with random start points and destinations are running on the map. For controlled experiments, we varied the zones sizes in a grid topology over a 1200 m * 1200 m area. Furthermore, we implement the two-way road in the system. The distance between two neighbor roadside replication nodes is set as 50 m. The value of the probability prob is set as 0.25 to stop the vehicle at the intersection when it reaches an intersection with an empty queue, and maximum value T for waiting is 10 s. The simulation results are averaged over repetitive runs repeated with at least ten times which each one takes 900 s (15 min) of simulation time. Table 1 lists the main parameters used in the simulations.

4.2   Simulation Results

The performance metrics used to evaluate the simulation results are as following:

- Packet delivery ratio: the ratio of originated data packets that are successfully delivered to their destinations to the original sent ones.
- Average end-to-end delay: the average time it takes for a packet to traverse the network from its source to destination.
- Routing overhead: the ratio of number of bytes of total control packets to those of total data packets is delivered to the destinations during the entire simulation.

Table 1. Parameters used in our simulation

Simulation setup/Scenario		MAC/Routing	
Network area	1200 m * 1200 m	Channel capacity	2 Mbps
Number of vehicles	200	Transmission range	100 m, 200 m, 300 m
Simulation time	900 s	Traffic type	CBR
λ_{vs}	0.1	Packet sending rate	4 packet/sec
prob	0.25		
Vehicle speed	0–60 km/hr	Data packet size	128 bytes
Stop time (T)	10 s		
Mobility	Dijkstra's algorithm, Manhattan mobility	Buffer size	Unlimited

In the data requesting and forwarding phases of MURD, the data source and destination are requesting vehicles and the base station, respectively. On the contrary, the requesting vehicles and the base station will perform as data destination and resource in the data responding and dissemination phase, respectively. Consequently, the measured three performance metrics are the average value of these phases. They are compared under various vehicle densities and various distributed event occurrence means. In order to observe the strongpoint of the MURD, we compare the performance of MURD, AODV [13] and GSR [14] in terms of those three metrics. In the AODV and GSR, entire vehicles communicate each other in a vehicle-to-vehicle ad hoc manner without using roadside replication nodes. The packets sources and destination are requesting vehicles and the base station, respectively. The packets are forwarded via the vehicles to the base station and send back to the requesting vehicles with multihop method. For the better performance, designed protocol should be tolerable to a small packet loss. We will show how the packet delivery is affected by the number of nodes and distributed event occurrence means.

This section compares MURD, AODV and GSR beyond varying numbers of nodes in a 1200 m * 1200 m grid topology with a zone size of 200 m * 200 m. The Figs. 4 and 5 show the packet delivery ratio and delay for the MURD over the AODV. The results indicate that MURD yields the better results than AODV. This is because the MURD through creating roadside routing backbone to send the data from the sources to the destinations. The impact of increasing moving nodes is relatively lower than the competitor. In the situation of AODV, the trend is that the delivery ratio increase with the number of nodes, up to 80 nodes, as the connectivity of the communication increases. However, the delivery ration starts decreasing as the number of nodes increases further. Typically, it drops quickly when the number of vehicles more than 120 where the data commutation collision is experienced frequently leading a high re-transmit rate in the system. The data delay of AODV displays the opposite trend with its delivery ratio plot. It first decreases as the number of nodes increases, and then there is a sharp increase thereafter, whereas MURD uses the clustering scheme that reduces the number of connections between vehicles and roadside sensors and set up an optimal routing path which yields an ideal end-to-end curve with the number of nodes increases. The Fig. 6 displays the routing overhead of MURD and its competitor. The results of AODV perform better than MURD because it only has three types of control messages (RREQ, RREP and RERR) [13] used for route discovery and route maintenance process. In the proposed MURD, more than three control messages are used for clustering, data requesting and dissemination phases. However, the AODV overhead plot will be higher than MURDs' when the system nodes more than 120 where a high re-transmit rate exists.

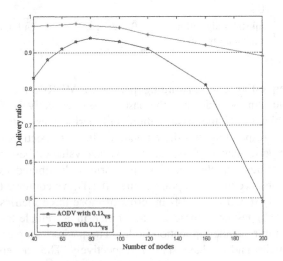

Fig. 4. Delivery ratio vs. number of nodes with 0.1 λ_{VS}

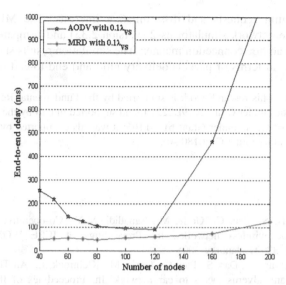

Fig. 5. End-to-end delay vs. number nodes with 0.1 λ_{VS}

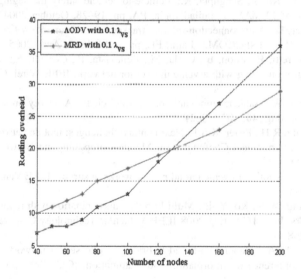

Fig. 6. Routing overhead vs. number of nodes with 0.1 λ_{VS}

5 Conclusion

In this paper, we have suggested an urban data communication scheme, named Multi-hop data Requesting and Dissemination (MURD) for proactive urban monitoring in VSNs. MURD constructs hybrid architecture in VSNs by using three categories system units. The vehicles request interests information from the base station through the roadside sensors. The data routing backbone is composed by roadside sensors, of which works by forwarding requests and disseminating interests. The MURD has been evaluated using the network simulator ns-2 (version 2.33) and compared with AODV which works in an ad hoc connection manner. The simulation results show that MURD outperform AODV in terms of packets delivery ratio and end-to-end delay.

Acknowledgements. This research work is supported by the Fundamental Research Funds for the Central Universities under Grant 2019B22214, and supported in part by the National Natural Science Foundation of China under Grant 61801166; it was also supported by Changzhou Sci. and Tech. Program under Grant CJ20180046.

References

1. Nandan, A., Das, S., Pau, G., Gerla, M., Sanadidi, M.Y.: Co-operative downloading in vehicular ad-hoc wireless networks. In: Proceedings of IEEE WONS, St. Moritz, Switzerland, pp. 32–41, January 2005
2. Nandam, A., Tewari, S., Das, S., Pau, G., Gerla, M., Kleinrock, L.: Ad-Torrent: delivering location cognizant advertisements to car network. In: Proceedings of IFIP WONS, Les Menuires, France, January 2006
3. Xu, Q., Mak, T., Ko, J., Sengupta, R.: Vehicle-to-vehicle safety messaging in DSRC. In: Proceedings of ACM VANET, Philadelphia, PA, pp. 19–28, October 2004
4. Ott, J., Kutscher, D.: A disconnection-tolerant transport for driven internet environments. In: Proceedings of IEEE INFOCOM, Miami, FL, pp. 1849–1862, April 2005
5. Lee, U., Magistretti, E., Zhou, B., Gerla, M., Bellavista, P., Corradi, A.: Mobeyes: smart mobs for urban monitoring with a vehicular sensor network. IEEE Wirel. Commun. **13**(5), 52–57 (2006)
6. Akyildiz, F., Su, W., Sankarasubramaniam, Y., Cayirci, E.: A survey on sensor networks. IEEE Commun. Mag. **40**, 102 (2002)
7. Kahn, J.M., Katz, R.H., Pister, K.S.J.: Next century challenges: mobile networking for smart dust. In: IEEE International Conference on Mobile Computing and Networking, pp. 270–278, August 1999
8. Li, F., Wang, Y.: Routing in vehicular ad hoc networks: a survey. IEEE Veh. Technol. Mag. **2**, 12–22 (2007)
9. Lim, K.W., Jung, W.S., Ko, Y.-B.: Multi-hop data dissemination with replicas in vehicular sensor networks. In: VTC Spring 2008 IEEE Vehicular Technology Conference, pp. 3062–3066, May 2008
10. Kong, F., Tan, J.: A collaboration-based hybrid vehicular sensor network architecture. In: International Conference on Information and Automation, ICIA 2008, pp. 584–589, June 2008
11. Takagi, H., Kleinrock, L.: Optimal transmission ranges for randomly distributed packet radio terminals. IEEE Trans. Commun. **32**(3), 246 (1984)
12. The Network Simulator – ns-2 (2010). http://www.isi.edu/nsnam/ns
13. Perkin, C., Belding-Royer, E., Das, S.: Ad Hoc On-Demand Distance Vector (AODV) routing. IETF Experimental RFC, MANET Working Group, RFC 3562, July 2003
14. Lochert, C., Hartenstein, H., Tian, J., Herrmann, D., Mauve, M.: A routing strategy for vehicular ad hoc networks in city environments. In: Proceedings of IEEE Intelligent Vehicles Symposium (IV2003), pp. 156–161, June 2003

Pervasive Sensing and Analysis

Searching the Internet of Things Using Coding Enabled Index Technology

Jine Tang[1] and Zhangbing Zhou[2,3($\boxtimes$)]

[1] School of Computer Science and Communication Engineering,
Jiangsu University, Zhenjiang, Jiangsu, China
tangjine2008@163.com
[2] School of Information Engineering,
China University of Geosciences, Beijing, China
zhangbing.zhou@gmail.com
[3] Computer Science Department, TELECOM SudParis, Evry, France

Abstract. With the Internet of Things (IoT) becoming a major component of our daily life, IoT search engines, which can crawl heterogeneous data sources and search in highly dynamic contexts, attract increasing attention from users, industry, and the research community. While considerable effort has been devoted to designing IoT search engines for finding a particular mobile object device, or a list of object devices that fit the query terms description, relatively little attention has been paid to enabling so-called spatial-temporal-keyword query description. This paper identifies an important efficiency problem in existing IoT search engines that simply apply a keyword or spatial-temporal matching to identify object devices that satisfy the query requirement, but that do not simultaneously consider the spatial-temporal-keyword aspect. To shed light on this line of research, we present a novel SMSTK search engine, the core of which is a coding enabled index called STK-tree that seamlessly integrates spatial-temporal-keyword proximity. Further, we propose efficient algorithms for processing range queries. Extensive experiments suggest that SMSTK search engine enables efficient query processing in spatial-temporal-keyword-based object device search.

Keywords: Internet of Things · Spatial-temporal-keyword query ·
SMSTK search engine · STK-tree · Range queries

1 Introduction

In recent years, the Internet of Things (IoT) [1, 2] becomes one of the hottest research issues over the world, which already penetrates to the fields of people's studying, working and life [3, 4]. For instance, the people can control the home, including curtain, refrigerator and air conditioner, in the office via intelligent terminals, such as cellphone; the logistics company can carry on real-time monitoring of transportation information through installing the sensors on the packages; the scientists can realize the timely and intelligent situation analysis of disaster-affected regions from sensory data transmitted to server-side. Typically, an IoT system is expected to connect and manage huge numbers of heterogeneous sensors and/or monitoring devices, called as "object device", which

© Springer Nature Switzerland AG 2019
S. Li (Ed.): GPC 2018, LNCS 11204, pp. 79–91, 2019.
https://doi.org/10.1007/978-3-030-15093-8_6

are capable of continuously providing real-time state information of real-world objects such as vehicles, buildings, airports, traffic networks, thus promoting the spatial-temporal-keyword-based object device search to be a promising service in the IoT [5, 6].

Due to increasing application demands and rapid technological advances in IoT systems, IoT search engine has been receiving a lot of attention from both industry and academia [7–10]. Same as the conventional search engines, an IoT search engine is required to quickly return moving object devices of high relevance in spatial-temporal and keywords aspects to a given query. Most of research on IoT search engine adopts hierarchical architectures [11–13] and is still very limited in query constraints, only supporting relatively simple keyword-based searches [11–13], static-location-based searches [12–14], or current state-based searches [15]. Snoogle [12] and Microsearch [7], a two-tier distributed hierarchical architecture, are based on keywords matching for searching. Dyser [15], a two-tier centralized hierarchical architecture, supports the keyword-based search for real-world entities with a user-specified current state. Recently, the research on IoT search engine develops the query constraints, including spatial-temporal and keywords aspects. IoT-SVKSearch [16], a multimodal search engine, is constructed on the basis of a two-tier distributed architecture of index master server - index node server, where each index node server is composed of a set of hierarchical trees indexing full-text keywords or the time-stamped, dynamically changing locations of moving monitored objects. In a word, existing search engines have been examined to fail in retrieving the dynamic real-time content-based object device measurements e effectively due to not fully considering the spatial-temporal-keyword searching constraint simultaneously.

Targeting at above problems, we develop a fine-grained IoT search engine (SMSTK) over moving object devices. To be specific, in this article, SMSTK search engine is constructed in four phases. We analyze the system structure. Then we investigate how to construct a safe region based on the frequent query regions. In the following, we aim to design an index structure called STK-tree, which considers to simultaneously concatenate the spatial-temporal-keywords of object devices in the safe regions. Finally, we explore the range search algorithms of the moving object devices. Comprehensive performance analysis is presented as well. We summarize our contributions as below.

1. We try to establish a three-tier search architecture in IoT system, where the moving object devices are organized into a form of key-based index tree to facilitate query and communication.
2. We develop the predictive safe regions of moving object devices to reduce the data transmitted within the IoT by frequent query grid cells and travel time structure.
3. On the basis of predictive safe regions, we further propose an index structure to improve the search performance through developing binary value concatenation of moving object devices' spatial-temporal-keywords simultaneously. Our range query are implemented on this index structure.
4. We provide experimental data on the searching performance. Our simulation results are consistent with the theoretical analysis and demonstrate the significant searching performance advantages over the competitive schemes.

The remainder of this paper is organized as follows. In Sect. 2, we present the system architecture. In the following, we present the construction process of predictive safe regions in Sect. 3. Section 4 explores the index construction process and searching algorithms. Section 5 contains the evaluation results. Section 6 provides a review of the related works and Sect. 7 concludes this article.

2 System Architecture

In this section, an overview of the three-tier hierarchical searching architecture is given in Fig. 1. We are now ready to describe the architecture and its tiers. Intuitively, the three tiers should be closely associated with each other to finish the search efficiently and accurately.

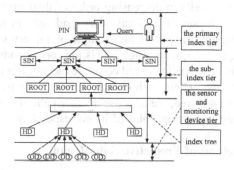

Fig. 1. Overview of SMSTK search system architecture.

1. The sensor and monitoring device tier. It is the lowest level of the hierarchy in the system architecture. This tier involves various categories of object devices (ODs), i.e., sensors and/or monitoring devices, which continuously sample the states of real-world objects.
2. The sub-index tier. Each sub-index node (SIN) manages the object devices in a certain area and is responsible for handing their text, space and time information in a single index structure for local search. Meanwhile, the object devices register themselves and transmit their textual description to the specific SIN via head nodes (HD) in each index hierarchy. We assume the object description data are either preloaded or incrementally uploaded by the object owner.
3. The primary index tier. It is the highest level of the hierarchy in the system architecture. The SINs forward the aggregated object information to the primary index node (PIN) so that the PIN can return a list of SINs that are most relevant to a certain user query. The PIN, considered as the sink of the network, holds the global object aggregation information, i.e., spatial-temporal-textual information, reported by each SIN in a single index structure for a distributed query. We can notice that the SINs route traffic among themselves or between the PIN and ODs.

3   Safe Region Construction

This section presents our construction process for predictive safe region. We first explore the method for frequent query regions mining. Then, we provide the specific construction process for predictive safe regions reachable to the frequent query regions within some time intervals.

3.1   Frequent Query Grid Cell Mining

We present the process for mining frequent query region/point using QFS-tree, which in turn is based on FP-tree [17, 18]. The FP-tree is a compact representation of all relevant frequency information in a database. Before a recursive mining, we partition the service area with equal or similar spatial grid cells, denoted as sg (e.g., the sg(1; 1) shown in Fig. 2(a)). The mining process contains two scans: (i) it is first required to find all frequent query grid cells in decreasing order of their query times; (ii) as each query region is scanned, the set of frequent query grid cells in it is inserted into the QFS-tree as a branch. Figure 2(a) shows an example of a query set, which consists of six query regions, i.e., q1, ... , q6, and Fig. 2(b) is the corresponding QFS-tree for that query set. The query grid cells written in red color are the multi-branch points with high query frequency in QFS-tree, where each branch represents the traversal path (denoted as) of a query region. In QFS-tree, we also use another path, the starting grid cell path (denoted as), to indicate the starting point of its corresponding traversal path.

3.2   Construction Method of Predictive Safe Region

In this section, we discuss how to represent the predictive safe regions. The safe region of a moving object device designates how far it can reach without affecting the results of any registered query. The basic idea is to use grid cells as the spatial division structure. A brief overview of each data structure is outlined below.

Definition 3.1. *Sequence Grid Cells List. For each object device OD, we keep track of an object device identifier and the sequence grid cells list as $CS = \{sg_1, sg_2, ... , sg_k\}$ that are traversed by OD in its current trip.*

Definition 3.2. *Travel Time Structure. This is a grid division structure where each cell $TTS[(i_1, i_2), (j_1, j_2), k]$ has the travel time interval between space cells $sg(i_1, i_2)$ and $sg(j_1, j_2)$ at time slot t_k.*
    In order to locate the predictive safe regions of the object devices in the grid cells reachable to the frequent query grid cells within a specific time interval, a key point is to support predictive location calculation that concerns a future reference time relevant to the specific time interval. Let us step through the formalization of predictive location calculation and safe region construction as follows.

Definition 3.3. *Predictive Position Calculation. Given a set of moving objects devices, a moving point MP represented by its position $MP(t_{ref})$ at reference time t_{ref} and its velocity MP_V, and a travel time interval $MP_T = [t_{min}, t_{max}]$, it is to find the predictive object position $MP(t_{fr}) = MP(t_{ref}) + MP_V (t_{fr} - t_{ref})$ at the future reference time $t_{fr} \in [t_{ref} + t_{min}, t_{ref} + t_{max}]$.*

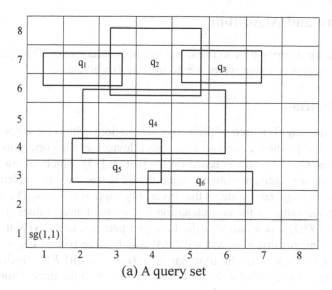

(a) A query set

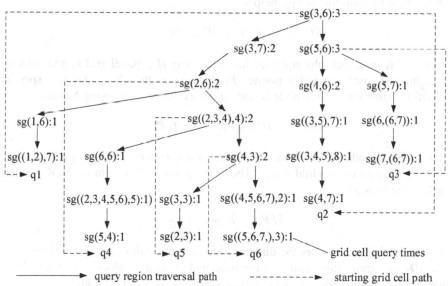

(b) The QFS-tree for the query set

Fig. 2. A QFS-tree example. (Color figure online)

Definition 3.4. *Predictive Safe Region Construction. Given a set of grid cells SQ_{rb} reachable to the frequent query grid cells within the travel time interval $[t_{min}, t_{max}]$, the predictive safe region for each object device $OD \in SG_{rb}$ is defined as the minimum bounding rectangle of the predictive positions within the travel time interval $[t_{min}, t_{max}]$.*

4 Structure and Algorithm

In this section, we describe how to construct the index structure of STK-tree. Then, we formalize the predictive range query we address in this article.

4.1 Index Structure

Note that constructing STK-tree of predictive safe regions within the range of a SIN is equal to index the predictive locations of object devices in the corresponding spatial grid cell regions. Our STK-tree is based on the B^x-tree [13], which in turn is based on the B^+-tree. This arrangement aims to make the STK-tree easy to implement in real SMSTK search management systems that invariably support B^+-trees. The value that is indexed, the STK-value, is the concatenation ($\oplus$) of the binary values ($[\cdot]_2$) of three components: (i) *TID*, which indicates the time grid partition in the STK-tree where an object device's information is stored; (ii) *HLV*, which is the Hilbert-Curve value of an object device's location as of the time partition *TID*; and (iii) *KY*, which is the keywords encoding value detailed in Sect. 4.2. After we obtain the three components, we combine them to form the STK_{key} below.

$$STK_{key} = [TID]_2 \oplus [HLV]_2 \oplus [KY]_2$$

In the STK-tree, each object device has an identity *ID*, spatial grid *sg* and velocity (v_x, v_y) at time instant t, and a pointer *PI* pointing to the object device's specific information. Therefore, a leaf node in the *STK*-tree has the following format:

$$<STK_{key}, ID_{OD}, sg, v_x, v_y, t, PI>$$

The internal nodes of the STK-tree serve as a directory that contains index key values and pointers to the child nodes. Therefore, a non-leaf node in the STK-tree has the following format:

$$<STK_{ckey}, ID_{HD}, PC>$$

Note here, STK_{ckey} denotes the directory that contains index key values of the child nodes. ID_{HD} indicates the identity of the head node which is selected from the object devices in the child nodes. *PC* is the pointer pointing to the child nodes.

4.2 Keywords Encoding

We make use of a context-free text mapping algorithm to encode keywords into a numeric system, or ids (See Definition 4.1). This allows textual data to be treated as just another numeric component of the index, enabling it to be efficiently constructed and queried over the data. Intuitively, an n-dimensional vector is actually a point in an n-dimensional Euclidean space. Thus, encoding keywords into a numeric component converts to mapping a point from n-dimension to 1-dimension. We consider to use Hilbert spatial filling curve, for this conversion and achieve the *KY*.

Definition 4.1. *Pair Coding Function. Given a function PC: $S_1 \rightarrow S_2$, where $S_1 = c_1 c_2 \ldots c_n$ is the keyword, $S_2 = b_0 b_1 \ldots b_{m-1}$ is the binary vector ($n < m$) and the initial value of b_i($i \in [0, m-1]$) is 0. The two adjacent characters (c_j, c_{j+1}) ($j \in [1, n-1]$) are hashed and mapped into a number ($\in (0, m-1)$) according to the hash function H, only when $H(c_j, c_{j+1}) = i$, $b_i = 1$. PC is called as a pair encoding function.*

4.3 Predictive Range Query

A predictive range query aims to retrieve all object devices whose locations fall within the rectangular range Q_r at time instant t, which is defined as follows.

Definition 4.2. *Predictive Range Queries. Given a set of object devices sequences DS, a space partitioned into a set of grid cells SQ, a predictive function F, a predictive query defined by a region Q_r, and a time period t, we need to find out the object devices predicted to be inside r after the time t.*

Since the spatial, temporal and keywords components of each object device are integrated in the STK-tree structure, the query processing algorithm can map any spatio-temporal-keyword aware range query into a key range search query regardless of search criteria. To reduce the query processing time, the algorithm reads and navigates the tree in the same order as it stored.

For each query, the user first queries the PIN. The PIN then returns the SINs that may contain the object devices in the search range. The user then performs a distributed range query from the returned list of SINs, i.e., the local root node of STK-tree, to find a satisfactory answer. The STK-tree starts from the SIN and moves down recursively, similar to the process used for index construction. At each level of the STK-Tree, the head node transmits the query to which child node depending on the STK_{ckey} of the current node and the concatenation value of the spatial, temporal and keywords constraints in the user query. When a leaf-node is reached, it is added to the result list of range query.

5 Implementation and Evaluation

In this section, we evaluate the performance of our spatial-temporal-keyword query scheme over the proposed SMSTK search engine. The goal of our experiment is mainly to evaluate the query response time, and the query accuracy.

5.1 Setup

The synthetic data of moving object devices are generated from the Network-based Generator of 10,000 moving object devices. A real road network map is extracted from the check-in records over 10,000 Beijing taxi cabs, which can be used for our experiment setting and for the generator as an input. The object devices generated are assumed to be randomly distributed over the spatial region, which is virtually partitioned into $N \times N$ squared grid cells of width relative to the minimum and the maximum step taken by any of the underlying moving object devices. The spatial grid structure stores a Hilbert value for each cell and updatable list of object devices moving within it. Therefore, all sampling values of the same mobile object device can generate a trajectory in the form of grid cells. As the check-in record of each taxi cab contains its *ID*, the grid cells of its geo-location (place of interest) and the check-in time, the travel time interval between any pair of cells can be obtained by taking the average minimum an average maximum time it takes from the underlying set of object devices to move between any pair of cells at time period *t*. Hence, the travel time data structure can be efficiently filled before starting the experiment and offline. Additionally, we apply a real-world dataset from newspaper such as China Daily [20], where we randomly select documents attached to each mobile object device. For the generation of spatial-temporal-keyword queries, the locations of the generated query regions are uniformly distributed over the space and the keywords are generated by randomly choosing consecutive words from the documents of object devices. All the index structures and algorithms are implemented in Java and performed on a machine with Intel Quad Core Processor 3.5 GHZ, 32G memory, 64 bits operating system. Table 1 lists the parameters used in our experiments.

Table 1. Parameters and their settings

Parameters	Setting
Number of object devices	1000, 3000, 5000, 7000, 9000
Query range length	10, 20, 30, 40, 50, 60
k	100, 200, 300, 400, 500
Number of keywords	10
Length of keywords	20
Number of frequent queries	100

For the sake of understanding the effectiveness of SMSTK search engine (SK) based on *ST K*-tree, we compare it with the following searching schemes.

– *Chained Structure* searching scheme (CS). A group of object devices are organized to a chained structure.
– *Snoogle* searching scheme (SL). This is an alternative information retrieval system built on sensor networks for the physical world [12].
– *IoT-SVK* searching scheme (IS). This is a real-time multimodal search engine mechanism implemented for the Internet of Things [16].

5.2 SMSTK Query Performance

To evaluate the query performance, we first test the query time of SMSTK search engine. In this experiment, we vary the total number of object devices from 1000 to 9000, the query range length from 10 to 60, and the *k* value from 100 to 600, to measure the range query and KNN query time. We give 10 keywords for each query with the length as 20.

Figures 3(a) and (b) report on range query time. We observe that it increases with the number of object devices and the query range length. As mentioned in the construction process of STK-tree, the index tree organizes object devices based on their spatial-temporal-keyword proximity. Hence, the spatial index needs to retrieve all object devices inside the searching key ranges constructed from the given query range, thus increasing the query costs. This accounts for the increase in query response time.

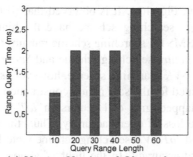

(a) Varying Number of Object devices

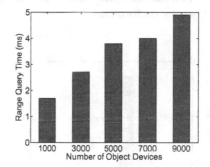

(b) Varying Number of Query Range Length

Fig. 3. SMSTK query time.

5.3 Query Performance Comparison

We proceed to evaluate the query performance of SMSTK search engine and the other three searching schemes on frequent query.

Figure 4 provides a comparison of the query time and query accuracy among the four searching schemes when given 100 frequent query ranges. With respect to the query processing time, SMSTK performs the best in this case, as shown in Fig. 4(a). Note that constructing the safe regions from the frequent query grid cells could guarantee the higher stability of the query results. Meanwhile, indexing the safe regions could guarantee the higher accuracy of the query results. Thus, SMSTK search engine costs the least query time.

As for the query accuracy, our estimation of the query accuracy for the frequent range query is:

$$QueryAccuracy = \frac{|QueryResult - ActualResult|}{ActualResult}$$

In this accuracy test, we let the number of object devices be as 1000 and k as 10, respectively. The estimated accuracy result of the 100 frequent range queries of the four searching schemes is shown in Fig. 4(b). It is observed that (i) the *chained structure* searching scheme and *Snoogle* searching scheme have the highest query accuracy; (ii) the query accuracy of our SMSTK searching scheme remains relatively higher, next only to that of the *chained structure* searching scheme and *Snoogle* searching scheme; (iii) the query accuracy of *IoT-SVK* searching scheme, however, is the lowest. Note that the prediction positions estimated for the safe region construction in SMSTK searching scheme and the index of mapped grid cell centers in *IoT-SVK* searching scheme contribute much weight in the loss of query accuracy. Thus, the test results show that SMSTK searching scheme and *IoT-SVK* searching scheme yield less query time at the expense of losing some query accuracy. In the light of almost one hundred percent query accuracy and more than two seconds query time, it proves our SMSTK searching scheme achieves much better query performance.

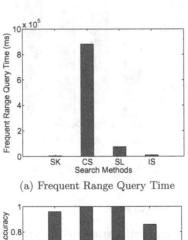

(a) Frequent Range Query Time

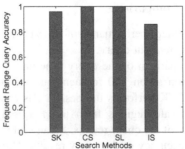

(b) Frequent Range Query Accuracy

Fig. 4. Query performance.

6 Related Work and Comparison

In this section, we discuss research in IOT search engine that are closely related to the problem studied in this article. Some of the notable works in this area are given as follows. Wang *et al.* [12] propose a search engine called Snoogle for pervasive environments, which describes the entity in the form of a set of keywords, stored in the sensor node. The users query the matched entities via keywords in the two-tier hierarchical architecture. However, the Snoogle system can not be applied to large-scale network environment because of inefficient data transmission mode. Therefore, it does not support the dynamic data search well. Yap *et al.* [13] design MAX system, in which the physical entities are sensed by tag instead of sensors. The tag stores the textual information of the physical entity, which is similar to Snoogle system. MAX system adopts pull mode to query, thus it is suitable for the mobile queries and queries with content frequently changed. However, the messages still need to be broadcasted to each sub station and tag, which results in high communication overhead. Therefore, it is inappropriate for large-scale network environments. Ostermaier *et al.* [15] design a real-time search engine Dyser, which abstracts the physical entities and sensors into Web pages. The predictive mechanism used not only improves the searching efficiency, but also reduces the searching overhead, which makes Dyser be applied to resource constrained networking environment. Ding *et al.* [16] propose a hybrid real-time search engine framework IoT-SVK for the Internet of Things based on spatial-temporal, value-based, and keyword-based conditions. IoT-SVK supports dynamic location changes of the sensors, and proposes to replace the original curve path as the grid regions with the same size. However, it cannot construct one integrated index structure based on spatial-temporal-keyword conditions, which affects the searching efficiency in some extent. Perera *et al.* [19] adopt cloud computing to resolve the architecture problem of IoT searching. This system constructs a cloud sensing the service on the basis of Internet of Things, infrastructure and services. It manages the sensors and sensing data by way of cloud computing. In a whole, most existing proposals can not directly enable effective searching of mobile object devices in the IoT. In this article, we propose SMSTK searching scheme, in which the single spatial indexing mechanism with spatial-temporal-keyword integrated into one key, can help it improve much more searching efficiency.

7 Conclusion

In this article, we propose a novel SMSTK search engine over mobile object devices. The focus of it is to construct a STK-tree index structure by incorporating the safe region and keywords encoding technique into the design mechanism of establishing the concatenated spatial-temporal-keyword key value. Further, we present predictive range query over our SMSTK search engine. In addition, our searching mechanism is developed to support frequent query operations. To evaluate the performance, we conduct a series of experiments and the results illustrate the searching efficiency of our scheme.

Acknowledgment. The authors gratefully acknowledge the financial support partially from the National Natural Science Foundation of China (No. 61702232, No. 61772479 and No. 61662021), and partially from the higher school research fund from Jiangsu University (No. 1291170040).

References

1. Li, S., Xu, L.D., Zhao, S.: The internet of things: a survey. Inform. Syst. Front. **17**, 243–259 (2015)
2. Pan, J., Jain, R., Paul, S., Vu, T., Saifullah, A., Sha, M.: An Internet of Things framework for smart energy in buildings: designs, prototype, and experiments. IEEE Internet Thing **2**, 527–537 (2015)
3. Park, E., Cho, Y., Han, J., Sang, J.K.: Comprehensive approaches to user acceptance of Internet of Things in a smart home environment. IEEE Internet Thing **4**, 2342–2350 (2017)
4. Zhang, F., Liu, M., Zhou, Z., Shen, W.: An IoT-based online monitoring system for continuous steel casting. IEEE Internet Thing **3**, 1355–1363 (2016)
5. Zhang, P., Liu, Y., Wu, F., Liu, S., Tang, B.: Low-overhead and high-precision prediction model for content-based sensor search in the Internet of Things. IEEE Commun. Lett. **20**, 720–723 (2016)
6. Shemshadi, A., Sheng, Q.Z., Qin, Y.: ThingSeek: a crawler and search engine for the internet of things. In: Proceedings of the 39th International ACM SIGIR Conference on Research and Development in Information Retrieval, pp. 1149–1152. ACM Press, New York (2016)
7. Tan, C.C., Sheng, B., Wang, H., Li, Q.: Microsearch: a search engine for embedded devices used in pervasive computing. ACM Trans. Embed. Comput. **9**, 1–43 (2010)
8. Shah, M., Sardana, A.: Searching in Internet of Things using VCS. In: International Conference on Security of Internet of Things, pp. 63–67. ACM Press, New York (2012)
9. Ma, H., Liu, W.: Progressive search paradigm for Internet of Things. IEEE Multimedia, 1–8 (2010). https://doi.org/10.1109/mmul.2017.265091429
10. Zhou, Y., De, S., Wei, W., Moessner, K.: Search techniques for the web of things: a taxonomy and survey. IEEE Sens. J. **16**, 1–29 (2016)
11. Aberer, K., Hauswirth, M., Salehi, A.: Infrastructure for data processing in large-scale interconnected sensor networks. In: 2007 International Conference on Mobile Data Management, pp. 198–205. IEEE Press, New York (2007)
12. Wang, H., Tan, C.C., Li, Q.: Snoogle: a search engine for pervasive environments. IEEE Trans. Parallel Distrib. **21**, 1188–1202 (2010)
13. Yap, K.-K., Srinivasan, V., Motani, M.: MAX: human-centric search of the physical world. In: Proceedings of the 3rd International Conference on Embedded Networked Sensor Systems, pp. 166–179. ACM Press, New York (2005)
14. Grosky, W.I., Kansal, A., Nath, S., Liu, J., Zhao, F.: Senseweb: an infrastructure for shared sensing. IEEE Multimedia **14**, 8–13 (2007)
15. Ostermaier, B., Römer, K., Mattern, F., Fahrmair, M., Kellerer, W.: A real-time search engine for the web of things. In: IEEE Internet of Things, vol. 9, pp. 1–8 (2010)
16. Ding, Z., Chen, Z., Yang, Q.: IoT-SVKSearch: a real-time multimodal search engine mechanism for the internet of things. Int. J. Commun. Syst. **9**, 1–8 (2010)

17. Han, J., Pei, J., Yiwen, Y.: Mining frequent patterns without candidate generation. ACM Sigmod Rec. **29**, 1–12 (2010)
18. Grahne, G., Zhu, J.: Fast algorithms for frequent itemset mining using FP-trees. IEEE Trans. Knowl. Data Eng. **17**, 1347–1362 (2005)
19. Perera, C., Zaslavsky, A., Liu, C.H., Compton, M., Christen, P., Georgakopoulos, D.: Sensor search techniques for sensing as a service architecture for the Internet of Things. IEEE Sens. J. **14**, 406–420 (2014)
20. http://www.chinadaily.com.cn/

Fuel Consumption Estimation of Potential Driving Paths by Leveraging Online Route APIs

Yan Ding[1], Chao Chen[1(✉)], Xuefeng Xie[2], and Zhikai Yang[1]

[1] College of Computer Science, Chongqing University, Chongqing, China
cschaochen@cqu.edu.cn
[2] School of Media and Communication, University of Leeds, Leeds, England

Abstract. Greenhouse gas and pollutant emissions generated by an increasing number of vehicles have become a significant problem in modern cities. Estimating fuel usage of potential driving paths can help drivers choose fuel-efficient paths to save more energy and protect the environment. In this paper, we build a fuel consumption model (FCM) for drivers based on their historical GPS trajectory and OBD-II data. FCM on a path only needs three parameters (i.e., the path distance, traveling time on the path and path curvature), which can be easily obtained from online route APIs. Based on experiment results, we can conclude that the proposed model can achieve high accuracy, with a mean fuel consumption error of less than 10% for paths longer than 15 km. In addition, the traveling time on paths provided by online route APIs is accurate and can be input into FCM to estimate the fuel usage of paths.

Keywords: GPS trajectories · OBD-II · Online route APIs · Fuel consumption model

1 Introduction

Vehicle fuel consumption has increased sharply in recent years, and it is widely recognized that over-consumed fuel is highly correlated with greenhouse gases and pollutant emissions [5,15]. On the other hand, high fuel consumption cost results in economic pressure for drivers. If we can provide drivers information of the fuel consumption on their potential driving paths, it can help them choose more fuel-economical paths to save more energy and protect the environment. Common mapping and navigation tools, such as Google Maps only provide users either the fastest or the shortest path from the source to the destination. Users have no idea about the fuel usage of these paths.

In general, the fuel consumption for a vehicle on a path depends strongly on three factors. The first one is the vehicle's parameters (e.g., vehicle weight), the second one is the driving behaviors, and the last one is the path shape (e.g., winding or straight paths). For example, a smooth acceleration or deceleration

© Springer Nature Switzerland AG 2019
S. Li (Ed.): GPC 2018, LNCS 11204, pp. 92–106, 2019.
https://doi.org/10.1007/978-3-030-15093-8_7

(i.e., small jerk) generally implies economical fuel consumption, while a high jerk corresponds more fuel consumption [1,4]. The first two factors have been investigated in lots of previous work [5–7,10]. In contrast, to our knowledge, the relationship between the path shape and the fuel consumption is rarely discussed. Vehicle that travels on *winding paths* usually consumes more fuel as compared with *straight paths*, because drivers have to speed up or down to traverse the path more carefully. This phenomenon has been partially captured by the state-of-art model (i.e., GreenGPS) [7,10]. GreenGPS points out that *the path containing left turns and right turns* increases the fuel usage. However, the path is just one typical example of winding paths, and 'left or right turns' are far from representing the path shape. Going a further step, we propose path curvature to define the path shape, and then make use of it to estimate fuel consumption more accurately.

Fuel consumption model is possible thanks to the popularization of GPS (Global Position System) and OBD-II (On-Board-Diagnostics) [3,9,13]. GPS trajectory consists of a sequence of points (i.e., latitude and longitude), and can be mapped on the road network to compute the 'path curvature' (more details in Sect. 3.2). OBD-II system is able to monitor users' driving behavior including instantaneous speed, traveling distance and fuel consumption via sensors. The data recorded by GPS and OBD-II provides rich sources to investigate the inherent relationship among vehicle's parameters, personalized driving behaviors, path shape and fuel consumption.

With the constructed fuel consumption model, we are able to estimate the fuel consumption on any path by plugging in three parameters (i.e., the path distance, traveling time on the path and the path curvature) extracted from drivers' historical GPS and OBD-II data. However, the model can only operate ex-postly (i.e., compute the fuel consumption after the trip is completed). In another word, because of the unavailability of drivers' GPS and OBD-II data, the fuel usage of potential driving paths cannot be estimated. The path distance and path curvature are easily to be obtained, once the driving path is identified. However, the traveling time on a path depends on the real-time traffic information. To migrate the problem, we need to obtain the real-time traffic information as input to the fuel-estimation model.

Some methods to obtain traffic information have been proposed in previous work. Authors in [11] estimate the traffic information on each segment using GPS trajectories received recently and contexts (e.g., point of interests) using a framework of collaborative filtering. In fact, this method can only obtain approximate results, as traffic condition varies dynamically due to unpredicted emergency events like road accidents. GreenGPS collects traffic information via Crowd Sensing [2,12], that relies on voluntary data collection and sharing within a community. Comparing to the data collected from recent GPS trajectories, Crowd Sensing can provide a more accurate traffic information on paths. However, many road segments are not traversed by volunteers drivers. Thus, it is difficult to obtain comprehensive traffic information merely relying on this method. In addition, the maintenance cost of such a community is high for many service providers. A good traffic information collection method ought to satisfy two essential requirements: providing accurate traffic information and low-cost. Traffic information with inaccurate information degrades the model accuracy, which causes incorrect recommendation result and damages users' economic interests. Many location-based service applications (e.g., Gaode Map and Google Maps) monitor traffic information on most road segments for many cities via multi-methods, such as induction coil and video detection [8,14]. Users have almost free access to obtain these traffic information data from Application Program Interfaces (APIs) provided by these applications. Hence, to meet requirements mentioned above, we collect the traffic information by leveraging route APIs.

To summarize, we make the following main contributions in this paper:

- We model drivers' fuel consumption (FCM), taking vehicle parameters, driving behaviors and the path shape into consideration. This model on a path only needs three parameters, i.e., path distance, traveling time on the path and path curvature, which can be easily obtained from online route APIs. To our knowledge, this study is the first to leverage route APIs for fuel consumption model.
- We leverage multi-sourced urban data including the GPS trajectory, OBD-II data and the open data extracted from OpenStreetMap to evaluate the framework extensively. Experimental results demonstrate FCM is accurate, and route APIs can provide accurate traffic information as input to the model.

2   Basic Concepts

Definition 1 (Road Network). *A road network $G(N, E)$ consists of a node set N and an edge set E. Each element n in N is an point of a pair of longitude and latitude coordinates (x, y), which represents its spatial location. Edge set E is a subset of the cross products $N \times N$. Each element in E is an edge e_i.*

Definition 2 (GPS Trajectory). *A GPS trajectory is a sequence of time-ordered spatial points $(p_1 \rightarrow p_2 \rightarrow \cdots \rightarrow p_9$, the blue dashed line with circle marks shown in Fig. 1). Each point is comprised of a timestamp, an instantaneous speed, and a geospatial coordinate, denoted by $p_i = (t_i, latitude_i, longitude_i)$.*

Definition 3 (Distance Between Two Positions). *Given two points $p_i = (x_i, y_i)$ and $p_j = (x_j, y_j)$, we use Eq. 1 to calculate the distance of them.*

$$dis(p_i, p_j) = R \times \sqrt{|x_i - x_j|^2 + |y_i - y_j|^2} \times \frac{\pi}{180} \qquad (1)$$

where R is the earth radius, and in this paper we set its value as $6370000\,m$.

Definition 4 (Mapped GPS Trajectory). *A mapped GPS trajectory is a sequence of sitting-on-edge GPS points $(p'_1 \rightarrow p'_2 \rightarrow \cdots \rightarrow p'_9$, the green solid line with square marks shown in Fig. 1), which can be obtained by map matching algorithms.*

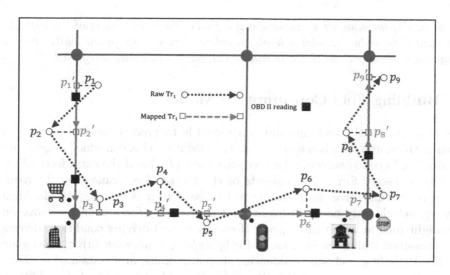

Fig. 1. Illustration of main concepts used in the paper. (Color figure online)

Definition 5 (OBD-II Data Reading). *A OBD-II Data Reading records the vehicle driving information at the sampling time, such as the distance that has been covered (in km), the instantaneous speed (in km/h), and the fuel usage (in liter), which can be denoted by* $obd_i = (t_i, v_i, distCov_i, fu_i)$.

Thus, the average fuel consumption (AFC, l) between any two given OBD-II data readings $(obd_j, obd_i, t_j > t_i)$ can be computed as:

$$AFC(obd_i, obd_j) = fu_j - fu_i \qquad (2)$$

Definition 6 (Average fuel consumption between two given GPS points). *The average fuel consumption between two given GPS points is not straightforward because we are not aware of the exact amount of fuel usage of the vehicle of a given GPS data point, which is due to the fact that the sampling time and rate of OBD-II data and GPS trajectory data are not consistent[1] (as illustrated in Fig. 1). To obtain the average fuel consumption (AFC) between any two given GPS data points, we first use the fuel usage value of the time-nearest OBD-II data reading to approximate that of each GPS point, then compute it according to Eq. 2.*

Definition 7 (Physical Features). *The physical features[2] of a segment contain a number of attributes: (1) the number of traffic lights/stop signs along the path; (2) the number of all neighboring edges that connect the edges within*

---

[1] The sampling time of the GPS trajectory data is 6 s; the sampling time of the OBD-II data is 10 s.

[2] Both road network and physical features are extracted from OpenStreetMap, with more details can be found at www.openstreetmap.org.

the path; (3) the number of some specific POIs within the range of the path. In this study, we only consider schools, residential zones, shopping malls and big companies. However, more POIs can be extracted and easily integrated.

3 Building Fuel Consumption Model

The objective of the fuel consumption model is to predict the fuel usage of a given path accurately when a driver has traveled it. Fuel is consumed to generate the driving force to overcome the necessary force (F_N) and the extra force (F_E).

The necessary force (F_N) consists of the friction force caused by the road, the gravitational force, and the air resistance, which is *unavoidable*. In ideal driving, vehicle only needs to overcome the necessary force if drivers move on a straight road at a constant speed. However, in real driving condition, drivers may encounter traffic congestions, traffic lights/stop signs while driving and some specific POIs (e.g., schools, residences, shopping malls, and restaurants). They may speed up or down and thus the extra force (F_E) is induced. In addition, drivers have to change their speed continually when they travel on winding paths, which also leads to an *increase* of the extra force.

As we known, the fuel consumption rate (fr) of a vehicle at time t is proportional to the power (P), generated by its engine at that time. Thus the fuel consumption rate at time t can be computed by Eq. 3.

$$fr(t) = \beta P(t) \tag{3}$$

The instantaneous power (P) can be calculated by Eq. 4, where $v(t)$ is the instantaneous speed at time t.

$$P(t) = (F_N + F_E)v(t) \tag{4}$$

By substituting Eq. 4 to Eq. 3, we can compute the fuel consumption rate as follows.

$$fr(t) = \beta(F_N + F_E)v(t) = \underbrace{\beta F_N v(t)}_{Necessary} + \underbrace{\beta F_E v(t)}_{Extra} \tag{5}$$

where the first part is defined as the *necessary fuel consumption rate* $(fr_n(t))$, the second part is defined as the *extra fuel consumption rate* $(fr_e(t))$; β is the specific fuel consumption (SFC), which refers to *the quantity of fuel consumed to produce one unit of power in one unit of time*, and varies under different travel and engine speeds. Figure 2(a) shows the details on how β changes with the travel speed at a given engine speed. β is much bigger when the vehicle travels at a low speed, which means that fuel is consumed more quickly because the fuel cannot be fully burned at that speed. Thus, we utilize two coefficients to distinct the slow and normal speed cases. To be more specific, $\bar{\beta}_1$ is used to approximate the SFC if the travel speed is less than v_{th} and $\bar{\beta}_2$ is used to approximate if the travel speed is above v_{th}, as shown in Fig. 2(b). Here, we set $v_{th} = 10\,\mathrm{km/h}$ as a matter of experience.

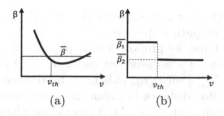

Fig. 2. The relation between β and the vehicle speed v at a given engine speed.

3.1 Necessary Fuel Consumption Rate

In our previous work *GreenPlanner*, the necessary fuel consumption rate has been computed, and here we just show the final result as follows. More details of the derivative process can be founded in [6].

$$fr_n(t) = \beta[\mu mg cos(\theta) + mg sin(\theta) + \frac{1}{2}\varphi AC_r v^2(t)]v(t) \tag{6}$$

where first three parts insides square brackets refer to the friction force caused by the road, the gravitational force and the air resistance, respectively; μ is the coefficient of friction, m is the mass of the vehicle, g is the gravitational acceleration, θ is the ground slope of the road, φ is the coefficient of the air resistance, A is the frontal area of the car, C_r is the air density, v is the speed of the vehicle at time t.

Note that in most cases, the slope (θ) of the road is very small and can be set to 0, thus $fr_n(t)$ can be simplified as follows.

$$fr_n(t) = \beta[\mu mg + \frac{1}{2}\varphi AC_r v^2(t)]v(t) \tag{7}$$

3.2 Extra Fuel Consumption Rate

In addition to the necessary fuel consumption rate, vehicle has to consume some extra fuel to overcome extra force. The extra force includes two parts F_{E1} and F_{E2}. F_{E1} is induced because drivers may speed up or down as they encounter traffic congestions, traffic lights/stop signs while driving and some specific POIs (e.g., schools, residences, shopping malls, and restaurants). F_{E2} is caused, as drivers have to change their speed continually when traveling on winding paths. In the previous work GreenPlanner, we have estimated the force F_{E1}, as shown in Eq. 8.

$$F_{E1} = m|a(t)|(\gamma + k_1'|path.ts| + k_2'|path.n| + k_3'|path.poi|) \tag{8}$$

where $a(t)$ and $v(t)$ are the acceleration and the speed of the vehicle at the time t, respectively; γ is a constant coefficient which is used to characterize the case of traffic congestions; $|path.ts|$, $|path.n|$ and $|path.poi|$ correspond to the number of traffic lights/stop signs, the number of neighboring edges, and the number of specific POIs of the given path respectively.

To define the path shape, we propose a concept *path curvature (rcur)*. Given a driving path consisting of a sequence of nodes n_i, $rcur$ can be computed in Eq. 9. In the equation, $\sum_{i=1}^{N-1} d(n_i, n_{i+1})$ is the total distance of the path, while $d(n_1, n_N)$ refers to the line distance between the source and the destination. If $rcur$ equals to its minimum value (i.e., 1), it means that the path is straight. On the other hand, if the value of $rcur$ is bigger than 1, it means that the path is more winding. Path curvature *can distinct straight and winding paths*. Note that the path can be obtained by mapping raw GPS trajectory on the road network.

$$rcur = \frac{\sum_{i=1}^{N-1} d(n_i, n_{i+1})}{d(n_1, n_N)} \tag{9}$$

where N is the number of nodes, and $d(n_i, n_{i+1})$ refers to the distance between two adjacent nodes n_i and n_{i+1}.

The accurate value of F_{E2} is difficult to get. Here, F_{E2} is simply estimated by Eq. 10.

$$F_{E2} = k_4' m (rcur - 1)|a(t)| \tag{10}$$

After F_{E1} and F_{E2} are calculated in Eqs. 8 and 10, the extra force F_E can be computed by adding them together, as follows.

$$\begin{aligned} F_E &= F_{E1} + F_{E2} \\ &= m|a(t)|[\gamma + k_1'|path.ts| + k_2'|path.n| + k_3'|path.poi| + k_4'(rcur - 1)] \end{aligned} \tag{11}$$

Thus, the extra fuel consumption model can be calculated, as follows.

$$\begin{aligned} fr_e(t) &= \beta F_E v(t) \\ &= \beta m|a(t)|[\gamma + k_1'|path.ts| + k_2'|path.n| + k_3'|path.poi| + k_4'(rcur - 1)]v(t) \end{aligned} \tag{12}$$

3.3   Fuel Consumption Model (FCM)

The total fuel consumption rate is simply the sum of the necessary and extra fuel consumption rates, as shown in Eq. 13.

$$\begin{aligned} fr(t) = fr_n(t) + fr_e(t) &= \beta v(t)\{[\mu mg + \frac{1}{2}\varphi AC_r v^2(t)] \\ &+ m|a(t)|[\gamma + k_1'|path.ts| + k_2'|path.n| + k_3'|path.poi| + k_4'(rcur - 1)]\} \end{aligned} \tag{13}$$

However, this fuel consumption rate $fr(t)$ cannot be applied directly, because we need develop a model on a path whose parameters can be easily measured by querying route APIs (i.e., path distance, traveling time on the path and path curvature). In another word, parameter $a(t)$ and $v(t)$ in Eq. 13 cannot be obtained by route APIs, which motivates us to simplify these two parameters. The path consists of several segments (i.e., sub-path). We assume that the speed

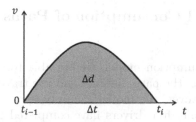

Fig. 3. Illustration of vehicle speed curve on a segment.

curve on each segment behaves like a parabola illustrated in Fig. 3. In addition, the speed value equals to 0 when vehicle enters and leaves the segment.

Given the ith segment distance Δd_i and the travelling time cost Δt_i, the speed curve on the segment can be well determined, as shown in Eq. 14.

$$v(t) = \frac{-6\Delta d_i}{\Delta t_i^3}t^2 + \frac{6\Delta d_i}{\Delta t_i^2}t, t \in [t_{i-1}, t_i] \tag{14}$$

where parameters t_{i-1} and t_i correspond to time when the vehicle enters and leaves the segment.

Considering the acceleration $a = dv/dt$, $a(t)$ on the segment can be calculated as follows.

$$a(t) = \frac{-3\Delta d_i}{\Delta t_i^3}t + \frac{6\Delta d_i}{\Delta t_i^2}, t \in [t_{i-1}, t_i] \tag{15}$$

By substituting Eqs. 14 and 15 to Eq. 13, we can derive the fuel rate $fr(t)$. The approximate fuel consumption of each segment can obtained by integrating the fuel rate $fr(t)$ during the time period $[t_{i-1}, t_i]$, as follows.

$$\hat{fc}_i = \beta\{k_1\Delta d_i + k_2\frac{\Delta d_i^3}{\Delta t_i^2} + \frac{\Delta d_i^2}{\Delta t_i^2}[k_3 + k_4|path.ts| \\ + k_5|path.n| + k_6|path.poi| + k_7(rcur - 1)]\} \tag{16}$$

Thus, the total fuel consumption of the path ($path_i$) is just the sum of fuel consumption on all segments, which can be computed based on Eq. 17.

$$\hat{fc}(path_i) = \sum_{i=1}^{M} \hat{fc}_i \tag{17}$$

where M is the number of segments determined by the number of sitting-on-edge GPS points (*Definition* 4) recorded along the path. Compared to the ground truth (fc_i) obtained according to Definition 6, each piece of $\hat{fc}_i$ may either be *overestimated or underestimated*. Thus, the modeling error could be *accumulated or reduced* as the path becomes longer and contains more GPS points.

4 Estimating Fuel Consumption of Paths by Leveraging Route APIs

With the built fuel consumption model, we are able to estimate the fuel consumption on a path, given the path distance and the traveling time on the path and path curvature. However, the model can only operate *ex-postly* (i.e., compute the fuel consumption after drivers have completed their trips). The path distance and path curvature are easily to be obtained, once the driving path is identified. However, the traveling time on a path depends on the real-time traffic information. To migrate the problem, we query route APIs to collect real-time traffic information on a path if the driver would travel it.

4.1 Real-Time Traffic Information Collection via Online Route APIs

Route APIs have access to real-time traffic information. It takes request (e.g., source and destination) as input and then returns a path along with its distance and traveling time cost in a XML document. We take an example below to illustrate the request and response information of Gaode Directions API. The request is an HTTP query string with user-specific parameters, including the origin and destination locations (in latitude-longitude), the travel mode (e.g., driving), as well as the strategy key. Gaode Directions API provides many travel strategies. For example, when the strategy key equals to 0, it indicates the path returned is the fastest path (minimal travelling cost). More details can be found in the homepage of Gaode API[3].

In the example, the origin is at $(106.4681, 29.5648)$, the destination is at $(106.4624, 29.5569)$, the user is at 'driving' mode, and the travel strategy is to choose the fastest path. According to the response, we can easily obtain the distance (in meter) and duration (in second) of the path from the source to the destination. In addition, the response also stores the a sequence of intermediate nodes of the path, that can be extracted from `"polyline"`. These nodes represent the path shape, which can be used to compute the parameter $rcur$ based on Eq. 9.

. . . .

```
                    HTTP request
http://restapi.amap.com/v3/direction/driving?key=?&
origin=106.4681,29.5648&destination=106.4624,29.5569&strategy=0
```

. . . .

```
                    XML response
"path" :{
          "origin" :"106.4681,29.5648",
          "destination" :"106.4624,29.5569",
          "distance" :"2202 (meter)",
```

---

[3] http://developer.amap.com/.

```
"duration" :"401 (second)",
"strategy" :"Fastestp",
"steps" :[
        "0" :{
            "polyline" :"106.4683,29.5651;
                        106.4682,29.5651;
                        ...",},
        "1" :{
            "polyline" :"106.4649,29.5654;
                        106.4647,29.5654;
                        ...",},
        ...
    ]
}
```
. . . .

The core of the real-time traffic information collection is to obtain the segment distance Δd_i and the traveling time cost Δt_i. According to the above explanation, Δd_i and Δt_i can be easily obtained as we input a pair of starting and ending points of the segment into route APIs.

4.2 Fuel Usage Estimation of the Shortest and the Fastest Path

Choosing the shortest or the fastest path from the source to the destination is the most common trip choice for drivers. However, the traditional navigation tools provide little information about the fuel economy of traveling paths. Thus, we estimate the fuel consumption of these two paths to provide more detail information (i.e., fuel usage) for users.

The steps can be summarized as follows.

- **Query** the shortest and the fastest paths from route APIs, for a given SD pair.
- **Divide** these two paths into a number of small segments (i.e., a sequence of points extracted from 'polyline') with a similar distance, which is calculated by summing up the distance between each pair of adjacent nodes.
- **Estimate** the fuel usage of each segment by inputting Δd and Δt into FCM. Δd and Δt of the segment can be obtained from route APIs, with inputting a pair of location of its first and the end points.
- **Measure** the total fuel consumption of the shortest and the fastest path by adding the fuel usage of all segments together.

Note that querying route APIs may raise the response time challenge. There are some potential methods to reduce the number of query requests. For example, traveling times on some segments nearly keep constant. If drivers travel on these path, there is no need to obtain the live traffic information on them. In the future, we will focus on how to reduce the number of query requests.

5   Evaluation

5.1   Experiment Setup

Data Preparation. In our experiments, we use several urban datasets in the city of Beijing, China. The datasets are gathered from multi-sources, i.e., the road network, the physical feature data extracted from OpenStreetMap, the GPS trajectory, and the OBD-II data generated by 595 taxis from September 15th to 21st, 2015. The mean travel distance and the mean working time for each taxi were 177.8 Km and 14.59 h, respectively. We use 30% of the total data to train the **FCM** and derive the parameters, and use the rest data to test the model.

In addition, to evaluate the quality of traffic information provided by route APIs, we conduct a small-scale experiment, as the limited human power. To be more specific, we recruit three volunteers to drive in the city of Chongqing from 14 pm to 22 pm on February 3th, 2018. They carry GPS devices implemented in their mobile-phone to collect the real traveling time (ground truth) on segments, denoted as t_{real}. Meanwhile, we collect the traveling time, denoted as t_{API}, returned from Gaode Directions API before trips. The total distance of segments is less than 100 km.

Baseline Model

- FCM_{rcur} – This model is a variant of the proposed FCM, in which we do not take the path shape into consideration.
- AFCM – The average fuel consumption model (AFCM) simply multiply the average fuel consumption (C_i) of the driver in history with the total distance of the path to approximate the total fuel consumed on the path.
- CFCM – The constant fuel consumption model (CFCM) first derives a constant average fuel consumption (C) which minimizes the following Eq. 18. Then, the estimation of the fuel consumption on a path is computed by multiplying C with the path distance, *regardless of the detailed path and the individual driving behaviors.*

$$\arg\min[C \times \sum_{i=1}^{n} td_i - \sum_{i=1}^{n} fc(i)] \tag{18}$$

where td_i is the total travel distance of the ith driver; $fc(i)$ is the true value of fuel consumption of the ith driver during the same whole day; n is the total number of taxis.

Evaluation Metrics. Two major metrics, i.e., the error and the mean error, are defined to evaluate the modeling accuracy. In addition, we define the *time cost error* as shown in Eq. 21 to evaluate the accuracy of traffic information collected from online route APIs.

$$error_i = \frac{\hat{fc}(path_i) - fc(path_i)}{fc(path_i)} \times 100 \tag{19}$$

$$m\_error = \frac{\sum_{i=1}^{m} error_i}{m} \qquad (20)$$

$$\#time\ cost\ error = \frac{|t_{real} - t_{API}|}{t_{real}} \times 100 \qquad (21)$$

where $\hat{f}c(path_i)$ is the fuel consumption on the given path estimated using the built model; $fc(path_i)$ is the true fuel consumed on that path, which is obtained by computing the fuel usage between the first GPS point entering and the last one leaving the given path; m is the total number of paths evaluated.

5.2 Evaluation on FCM

We compute the model accuracy of the proposed **FCM** and the baseline, with results shown in Fig. 4. The x-axis refers to the absolute error value of the model accuracy and y-axis refers to the corresponding Cumulative Distribution Functions (CDFs) of the number of segments. As shown in Fig. 4, **FCM** is able to obtain a good accuracy on the estimation of fuel consumption. For example, about 80% of segments are with an absolute error less than 30%. Moreover, comparing with the proposed **FCM**, **FCM**$_{rcur}$ always performs poorly, which demonstrates the path shape has effect on the fuel usage when driving. The two simple models (i.e., **AFCM** and **CFCM**), however, fail to provide a meaningful accuracy. Because these two models only consider the length of the path, information such as real-time traffic conditions and driving behaviors, which has important effect on the fuel consumption, is not included in the modeling process.

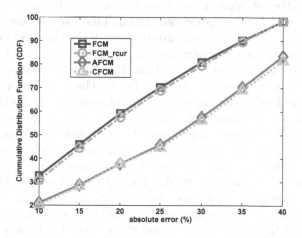

Fig. 4. Comparison results among FCM and three baseline models.

In order to better understand how models perform on paths of different lengths from 10 Km to 30 Km with an interval of 5 km qualitatively, we bin the paths based on their length and compute the mean errors. The corresponding results are shown in Fig. 5. The model accuracy gets higher as path length increases. In more details, **FCM** achieves a mean fuel consumption error less than 10% when the path length is 15 km.

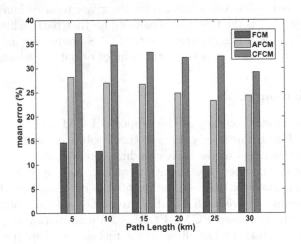

Fig. 5. Model performance of FCM and other two baselines by varying the length of the path.

5.3   Evaluation on Traffic Information Collection

The real-time traffic information collected from route APIs has a great impact on the accuracy of fuel consumption estimation. Thus, we conduct a experiment to evaluate the accuracy of the traffic information. If the time cost error between t_{real} and t_{API} on the same segment is small, it is safe to claim that the real-time information collected from route APIs is accurate. We compute the time cost error and the results are presented in Fig. 6. The x-axis refers to the time cost error and the y-axis refers to the number of segments. From the figure, we can see that most of errors have an acceptable accuracy, with a mean absolute error less than 20%. However, some cases has relatively big error (more than 30%), because there are two sudden traffic accidents after obtaining the query result. Based on the result, the traveling time obtained from online route APIs is demonstrated to be accurate. It also infers that the output of FCM (i.e., fuel consumption estimation) will be also accurate, as the input of FCM (i.e., traveling time) of FCM is accurate, which infers that the estimation calculated by the model will be also accurate. Limited by current human power, in the future, we will conduct more experiments to further evaluate the accuracy of fuel consumption estimation on paths, comparing with real fuel usage.

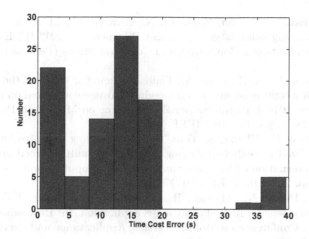

Fig. 6. The time cost error between the t_{real} and t_{API}.

6   Conclusion

In this paper, we propose FCM to estimate the fuel consumption on paths for drivers, which only needs three parameters (i.e., path distance, traveling time on the path and path curvature). And in the later estimating fuel usage of potential paths, these parameters can be easily obtained from route APIs. The proposed model and the traffic information collection method (i.e., route APIs) have been evaluated using real-world datasets, which consists of the road network, POI, the GPS trajectory and OBD-II data. Experimental results demonstrate that, compared to the baseline models, our model achieves good accuracy. In addition, the traffic information provided by route APIs is accurate.

Acknowledgments. The work was supported by the National Science Foundation of China (No. 61602067), the Fundamental Research Funds for the Central Universities (No. 106112017cdjxy180001), Chongqing Basic and Frontier Research Program (No. cstc2015jcyjA00016), Open Research Fund Program of Shenzhen Key Laboratory of Spatial Smart Sensing and Services (Shenzhen University).

References

1. Castro, P.S., Zhang, D., Chen, C., Li, S., Pan, G.: From taxi GPS traces to social and community dynamics: a survey. ACM Comput. Surv. (CSUR) **46**(2), 17 (2013)
2. Chen, C., Jiao, S., Zhang, S., Liu, W., Feng, L., Wang, Y.: Tripimputor: real-time imputing taxi trip purpose leveraging multi-sourced urban data. IEEE Trans. Intell. Transp. Syst. **99**, 1–13 (2018)
3. Chen, C., et al.: Crowddeliver: planning city-wide package delivery paths leveraging the crowd of taxis. IEEE Trans. Intell. Transp. Syst. **18**(6), 1478–1496 (2017)

4. Chen, C., Zhang, D., Zhou, Z.-H., Li, N., Atmaca, T., Li, S.: B-planner: night bus route planning using large-scale taxi GPS traces. In: 2013 IEEE International Conference on Pervasive Computing and Communications (PerCom), pp. 225–233. IEEE (2013)

5. Chen, H., Guo, B., Yu, Z., Chin, A., Tian, J., Chen, C.: Which is the greenest way home? A lightweight eco-route recommendation framework based on personal driving habits. In: 2016 12th International Conference on Mobile Ad-Hoc and Sensor Networks (MSN), pp. 187–194. IEEE (2016)

6. Ding, Y., Chen, C., Zhang, S., Guo, B., Yu, Z., Wang, Y.: Greenplanner: planning personalized fuel-efficient driving routes using multi-sourced urban data. In: 2017 IEEE International Conference on Pervasive Computing and Communications (PerCom), pp. 207–216. IEEE (2017)

7. Ganti, R.K., Pham, N., Ahmadi, H., Nangia, S., Abdelzaher, T.F.: GreenGPS: a participatory sensing fuel-efficient maps application. In: Proceedings of the 8th International Conference on Mobile Systems, Applications, and Services, pp. 151–164. ACM (2010)

8. Li, Y., Yiu, M.L.: Route-saver: leveraging route apis for accurate and efficient query processing at location-based services. IEEE Trans. Knowl. Data Eng. 27(1), 235–249 (2015)

9. Liu, L., Andris, C., Ratti, C.: Uncovering cabdrivers behavior patterns from their digital traces. Comput. Environ. Urban Syst. 34(6), 541–548 (2010)

10. Saremi, F., et al.: Experiences with greengps—fuel-efficient navigation using participatory sensing. IEEE Trans. Mob. Comput. 15(3), 672–689 (2016)

11. Shang, J., Zheng, Y., Tong, W., Chang, E., Yu, Y.: Inferring gas consumption and pollution emission of vehicles throughout a city. In: Proceedings of the 20th ACM SIGKDD International Conference on Knowledge Discovery and Data Mining, pp. 1027–1036. ACM (2014)

12. Wang, L., Zhang, D., Wang, Y., Chen, C., Han, X., M'hamed, A.: Sparse mobile crowdsensing: challenges and opportunities. IEEE Commun. Mag. 54(7), 161–167 (2016)

13. Yu, Z., Xu, H., Yang, Z., Guo, B.: Personalized travel package with multi-point-of-interest recommendation based on crowdsourced user footprints. IEEE Trans. Hum. Mach. Syst. 46(1), 151–158 (2016)

14. Zhang, D., Chow, C.-Y., Li, Q., Zhang, X., Xu, Y.: Efficient evaluation of k-NN queries using spatial mashups. In: Pfoser, D., et al. (eds.) SSTD 2011. LNCS, vol. 6849, pp. 348–366. Springer, Heidelberg (2011). https://doi.org/10.1007/978-3-642-22922-0_21

15. Zhang, J., Zhao, Y., Xue, W., Li, J.: Vehicle routing problem with fuel consumption and carbon emission. Int. J. Prod. Econ. 170, 234–242 (2015)

Large-Scale Semantic Data Management
For Urban Computing Applications

Shengli Song[1](✉), Xiang Zhang[2], and Bin Guo[3]

[1] Software Engineering Institute, Xidian University, Xi'an, China
shlsong@xidian.edu.cn
[2] School of Computer Science and Technology, Xidian University, Xi'an, China
[3] School of Computer Science, Northwestern Polytechnical University,
Xi'an, China

Abstract. Due to the current lack of effectiveness on perception, management, and coordination for urban computing applications, a great number of semantic data has not yet been fully exploited and utilized, decreasing the effectiveness of urban services. To address the problem, we propose a semantic data management framework, RDFStore, for large-scale urban data management and query. RDFStore uses hashcode as the basic encoding pattern for semantic data storage. Based on the characteristics of strong connectedness of the data clique with different semantics, we construct indexes through the maximum clique on the whole semantic data. The large-scale semantic data of urban computing is organized and managed. On the basis of clique index, we adopt CLARANS clustering to enhance the accessibility of vertexes, and the data management is fulfilled. The experiment compares RDFStore to the mainstream platforms, and the results show that the proposed framework does enhance the effectiveness of semantic data management for urban computing applications.

Keywords: Urban computing · Semantic data management ·
Encoding pattern · Data clustering

1 Introduction

Nowadays, people's lives rely on two worlds. One is the virtual world connected by the Internet, and the other is the physical world in reality. In spite of the high degree of connectivity and digitization of the former, the physical world contains a wealth of information and knowledge, but due to the current lack of effective perception, management, and coordination, these large-scale semantic data have not yet been fully exploited and utilized. At the same time, there is also a lack of effective communication between the real world and the virtual world, resulting in the isolation of human-owned information and thus failing to maximize its energy efficiency. Therefore, in today's highly developed Internet, in-depth exploration of the knowledge and intelligence contained in the physical world has become imperative.

Urban computing [1] was proposed and soon received a great deal of attention, in which concept any sensor, road, house, vehicle, and person in the city can be used as a computing unit to collaborate on a city-level calculation. Urban computing is a process

© Springer Nature Switzerland AG 2019
S. Li (Ed.): GPC 2018, LNCS 11204, pp. 107–123, 2019.
https://doi.org/10.1007/978-3-030-15093-8_8

of acquisition, integration, and analysis of big and heterogeneous data generated by diverse sources in urban spaces, such as sensors, devices, vehicles, buildings, and humans, to tackle the major issues that cities face. Urban computing applications connect unobtrusive and ubiquitous sensing technologies, advanced data management and analytic models, and novel visualization methods to create win-win-win solutions that improve the urban environment, human life quality, and city operation systems. Urban computing can be divided into four contents: urban sensing, data management, data analytics, service providing [2].

We put our emphasis on large-scale urban semantic data management and service providing. In service proving step, it aims to deliver the information to the people who ask for it. In data management step, the urban data information is well organized by indexing structure that simultaneously incorporates spatiotemporal information and texts for supporting efficient data analytics. This step is involved with three common data structures, i.e., stream, trajectory, and graph data. This paper mainly deals with graph data via the Resource Description Framework (RDF) that was proposed by the WWW for describing the information on the World Wide Web.

In this paper, we propose a large-scale semantic data management framework, named RDFStore, for RDF data storing and query. We use hashcode as the basic encoding pattern, the process of encoding is involved with semantic data and query subgraphs. Then we construct the index structure through the maximum clique on the whole semantic dataset. After that, by the maximum clique index, we adopt CLARANS clustering technique to enhance the accessibility of vertexes. According to the different query situations, RDFStore can effectively meet a variety of different query requests with the best performance.

2 Related Works

Many RDF storage systems [3] have been developed in the past decade, such as Sesame [4], Jena [5], RDF-3X [6], Hexastore [7], BitMat [8] and gStore [22]. These systems can be divided into four categories: property table, triples table, column storage with vertical partitioning and RDF graph-based storage. We will analyze these four categories of RDF stores [9].

Property Table. In this method, triples are classified according to attributes (predicates) and each type is stored as a relational table. It can effectively reduce the self-connection among subjects and improve the query efficiency. However, in practice, a query often involves multiple property tables, especially a query with unknown properties, and may require multiple table joining or merging operations. Meanwhile, due to the structural weaknesses of RDF data and the classification of properties, it is not possible to avoid having a large number of null values in many cases. Therefore, this storage scheme is only used in special applications. BitMat [10] represents an alternative design of the property table approach in which RDF triples are represented as a 3D bit-cube that represents subjects, predicates, and objects, and slicing along the dimensions yields 2D matrices SO, PO and PS.

Triple Table. Based on the basic triple means, through which the RDF triples are stored in the original ecological mode, the triples are used to store the RDF data directly in different orders of S, P, and O on the index structures of B+-trees and Hash, on which indices the query of RDF data is realized. A popular approach to improving the performance of queries over a triple table is to use an exhaustive indexing method that creates a full set of (S, P, O) permutations of indices [11]. For example, RDF-3X, which is currently one of the best RDF storage formats, builds clustered B+-trees on all six (S, P, O) permutations (SPO, SOP, PSO, POS, OSP, OPS). Thus, each RDF dataset is stored in six duplicates, with one per index.

Column Storage with Vertical Partitioning. An RDF triple table vertical partition (vertical partition) that is divided according to the predicate has the same predicate triplet stored in the same table. Then, the predicate can be used as the table name, just keep the S and O columns and each table is sorted by subject, which can quickly be used to query on the subject. However, this approach may suffer from scalability problems when the sizes of the tables differ significantly [12]. Furthermore, processing join queries with multiple join conditions and unrestricted properties can be extremely expensive due to the need to access all 2-column tables and the possibility of generating large intermediate results.

Graph-Based Storage. Graph-based approaches represent an orthogonal dimension of RDF storage research [13] and aim at improving the performance of graph-based manipulations on RDF datasets beyond RDF SPARQL queries. However, most graph-based approaches focus on improving the performance of specialized graph operations rather than on the scalability and efficiency of RDF query processing [14]. Large-scale RDF data are represented as a very large sparse graph, which poses a significant challenge to efficient storage and query. gStore proposes VS-tree and VS*-tree index for processing both exact and wildcard SPARQL [15] queries by efficient subgraph matching. In comparison, TripleBit advocates two important design principles: First, to truly scale the RDF query processor, we should design a compact storage structure and minimize the number of indices that are used in query evaluation. Second, we require a compact index structure as well as efficient index utilization techniques to minimize the size of the intermediate results that are generated during query processing and to process complex join operations efficiently.

3 Encoding Pattern

In this section, we first introduce the definition of hash function and then give an example to illustrate it. Next, we describe how to apply HashCode in our RDFStore framework. At last, we show an example in the following sections to vividly describe the process flow of RDFStore.

Definition 1. A *Hash Function* is any function that can be used to map data of arbitrary size to data of fixed size. The values that are returned by a hash function are called hash values, hash codes, digests, or simply hashes.

A hash function H projects a value from a set with many members to a value from a set with a fixed number of members. Hash functions are not reversible. A hash function H might, for instance, be defined as $y = H(x) = 10x(\text{mod}1)$, where $x \in R$ and $y \in [0,9]$.

Hash functions can be used to determine whether two objects are equal. Other common uses of hash functions are checksums over a large amount of data and to find an entry in a database based on a key value.

Definition 2. *HashCode* digests the data that are stored in an instance of a class into a single hash value. This hash value is used by other code when storing or manipulating the instance. The values are intended to be evenly distributed for varied inputs for use in clustering.

In an attempt to provide a fast implementation, early versions of the Java String class provided a hashCode implementation that considered at most 16 characters picked from the string. For some common data, this worked very poorly, delivering unacceptably clustered results and consequently slow hashtable performance.

HashCode uses a product sum algorithm over the entire text of the string. An instance of the java.lang.String class would have a hash code h(s) defined in Eq. (1).

$$h(s) = \sum_{i=0}^{n-1} s[i] \bullet 31^{n-1-i} \tag{1}$$

where terms are summed using Java 32-bit, $s[i]$ denotes the UTF-16 code unit of the ith character of the string, and n is the length of s.

The coefficient of 31 is adopted for two reasons: First, 31 is a prime number. A prime number (or a prime) is a natural number that is greater than 1 and cannot be formed by multiplying two smaller natural numbers. According to the characteristics of a prime number, the result of multiplication with a prime number is much more likely to be unique than with other methods, namely, the repeatability of the obtained hash value is relatively small. Second, this specific prime number was chosen because when computing a hash address, we want to minimize the amount of data with the same hash address, which is the so-called "conflict". If a large amount of data has the same hash address, the hash chain of these data will be too long, thereby reducing the query efficiency. Consider the above two points synthetically, under the premises of no overflow and the coefficient being a prime number. With the higher coefficient, the so-called "conflict" will decrease and the search efficiency will increase.

We complete two tasks in our framework. First, each entity needs to be encoded. In this task, the mapping table IDString is constructed. IDString stores the URI of each entity and the corresponding hashcode. The main aim of IDString is the interconversion between URI and hashcode. Second, the whole set of RDF data must be encoded, i.e., each SPO must be encoded. We encode Subject and Object for each SPO and retain the association relationship among entities. In this task, the table CodedGraph is constructed. CodedGraph stores the hashcode of Subject and Object for each SPO. The main aim of CodedGraph is to represent the whole structure and the association relationships among the entities in the RDF dataset.

For instance, given an example of SPARQL query involved with an urban data query, and Fig. 1 is the corresponding query subgraph. Figure 2 represents the part of the RDF dataset involved with urban data. The query asks for the facts that "Which place established in 1965 is Alice born in?", "Which unit of USA dose Alice live next to?", "Which place dose Alice purchase fruits at?", and "Which college does Alice teach philosophy at?".

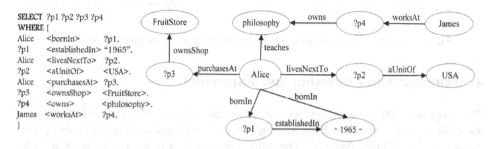

```
SELECT ?p1 ?p2 ?p3 ?p4
WHERE {
Alice    <bornIn>           ?p1.
?p1      <establishedIn>    "1965".
Alice    <livesNextTo>      ?p2.
?p2      <aUnitOf>          <USA>.
Alice    <purchasesAt>      ?p3.
?p3      <ownsShop>         <FruitStore>.
?p4      <owns>             <philosophy>.
James    <worksAt>          ?p4.
}
```

Fig. 1. A sample query graph over given RDF dataset.

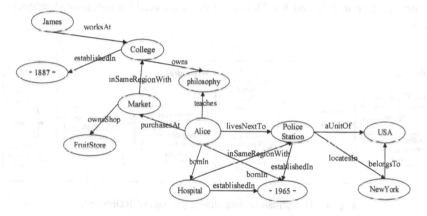

Fig. 2. A sample RDF dataset graph.

Encode the entities in the query subgraph and the whole RDF dataset, and we obtain two encoded graphs: encoded query graph and encoded dataset graph, shown in Fig. 3.

In the encoded query subgraph, each entity corresponds to a unique hashcode. We encode the known entities in the query subgraph because the unknown entities are obtained from the known ones. In the encoded RDF dataset, after being encoded, the association relationships among entities have been retained.

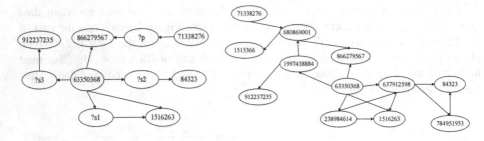

Fig. 3. A sample encoded query graph and RDF dataset over given RDF dataset.

4   Semantic Data Indexing and Expanding

The proposed RDFStore framework can be divided into two phases: the first phase is the Offline phase "Semantic Data Storing Process" and the second is the Online phase "Semantic Data Query Process". The offline phase, which is shown in Fig. 4, aims to build indexes for vertexes and on the basis of the index to enhance the accessibility of vertexes through clustering technique. Thus, the Offline phase includes two processes: Indexing and Expanding and the Online phase is finished by semantic data querying process.

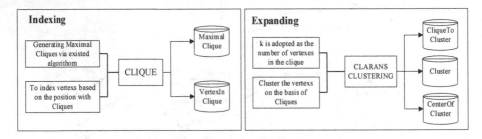

Fig. 4. The semantic data storing process of RDFStore.

4.1   Indexing

Definition 3. A *Clique* is a subset of vertices of an undirected graph such that every two distinct vertices in the clique are adjacent; that is, its induced subgraph is complete.

A clique, C, in an undirected graph $G = (V, E)$ is a subset of the vertices, $C \subseteq V$, such that every two distinct vertices are adjacent. This is equivalent to the condition that the induced subgraph of G induced by C is a complete graph.

Definition 4. A *Maximal Clique* is a clique that cannot be extended by including one more adjacent vertex, that is, a clique which does not exist exclusively within the vertex set of a larger clique.

In this paper, we propose mining maximal cliques to build the index. An approach to enumerating all the maximal cliques in a graph is not our contribution. Thus, we implement a method that was proposed in an existing study [16]. We introduce its main strategy here [17]. The search process maintains two sets: the Current Maximal Clique (*CMC*) and the candidate vertices to be expanded (SUBG). Initially, *CMC* is equal to Ø and SUBG is equal to V. In the expansion process, choose a vertex $v \in SUBG$. Then, $CMC = CMC \cup v$ and $SUBG = SUBG \cap NB(v)$, where $NB(v)$ is the neighbor set of vertex v. *CMC* is expanded iteratively until SUBG is equal to Ø. In addition, pruning strategies are used to reduce the size of the search space [18]. The corresponding algorithm is shown in Algorithm 1.

After obtaining all maximal cliques, we modified the source code to encode each maximal clique with a numerical ID, i.e., CliqueID. We adopt each CliqueID as a key and the hashcodes of vertices, which are denoted as vertexID, that are contained in each clique as values that generate a mapping table between cliques and vertices, which are denoted as MaximalCliques. We also store the ID of each vertex and the CliqueIDs of the maximal cliques that contain the vertex that generates the mapping table between vertices and cliques, which is denoted as VertexInClique. The structures of MaximalCliques and VertexInClique are as shown in Tables 1 and 2. The algorithm is described in Algorithm 2.

Algorithm 1. Maximal cliques mining algorithm.

```
    procedure CLIQUES(G)
    begin
1   EXPAND(V, V)
    end of CLIQUES
    procedure EXPAND(SUBG, CAND)
    begin
2   if SUBG = Ø
3     then print("clique,")
4   else
5     u:= a vertex u in SUBG that maximizes |CAND ∩ Γ(u)|
6     while CAND − Γ(u) ≠ Ø
7     do q:= a vertex in (CAND − Γ(u))
8       print(q, ",");
9       SUBG_q := SUBG ∩ Γ(q);
10      CAND_q := CAND ∩ Γ(q);
11      EXPAND(SUBG_q, CAND_q);
12      CAND := CAND − {q};
13      print("back,");
14    end if
15  end of EXPAND
```

Table 1. Table MaximalCliques

CliqueID	VertexIDs
C1	v1, v2, v3, ...
C2	v4, v5, v6, ...
...	...

Table 2. Table VertexInClique.

VertexID	CliqueIDs
v1	C1, C2, C3, ...
v2	C4, C5, C6, ...
...	...

Table 3. MaximalCliques of the sample data.

CliqueID	VertexIDs
1	63350368, 637912598, 23898461, 1516263
2	637912598, 784951953, 84323

Table 4. VertexInClique of the sample data.

VertexID	CliqueIDs
63350368	1
23898461	1
1516263	1
637912598	1, 2
784951953	2
84323	2

For the sample RDF data given in Sect. 3, MaximalCliques and VertexInClique are generated in Tables 3 and 4.

Algorithm 2. Index construction algorithm.

Input: All maximal cliques *CLIQUES*
Output: MaximalCliques and VertexInClique
1 cliqueid = 1;
2 **for** each clique *clique* in *CLIQUES*
3   insert cliqueid into MaximalCliques
4   **for** each vertex *vertex* in *clique*
5   insert *vertex* with key cliqueid in MaximalClique
6   **if** *vertex* exists in VertexInClique
7   insert cliqueid with key *vertex* in VertexInClique
8   **else**
9   insert < *vertex* , cliqueid> into VertexInClique as <key, value>
10   **end if**
12   **end for**
13 **end for**
14 **end**

As can be seen from the algorithm, due to that the two for loops are nested, its time complexity is $O(n^2)$, in which n is the number of vertex.

4.2 Expanding

Through the clique index, we can easily access all vertices in the same clique via one vertex. However, due to the strict characteristics of strong connectivity and accessibility, the ratio of the number of nodes in cliques to the number of nodes in the whole RDF dataset is not satisfactory (the ratio is proved to be in the range from 13.1% to 21.5% in a later experiment). The low accessibility will lead to a low recall ratio, thereby eventually decreasing the overall effectiveness. Thus, we must substantially increase the number of accessible nodes. On the basis of the clique index, we introduce a clustering technique for solving the problem.

We use CLARANS as clustering technique in RDFStore framework. We must figure out how to apply CLARANS to solve the problem of low accessibility. On the basis of the clique index, we take the vertices in cliques as the input of the clustering algorithm, i.e., we choose the vertices that are contained in a clique as the initial cluster centers. For datasets of different sizes, we select the corresponding *maxneighbor* and *numlocal*. CLARANS clustering is performed on the dataset, and three mapping tables are obtained during the clustering process: CliqueToCluster, Cluster, and CenterOfCluster. While clustering, we encode each cluster as ClusterID and save the ID of the final center vertex of each cluster.

Table CliqueToCluster stores the ID of the vertex that is inside the clique as the key, the ID of the cluster that contains the vertex as value1 and the adjacent vertices that are inside the cluster as value2. The aim of Table CliqueToCluster is to associate clusters with cliques on the basis of the clique index. We can access each cluster via the vertices that are inside the cliques.

Table Cluster stores the ID of each cluster as the key and the IDs of the corresponding vertices that are contained in the cluster as the value. Table Cluster aims at making the scope of each cluster explicit. Once we have the ID of a cluster, we can access all the vertices that are inside it.

Table CenterOfCluster stores the ID of each cluster as the key, the ID of the corresponding final center vertex as value1 and the IDs of the adjacent vertices as value2. The aim of table CenterOfCluster is to provide support for the query inside clusters. Since we record the center vertex of each cluster and its adjacent vertices, we can access the majority of each cluster.

The core process of applying CLARANS to enhance the accessibility is described in Algorithm 3.

Algorithm 3. Clique-based CLARANS clustering algorithm.

Input: A training set G; $G=\{V_h\}_{h=1,...,n}$; n is the size of G
Output: Table CliqueToCluster, Table Cluster and Table CenterOfCluster
Initialize: $COST_{min}$=Infinity; iteration = 0;
Repeat
1 **Set** C a node from G; $(C = [R_1, R_2, ..., R_k]$, R_i is the vertex that is contained in each clique)
2 **Set** j=1, clusterid = 1;
3 **Repeat**
   Consider a random neighbor C* of C;
   Compute TS_{ih} of C^* and TS'_{ih} of C
   if $TS_{ih} > TS'_{ih}$ **then**
      $C=C^*$
      $j = 1$
   else
      $j = j+1$;
   end if
   Until j = Maxneighbor;
   clusterid = clusterid + 1;
4 **for** each vertex V_h do
   Compute the similarity of V_h with each R_i;
   Assign V_h to the cluster with the nearest R_i;
   Insert R_i into table CliqueToCluster with the key clusterid;
   Insert the adjacent vertices inside the cluster into table CliqueToCluster with key clusterid
   Insert V_h into table Cluster with the key clusterid the ID of which cluster contains R_i
5 Compute COST
6 **if** COST > COSTmin **then**
      iteration = iteration + 1;
   else
      COSTmin = COST;
      BestSets = CurrentSets;
      Insert the final center vertex into table CenterOfCluster as value1 and its adjacent vertices
as value2
      Go back to Step3;
      Until iteration = q;
   end if
end

The algorithm adopts CLARANS as the clustering technique and it has the same time complexity as CLARANS. Therefore, its time complexity is $O(n^2)$, in which n is the number of vertex.

The structure of Table CliqueToCluster, Cluster and CenterOfCluster are shown in Table 5.

For the sample query described in Sect. 3, the Table CliqueToCluster, Table Cluster, and Table CenterOfCluster are shown in Tables 6, 7 and 8, respectively.

The main task of this section is to filter the maximum clique on the encoded whole RDF dataset to construct the clique index. Then, on the basis of the clique index, the CLARANS clustering technique is applied to enhance the accessibility of the vertices.

Table 5. The structure of Table CliqueToCluster, Cluster and CenterOfCluster.

VertexID	ClusterID	AdjacentVertexIDs	ClusterID	VertexIDs
v1	C1	v3, v4, ...	C1	v1, v2, v3, ...
v2	C2	v5, v6, ...	C2	v4, v5, v6, ...
...	...	...	...	...

ClusterID	CenterVertexID	AdjacentVertexIDs
C1	v1	v3, v4, ...
C2	v2	v5, v6, ...
...	...	...

Table 6. Table CliqueToCluster for sample data

VertexID	ClusterID	AdjacentVertexIDs
63350368	1	1997438884, 866279567

Table 7. Table Cluster for sample data.

ClusterID	VertexIDs
1	1680869001, 1515366, 71338276, 866279567, 63350368, 1997438884, 912237235

Table 8. Table CenterOfCluster for sample data.

ClusterID	CenterVertexID	AdjacentVertexIDs
1	1680869001	1515366, 71338276, 866279567, 1997438884

In the Online phase, query process is corresponding to the step of service providing in Urban Computing.

Given a SPARQL query, there are four scenarios regarding the positions of the known entities and the unknown entity in an <S, P, O> triple, such as <?x, P, O> or <S, P, ?x>. (1) The known entities and the unknown entity are in the same clique; (2) The known entities are in one clique, while the unknown entity is in the adjacent clique; (3) The known entities are in one clique, and the unknown entity is in the cluster obtained from the clique; (4) The known entities and the unknown entity are in the same cluster.

For these four situations, we propose four corresponding query modes: Query Inside Cliques, Query Among Cliques, Query Between Cliques and Clusters, and Query Inside Clusters. Given a SPARQL query, selecting the corresponding query mode is vital to query effectiveness. The complete query processing approach is described in the next section.

5   Experiment

5.1   Dataset and Metrics

Dataset. We choose DBpedia 3.9 as the experimental RDF dataset. DBpedia 3.9 is an open-domain knowledgebase, which is constructed by using structured information that is extracted from Wikipedia 6. It contains 4.0 million objects, of which 3.22 million are classified in a consistent Ontology, including 832,000 persons, 639,000 places (including 427,000 populated places), 372,000 creative works (including 116,000 music albums, 78,000 films and 18,500 video games), 209,000 organizations (including 49,000 companies and 45,000 educational institutions), 226,000 species and 5,600 diseases. DBpedia 3.9 contains 5,040,948 vertices and 61,481,483 edges and occupies 3.67 GB of storage space.

SPARQL Queries. We use QALD-4 [19], which is a benchmark that is delivered in the fourth evaluation campaign on question answering over linked data. It contains 236 SPARQL queries, each of which has an answer. QALD-4 is the fourth in a series of evaluation campaigns on multilingual question answering over linked data, which has a strong emphasis on interlinked data and hybrid approaches using information from both structured and unstructured data. It is aimed at all kinds of systems that mediate between a user, who expresses his or her information needs in natural language, and semantic data.

Metrics. Since QALD-4 provides the gold standard, we adopt two classical metrics, namely, *precision* (the ratio of the number of correctly discovered matches over the total number of discovered top-k matches, which is denoted by P) and *recall* (the ratio of the number of correctly discovered matches over the total number of correct matches, which is denoted by R), and *F1* (the measure that combines precision and recall and coincides with the square of the geometric mean divided by the arithmetic mean) to evaluate the effectiveness of our approach.

All experiments are conducted on an Intel(R) Core(TM) i5-6400 CPU@2.70 GHz and 16G RAM, on Windows 7. Programs are implemented in Java.

5.2   Effectiveness Evaluation

As presented in latest study, gStore, NeMa [20], SLQ [21] and S^4 [22] are state-of-the-art methods dealing with RDF data query over semantic data sets. We take these query platforms as the experimental comparison to our method RDFStore.

We adopt 7 SPARQL queries from QALD-4 as testing queries. These queries are all involved in urban RDF data. These queries are shown in Table 9.

We input these seven SPARQL queries into NeMa, SLQ, and our method RDFStore. For each query, we obtain the answers from each query platform, and according to the standard answer sets we can obtain Precision, Recall, and F1-measure for each platform via calculation. As shown in Table 10.

Since the dataset that we adopt is the same as the one that was adopted in paper [22], in terms of effectiveness, we can use its experimental data for our comparison.

Table 9. SPARQL Queries from QALD-4.

	Description	SPARQL Query

1	Give me all units that locate in NewYork.	SELECT DISTINCT ?uri
		WHERE {
		?uri rdf:type dbo:Unit .
		?uri dbo:locateIn res:NewYork .
		}

2	Give me the negative comment for the restaurants next to police station.	SELECT DISTINCT ?k
		WHERE {
		?uri rdf:type dbo:Restaurant
		?uri dbo:nextTo dbo:PoliceStation
		?uri <http://www4.wiwiss.fu-ber-
		lin.de/sider/resource/negativeComment> ?k
		}

3	List the markets in NewYork which have a fruit store?	SELECT DISTINCT ?x
		WHERE {
		?uri rdf:type dbo:Market .
		?uri dbo:locateIn res:NewYork .
		?uri dbo:has dbo:FruitStore.
		}

4	List the name of streets that passe a hospital and a restaurant?	SELECT DISTINCT ?uri
		WHERE {
		?uri rdf:type dbo:Street .
		?uri dbo:pass dbo:Hospital .
		?uri dbo:pass dbo:Restaurant.
		}

5	List restaurants that are in the same region with a hospital in Chicago.	SELECT DISTINCT ?uri
		WHERE {
		?uri rdf:type dbo:Restaurant .
		?uri dbo:inSameRegionWith ?x .
		?x rdf:type dbo:Hospital .
		?x dbo:locateIn res:Chicago .
		}

6	Which stores have no negative comment?	SELECT DISTINCT ?z
		WHERE {
		{?z a <http://www4.wiwiss.fu-ber-
		lin.de/sider/resource/store>} MINUS
		{?z <http://www4.wiwiss.fu-ber-
		lin.de/sider/resource/sider/negativeCom-
		ment> ?y.
		}

7	List restaurants which have no negative comment.	SELECT DISTINCT ?k
		WHERE {
		{?k rdf:type dbo:Restaurant . } MINUS
		{?k <http://www4.wiwiss.fu-ber-
		lin.de/sider/resource/sider/negativeCom-
		ment> ?y.}
		}

Table 10. Query efficiency of NeMa, SLQ and RDFStore

Query	NeMa			SLQ			RDFStore		
	Precision	Recall	F1	Precision	Recall	F1	Precision	Recall	F1
Q1	0.517	0.675	0.586	0.575	0.749	0.651	0.652	1.000	0.789
Q2	0.515	0.685	0.588	0.571	0.753	0.650	0.667	0.971	0.791
Q3	0.519	0.691	0.593	0.581	0.751	0.656	0.701	0.963	0.811
Q4	0.524	0.668	0.587	0.579	0.741	0.650	0.689	0.967	0.805
Q5	0.527	0.670	0.590	0.583	0.746	0.654	0.701	0.961	0.811
Q6	0.511	0.691	0.588	0.581	0.743	0.652	0.700	0.953	0.807
Q7	0.522	0.681	0.591	0.578	0.755	0.655	0.684	0.956	0.798
Average	**0.519**	**0.680**	**0.589**	**0.578**	**0.748**	**0.652**	**0.685**	**0.967**	**0.802**

Table 11. Experimental performance of gStore, NeMa, SLQ, S^4 and RDFStore

Method	Precision	Recall	F1
gStore	1	0.332	0.496
NeMa	0.519	0.680	0.589
SLQ	0.578	0.748	0.652
S^4	0.712	0.866	0.781
RDFStore	**0.685**	**0.967**	**0.802**

We average the precision and recall ratios for seven queries and recalculate the F1-measure for each platform. The final comparison results are shown in Table 11.

Table 11 shows the comparison results among other mainstream platforms and our RDFStore on the metric of F1-measure. The measure can intuitively reflect the effectiveness of a query platform. We find that the F1-measure gradually rises in the order of gStore, NeMa, SLQ, S^4, and RDFStore. gStore is in the lowest place with an F1-measure of 49.6%, while NeMa has an F1-measure of 58.9%. SLQ is in the middle place with an F1-measure of 65.2%, and S^4 has an F1-measure of 78.1%. Our method RDFStore obtains the best F1-measure of 80.2%.

We can obtain the fact that our method RDFStore has the highest F1-measure among all the query platforms. Namely, RDFStore is the most effective method in the comparison experiment. RDFStore outperforms NeMa and SLQ in terms of both precision and recall ratio. Therefore, RDFStore has an absolute advantage over NeMa and SLQ on effectiveness. The precision of gStore is 31.5% higher than that of RDFStore, but the recall ratio of RDFStore is far higher than that of gStore, with a gap of 63.5%. Consequently, RDFStore is more effective than gStore. In terms of effectiveness, S^4 is the closest to RDFStore. With a precision of 71.2%, it surpasses RDFStore by 2.7%, while the recall ratio of RDFStore surpasses that of S^4 by 10.1%. The recall ratio of S^4 is 86.6%, which is high, while RDFStore adopts a clustering technique to improve the accessibility of the vertices. As a result, RDFStore has higher effectiveness than S^4.

For the RDF graph dataset involved with urban information, RDFStore adopts HashCode as encoding technique, decreasing storage space for graph dataset. For data management in urban computing, encoding the urban information into hashcode enhances the frequency of I/O operation, indirectly enhancing the whole effectiveness of urban computing.

In the process of Indexing, we adopt maximal cliques as an index on the whole RDF graph dataset. According to the characteristics of strong connectivity, the urban data information is organized in clique index, which enhances the association among the vertexes in graph dataset. The information in data management of urban computing is organized.

In the process of Expanding, on the basis of clique index, CLARANS clustering technique is adopted to enhance the accessibility of the vertexes in graph dataset, which means most of the urban information is well organized within the clique index and clusters. In data management of urban computing, almost every vertex can be accessed in graph dataset.

In the online phase, namely, service providing step of urban computing, four query modes are proposed in accordance with the storage structure. For the vertexes in graph dataset, depending on its position, different query modes are corresponded, owning a great effect on the output of urban computing. Taking gStore, NeMa, SLQ, and S^4 as a comparison, RDFStore outperforms these platforms on effectiveness. Due to adopting clustering technique, RDFStore is much more suitable for dealing with massive urban data information. The experimental results show that RDFStore does enhance the effectiveness of urban computing compared to other mainstream platforms.

6 Conclusion

In this paper, due to the current lack of effectiveness on urban data perception, management and coordination for urban computing applications, a great of number knowledge has not yet been fully exploited and utilized, decreasing the effectiveness of urban computing. To address the problem, we propose a large-scale semantic data management framework for knowledge management and query for urban computing applications. We use hashcode as the encoding pattern for data management in urban computing. Then based on the characteristics of strong connectedness of the clique, we construct index through the maximum clique on the whole semantic dataset. The data information in data management of urban computing is organized. On the basis of clique index, we adopt CLARANS clustering technique to enhance the accessibility of vertexes and the data management step is fulfilled. Then in the step of service providing, four corresponding query modes are proposed. Experiment compares RDFStore to the mainstream platforms, and the results show that compared to other mainstream platforms RDFStore does enhance the effectiveness of urban computing. In the future, the main research direction is involved with two tasks. (1) How to sense urban dynamics more reasonably and effectively, including people's mobility, traffic flow, environment, and energy consumption in cities. (2) How to effectively express the acquired knowledge and extract the intelligence that can be used to make the decision. Current urban computing is still in its infancy, and the space to explore is huge and not far behind.

References

1. Zheng, Y., Liu, Y., Yuan, J., et al.: Urban computing with taxicabs. In: Proceedings of the 13th International Conference on Ubiquitous Computing, pp. 89–98. ACM (2011)
2. Jiang, S., Fiore, G.A., Yang, Y., et al.: A review of urban computing for mobile phone traces: current methods, challenges and opportunities. In: Proceedings of the 2nd ACM SIGKDD International Workshop on Urban Computing, p. 2. ACM (2013)
3. Yuan, P., Liu, P., Wu, B., et al.: TripleBit: a fast and compact system for large scale RDF data. Proc. VLDB Endow. **6**(7), 517–528 (2013)
4. Broekstra, J., Kampman, A., van Harmelen, F.: Sesame: a generic architecture for storing and querying RDF and RDF schema. In: Horrocks, I., Hendler, J. (eds.) ISWC 2002. LNCS, vol. 2342, pp. 54–68. Springer, Heidelberg (2002). https://doi.org/10.1007/3-540-48005-6_7
5. Wilkinson, K., Wilkinson, K.: Jena property table implementation (2006)
6. Neumann, T., Weikum, G.: The RDF-3X engine for scalable management of RDF data. VLDB J. Int. J. Very Large Data Bases **19**(1), 91–113 (2010)
7. Weiss, C., Karras, P., Bernstein, A.: Hexastore: sextuple indexing for semantic web data management. Proc. VLDB Endow. **1**(1), 1008–1019 (2008)
8. Atre, M., Chaoji, V., Zaki, M.J., et al.: Matrix Bit loaded: a scalable lightweight join query processor for RDF data. In: Proceedings of the 19th International Conference on World Wide Web, pp. 41–50. ACM (2010)
9. Modoni, G.E., Sacco, M., Terkaj, W.: A survey of RDF store solutions. In: 2014 International ICE Conference on Engineering, Technology and Innovation (ICE), pp. 1–7. IEEE (2014)
10. Chambi, S., Lemire, D., Kaser, O., et al.: Better bitmap performance with roaring bitmaps. Softw. Pract. Exp. **46**(5), 709–719 (2016)
11. Yan, Y., Wang, C., Zhou, A., et al.: Efficiently querying RDF data in triple stores. In: Proceedings of the 17th International Conference on World Wide Web, pp. 1053–1054. ACM (2008)
12. Sidirourgos, L., Goncalves, R., Kersten, M., et al.: Column-store support for RDF data management: not all swans are white. Proc. VLDB Endow. **1**(2), 1553–1563 (2008)
13. Bonstrom, V., Hinze, A., Schweppe, H.: Storing RDF as a graph. In: Proceedings of Web Congress. First Latin American, pp. 27–36. IEEE (2003)
14. Kim, J., Shin, H., Han, W.S., et al.: Taming subgraph isomorphism for RDF query processing. Proc. VLDB Endow. **8**(11), 1238–1249 (2015)
15. Peng, P., Zou, L., Özsu, M.T., et al.: Processing SPARQL queries over distributed RDF graphs. VLDB J. **25**(2), 243–268 (2016)
16. Tomita, E., Tanaka, A., Takahashi, H.: The worst-case time complexity for generating all maximal cliques and computational experiments. Theoret. Comput. Sci. **363**(1), 28–42 (2006)
17. Zheng, W., Zou, L., Lian, X., et al.: SQBC: an efficient subgraph matching method over large and dense graphs. Inf. Sci. **261**, 116–131 (2014)
18. Grosso, A., Locatelli, M., Pullan, W.: Simple ingredients leading to very efficient heuristics for the maximum clique problem. J. Heuristics **14**(6), 587–612 (2008)
19. Unger, C., Forascu, C., Lopez, V., et al.: Question answering over linked data (QALD-4). In: Working Notes for CLEF 2014 Conference (2014)

20. Khan, A., Wu, Y., Aggarwal, C.C., et al.: Nema: fast graph search with label similarity. Proc. VLDB Endow. **6**(3), 181–192 (2013)
21. Yang, S., Wu, Y., Sun, H., et al.: Schemaless and structureless graph querying. Proc. VLDB Endow. **7**(7), 565–576 (2014)
22. Zheng, W., Zou, L., Peng, W., et al.: Semantic SPARQL similarity search over RDF knowledge graphs. Proc. VLDB Endow. **9**(11), 840–851 (2016)

Parking Availability Prediction with Long Short Term Memory Model

Wei Shao[1], Yu Zhang[1,2]([email]), Bin Guo[2], Kai Qin[1], Jeffrey Chan[1], and Flora D. Salim[1]

[1] School of Science, RMIT University, Melbourne, Australia
[2] School of Computer Science, Northwestern Polytechnical University, Xian, China
zhangyu@nwpu.edu.cn

Abstract. Traffic congestion causes heavily energy consumption, carbon dioxide emission and air pollution in cities, which is usually created by cars searching on-street parking spaces. Drivers are likely to move slowly and waste time on the road for an available on-street parking space if parking slot availability information is not revealed in advanced. Therefore, it is necessary for city councils to provide a car parking availability prediction service which could inform car drivers vacant parking slots before they start the journey. In this paper, we propose a novel framework based on recurrent network and use the long short-term memory (LSTM) model to predict parking multi-steps ahead. The core idea of this framework is that both the occupancy rate of on-street parking in a specific region and car leaving probability are exploited as prediction performance metric. A large real parking dataset is used to evaluate the proposed approach with extensive comparative experiments. Experimental results shows the proposed model outperform the state-of-art model.

Keywords: Internet of Things · Recurrent neural networks · Parking occupancy · Parking sensors · Smart city

1 Introduction

Traffic congestions leads to air pollution, green house gas emission and energy consumption in central business district (CBD) areas. More than one-third of congestions are caused by parking space searching tasks [1,10]. A data-driven and robust solution for parking availability prediction can guide the drivers, thereby reducing traffic congestions and the time cost. However, such solutions cannot be proposed without network infrastructure in the past.

The concept of smart cities becomes possible with the emergence of the Internet of Things (IoT), which integrated the networks into interconnected objects such as sensors. Recently, many cities placed Internet of Things (IoT) devices in parking spaces around CBD areas which can record parking event and then send

S. Li (Ed.): GPC 2018, LNCS 11204, pp. 124–137, 2019.
https://doi.org/10.1007/978-3-030-15093-8_9

these events to control centre in real-time way. The data collected by those IoT sensors consists both spatial and temporal information, which raise the necessity for a spatio-temporal data processing system. Various systems developed for providing parking information ahead have been proposed by researchers in last few decades [5–7,9–12].

It is challenging to establish a smart parking prediction system with historical time-series sensor data. Firstly, parking events depends on many factors such as time, day of week, weather, special events, holidays, and etc. There is almost no real time information about free parking spots with those factors. Secondly, although in areas with internet-connected parking meters providing information on availability, those data are not well organized for querying or searching. Finally, with the constructions and new plans proposed by city councils, the location and demand of parking changes rapidly, so the new and well-equipped system is at risk of being outdated as soon as it has been built.

Existing works usually applied conventional machine learning methods and time-series models to predict both parking occupancy and duration with collected sensor data. With increasing amount of time-series sensor data, the performance of traditional regression model cannot catch up with deep neural networks as deep neural network can approximate any linear or non-linear complex functions with enough samples. Comparatively speaking, the traditional regression model needs to choose the right kernel or features [13]. Hence deep learning techniques could be employed to predict the occupancy especially for feed-forward networks [5,9]. However, simple feed-forwards neural networks are not able to incorporate temporal domain information, which is important to parking prediction especially for duration problem.

Recurrent neural networks (RNN) [14] is a class of neural networks that exploit the sequential nature of their input, which is widely applied to many time dependent problems such as electricity consumption, text prediction and POS tagging. The input of all of these problems are time dependent. That is, the value of occurrence of an element in the sequence is depended on the elements which appears before. Parking duration and occupancy data are both time series data [5,9]. The occurrence of parking events are highly depended on the end time of the last event and the time in a day [15]. Therefore, the parking occupancy or duration time can be estimated by RNN and its variants if we can apply the different training model to different time spans and regions.

In this paper, we propose a novel two-steps approach to estimate the parking occupancy and duration time with recurrent neutral networks. Firstly, we divide the complete temporal dimension to a couple of time slots with temporal features. By using temporal clustering methods, these features has significant impacts on the parking event model. Secondly, we train the LSTM [16] model which are variants of simple recurrent neural networks to fit for each cluster. Finally, we predict the parking occupancy and duration time with temporal information and the corresponding trained deep learning model.

In this paper, we make the contribution as follows:

- We analyze the duration problem with hazard-based free parking duration modeling and fit the distribution with the regression model.
- We transform the parking events data to time-series data and apply a novel two-step approaches to predict parking occupancy and duration time.
- We first use the LSTM model to predict the parking availability to our best knowledge.
- We evaluate the performance of our proposed model and compare it with state-of-art approach with a large real-world parking dataset.

The rest of this paper is organized as follows. The related work is discussed in Sect. 2. We present the novel framework of the model and details of the two-step algorithm in Sect. 3. This is followed by an evaluation of the approach in Sect. 4. Section 5 provides a discussion and possible future works.

2   Related Work

There is an extensive body of research on helping drivers in the CBD areas to find an available car slot. The main idea of those works is to establish a probabilistic model or apply machine learning algorithms to video-based [11] or sensor-based time series data [5] and train a predictive model for parking occupancy or duration.

Traditional regression model are most obvious option in time-series prediction [2,3]. Zheng et al. proposed many strategies include polynomial fitting, Fourier series and k-means clustering and tested on parking occupancy data published by the Birmingham city council [6]. The result is acceptable if we consider the model only need five parameters. Compared with machine learning techniques, those works does well in the case with limited amount of data or time series data are easy to fit by a kernel based model.

Markov model is a popular probabilistic model in parking prediction area. Ford Motor company [10] built a on-street parking system to predict the parking occupancy level based on queuing theory model which take advantage of a transient probability model and K-means clustering algorithm. Tilahun et al. proposed a cooperative multi-agent system to predict the parking availability with dynamic Markov chains [1]. They constructed a data transition matrices based on previous data and learning the transition matrix with each iterations. A continuous time Markov chain was used in [23] to predict the available parking space by connecting the parking garage and the navigation system of cars. Markov chain based model exploits temporal information in the parking events include arriving and departure. However, transition matrices are not fixed with different period and locations. It also cannot approximate some complex functions mapping from time and events.

Neural networks are also widely used in time series event prediction model [4,5,17,18,26]. Reference [5] predicts the occupancy rate with a static MLP model which was used in traffic flow estimation [24]. It is accomplished with adding a memory structure to classical MLP. Hence, such static MLP can also predict short-term time-series event. Pengzi et al. [9] firstly analyzes the parking dataset collected by sensor located in Xi'an, China. It uses BP neural network which belongs to nonlinear dynamic system and GA-BP neural network which have the optimized initial weights and thresholds to carry on the data analysis to the above five parking lots.

Recurrent neural networks, or RNNs [14], are a family of deep learning neural networks which is mainly to process the sequential data. Unlike the feed-forward neural networks, such as the multilayer perceptron (MLP) which is most popular neural networks used in parking prediction, RNNs share the same set of parameter across the different parts of the model. With powerful prediction ability, RNNs have been applied to many sequential data based research areas such as text generation [19], power forecasting [20] and dynamic mortality risk predictions [21]. It has been proved to be a powerful sequential event prediction neural networks in the last few years. However, they are never used in parking availability prediction. LSTM [16] architecture is one of the most popular variant of RNNs. It is capable of learning long term dependencies and widely used in a large variety of problems as it addresses the problems such as vanishing and exploding gradients in the training stage of simple RNNs [22]. Therefore, we firstly use this model to predict both occupancy and duration for parking spaces.

3 Parking Availability Prediction Model

3.1 Overview of Parking Prediction Architecture

Parking occupancy and duration time are two main indicators to define the parking efficiency [5]. Occupancy O_t indicates the percentage of parking spaces in a selected region during a predefined time period. Duration D_t is used to measure the average time duration that a parking slot is free, over a certain time period.

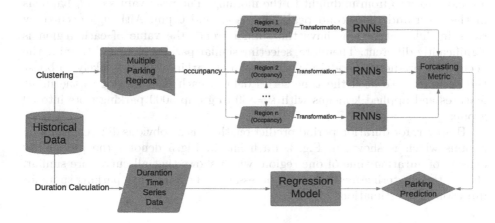

Fig. 1. Functional architecture of the proposed parking prediction scheme.

Figure 1 illustrates the framework of proposed parking prediction system. It mainly consists of two modules: (1) occupancy prediction module and (2) duration estimation module. We firstly investigated the parking data and found that occupancy rates for different regions are significantly different even at the same time. However, the duration time for different areas are similar. Therefore, we applied the clustering algorithm to spatio-temporal parking data and grouped around 3000 on-street parking spaces into regions with similar patterns. For each region, occupancy was calculated and transformed to the input data of neural networks. An unique models has been trained for corresponding area and a standard evaluation metric is used to measure the performance of the model. Once model trained, we can apply current time t and previous occupancy records $O_t, O_{t-1}, ..., O_{t-n}$ in the specific region i to corresponding LSTM model and get the predicted occupancy O_{t+k} at time $t + k$.

For the other part, we did not use clustering method for different regions of data as all duration model are similar. We directly applied the regression model to learn the functions between time and density probability. The output of the regression model can be visualised in the specific region with heatmap.

Using both predicted duration time and occupancy rate, we can offer drivers a most likely empty parking space ahead.

3.2   Clustering Methodology

As the distribution of occupancy rate with time is different for each region in the CBD area, it is necessary to group similar parking slots before learning process. K-means clustering is a popular clustering method in data mining area. It aims to partition samples into k clusters in which each sample belongs to the cluster with the highest similarity [25]. In the occupancy rate estimation problem, each group of parking slots needs to have the similar trend. However, we can observe that occupancy rates for different regions are not similar in Fig. 2.

Figure 2 shows that occupancy rate varies with time. The occupancy rates are close to zero from midnight to the morning. The peak values of O_t happens in the noon and the second peak time is around 6 pm. Although occupancy rates in different regions have the similar trend, the value of each region is significantly different. Therefore, selecting similar parking slots and training the occupancy prediction model for each group of parking spaces is likely to boost the accuracy. We used the time series data of each parking slot as the input features and applied k-means with $k = 30$ to group 3000 parking slots into 30 groups.

However, for duration period prediction, there is no obvious difference among regions which is shown in Fig. 3. Each line in Fig. 3 denotes the probability density of duration time of one region, which shows that all curves are similar. Hence, the clustering method is not necessary to be applied to parking spots for duration time estimation.

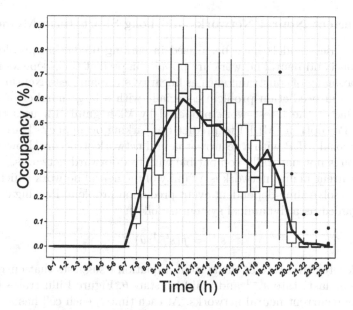

Fig. 2. Hourly evolution of parking occupancy for 30 regions (%). The line graph indicates the mean value of occupancy for all regions.

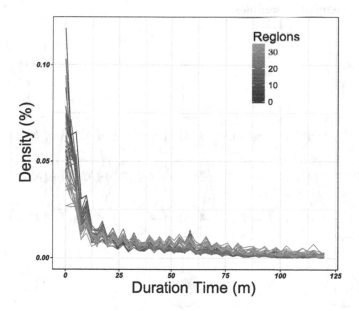

Fig. 3. Probability density function of duration time for each region.

3.3   Recurrent Neural Networks and Long Short Term Memory

Many time-series models have been used in parking prediction problem. Both Markov model and neural network are popular in predict time-dependent events as they relax many of constraint such as linearity, stationariness. Neural networks perform well in prediction problem especially with a large number of available samples. One of the latest work proposed by Vlahoginni [13] uses the static Multilayer Perceptron (MLP) to predict the parking occupancy. They modified the conventional MLP with a recurrent neural network structure. However, such modification is not naturally compatible with feed-forward networks.

The recurrent neural network is a family of the neural network which is naturally used to solve time-dependent event prediction problem. Parking occupancy can be regarded as a dynamical system as follow:

$$s^{(t)} = f(s^{t-1}; \theta) \tag{1}$$

where s^t denotes the state of the system. Equation 1 suggests that current state s^t depends on last state s^{t-1} and a hidden state θ. Figure 4 illustrates the basic structure of recurrent neural networks. At each time t, each cell has an input v_t and an output y_t. Part of y_t is fed back into the cell for use for next step $t+1$, and h_t denotes the hidden state. The input v_t, output y_t and hidden state h_t are associated with a weight matrix. The training process is to adjust the weight matrix with sequential samples.

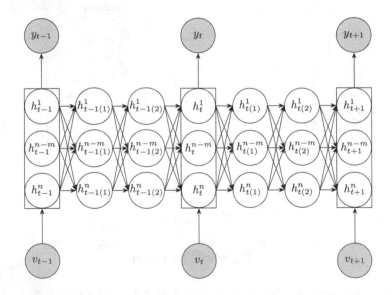

Fig. 4. Recurrent neutral network structure.

The LSTM is a variant of RNN that is capable of learning long-term dependencies. LSTM also implement the recurrent structure as what simple RNN does. However, LSTM adds four gates to inoperative all cell states. There are three main gates called input gate, forget gate and output gate for input state, hidden state and output state, respectively. The other gate is a sigmoid function which is used to modulates the output of these gates. The basic stricture of LSTM is illustrated in Fig. 5.

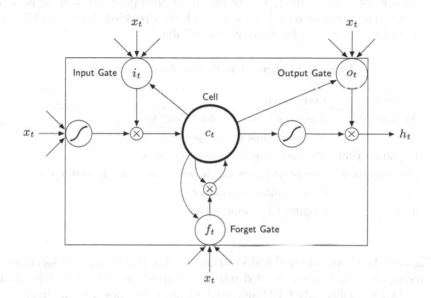

Fig. 5. Long short term memory structure.

In this paper, we use the similar configuration of static MLP [5]. The input of neural networks is a sequential value of occupancy rate $\mathbf{O(t)} = \{O(t-\tau), ..., O(t-(m-1)\tau)\}$. Where τ is the delay and m denotes the time step. The output is a sequence of occupancy value as well. Therefore, the parking occupancy prediction problem is a classical many to many LSTM based problem.

3.4   Regression Model

Duration time can be estimated with a non-linear regression model. As we shown above, this model does not rely on the location of parking slots. Therefore, we apply a unified model to all parking slots. However, some works [5,18,27] showed that the distribution of duration time changed with the temporal information such as weekdays and weekends, morning or evening. In order to keep the model as simple as possible, we only consider three factors: the day in a week, occupancy rate and time in a day. We use the nonlinear least square to estimate the probability density function of duration time with the corresponding regression model based on three factors above.

4   Experiments

4.1   Data Set

Parking events data were collected from more than in-ground 3000 sensors which were placed in each on-street parking slot around Melbourne CBD area. The sensors is able to detect the parking sites availability and report parking events to the central system. A total number of 12, 208.178 parking events have been recorded in a year (2011–2012). Melbourne city council has published the parking dataset, and many works have been done with such parking dataset [6,15]. The relevant features in this dataset are listed in Table 1.

Table 1. Features of Melbourne parking dataset.

Features	Description
Area name	The region parking slot belong to
Arrive time	The start time of car parking event
Departure time	The end time of car parking event
Duration time	The period between the arrival and the departure event
Longitude	Geographical information
latitude	Geographical information

Figure 6 shows all geographical locations of placed sensors. The locations of all parking slots have been divided into many areas and labelled with colour. Limited by the space, only CBD area parking slots are shown in the figure.

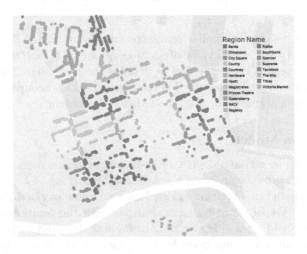

Fig. 6. Parking sensors located in different areas of Melbourne. (Color figure online)

4.2   Parking Occupancy Prediction

In this section, we conducted two experiments on occupancy prediction. In the first experiment, we compared the statistical results from prediction models with clustering and without clustering. In the second experiment, we applied our approaches to parking dataset and compared it with one of the state-of-art approaches called static MLP which is also used to predict the occupancy of parking [5].

A unique LSTM for each parking region is trained to predict overall occupancy of parking slots using historical records which were represented as a couple of time-series data. The input space of each LSTM was a sequential temporal data. In both experiments, we use a 10 min time window data as the input features. The output spaces were from 1 min to 30 min. That is, all models is able to predict parking occupancy from 1 to 30 min ahead. In the train-and-test session, we use 80–20 train and test split.

In order to compare with other works, we use the same evaluation metric which are used for occupancy prediction in work [5] as below:

- Mean absolute error (MAE): $\frac{1}{N} \sum_{i=1}^{N} |\widehat{y}_{i+\tau} - y_{i+\tau}|$
- Root mean squared error (RMSE): $\sqrt{\frac{1}{N} \sum_{i=1}^{N} (\widehat{y}_{i+\tau} - y_{i+\tau})^2}$
- Mean absolute percentage error (MAPE): $\frac{1}{N} \sum_{i=1}^{N} \frac{(\widehat{y}_{i+\tau} - y_{i+\tau})}{y_{i+\tau}}$
- Root relative squared error (RRSE): $\sqrt{\frac{1}{N} \sum_{i=1}^{N} \frac{(\widehat{y}_{i+\tau} - y_{i+\tau})^2}{y_{i+\tau}}} \Big/$

$$\sqrt{\frac{1}{N} \sum_{i=1}^{N} \frac{(\widehat{y}_{i+\tau} - y_{i-\tau})^2}{y_{i+\tau}}}$$

where $y_{i+\tau}$ denotes the actual value and $\widehat{y}_{i+\tau}$ denotes the predicted value with sample i from 1 to N. τ is the timestep and N is the number of samples. $y_{i-\tau}$ denotes the last known occupancy rate.

Clustering Comparison. In the clustering stage, we applied the k-mean (k = 30) to all parking slots and used the parking availability state 1 or 0 at every minute t as features.

Figure 7 shows the hourly occupancy rate for three different clusters. It shows that the mean value of parking occupancy of each region is significantly different from each other, which also support our assumption before that the shape of occupancy rate curses vary with the locations of parking space. Therefore, we train each cluster or region with a unique LSTM model.

Table 2 compares the parking occupancy prediction result from non-clustering approach and clustering based approach. Obviously, prediction result with the specific model for each region divided by the cluster method is better than using the original region division. In this table, raw data means the using data were not divided by clustering methods but original region division which is given by city council. We use the mean value of parking occupancy rate of different regions and four prediction result (1, 5, 15, 30 min).

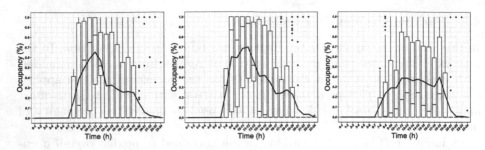

Fig. 7. Hourly evolution of parking occupancy for 3 random selected regions (%). The line graph indicates the mean value of parking occupancy rate for each region.

Table 2. Results for prediction horizons for parking regions (Mean Value) with clustering and without clustering.

	Prediction horizon							
	1 min		5 min		15 min		30 min	
	raw data	k-means	raw data	k-means	raw data	k-means	raw data	k-means
MAE	0.025	0.019	0.041	0.033	0.042	0.048	0.068	0.065
RMSE	0.030	0.023	0.051	0.040	0.051	0.058	0.078	0.075
MAPE	0.047	0.036	0.080	0.062	0.081	0.092	0.130	0.124
RRSE	2.148	1.640	3.618	2.874	3.562	4.010	5.295	5.135

Prediction Result Comparison with Static MLP. In the second experiment, we test LSTM model and static MLP with clustering results. Table 3 shows LSTM model outperforms the static MLP for predicting the occupancy in 1 min, 5 min and 30 min in all metrics. Static MLP performs slight better in 15 min prediction. It suggests that LSTM is better at parking occupancy prediction in general as LSTM is developed with leveraging temporal dependency and static MLP cannot fully use all dependency.

Table 3. Results for prediction horizons for parking regions (Mean Value) by proposed method and static MLP.

	Prediction horizon							
	1 min		5 min		15 min		30 min	
	LSTM	Static-MLP	LSTM	Static-MLP	LSTM	Static-MLP	LSTM	Static-MLP
MAE	0.018	0.025	0.034	0.042	0.047	0.042	0.062	0.064
RMSE	0.022	0.030	0.041	0.053	0.057	0.051	0.072	0.076
MAPE	0.035	0.047	0.065	0.082	0.090	0.081	0.117	0.122
RRSE	1.524	2.062	2.884	3.657	3.893	3.518	5.027	5.260

4.3   Duration Time Factors

We develop the free parking space duration time models with nonlinear least square function. In this experiment, we evaluate the influence of three factors on duration time probability distribution: the day in a week, occupancy rate and the time in a day on duration time.

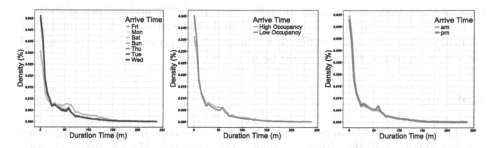

Fig. 8. Probability density function of duration time based on factors (day in a week, occupancy rate and time in a day).

Figure 8 shows the probability density function of duration time grouped by three different factors. The horizontal axis denotes the duration time by minutes and vertical axis denotes the probability density. We observe that none of three factors significant influences the probability density of duration time. Therefore, we think it is reasonable to apply one unified duration regression model to estimate the probability of car leave with a certain duration time.

5   Conclusion and Future Work

In this paper, we exploited a large real-world dataset and proposed a framework to predict the parking availability with LSTM and clustering techniques. The framework consists of two modules: parking occupancy prediction and duration time estimation. In the parking occupancy module, we introduce a popular recurrent neural networks called LSTM to learn the pattern of the occupancy rate of each region clustered by k-means. For duration time estimation module, we use the regression analysis method to estimate the probability of car leaving with time t and evaluate the potential influential factors. The proposed framework has been used for comparison with the state-of-art work and perform better, in general, using a large real-world parking dataset.

There are some weakness of this paper and needed to be solve in the future. Firstly, estimating the duration time should be integrated with occupancy prediction. In this paper, both occupancy prediction and duration time are shown separately to drivers. In the future, we hope we can develop an intelligent system which can employ context information and guide the drivers with an optimised solution for parking and route planning.

From methodology view, proposed approaches could be improved. k-means is good at solving the clustering problem with a certain number of cluster. However, in parking problem, the best number of regions is uncertain, which is needed to be decided by the size of the specific parking area and nearby context information. A more suitable clustering method is needed for cluster spatio-temporal interval-based data such as parking data [27]. From the learning perspective, we only use the basic LSTM to learn the pattern of time-series events. With recent rapid evolution for deep learning techniques, more powerful and suitable recurrent neural networks are proposed to solve such problems with faster speed and higher accuracy. Therefore, using different techniques to improve the effectiveness and the efficiency of the system should be in the future plan.

References

1. Tilahun, S.L., Di Marzo Serugendo, G.: Cooperative multiagent system for parking availability prediction based on time varying dynamic Markov chains. J. Adv. Transp. **2017** (2017)
2. Rahaman, M.S., Hamilton, M., Salim, F.D.: Queue context prediction using taxi driver knowledge. In: Proceedings of the Knowledge Capture Conference, Austin, TX, USA, pp. 35:1–35:4 (2017)
3. Arief-Ang, I.B., Salim, F.D., Hamilton, M.: DA-HOC: semi-supervised domain adaptation for room occupancy prediction using CO_2 sensor data. In: The Proceedings of the Fourth ACM International Conference on Systems for Energy-Efficient Built Environments (BuildSys), Delft, The Netherlands (2017)
4. Abdullah, S.S., Rahaman, M.S.: Stock market prediction model using TPWS and association rules mining. In: 15th International Conference on Computer and Information Technology (ICCIT), Chittagong, pp. 390–395 (2012)
5. Vlahogianni, E.I., Kepaptsoglou, K., Tsetsos, V., Karlaftis, M.G.: A real-time parking prediction system for smart cities. J. Intell. Transp. Syst. **20**(2), 192–204 (2016)
6. Zheng, Y., Rajasegarar, S., Leckie, C.: Parking availability prediction for sensor-enabled car parks in smart cities. In: 2015 IEEE Tenth International Conference on Intelligent Sensors, Sensor Networks and Information Processing (ISSNIP), pp. 1–6. IEEE (2015)
7. Caliskan, M., Barthels, A., Scheuermann, B., Mauve, M.: Predicting parking lot occupancy in vehicular ad hoc networks. In: IEEE 65th Vehicular Technology Conference, VTC2007-Spring, pp. 277–281. IEEE (2007)
8. Caicedo, F., Blazquez, C., Miranda, P.: Prediction of parking space availability in real time. Expert Syst. Appl. **39**(8), 7281–729 (2012)
9. Pengzi, C., Jingshuai, Y., Li, Z., Chong, G., Jian, S.: Service data analyze for the available parking spaces in different car parks and their forecast problem. In: Proceedings of the 2017 International Conference on Management Engineering, Software Engineering and Service Sciences, pp. 85–89. ACM (2017)
10. Ma, J., Clausing, E., Liu, Y.: Smart on-street parking system to predict parking occupancy and provide a routing strategy using cloud-based analytics. No. 2017-01-0087. SAE Technical Paper (2017)
11. Shi, C., Liu, J., Miao, C.: Study on parking spaces analyzing and guiding system based on video. In: 2017 23rd International Conference on Automation and Computing (ICAC), pp. 1–5. IEEE (2017)

12. Tamrazian, A., Qian, Z., Rajagopal, R.: Where is my parking spot? Online and offline prediction of time-varying parking occupancy. Transp. Res. Rec. J. Transp. Res. Board **2489**, 77–85 (2015)
13. Goodfellow, I., Bengio, Y., Courville, A.: Deep Learning, vol. 1. MIT Press, Cambridge (2016)
14. Rumelhart, D.E., Hinton, G.E., Williams, R.J.: Learning representations by back-propagating errors. Nature **323**(6088), 533 (1986)
15. Shao, W., Salim, F.D., Gu, T., Dinh, N.-T., Chan, J.: Travelling officer problem: managing car parking violations efficiently using sensor data. IEEE Internet Things J. (2017)
16. Hochreiter, S., Schmidhuber, J.: Long short-term memory. Neural Comput. **9**(8), 1735–1780 (1997)
17. Chen, X.: Parking occupancy prediction and pattern analysis. Department of Computer Science, Stanford University, Stanford, CA, USA, Technical report CS229-2014 (2014)
18. Vlahogianni, E.I., Kepaptsoglou, K., Tsetsos, V., Karlaftis, M.G.: Exploiting new sensor technologies for real-time parking prediction in urban areas. In: Transportation Research Board 93rd Annual Meeting Compendium of Papers, pp. 14–1673 (2014)
19. Vinyals, O., Toshev, A., Bengio, S., Erhan, D.: Show and tell: a neural image caption generator. In: 2015 IEEE Conference on Computer Vision and Pattern Recognition (CVPR), pp. 3156–3164. IEEE (2015)
20. Barbounis, T.G., Theocharis, J.B., Alexiadis, M.C., Dokopoulos, P.S.: Long-term wind speed and power forecasting using local recurrent neural network models. IEEE Trans. Energy Convers. **21**(1), 273–284 (2006)
21. Aczon, M., et al.: Dynamic mortality risk predictions in pediatric critical care using recurrent neural networks. arXiv preprint arXiv:1701.06675 (2017)
22. Rumelhart, D.E., Hinton, G.E., Williams, R.J.: Learning internal representations by error propagation. No. ICS-8506. California Univ San Diego La Jolla Inst for Cognitive Science (1985). Harvard
23. Klappenecker, A., Lee, H., Welch, J.L.: Finding available parking spaces made easy. Ad Hoc Netw. **12**, 243–249 (2014)
24. Vlahogianni, E.I., Karlaftis, M.G., Golias, J.C.: Optimized and meta-optimized neural networks for short-term traffic flow prediction: a genetic approach. Transp. Res. Part C Emerg. Technol. **13**(3), 211–234 (2005). Harvard
25. Kanungo, T., Mount, D.M., Netanyahu, N.S., Piatko, C.D., Silverman, R., Wu, A.Y.: An efficient k-means clustering algorithm: analysis and implementation. IEEE Trans. Pattern Anal. Mach. Intell. **24**(7), 881–892 (2002)
26. Song, H., Qin, A.K., Salim, F.D.: Multivariate electricity consumption prediction with extreme learning machine. In: 2016 International Joint Conference on Neural Networks (IJCNN), pp. 2313–2320. IEEE (2016)
27. Shao, W., Salim, F.D., Song, A., Bouguettaya, A.: Clustering big spatiotemporal-interval data. IEEE Trans. Big Data **2**(3), 190–203 (2016)

Time-Based Quality-Aware Incentive Mechanism for Mobile Crowd Sensing

Han Yan and Ming Zhao[(✉)]

School of Software, Central South University, Changsha 410075, China
{yanhan,meanzhao}@csu.edu.cn

Abstract. Recent years have witnessed the advance of mobile crowd sensing (MCS) system. How to meet the demands of task time requirements and obtain high-quality data with little expense has become a critical problem. We focus on exploring incentive mechanisms for a practical scenario, where the tasks are time window dependent. An important indicator, "quality of user's data (QOD)" is also considered. First, we design a prediction model based on user history data (p-QOD), to calculate the next time of the user's QOD. Second, we design a dynamic programming algorithm based on time windows and p-QOD, to ensure all of the task time windows are covered, as well as minimizing the platform's cost. Finally, we determine the payment for each user through a Vickrey–Clarke–Groves auction (VCG) considering the user's true data quality (t-QOD), which is based on their submission time. Through both rigorous theoretical analysis and extensive simulations, we demonstrate that the proposed mechanisms achieve high computation efficiency, fairness, and individual rationality.

Keywords: Mobile crowd sensing · Incentive mechanism · VCG auction · Quality of user's data · Time window

1 Introduction

Recently mobile crowd sensing systems (MCS) have emerged as a novel networking model (e.g., smartphones and smart watches), giving rise to extensive concerns for solving the complex sensing applications from the significant demands of people's lives, including Haze Watch for pollution monitoring, NoiseTube and Ear-Phone for creating noise maps, which leverage the ubiquity of sensor-equipped mobile devices to collect data at low costs. These provide a new compelling paradigm for solving the problem of large-scale sensing data collection.

The key challenge for the MCS system is how to design an effective incentive mechanism. The incentive mechanism can motivate more participation in MCS with minimum cost, to ensure the reliability of the data quality. For the participants, incentive mechanisms ensure individual rationality and make the payment fair. However, most of MCS applications are based on voluntary user participation or lack effective incentive mechanisms, especially for the time-sensitive crowd sensing system.

Recent studies have focused on time-sensitive systems. These studies include the continuous time interval coverage tasks that require completing sensing data in the entire time interval publicized by the platform [1]. However, most of the existing

© Springer Nature Switzerland AG 2019
S. Li (Ed.): GPC 2018, LNCS 11204, pp. 138–151, 2019.
https://doi.org/10.1007/978-3-030-15093-8_10

mechanisms fail to incorporate users' quality of data (QOD) and the impact of user submission time on the platform. The meaning of QOD varies for different applications. For example, in the Med Watcher system QOD refers to the quality of up-loaded photos (e.g., resolution, contrast, sharpness, etc.). In air quality monitoring MCS systems, QOD means a user's estimation accuracy of air quality.

In order to solve these problems, we propose a time-sensitive incentive mechanism for a practical scenario. Unlike most of the previous work, we use an important indicator known as "quality of user's data" (QOD). First, we design a prediction model based on the user history data (p-QOD). We use the time attenuation factor (TAF) to denote the affect weight of QOD, to calculate user's p-QOD. Next, considering users' strategic behaviors, our system uses p-QOD to calculate user bids. We design a dynamic programming algorithm based on the time window and user's bid to ensure the continuous time-interval coverage while satisfying the minimum social cost. Finally,considering the impact of users' submit time on QOD, we weight the true quality of the data (t-QOD) based on the submission time and determine the payment for each user through VCG auction while considering the t-QOD. Similar to traditional VCG mechanisms, ours maximizes the social welfare.

- Apart from other reverse combinatorial auctions, our mechanism also satisfies fairness between users while approximately maximizing the social efficiency and reducing the platform costs in time window case. The key contributions of our work are the following: A simple yet representative formulation based on the social optimization user selection (SOUS) problem to address the strong requirement of continuous time-interval coverage. Using the time attenuation factor (TAF) to denote the affect weight of history QOD. Next, we integrate p-QOD into the time-window selection phase, and select winners that meet the needs of the platform.
- We design a payment incentive mechanism t-QOD VCG considering the influence of the user submit times. Apart from other reverse combinatorial auctions, our mechanism also satisfies fairness between users while approximately maximizing the social welfare and reducing the platform costs in the time window case.
- We perform extensive simulations to evaluate the effectiveness of our incentive mechanism.

The rest of the paper is organized as follows. Section 2 reviews related work. The system model of our mechanism is introduced in Sect. 3. We evaluate the performance of the user selection and user payment mechanism via simulations in Sect. 4. Finally, we conclude our paper in Sect. 5.

2   Related Work

At present, a number of mobile crowd sensing applications have been designed and implemented. For example, Ahnn [2] designed a system named GeoServ as a distributed sensing platform, where millions of participants can take part in urban sensing and share information using always-on cellular data connections. Lee et al. proposed a reverse auction by use of mechanisms such as virtual participation credit (VPC) and recruitment credit (RC) [3] for collecting users' sensing data. They designed a novel reverse auction-based dynamic price (RADP) incentive mechanism, in which the service provider publicized time tasks in each round and users sold their sensing data to the provider with users' claimed bid prices. However, the data collected did not meet the task time requirements. The Vickrey Clark Groves (VCG) reverse auction [4] is a payment model where each bidder's compensation is the damage caused by the bidder's accession to all other bidders. In [5], the VCG auction updated rules to adjust user distribution based on the online mechanisms. Koutsopoulos [6] introduced the VCG auction. Each time the service provider receives a request, it publishes the sensing task and a reverse auction is opened to complete the task. A Bayesian game is used on the participants once the auction is open. The mechanism maximizes the revenue of each user and proves that the game achieves a Bayesian Nash equilibrium. Yang et al. considered two system models [7]: the platform-centric model design and incentive mechanism using a Stackelberg game; and the user-centric model design and auction-based incentive mechanism that allows users to control their payment more.

Guo in [8] introduced a formal concept model to characterize group activities and classify them into four organizational stages. This paper presents a group-aware, mobile, crowd-sensing system called MobiGroup, which supports group activity organization in real-world settings. In [9], the consideration of data quality is included into the design of incentive mechanisms for crowd sensing, and participants are paid for how well they perform, to motivate the rational participants to perform data sensing efficiently. This mechanism estimates the quality of sensing data, and offers each participant a reward based on effective contribution. Pouryazdan [10] adopt vote-based approaches, and presented a thorough performance study of vote-based trustworthiness with trusted entities that are a subset of the participating smartphone users. The reputations of regular users are determined based on vote-based (distributed) reputations.

Liu et al. [11] introduced four key design elements. The QoI is quantified in relation to the level they require. The credits are quantified based on the degree of satisfaction. The Gur Game used the two above indexes in the mathematical framework of the Gur Game for distributed decision-making, and the dynamic pricing scheme allocated credits to participants while minimizing the necessary adaptation of the pricing scheme from the network operator. A common feature of the existing work is that they do not consider the QOD may change over time. This is the major difference with our mechanisms.

3   Problem Formulation and Proposed Solution

3.1   System Overview

The MCS in this paper consists of a cloud platform and a set of N users $U = \{1, \ldots, n\}$. In real scenarios of MCS, there are many applications based on the time window, such as monitoring of real-time vehicle flow, continuous measurement of air quality, and the long-term observation of specific regional noises. The MCS platform must collect continuous data at a specific time window. In this paper, we assume the MCS is designed for these practical and universal time-window scenarios. For the platform, these time-sensitive tasks can't be done by a single user. The sensing data that all users must submit must meet the full coverage of the task time, to ensure the data integrity of the system.

Our platform first publishes the time-sensing tasks. Let G represent the length of task time requirements, $G = [T_S, T_E]$, T_S denote the task start time, and T_E denote the task end time. The platform requests participants executing sensing tasks from T_S to T_E to submit their sensing data. Thus, participants should submit their sensing data in the required time window. In addition, the sensing data is invalid if users submit data that is not in the valid time G. Assume that a crowd of smartphone users $U = \{1, \ldots, n\}$ are the participants interested in our sensing tasks. Each participant uploaded their own free time windows set $\Phi_i = \{[s_i(1), e_i(1)], \ldots, [s_i(k), e_i(k)]\}$, $s_i(k)$ and $e_i(k)$ are the start time and end time, respectively. Any $s_i(k) < T_S$ or $e_i(k) > T_E$ is invalid data and can't bring extra revenue for any user i. The problem of winner determination and payment can be decoupled into two separate problems. We formulate the winner determination phase as Social Optimization User Selection (SOUS) problem. Considering each user's bid and time window, our MCS system maximizes the social efficiency of the platform as given in Definition 1.

Definition 1. (SOUS Model) The expected social efficiency maximization problem

$$Y(H) = \max\left[y(G) - \sum B_i\right] \qquad (1)$$

$$\text{s.t.} G \sqsubseteq \cup_{i \in U, \theta \in \{1, \ldots, k\}} [s_i(\theta), e_i(\theta)] \qquad (2)$$

where $y(G)$ is the utility function of the platform when the entire time window tasks were performed, which are computed as follows. Each user's bid B_i is calculated by the platform based on the user's history quality of data. H is the set of winners, $H \in U$.

Definition 2. (Platform Utility) The utility of the platform is

$$y(G) = v(G) - \sum_{i \in H} P_i \qquad (3)$$

which denotes if all of the data obtained by the platform meets the coverage of time window G. The data value obtained by the platform is $v(G)$. Each user's paid is P_i, which is computed by the platform.

The workflow of the system is described as follows

1. First, the platform publishes the sensing time window, $G = [T_S, T_E]$, to users.
2. Each user i submits its set of tasks Φ_i, consisting of the set of tasks that user i wants to execute.
3. Based on the user's p-QOD and time windows Φ_i, the platform determines the set of winners H ($H \in U$).
4. Based on user's true data quality (t-QOD) and the submission time, the platform pays the winners. Specifically, a loser does not execute any task and receives zero payment.

3.2   The Predict Quality of Data (p-QOD)

In general, the definition of QOD is as a metric to measures the quality of sensed data from participants. QOD determines user bids and the choice of the platform. In our user selection phase, we use historical records to predict the next time of quality that the user can achieve, p-QOD. First, we use the time attenuation factor (TAF) to denote the affect weight of p-QOD. Second, the DNC algorithm [12, 13] is used to calculate users' p-QOD. Assuming that each historical data's quality is arranged in a queue Q, and each history execution time $t(Q)$ satisfies:

(1) $t(Q) \geq t - T$, where t represents the current time and T represents the span of time (e.g. a week or a mouth).
(2) The header of the queue was the QOD for the last time, and the tail of the queue was the QOD for the earliest time, as shown in Fig. 1:

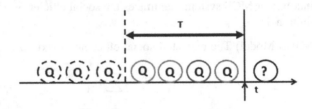

Fig. 1. Historical data's quality.

Different colors represent different users' QOD. The blue one in Fig. 1 was the QOD for the last time and the tail of the queue with the dotted line was the QOD for the earliest time. As shown in Fig. 1, we only consider the historical quality of data in t-T (e.g. the data for the last month) when calculating the quality of the next time. Thus, the TAF λ_i of the first h data of user i is

$$\lambda_i = \begin{cases} 0 & t(Q) < t - T \\ 1 - \frac{t - t(Q)}{T} & t(Q) \geq t - T \end{cases} \tag{4}$$

According to the TAF, the p-QOD p_q_i of this time is

$$p\_q_i = \frac{\sum \lambda_i Q_h}{\sum \lambda_h} \tag{5}$$

Q_h indicates the first h data's quality for user i, and λ_h is the first h data's TAF, calculated by (4). The TAF indicates the current task time had a higher impact on user's p-QOD, while the earlier task time had a lower influence on user's p-QOD.

3.3   Incentive Mechanism in SOUS Model

3.3.1   Mechanism Design

In our SOUS model, the platform first publicizes the sensing time window $G = [T_S, T_E]$. Next, it determines the winners based on their $\widehat{c}_i(p\_q_i)$, which represents the user's cost function in this round that is calculated by the platform. The users' time windows are independent of each other, which means there is no contact between different users' time windows Φ_i and Φ_j. According this, we propose an algorithm based on dynamic programming to solve the SOUS problem illustrated in Algorithm 1.

Algorithm 1. User Selection algorithm based on the Time Window

---

Input: Time Window G, Set of Users U

1:    $H \leftarrow \{\phi\}$;

2:    **for** all $i \in U$ **do**

3:        $C(i) \leftarrow \infty$;

4:    **end for**

5:    **Sorting** all time windows Φ_i based on $e_i(k)$ in descending order for $\forall i \in U$ and the sequence is denoted by $\{\Phi_1, \Phi_2, \dots, \Phi_n\}$;

6:    **for** i = 1 to n **do** :

7:        **if** $T_S \in [s_i(k), e_i(k)]$ **then**

8:            front(i) $\leftarrow$ (-1) ;

9:            $C(i) = \widehat{c}_i(p\_q_i)$;

10:       **else**

11:           front(i) $\leftarrow \arg\min_{e_j(k) \geq s_1(k), j < i} C(j)$;

12:           $C(i) \leftarrow C(\text{front}(i) + \widehat{c}_i(p\_q_i))$;

13:       **end if**

14:   **end for**

15:   $i \leftarrow \arg\min_{T_E \in [s_j(k), e_j(k)], j \in U} C(j)$;

16:   $y \leftarrow C(i)$;

17:   **while** $i \neq -1$

18:       $H \leftarrow H \cup \{i\}$; $i \leftarrow \text{front}(i)$;

19:   **end while**

20:   **return** H;

The SOUS problem is equivalent to the problem of maximizing social efficiency, as shown in Definition 1: $Y(H) = \max[y(G) - \sum B_i]$, the value of $y(G)$ is continuous, since the platform publishes the sensing time window G until all time window tasks are performed. Therefore, the problem can be treated as a minimizing social cost problem.

$$min_{i \in U}\{C(i)|T_E \in [s_i(k), e_i(k)]\} \tag{6}$$

$$s.t. G \subseteq \bigcup_{i \in \{1,...,n\}[s_i(k),e_i(k)]} \tag{7}$$

The constraint in formula (7) means summing all users' time windows should cover the platform requirements from T_S to T_E. We assume that there are enough users satisfying the constraint involved in the sensing tasks. Users are sorted according to the end time of the time windows, such as $e_1(1) \leq e_2(2) \leq ... \leq e_n(n)$. $C(i)$ is the minimum social cost covering $[T_s, e_i(k)]$. Considering all $C(i)$ in the above order have been computed, the state transition is

$$C(i) = \begin{cases} \arg min_{e_j(k) > s_i(k)_{j<i}} C(j) + \widehat{c}_i(p\_q_i), & T_S \notin [s_i(k), e_i(k)] \\ \widehat{c}_i(p\_q_i), & T_S \in [s_i(k), e_i(k)] \end{cases} \tag{8}$$

$C(i)$ is the minimum value of the sum of $C(j)$ which satisfies $e_j(k) > s_i(k)(\forall j < i)$ and the bid $B_i = \widehat{c}_i(p\_q_i)$. We obtain the winners H that satisfy the minimum social cost.

3.4   QOD-VCG Auction in User Payment

3.4.1   Weight the True Quality of Data (t-QOD)

After the winners submit their data, our platform obtains users' true data quality. The users' submission times affect the performance of the task execution (e.g. the user submits its data after the end of the task, and can't collect sufficient data). Therefore, in order to ensure the fairness of the payment mechanism, we design an incentive mechanism based on QOD-VCG auction, using user's true quality of data (called t-QOD) t_q_i. We weight the true quality of the data (t-QOD) based on the submission time and determine the payment for each user through VCG auction while considering the t-QOD.

In this paper, we reasonably assume that the platform uses weighted aggregation to calculate Q (the sum of the QODs of the winners that execute these tasks) as $Q = \sum_{i:i \in H} \alpha_i * t\_q_i$, where α_i is the weight of i submits its quality t_q_i. The value of α_i changes over time. Thus, α_i is considered as a time factor [9]. Suppose that the user's data submits the time as t. Let $T_S = s$ and $T_E = e$ for convenience. If i submits his data in $\Phi_i(s \leq t \leq e)$, then the time factor $\alpha_i = 1$. However, with the delay of i submits it's data $(t > e)$, the negative effect of α_i on q_i is greater. Formula (10) is the function of α_i, $\alpha_i = g(t - e)$ (9). In addition, $f(x)$ is a sigmoid function (10) and $sgn(x)$ is a sign function (11).

$$g(t - e) = 2 \times sgn(t - e) \times f(e - t) + sgn(e - t) \tag{9}$$

The expression of sigmoid function $f(x)$:

$$f(x) = \frac{1}{1+e^{-x}} \tag{10}$$

The expression of sign function $sgn(x)$:

$$sgn(x) = \begin{cases} 1, & x > 0 \\ \frac{1}{2}, & x = 0 \\ 0, & x < 0 \end{cases} \tag{11}$$

The functional model of α_i is shown in Fig. 2, where $x = t - e$, $\alpha_i = g(t - e)$.

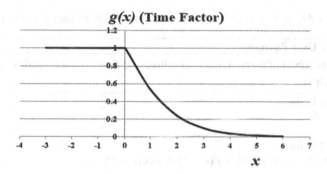

Fig. 2. Time factor function model.

The MCS system accounts for the dynamic changes of users' behavior patterns when they submit their data. The aggregation error [14] of the sum of the QODs (Q) submitted by the winners and the task demand QODs ($\overline{Q}$) is upper bounded by a predefined threshold ϑ, $\overline{Q} - Q \leq \vartheta$. Intuitively, the larger ϑ is, the more QOD will be available to the platform.

3.4.2 Mechanism Design
In this paper, we introduce QOD-VCG extended from the payment mechanism proposed in [15]. The QA-VCG payment mechanism relied on data quality to pay for users. However, the time that winners submit their sensing data in our time window system can't be predicted accurately. Since the QOD varies with time, we propose a time-based QOD-VCG payment mechanism based on the QA-VCG. We formulate the payment problem in (12), and set t_q_i to be controlled by time factor α_i, to eliminate the unfairness caused by time delay.

$$P_i(\widehat{c}_i|\overline{c}_i) = Y(H) + \widehat{c}_i(t\_q_i) - Y(H_{-i}, \overline{c}_i) \tag{12}$$

$P_i(\widehat{c}_i|\overline{c}_i)$ in (12) represents the platform's payment to user i when the bid of i is $\overline{c}_i$. The cost function of i is $\widehat{c}_i(\cdot)$. Thus, under the limitation of the aggregation error, the objective of QOD-VCG can be formulated as

$$Y(H) = \max_{q_i \in Q} \sum_{i=1}^{n} [y_i(\alpha_i * t\_q_i) - \widehat{c}_i(t\_q_i)],$$

$$s.t. Q - \overline{Q} \leq \vartheta$$

(13)

We use $Y(H_{-i}, \overline{c}_i)$ (H_{-i} denotes all the winners except user i) to denote the social surplus, in case of the bid of i is $\overline{c}_i$ and the cost function of i is $\widehat{c}_i(\cdot)$ while other users' k (except user i) bid is $\widehat{c}_k$.

$$Y(H_{-i}, \overline{c}_i) = \sum_{k \neq i}^{n} [y_k(\alpha_k * t\_q_k) - \widehat{c}_k(t\_q_k)] + [y_i(\alpha_i * t\_q_i) - \overline{c}_i(t\_q_i)] \quad (14)$$

Both t_q_i and t_q_k in Eq. (14) are the different QODs of user i and k, respectively.

Algorithm 2. User Payment
Input: Optimal subset of users H, User submitted task time t, Data of quality q_i
Output: Payoff P
1:   for all $i \in U$ do
2:       $P_i \leftarrow 0$;
3:   end for
4:   for all $i \in H$ do
5:       $\alpha_i = 2 \times sgn(t - e) \times f(e - t) + sgn(e - t)$
6:       $Y(H) = \max_{q_i \in Q} \sum_{i=1}^{n} [y_i(\alpha_i q_i) - \hat{c}_i(q_i)]$
7:       $P_i \leftarrow Y(H) + \hat{c}_i(q_{s_i}) - Y(H_{-i}, \overline{c}_i)$;
8:   end for
9:   return P

If user i reports it's true cost function $c_i(\cdot)$, the profit of i can be expressed as

$$u_i(c_i|\overline{c}_i) = P_i(\widehat{c}_i|\overline{c}_i) - c_i(t\_q_i)$$
$$= \sum_{k \neq i}^{n} [y_k(\alpha_k * t\_q_k) - c_k t\_(q_k)] + [y_i(\alpha_i * t\_q_i) - c_i(t\_q_i)] - Y(H_{-i}, \overline{c}_i)$$

(15)

Based on the above definitions, (15) is maximized when the bid of i is the true cost $c_i(\cdot)$. As a result, only users reporting their real cost function would maximize their profit $u_i(c_i|\overline{c}_i)$.

4 Performance Evaluation

4.1 Data Description

The experiment in this paper is based on a noise model, where the intensity of the noise is varies at different times in the same place. The noise model provides help with travel and purchases. A number of college students were selected as experimental participants. The participants in this experiment were equipped with a timer, accelerometer, and the audio signal sampling microphone device that supports 16 bits 44.1 Hz. The mobile devices of the participants are not restricted. First, we publicize different tasks with different sensing time windows in many different locations. Next, participants collect sensing data with different devices in different ways (i.e., walking, taxi, or bus). Finally, the platform measures the performance of the participants according to the time and QOD submitted by the user and pays the participant.

4.2 Baseline Method

We first compare the platform's total payment of the QOD-VCG auction with the traditional VCG auction. We integrate the concept of t-QOD defined in Sect. 3 into the VCG payment problem. Next, we use a baseline method MSG-greed to select N users as winners, according to the descending order of the value q_i, until the error-bound constraints of all tasks are satisfied. Like our QOD-VCG mechanism, the baseline auction also satisfies individual rationality.

4.3 Experiment Settings

In order to measure the performance of the system, we set different time windows in different areas. We chose three different locations and four different time periods. The start times and end times are different for each. Because the noise collection experiment is conducted in colleges, it accounts for the class times and rest times of the users. The participation may be different, so the time window settings are shown as Table 1.

Table 1. Settings for areas and sensing time windows

Area	Time windows
1	[10:20:37, 13:25:40]
2	[18:23:45, 22:34:05]
3	[07:56:32, 10:43:21]

In our QOD-VCG auction, we consider the two settings described in Table 2. In setting I, we fix the number of tasks as K = 30, and vary the number of users from 50 to 150. In setting II, we fix the number of users as N = 130 and vary the number of tasks from 30 to 60. We assume that participants in the SOUS model calculated the same QOD through their historical data. The values of ϑ and α_i for any user $i \in N$ are based on their true behavior.

Table 2. Settings for QOD-VCG auction

Setting	Area	α_i	ϑ	N	K
I	1	[0, 1]	[0.1, 0.2]	[50, 150]	30
II	2	[0, 1]	[0.1, 0.2]	130	[30, 60]

4.4 Performance Evaluation

A: Time Window

Since the platform needs to collect data from different areas and time windows, the following figures plot the number of users and number of winners in different time windows. We determine the different end times, which means the different time windows G are an index to measure the system.

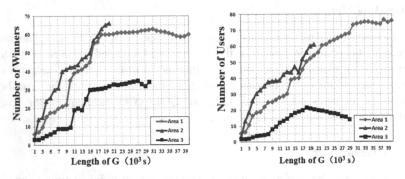

Fig. 3. Performance of SOUS with various end time of the sensing time window

For our simulation of the SOUS mechanism, which is illustrated in Fig. 3 shows that, when the length of |G| increases, not only do the number of users increase, but the number of winners also increase. Because of the increase of the sensing time, there are more participants in each area to perform tasks. Therefore, the platform needs to collect more participants to perform the sensing task. As the time increases, the winners of the auction will also grow. The unsmoothed curves in Fig. 3 are due to the experiment that was conducted at the school and affected by the class time, resulting in the fluctuation of Fig. 3.

B: Platform's Total Payment

We compare the platform's total payment generated under setting I (a) and II (b) using the QOD-VCG auction and the baseline auction MSG-greed mechanism, in both setting I and II as shown in Fig. 4. The platform's total payment of the QOD-VCG auction is far less than that of the baseline and VCG auction. The unsmoothed curves in Fig. 4 as well as in the forthcoming Fig. 5 are due to the parameter α_i, which varies with time, as shown in Fig. 2. We conclude that the platform's total payment of the QOD-VCG auction is close to optimal and far better than that of the baseline auction. Compared with the traditional VCG auction, the data of QOD-VCG is varied, due to the

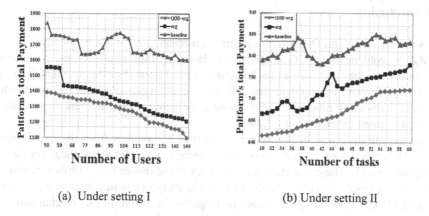

(a) Under setting I   (b) Under setting II

Fig. 4. Performance of QOD-VCG in platform's payment comparison with other mechanisms under setting I and II

introduction of the time factor. Hence, the payment to the winner will be more reasonable than other payment mechanisms.

C. Social Efficiency

In Fig. 5, we show the comparison of the social efficiency under setting I (c) and II (d) between the QOD-VCG auction, VCG auction, and the baseline auction mechanism in both settings I and II. It is obvious from these two figures that the social efficiency of the baseline auction is significantly less than the QOD-VCG auction. By increasing the number of users and tasks, the social efficiency is also increased. Hence, our social efficiency is closely related to the social cost; the cost will increase with the increase of time window. The bid price may decline with the increase of the number of bidders over time, and the social cost unexpectedly decreased. From the two figures above, we also know that the QOD-VCG auction obtains close-to-optimal social welfare, which is closer to the optimal social welfare.

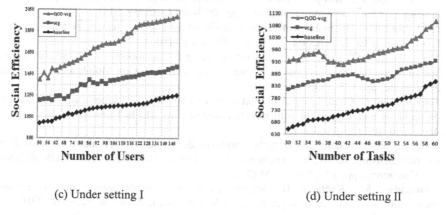

(c) Under setting I   (d) Under setting II

Fig. 5. Performance of QOD-VCG in social efficiency compare with other mechanisms under setting I and II.

5 Conclusion

In this paper, we design QOD-aware incentive mechanisms for MCS systems based on the SOUS problem and VCG auction. We design an individual rationally and computationally efficient mechanism for selecting optimal users. For the payment stage, we design a QOD-VCG mechanism that achieves close-to-optimal social efficiency while satisfying individual rationality and fairness. Moreover, our theoretical analysis is validated through extensive simulations.

The system designed in this paper has certain advantages from the experimental results. However, there are still some shortcomings. The next steps focus on the following aspects: (1) The computational efficiency of the model being further optimized; (2) Designing online incentive model to meet the participants in real-time dynamics to join; and (3) Using the long-term user participation incentive in our mechanism.

Acknowlededment. The research is supported by "National Natural Science Foundation of China" (No. 61572526) and "Innovation Project for Graduate Students in Central South University" (No. 502211708).

References

1. Zhou, P., Zheng, Y., Li, M.: Demo: how long to wait?: predicting bus arrival time with mobile phone based participatory sensing. IEEE Trans. Mob. Comput. **13**(6), 1228–1241 (2014)
2. Ahnn, J.H., Lee, U., Moon, H.J.: GeoServ: a distributed urban sensing platform. In: IEEE/ACM International Symposium on Cluster, Cloud and Grid Computing, pp. 164–173. IEEE (2011)
3. Lee, J.S., Hoh, B.: Sell your experiences: a market mechanism based incentive for participatory sensing. In: IEEE International Conference on Pervasive Computing and Communications, pp. 60–68. IEEE (2010)
4. Nisan, N., Ronen, A.: Computationally feasible VCG mechanisms. In: ACM Conference on Electronic Commerce, pp. 242–252. ACM (2011)
5. Gao, L., Hou, F., Huang, J.: Providing long-term participation incentive in participatory sensing. In: Computer Communications, pp. 2803–2811. IEEE (2015)
6. Koutsopoulos, I.: Optimal incentive-driven design of participatory sensing systems. In: 2013 Proceedings IEEE INFOCOM, pp. 1402–1410. IEEE (2013)
7. Yang, D., Xue, G., Fang, X., et al.: Crowdsourcing to smartphones: incentive mechanism design for mobile phone sensing. In: International Conference on Mobile Computing and Networking, pp. 173–184. ACM (2012)
8. Guo, B., Yu, Z., Chen, L., et al.: MobiGroup: enabling lifecycle support to social activity organization and suggestion with mobile crowd sensing. IEEE Trans. Hum. Mach. Syst. **46**(3), 390–402 (2016)
9. Peng, D., Wu, F., Chen, G.: Pay as how well you do: a quality based incentive mechanism for crowdsensing. In: ACM International Symposium on Mobile Ad Hoc NETWORKING and Computing, pp. 177–186. ACM (2015)
10. Pouryazdan, M., Kantarci, B., Soyata, T., et al.: Anchor-assisted and vote-based trustworthiness assurance in smart city crowdsensing. IEEE Access **4**, 529–541 (2016)

11. Liu, C.H., Hui, P., Branch, J.W., et al.: Efficient network management for context-aware participatory sensing, pp. 116–124 (2011)
12. Chang, J.S., Wang, H.M., Yin, G.: DyTrust: a time-frame based dynamic trust model for P2P systems. In: International Conference on Information Security, pp. 1301–1307. Springer, Heidelberg (2006)
13. Su, L., et al.: Generalized decision aggregation in distributed sensing systems. In: IEEE Real-Time Systems Symposium, pp. 1–10 (2014). IEEE
14. Chen, J., Huang, H., Kauffman, R.J.: A public procurement combinatorial auction mechanism with quality assignment. Decis. Support Syst. **51**(3), 480–492 (2011)
15. He, H., Jian, C.: On revised QA-VCG mechanism in procurement combinatorial auction. Syst. Eng. Theory Pract. **27**(11), 43–47 (2007)

Cloud Computing, Mobile Computing and Crowd Sensing

Container-Based Customization Approach for Mobile Environments on Clouds

Jiahuan Hu, Song Wu$^{(\boxtimes)}$, Hai Jin, and Hanhua Chen

Services Computing Technology and System Lab, Cluster and Grid Computing Lab,
Huazhong University of Science and Technology, Wuhan 430074, China
{zjsyhjh,wusong,hjin,chen}@hust.edu.cn

Abstract. Recently, mobile cloud which utilizes the elastic resources of clouds to provide services for mobile applications, is becoming more and more popular. When building a *mobile cloud platform* (MCP), one of the most important things is to provide an execution environment for mobile applications, e.g., the Android mobile *operating system* (OS). Many efforts have been made to build Android environments on clouds, such as Android *virtual machines* (VMs) and Android containers. However, the need of customizable Android execution environments for MCP has been ignored for many years, since the existing OS customization solutions are only designed for hardware-specific platforms or driver-specific applications, and taking little account of frequently-changing scenarios on clouds. Moreover, they lack a unified method of customization, as well as an effective upgrade and maintenance mechanism. As a result, they are not suitable for varied and large-scale scenarios on clouds. Therefore, in this paper, we propose a unified and effective approach for customizing Android environments on clouds. The approach provides a container-based solution to custom-tailor Android OS components, as well as a way to run Android applications for different scenarios. Under the guidance of this approach, we develop an automatic customization toolkit named AndroidKit for generating specific Android OS components. Through this toolkit, we are able to boot new Android VM instances called AndroidXs. These AndroidXs are composed of OS images generated by AndroidKit, which can be easily customized and combined for varied demands on clouds.

Keywords: Mobile cloud · Execution environment · Android · Container-based customization approach · AndroidKit · AndroidX

1 Introduction

Currently, mobile cloud, which leverages elastic resources of cloud platform to provide services for mobile applications, is becoming more and more attractive. There are many scenarios depend on MCP for different requirements. Mobile computation offloading [1–3], which is able to offload parts of workloads to cloud, exploits rich computing resources of cloud platform to enhance performance of

© Springer Nature Switzerland AG 2019
S. Li (Ed.): GPC 2018, LNCS 11204, pp. 155–169, 2019.
https://doi.org/10.1007/978-3-030-15093-8_11

mobile applications and reduce power consumption of mobile devices. And cloud-based mobile testing [4], which uses elastic cloud infrastructure to test mobile applications for different application requirements, has been widely used.

Many challenges need to be faced when building a practical MCP. One of the typical challenges is how to build a mobile execution environment, e.g., the Android mobile OS. Technically speaking, Android OS is not an ordinary Linux distribution although it is built on top of the standard Linux kernel. There are many differences between Linux and Android. For example, Linux uses X11 or Wayland to run a GUI, but Android uses SurfaceFlinger. What's more, Android relies on the special kernel features (e.g., Binder IPC subsystem [5]), which simply do not exist on Linux. Therefore, it is a challenging job to launch Android applications on MCP, since the vast majority of MCPs are currently based on ordinary GNU/Linux distributions.

Using VMs as the MCP execution environments is a practical solution. Projects like shashlik [6] or genymobile [7] use an emulator, which is essentially a VM, to run the Android environment. The emulator creates an entire emulated system with independent kernel to provide rich functionality. By targeting an emulator, they avoid the hardware compatibility problems, but it causes a lot of resource costs since each VM runs a full copy of an OS and suffers high virtualization overhead. In contrast to this, containers (e.g., Docker [8], LXC [9], and rkt [10]), as the center of lightweight virtualization technologies, have significantly lower overhead when compared to VMs. However, because of the shared kernel mechanism, many efforts need to be made when using containers as MCP runtime environments for running mobile applications. Anbox [11] puts a full Android-x86 [12] OS into a LXC container and runs the Android-x86 OS under the same kernel as the host OS does. It communicates with the host system by using different pipes and sends all hardware access commands by reusing what Android implements within the QEMU-based emulator. Rattrap [13] is a container-based cloud platform. It provides an on-demand execution environment for running Android codes through *Cloud Android Container* (CAC), which runs an Android-x86 OS inside a LXC container with dynamic driver extensions. However, the common drawback of Anbox and Rattrap is their hardware compatibility problem, which results in only a handful of Android applications can be run normally.

Another challenge is the unified customization model for frequently-changing usage scenarios on mobile clouds. The need for customizable OS arises when the existing OS components can not match the specific use cases (such as resource-constrained hardware platforms and driver-specific applications). Many research works have been made on exploring the customization of OS. Exokernel [14] and Nemesis [15] customize OS by restructuring the OS with the target application into a set of libraries. Think [16] provides a highly flexible programming model for building flexible OS kernels from components, and [17] designs a system that has the ability to guide Linux images customization for scientific applications. The common problem of these OS customization works [14–17] is the lack of a unified customization model and an effective upgrade and maintenance mechanism, since

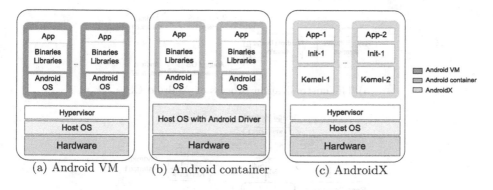

Fig. 1. The difference between Android VM, Android container, and AndroidX

they are only designed for a particular type of platform or device and do not focus on the various frequently-changing scenarios on clouds. In other words, they are inappropriate for varied demands on clouds.

According to the above analysis, in this paper, we mainly address two challenges mentioned above. Inspired by the previous OS customization works, we present a customizable Android execution environment called AndroidX. As Figs. 1(c) and 2(b) show, AndroidXs are built from system images and run with them. From Fig. 1(c), we can find that one of the AndroidX instances is launched with the Kernel-1 image, and another is launched with the Kernel-2 image, but they share the Init-1 image. This kind of combination effectively improves the utilization of images and simplifies the maintenance and update process of the whole system. By leveraging the customizability and portability of Docker images [8], AndroidXs can be easily customized for varied scenarios. In addition, from Fig. 1(c), it can be also observed that AndroidXs combine the strong hardware-enforced isolation and compatibility of VMs and the flexibility of containers.

In particular, the main contributions of this paper are summarized as follows:

1. We present a unified approach for customizing Android OS. The approach provides a container-based solution for building Android OS system component images. Following the guidance of this approach, we develop an automatic customization toolkit called AndroidKit, which is much suitable for varied demands on clouds by leveraging the customizability and portability of Docker images.
2. We give a prototypical implementation called AndroidX with the help of the toolkit. The AndroidXs combine the great hardware-enforced isolation and compatibility of VMs and the flexibility and portability of containers. To demonstrate the usability and practicability of the toolkit described here, we use it to implement a type of lightweight AndroidX specifically customized for mobile computation offloading.

The rest of this paper is organized as follows: Sect. 2 explains the research motivation, which gives us an important guideline on designing AndroidX.

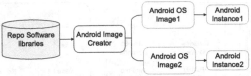

(a) The customization process of Android OS

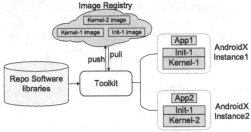

(b) The customization process of AndroidX

Fig. 2. The comparison of customization process for Android and AndroidX

Then in Sect. 3, we describe the architecture and design of AndroidX in detail, as well as AndroidKit that we developed. Section 4 shows an implementation of AndroidX. After that, in Sect. 5, we evaluate the usability of the toolkit by using a case study. In Sect. 6, we describe and review some related works. Finally, a conclusion is summarized in Sect. 7.

2   Research Motivation

First of all, when building a MCP, the cloud platform should provide a runtime environment that has the ability to run mobile applications, e.g., the Android mobile OS. The need for customizing Android OS exists because varied cloud-based scenarios have their own demands for runtime environments. For example, since mobility and interactivity are the keys to the success of computation offloading, the cloud execution environment, e.g., the tailor-made, trimmed-down Android OS, should be as lightweight as possible and only contain the functionality that computation offloading depends on. By contrast, cloud-based mobile testing, which needs MCP to provide the full integrated testing environments, requires the cloud platform has multiple versions of mobile OS for testing different application requirements, e.g., Android OS with a particular architecture or kernel. Therefore, there is a demand for custom-tailored Android OS for varied scenarios on clouds.

Second, many efforts [18,19] have been made to implement the customization of Android OS, but they only focus on the hardware-specific devices or driver-specific applications. From Fig. 2(a), we can clearly observe that traditional customization approaches of Android OS have the following problems. On one hand, the image creator needs to customize complete system images for different hardware devices and platforms, which results in low reuse of OS image components. On the other hand, traditional approaches require developers have a deep understanding of Android OS and be familiar with each customization step, which is a challenging job for developers. In addition, due to the lack of a unified authentication and update mechanism, it is difficult to ensure the maintainability and security of OS components.

Finally, as known to all, a lot of scenarios use Android VMs and Android containers as on-demand execution environments for running Android applications on clouds. The Android VMs, as shown in Fig. 1(a), have a full copy of an Android OS and run on top of the specific hypervisor. The main defects of Android VMs are their heavy overhead of virtualization and slow startup speed, which are unable to meet the requirements for some scenarios on clouds that require low time-delay, high interactivity, and mobility [20, 21]. The Android containers (shown in Fig. 1(b)), by contrast, are much more lightweight than Android VMs because they share kernel with host OS and suffer small overhead [13]. However, as an alternative solution to Android VMs, Android containers are not appropriate for multi-tenant deployments on clouds because of their poor compatibility and weak isolation. Therefore, it is necessary to customize such an execution runtime environment with great isolation and compatibility like VMs, as well as the flexibility like containers.

3   System Design

Through the above analysis and discussion, we conclude that the current execution environments can not meet the needs of varied scenarios on mobile clouds. In this section, we introduce AndroidX, a customizable Android runtime environment for running Android applications on clouds, as well as AndroidKit, a toolkit that we have developed to create Android VM images around specified applications.

3.1   Overview

Based on the previous analysis of the deployment of Android on GNU/Linux platform and the traditional OS customization approaches, and taking into account the current demands for varied scenarios on clouds, we design AndroidX with the following primary targets:

1. **Customizable components:** The customizability of OS components is important when faced with the varied demands on clouds. Our AndroidX is designed for the purpose of customizable OS.
2. **Easy tooling and easy iteration:** The previous research works do not consider the iterability and maintainability of system components, since they are only designed for a certain hardware-specific platform or driver-application, and without taking into account frequently-changing usage scenarios on clouds.

3. **Immutable infrastructure:** The immutable infrastructure, as the term suggests, is comprised of immutable components. It makes service maintenance as simple as installing fresh copy of applications and removing the old versions. With the advantages of repeatable deployments and scalability, it is widely used on cloud environments.

We use Linux platform as the running and testing environments since our prototype implementation is based on it. Figure 3 provides an overview of AndroidX's architecture. It can be clearly observed that AndroidX has two core components. One is a set of tools with command line interfaces called Android-Kit that is used to parse input *yaml* configuration file and generate Docker images. The other is a hypervisor-agnostic runtime, which is able to launch Docker images with a new VM instance.

According to the above targets that we presented, the following four primary challenges have to be overcome: (1) keep the customization process of Android OS simple enough and make it iterate as quickly as possible on the development of system components, (2) be compatible with the *Open Containers Initiative* (OCI) specification for Docker containers, (3) minimize performance overhead as far as possible, and (4) have the speed of containers and excellent isolation of VMs.

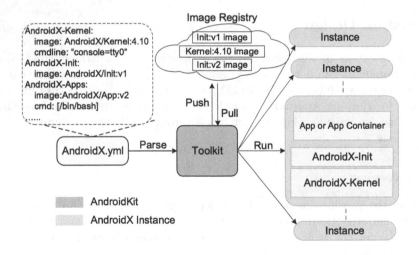

Fig. 3. Overview of AndroidX architecture

3.2   AndroidKit

One of our targets is to achieve easy tooling and easy iteration. To achieve this target, we provide a set of tools called AndroidKit for images building. The toolkit contains an automatic build system that has the ability to create minimalistic Docker images for specific applications. In general, it builds the images according to the corresponding package source that consists of a directory containing a Dockerfile, which contains the steps to build the package. From Fig. 3, we can find that the toolkit uses a *yaml* configuration template as the input file, which contains the specific system modules that need to be customized.

As we heavily exploit Docker containers, which allow users to package up an application with all of the parts it needs, such as libraries and other dependencies, and ship it all out as one package, for building tailor-made Android OS, we are able to ensure the freshness and integrity of generated images by Docker content trust. Moreover, benefit from the immutable and self-contained features of the images, we can test them for continuous development and delivery. Because of the rich tools provided by Docker, it is very easy for users to add or update the existing system components for different application requirements, which is much suitable for varied demands on clouds.

After building or pulling the images mentioned in the *yaml* configuration file, AndroidKit is able to run the generated images with a new VM instance called AndroidX, which is designed to be architecture agnostic and compatible with the OCI specification that allows users to run Docker images on any hypervisor. Internal to the AndroidX instance in Fig. 3, a minimalist Android Kernel called AndroidX-Kernel is booted directly by hypervisor, the AndroidX-Kernel employs a tiny Android initialization service called AndroidX-Init to load the Docker images from host and then launch them. Through containing applications within separate VM instances and kernel spaces, AndroidXs are able to provide more excellent workload isolation than containers, and security advantages like VMs, which are much suitable for multi-tenant cloud environments.

3.3 AndroidX

Another of our targets is able to launch Docker images with a new Android VM instance, which is called AndroidX. After pulling user-defined Docker images, AndroidKit assembles them into bootable images, then they are launched directly by the toolkit on plain hypervisor, e.g., *qemu*. AndroidX promises immutable infrastructure by eliminating the middle layer of Guest OS. Since the image is run as an *initramfs*, upgrades are done by updating the system components externally. This makes AndroidX immutable, but persistent storage can be attached by adding -drive parameter when booting up.

The customization process of AndroidX is shown in Fig. 3. It can be clearly observed that the customization is processed by the order of AndroidX-Kernel, AndroidX-Init, and AndroidX-Apps, which are defined in the input *yaml* configuration file.

(1) **AndroidX-Kernel:** The AndroidX-Kernel section defines the kernel configuration. One of the features of AndroidX is the customizable kernel. As the previous works shown [11,13], Anbox and Rattrap use LXC to run an entire Android-x86 OS in a container. However, since containers share kernel with host, if users want to add extra new kernel features, they have to modify the host kernel configuration and recompile the kernel, which is a big challenge for

non-kernel developers. What's more, it is not suitable for multi-tenant environments on clouds because of the weak isolation of containers. In contrast to this, AndroidX has its own independent kernel, and we provide a set of useful tools to simplify the customization process of kernel.

To build the AndroidX-Kernel, we divide the kernel configuration file into two parts according to the content of configuration options. One is the general-purpose configuration of Android that acts as a baseline, and the other is the user-optional kernel configuration, which can be replaced or added according to different hardware platforms or application requirements. When creating tailor-made AndroidX-Kernel images, AndroidKit can take a set of user-provided kernel options, and then uses the merge_config script that we provided to generate a minimalist configuration file, which can be used to build the device-specific kernel with user-provided features enabled. This helps AndroidKit create more streamlined kernel images.

(2) AndroidX-Init: The AndroidX-Init section consists of basic init process image called *initrd*, which is unpacked directly into the root filesystem and contains an Android init program. This and the AndroidX-Kernel can be booted directly on plain hypervisor. Unlike other Linux based systems, which use combinations of `/etc/inittab` and `init` programs included in busybox, Android uses its own initialization program, which parses an *init.rc* script that including actions to mount the basic filesystem, set system properties, and start the specified Android system services.

To bring up containerd [22], an industry-standard container runtime, and use runC [23] to run application containers, we migrate containerd and runC to Android-x86 runtime environment. The original runC program has no ability to launch application containers since the `pivot_root` system call, which is used to change root filesystem by runC, does not work on a *ramfs* or *tmpfs* root filesystem. Instead, AndroidX first creates a new *tmpfs* as root filesystem and then uses `switch_root` system call to change root filesystem in an *initrd* shell script provided by Android-x86. Since this is done before starting Android init program, runC is able to support `pivot_root` system call without errors. In addition, as runC provides a native Go implementation for creating containers with a few Linux kernel features (e.g., *namespaces* and *cgroups*), the options of *namespaces* and *cgroups* must be selected in AndroidX-Kernel configuration file except the IPC *namespace* since Android uses binder for interprocess communication. Through our efforts, AndroidX-Init can finally bring up containerd and use runC to run application containers (rootless containers can be launched successfully we tested).

(3) AndroidX-Apps: The AndroidX-Apps section shows a list of images for running applications and services. It contains the specific apps and services that need to be launched when AndroidX starts. The final goal of AndroidX-Apps is that the specific apps and services can be emitted directly both by AndroidX-Init and runC. Since AndroidX-Init has the native ability to run Android applications and services, there is no need to do extra efforts when exploiting AndroidX-Init to run Android applications and services. By contrast, as we use runC to launch application containers, some kernel features (e.g., *cgroups* and *namespaces*) and the Go runtime environment must be supported. In our efforts, rootless containers currently can be instantiated successfully by runC.

As we hope that more than one AndroidX instance could share the system components as much as possible, we exploit the read-only feature of the Android system partition, which can be attached by adding `-drive` parameter and shared with other instances when booting AndroidX instances. Benefit from this, the average disk usage size of each AndroidX instance decreases and gets close to Android containers when more and more AndroidX instance are started. Since we design AndroidX for immutable infrastructure, AndroidX-Apps can be attached by using SD card and data partitions for persistent storage, which will be identified by tailor-made AndroidX initialization program when booting up.

4   Implementation

We have implemented the prototype of AndroidX on our machine. The machine contains an Intel Core i5-7200 2.50 GHz CPU (2 cores) with 16 GB of DDR4 RAM and 256 GB HDD, running Ubuntu 16.04. In our current version, the implementation of AndroidX consists of two parts. One of them provides a rich set of tools named AndroidKit for images building. We develop the toolkit with the guidance of the container-based customization approach. Another is the execution runtime environment, which can be easily customized for varied scenarios on clouds.

As we mentioned above, the build process of AndroidX heavily leverages Docker images for packaging, and all intermediate images are referenced by digest to ensure reproducibility across its build process. To guarantee the freshness and integrity of the images, all of the generated images will be signed using Docker content trust. After building the Docker images, AndroidX instance can be booted directly on plain hypervisor by targeting with AndroidX-Kernel+AndroidX-Init. Through combining with the portability of app container images, AndroidX is able to allow users to build, ship, and run apps anywhere, without considering the underlying technology stack.

The construction of AndroidX is based on a series of Android-x86_6.0_r3 components, which can be customized for varied demands. In our prototype implementation, we provide a lot of templates, e.g., Dockerfiles, for users customize OS components. The source code of AndroidX is publicly available online at https://github.com/CGCL-codes/AndroidX.

5   A Case Study

To demonstrate the usability and practicability of AndroidKit, we use it to implement a type of lightweight AndroidX, which is based on Android-x86 and specifically customized for mobile computation offloading. These AndroidXs provide the excellent workloads isolation like Android VMs, as well as the extremely fast

Table 1. Performance comparison for Android container, Android VM, and AndroidX. For the experiment we allocate a single core and 1024 MB of memory to each test instance.

Runtime	Boot time	Memory footprint	Disk usage	CPU allocation	Memory allocation
Android container	1.8 s	96 MB	1044 MB	1vCPU	1024 MB
Android VM	31.8 s	493 MB	1728 MB	1vCPU	1024 MB
AndroidX	3.9 s	270 MB	1101 MB	1vCPU	1024 MB

instantiation time like Android containers. In this section, we present an evaluation of these customized AndroidXs, including the comparisons of startup time, memory footprint, and disk usage with standard Android VMs and Android containers. All experiments are run on a machine mentioned in Sect. 4.

5.1   Boot Time

Boot time is a critical performance evaluation point in many cloud computing scenarios. Since we evaluate the toolkit by using it to implement a type of lightweight AndroidX specifically customized for mobile computation offloading (which offloads computational codes to clouds and requires low time-delay), we want to measure how long AndroidKit takes to create and boot such an AndroidX instance.

The main limiting factors of instantiation time are the image size of VMs and the number of processes that need to be started. LightVM [24] has demonstrated that startup times grow linearly with VM image size by booting the same unikernel VM from images of different sizes in the experiment. The reason why large VMs slow down instantiation time can be summarized as follows: launching a large VM instance needs time to read the image from disk, parse it and finally run it in memory. Inspired by this, we have made a great effort on optimizing AndroidX for mobile computation offloading, and in our efforts, AndroidKit can generate compact AndroidXs which have the speed of containers. Table 1 compares boot times for a noop AndroidX instance against a noop Android VM, as well as a noop Android container. Time is measured form booting to the point where instantiation is finished.

In order to find the most time-consuming process during the whole startup of standard Android VM instances, we try to take a look at the CPU usage when starting a noop Android VM. For the measurement we use bootchart [25] and adb [26] tools to get a noop Android VM's CPU utilizations. The bootchart is a tool for performance analysis and virtualization of the GNU/Linux boot process, and the adb tool which is specially designed for Android, has the ability to analyze the startup log of each process. As shown in Fig. 4, the Android VM reaches a maximum CPU utilization of about 100% after 7 s of startup and lasts for a period of time. To find out which processes occupy a large amount of CPU resources at that time, we try to get the time consumption of each process in the whole startup process of the VM by leveraging bootchart and adb tools, and find that dex2oat, package scanning, and class preloading take up about 72.7% of the total system boot time and consume a lot of CPU resources at that time. To shorten the time taken by these processes, we reduce the number of preloaded packages, as well as the preloaded classes and resources when customizing AndroidX for mobile computation offloading. The optimized result is shown in Fig. 5. This figure shows the time consumption comparison of class preloading, package scanning and dex2oat during the whole startup for Android VM, AndroidX, and Android container. From this figure, we can conclude that the time consumption of optimized processes is only 15.7% of the non-optimized. In addition, after using a pre-prepared data partition, we almost eliminate the time taken up by the dex2oat process. In order to minimize the size of VM image, we reduce the functionality of AndroidX, such as Camera and Bluetooth that will not be used on cloud environments. In our efforts, the size of the final image generated by AndroidKit is reduced from the original 1728 MB to 1101 MB. As a result, the startup time of an optimized AndroidX instance can be shortened to 3.9 s (shown in Table 1), which is much lightweight than Android VM, and as fast as Android container.

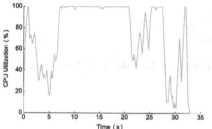

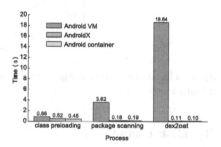

Fig. 4. CPU usage during the whole startup of a noop Android VM

Fig. 5. Time consumption comparison of class preloading, package scanning, and dex2oat during the whole startup for Android VM, AndroidX, and Android container

5.2 Memory Footprint

As known to all, in order to concurrently launch multiple instances on a single host, the most effective solution is to reduce per-instance memory footprint. One of the advantages of Android containers is that they typically need less memory than Android VMs because they use a common kernel with host OS and have smaller root filesystems. By contrast, Android VMs, where each instance has their own independent kernel and runs an entire Android OS, suffer more resource overhead than Android containers when concurrently running multiple instances on a single machine. In our next analysis, we try to find if the memory footprint of each compact AndroidX is close to Android container.

We observe, as others [27], that most VMs and containers run a single application on clouds. By reducing the functionality of AndroidX to include only what is necessary for that specified application, we are able to reduce the memory footprint of each AndroidX instance. With the help of `adb shell procrank` command, we get exact memory footprint of a noop AndroidX instance. Table 1 shows the memory usage of a noop AndroidX instance against a noop Android VM and a noop Android container (which is essentially a Rattrap instance). From the table, we can conclude that the memory usage of a noop AndroidX instance is only 54.7% of a noop Android VM. The reason why we use a noop AndroidX instance is that the cloud execution environment for mobile computation offloading has little association with applications, and AndroidKit has the ability to generate AndroidX for different application requirements since we provide a lot of templates for customization.

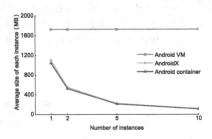

Fig. 6. The average disk usage size of each instance for Android VM, AndroidX, and Android container

5.3   Disk Usage

As we mentioned above, one of the main limiting factors of instantiation time is the image size of instance. In order to shorten the startup time and improve the disk utilization of the offloading code execution environment, Rattrap analyzes the entire Android-x86 OS files and finds out that 68.4% of them are never accessed by offloaded codes, which are composed of unnecessary libraries and modules. By removing the unnecessary parts and sharing the system libraries, an optimized Rattrap instance only takes up 7.1 MB space and has faster startup speed.

Inspired by this, we put forward the idea of sharing partitions. AndroidX instance has the ability to share the data and system partitions by the predefined `-drive` parameter, which can be identified by AndroidKit. In this way, the disk usage of each AndroidX instance is close to a Rattrap instance, which is less than 10 MB. From Table 1 and Fig. 6, we can find that an AndroidX instance takes up about 1101 MB of disk. However, the average disk usage size of each AndroidX instance decreases and gets close to Android containers when more and more Android instances are launched. By contrast, as Fig. 6 shows, the average disk usage size of each Android VM instance always remains the same when the number of running instances increases, since Android VMs lack a sharing mechanism. As a result, AndroidX has a great disk utilization when running multiple instances, which is more suitable for large-scale scenarios on clouds.

6   Related Work

A lot of container-based virtualization technologies (e.g., Docker, LXC, and rkt) have been widely deployed on cloud platforms because of their low resource overhead and great scalability. However, as known to all, the weak isolation of containers has caused some security problems on multi-tenant cloud environments [28]. Unlike containers, VMs which are based on hypervisor technologies, have excellent hardware-enforced isolation, but they cause high resource overhead since each VM runs a full copy of an OS. Intel sets out to build hypervisor-based container named *Intel Clear Containers* (ICC) [29] by combining the best benefits of VMs and Linux containers. Kata Containers [30], which combines technology from ICC and Hyper [31], tries to run Docker containers on agnostic hypervisors to provide the workload isolation like VMs, as well as the portability like containers. A part of our design idea comes from ICC and Hyper, but as we have already shown, AndroidXs are specifically designed for the various frequently-changing scenarios on mobile clouds.

Traditional operating systems, e.g., Linux, focus on the versatility and integrality of system functions and contain the entire software stack with the tradeoff of overhead and efficiency. By contrast, unikernels [27], which are designed for supporting cloud services rather than desktop applications, are tiny VMs that pack the minimalistic OS with the target application into a single bootable VM image. Many research works have been made on constructing unikernels (e.g., OSv [32], MirageOS [33], and ClickOS [34]), and the common goal of these unikernels is to run single application on a single machine to eliminate the redundancy and provide great performance. Through booting directly on plain hypervisor, they are able to avoid the hardware compatibility problems suffered by traditional library operating systems (e.g., Exokernel [14] and Nemesis [15]).

7   Conclusion

This paper presents AndroidXs as customizable execution environments for Android applications on clouds, as well as AndroidKit as a toolkit for building customizable Android OS images. Our idea comes from the experience of running Android applications on Linux platform, as well as *Intel Clear Containers* and Hyper open source projects. After investigating the relevant research works of OS customization, and inspired by the design of unikernels, we propose a container-based approach for customizing mobile cloud execution environment. Under the guidance of this approach, we have developed a set of tools named AndroidKit, which is able to generate images for customizable AndroidXs. To demonstrate the practicability of the AndroidKit, we use it to implement a type of lightweight AndroidX specifically customized for mobile computation offloading. The AndroidXs are composed of OS images generated by AndroidKit, and in our efforts, rootless containers are now able to be brought up by runC in AndroidX instances.

Acknowledgements. This research is supported by National Key Research and Development Program under grant 2016YFB1000501, and National Science Foundation of China under grants No. 61732010 and 61872155.

References

1. Cuervo, E., et al.: MAUI: making smartphones last longer with code offload. In: Proceedings of MobiSys, pp. 49–62. ACM (2010)
2. Chun, B., Ihm, S., Maniatis, P., Naik, M., Patti, A.: Clonecloud: elastic execution between mobile device and cloud. In: Proceedings of EuroSys, pp. 301–314. ACM (2011)
3. Kosta, S., Aucinas, A., Hui, P., Mortier, R., Zhang, X.: Thinkair: Dynamic resource allocation and parallel execution in the cloud for mobile code offloading. In: Proceedings of INFOCOM, pp. 945–953. IEEE (2012)
4. Mobile testing. https://en.wikipedia.org/wiki/Mobile_application_testing
5. Android binder. https://elinux.org/Android_Binder
6. Shashlik. http://www.shashlik.io/
7. Genymobile. https://www.genymobile.com/
8. Docker. https://www.docker.com/
9. Lxc. https://en.wikipedia.org/wiki/LXC
10. Rkt. https://coreos.com/rkt/
11. Anbox. https://anbox.io/
12. Android-x86. http://www.android-x86.org/
13. Wu, S., Niu, C., Rao, J., Jin, H., Dai, X.: Container-based cloud platform for mobile computation offloading. In: Proceedings of IPDPS, pp. 123–132. IEEE (2017)
14. Engler, D.R., Kaashoek, M.F., O'Toole, J.: Exokernel: An operating system architecture for application-level resource management. In: Proceedings of SOSP, pp. 251–266. ACM (1995)
15. Leslie, I.M., et al.: The design and implementation of an operating system to support distributed multimedia applications. IEEE J. Sel. Areas Commun. **14**(7), 1280–1297 (1996)
16. Fassino, J., Stefani, J., Lawall, J.L., Muller, G.: Think: a software framework for component-based operating system kernels. In: Proceedings of ATC, pp. 73–86. ACM (2002)
17. Krintz, C., Wolski, R.: Using phase behavior in scientific application to guide linux operating system customization. In: Proceedings of IPDPS. IEEE (2005)
18. Shanker, A., Lai, S.: Android porting concepts. In: Proceedings of ICECT, vol. 5, pp. 129–133. IEEE (2011)
19. Yaghmour, K.: Embedded Android: Porting, Extending, and Customizing. O'Reilly Media Inc., Sebastopol (2013)
20. Duan, Y., Zhang, M., Yin, H., Tang, Y.: Privacy-preserving offloading of mobile app to the public cloud. In: Proceedings of HotCloud, pp. 18–18. ACM (2015)
21. Shiraz, M., Abolfazli, S., Sanaei, Z., Gani, A.: A study on virtual machine deployment for application outsourcing in mobile cloud computing. J. Supercomput. **63**(3), 946–964 (2013)

22. Containerd. https://containerd.io/
23. Runc. https://blog.docker.com/2015/06/runc/
24. Manco, F., et al.: My VM is lighter (and safer) than your container. In: Proceedings of SOSP, pp. 218–233. ACM (2017)
25. Bootchart. http://www.bootchart.org/
26. Android debug bridge. https://en.droidwiki.org/wiki/Android_Debug_Bridge
27. Madhavapeddy, A., Scott, D.J.: Unikernels: the rise of the virtual library operating system. Commun. ACM **57**(1), 61–69 (2014)
28. Container security. https://arxiv.org/abs/1507.07816
29. Intel clear container. https://clearlinux.org/containers
30. Kata container. https://katacontainers.io/
31. Hyper. https://hypercontainer.io/
32. Kivity, A., et al.: Osv - optimizing the operating system for virtual machines. In: Proceedings of ATC, pp. 61–72 (2014)
33. Madhavapeddy, A., et al.: Unikernels: library operating systems for the cloud. In: Proceedings of ASPLOS, pp. 461–472. ACM (2013)
34. Martins, J., et al.: Clickos and the art of network function virtualization. In: Proceedings of NSDI, pp. 459–473. ACM (2014)

A Dynamic Resource Pricing Scheme for a Crowd-Funding Cloud Environment

Nan Zhang, Xiaolong Yang$^{(\boxtimes)}$, Min Zhang, and Yan Sun

School of Computer and Communication Engineering,
University of Science and Technology Beijing, Beijing, China
yangxl@ustb.edu.cn

Abstract. With the rapid development of cloud computing and the exponential growth of cloud users, federated clouds are becoming increasingly prevalent based on the idea of resource cooperation. In this paper, we consider a new resource cooperation model called "Crowd-funding", which is aimed at integrating and uniformly managing geographically distributed resource-limited resource owners to achieve a more effective use of resources. The resource owners are rational and maximize their own interest when contributing resources, so a reasonable pricing scheme can incentivize more resource owners to join the Crowd-funding system and increase their service level. Therefore, we propose a dynamic pricing scheme based on a repeated game between the "Crowd-funding" system and the resource owners. The simulation results show that our resource pricing scheme can achieve more effective and longer-lasting incentivizing effects for resource owners.

Keywords: Cloud computing · Resource pricing · Resource crowd-funding

1 Introduction

In traditional cloud computing, a data center acts as the only resource provider, performing hardware maintenance and managing task execution and network traffic [1], and users purchase resources from one resource provider using a fixed pricing scheme. However, with the rapid development of cloud computing, the number of cloud users has grown exponentially and federated clouds have become more prevalent [6]. The aim of federated clouds is to integrate resources from different providers [7, 8]. Therefore, in this paper, we consider a Crowd-funding cloud system [9], which can aggregate the resources of geographically distributed resource owners to provide cloud services.

Cloud providers are rational, self-interested parties who exercise their partial or complete autonomy to maximize their benefit [4–6]. A dynamic pricing scheme is significantly affected by rational resource contributors, and they may join or leave the system at any time by measuring their earnings [7]. In the market-oriented environment, both the resource contributors' revenue and the users' quality-of-service (QoS) assurance are important. A trust and reputation system has been proposed to

© Springer Nature Switzerland AG 2019
S. Li (Ed.): GPC 2018, LNCS 11204, pp. 170–181, 2019.
https://doi.org/10.1007/978-3-030-15093-8_12

differentiate the service providers [3], and some techniques to evaluate the service quality of the cloud vendors are introduced based on parameters such as response time, availability and elasticity [9]. Therefore, the price process should consider the degree of reliability of resource contributors to ensure QoS. A reasonable price scheme should be able to incentivize more resource owners to join the Crowd-funding cloud system and to increase their degree of reliability. In this paper, we design a new resource price scheme based on the repeated game between the resource owners and the Crowd-funding system. In our scheme, the resource prices are set differently according to the degree of reliability of resource owners, and the degree of reliability will be updated in each new task cycle.

The rest of this paper is organized as follows: Sect. 3.1 describes the application scenarios of our price scheme and the modeling process. The detailed pricing process is introduced in Sect. 4. Section 5 presents the performance simulation results. Finally, Sect. 6 concludes the paper.

2 Related Works

Resource sharing is not a new concept. This idea is applied in many fields, which can inspired us. P2P (peer-to-peer) architectures and systems are characterized by resource sharing and direct access between peer computers, rather than through a centralized server [10]. Therefore, we considered that the idea of resource sharing can be used for the resource providing in cloud computing. In addition, The authors in [11] propose a "Crowd-Cloud" architecture by integrating the sensing and processing capabilities of the dynamic mobile cloud. What is more, the concept of a "local crowd" is proposed, which can realize a lower delay of offloading data than can be realized from a remote cloud [12]. Inspired by these works, in this paper, we consider realize the resource sharing among various devices in a network, which is named "Crowd-funding". For "Crowd-funding" model, all kinds of resource owners in the network that have idle resources could be considered a participator, which allows different cloud providers to share resources for increased scalability and reliability. In addition, the authors in [13] claim that the more decentralized a system is, the less energy is consumed. Therefore, in "Crowd-funding" model, the high energy consumption of relatively concentrated large servers in traditional data centers can be converted to the low energy consumption of the relatively small devices that are widely distributed in the network.

For the resource providers in Crowd-funding model, an impartial and reasonable pricing scheme is important. In [2], the authors propose a new resource pricing and allocation policy where users can predict the future resource price as well as satisfy budget and deadline constraints. The authors in [14] propose a reverse auction-based pricing and allocation scheme, which they prove formally to be individual rational, incentive compatible and budget balanced. In addition, they present a dynamic pricing scheme suitable for rational provider requests containing multiple resource types in federated cloud [6]. In federated cloud, the resource come from multiple cloud providers, so the trust and reputation of multiple cloud providers need to be considered [3]. However, these works do not consider the capacity variance of the resources, which can be represented by some parameters, such as the task completion rate. What is more, the price will be updated along with the change of task cycle in this paper.

3 Problem Statement

3.1 Description of Application Scenarios

In the network, some small cloud providers or private devices could not afford certain computing-intensive tasks because of limited resources. Nevertheless, the resources could be considerable if all the resources within resource-limited devices were integrated. Therefore, in this paper, we consider a Crowd-funding cloud system, which can integrate the cloud resources of small resource providers and generate a larger federated cloud as shown in Fig. 1. The "cloud Broker" is the initiator of the resource crowd-funding, and will manage the Crowd-funding system. The computing tasks from the users will be split into some small sub-tasks and reorganized by cloud brokers. And then the cloud broker will assign these sub-tasks to the resource supporters in a certain order. What is more, the cloud broker will determine whether the cloud supporter has completed the task at a satisfactory level. Accordingly, the cloud broker will pay the resource supporters for their services.

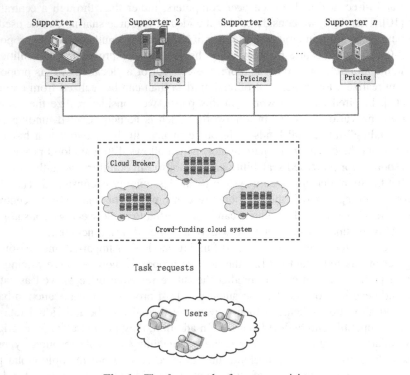

Fig. 1. The framework of resource pricing

3.2   Modeling and Related Definitions

In the network, all the small cloud providers and private devices are called "resource owners". The resource owners can gain revenue in two ways: be self-employed (complete simple tasks alone) or rent their resources to the Crowd-funding system. In this paper, every resource owner is considered rational with a preference to achieve more revenue.

In the Crowd-funding system, the resources belong to different owners, whose service abilities and reliabilities are different. Some resource owners may agree to join the Crowd-funding system initially, but then fail to complete the tasks because of their dishonesty or poor ability. Therefore, the resource prices should be different among different resource owners according to their service levels. The resource price includes two parts, which means the cloud broker will pay the bonus to the resource owners in two steps. In the first step, one part of the bonus will be paid according to the strategy of resource owners if they participate in the Crowd-funding system. In the second step, another part of the bonus will be paid if the resource owners can complete the task assigned to them.

In our Crowd-funding system, the degree of service abilities and reliability of resource owners can be represented as their "completion rate", which can be expressed by the probability that the resource owner can successfully complete the assigned tasks and the average completion rate of Crowd-funding system, which is defined as follows:

Definition 1. Completion Rate—The completion rate of one resource owner can represent the service level and reputation of its resources, which is described as follows:

$$
P = \begin{cases} \alpha * \frac{N_{Com}}{N_{Tot}} + \beta * \overline{P}, & N_{Tot} < N_{TH} \\ \frac{N_{Com}}{N_{Tot}}, & N_{Tot} \geq N_{TH} \end{cases} \tag{1}
$$

where $\frac{N_{Com}}{N_{Tot}}$ is the proportion of tasks completed by a resource owner out of the total number of tasks assigned to that resource owner, expressed by P. When the total number of tasks assigned to one resource owner is too small to reflect the actual service level ($N_{Tot} < N_{TH}$, N_{TH} is the threshold value, which can be set as a constant), the average completion rate of the current Crowd-funding system ($\overline{P}$) is considered as shown in Eq. (1). In addition, α and β are the weighted values of an owner's completion rate and the average completion rate of the system.

To encourage more resource owners to participate in the Crowd-funding system, the resource revenue achieved by the cloud broker from renting resources should be greater than the revenue from self-employing. In this paper, the main issue to be solved is described as:

- How to encourage the resource owners to participate in the Crowd-funding system by paying a reasonable bonus to them according to their completion rate.

4 Resource Pricing Model

In this section, we introduce the game process between the broker and the resource owner. Then, the solutions in different situations are analyzed in detail to achieve a reasonable pricing process.

4.1 The Game Process Between the Broker and Resource Owner

We know that we can obtain sufficient resources when needed only if a sufficient number of resource owners are willing to become the resource supporter. Therefore, a bonus should be paid to the resource owners if they are willing to participate in the Crowd-funding system as the resources reserve. For every task cycle, the resource owner can choose whether to join the Crowd-funding system, and the broker will re-calculate their values of completion rate. The pricing process can be seen as a repeated game between the broker and the resource owner. The detailed game process is shown in Fig. 2.

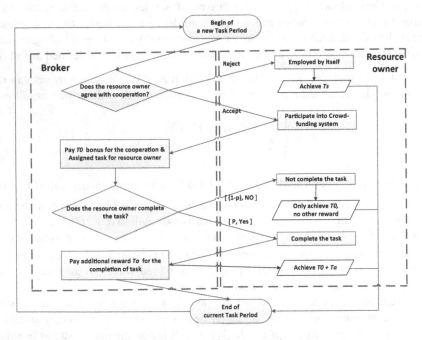

Fig. 2. Flowchart of the repeated game

A. As shown in Fig. 2, there are three strategies for the resource owners: "non-cooperation", "cooperation and completing the task", and "cooperation but not completing the task". First, if the resource owner participates in the Crowd-funding system and completes the tasks assigned by the broker, the broker will pay the resource not only T_0 but also an additional Ta as the reward for completing tasks. Secondly, if the resource owner participates in the Crowd-funding system but does not complete the assigned tasks in this task cycle, the broker will pay only T_0, as shown in Fig. 2. Furthermore, in this situation, the completion rate of the resource will decline according to Eq. (1). Third, if the resource owner refuses to participate in the Crowd-funding system, they can achieve Ts through self-employment (providing services alone). For every task cycle, the resource owners will choose their strategy again from the abovementioned choices and the cloud broker will pay the corresponding rewards. Therefore, the relationship between the resource owner and the cloud broker can be regarded as an infinite repeated game model. For an infinite repetition process, the revenue up to the current period can be considered equal to the revenue up to the previous cycle in mathematics, as shown in Eqs. (2) and (3). In addition, if the resource owner always choose the strategy of "completing the task" (i.e., P = 1), some of his own tasks must be affected, which is called as "disutility". If the resource owners choose cooperation but only completing the tasks when their resources are idle (P < 1), there is no disutility, but the value of Completion Rate will decrease. We consider three situations of unit rewards for the resource owner, which are described as follows:

(1) If one resource owner can always cooperate and complete all the tasks assigned by the broker, i.e., P = 1, then the total revenue of a unit resource after infinite task cycles should be:

$$V_e = \delta * V_e + T_0 + Ta - e \qquad (2)$$

where e is assumed to be the cost or the disutility for maintaining P = 1. In economics, the values of unit money are not the same in different periods. In game theory, the time value of money is described as the discount factor. δ is the discount factor in this paper, which can be described as the exchange rate of a unit reward between the current task cycle and the previous cycle. Equation (2) can be explained as the total revenue up to the current task cycle is equal to the sum of the total revenue up to the previous cycle and the rewards achieved in the current cycle.

(2) If one resource owner chooses to cooperate and completes tasks with a probability P(P < 1), then the total revenues of a unit resource after infinite task cycles should be:

$$V_s = \delta * V_s + P * (T_0 + Ta) + (1 - P) * T_0 \qquad (3)$$

(3) If one resource owner refuses to join the Crowd-funding system, he will be self-employed with a resource price of T_s. Then, the total revenue of a unit resource after infinite task cycles should be:

$$Vn = T_s + T_s * \delta + T_s * \delta^2 + T_s * \delta^3 + \ldots = \frac{T_s}{1 - \delta} \qquad (4)$$

B. To achieve a reasonable pricing scheme and incentivize more resource owners to cooperate, some relationships should be met: (1) The revenue of a resource owner that completes the tasks with $P = 1$ should be greater than owners that complete tasks with $P < 1$. (2) The revenue of a resource owner that chooses to cooperate should be greater than an owner that refuses to cooperate. (3) The pseudo-cooperation of a resource owner should be avoided. The relationships can be described as follows:

$$\begin{cases} V_e \geq V_s \\ V_s \geq Vn \\ T_0 \leq T_s \end{cases} \tag{5}$$

C. To incentive more resource owner participate the Crowd-funding system and complete the tasks actively, the price should be paid at the price of the equation solutions.

4.2   Analysis of the Solutions

(1) To encourage more resource owners to complete the assigned tasks, the revenue V_e should be greater than V_s, as shown in Eq. (5). Solving the equations:

$$T_a \geq \frac{e}{1 - P} \tag{6}$$

The value of $\frac{e}{1-P}$ increases as P increases. Therefore, in this situation, the resource owners would prefer to increase their performance, which would lead to increased completion rate and increased rewards.

(2) To incentivize more resource owners to join the Crowd-funding system, the revenue when cooperating should be greater than the revenue of owners that are self-employed. Therefore, the relationship $V_s \geq Vn$ should be met as shown in Eq. (5). Solving the equations:

$$T_0 + P * T_a \geq T_s \tag{7}$$

(3) If the bonus T_0 for agree simply cooperating with the Crowd-funding system is already greater than the bonus for choosing self-employment (i.e., $T_0 > T_s$), the resource owner may choose pseudo-cooperation but not complete tasks. This means that the resource owner can achieve higher revenue without completing any tasks. To avoid this situation, the revenue for cooperation (T_0) should be less than the revenue for self-employment:

$$T_0 \leq T_s \tag{8}$$

5   Simulations

In this section, we consider a Crowd-funding system with an average completion rate value of approximately 0.6. In addition, α and β are set equal to 0.2 and 0.8, respectively. The situations of alternative resource owners are shown in Table 1. "MUs" is the abbreviation of "monetary units", which is the unit of resource revenue. For simplicity, each application request consists of one resource type. We consider the application of speech recognition as the cloud task, which needs 5 resources in one request. The resource numbers provided by the resource owners are in the range of [50,100]. In this paper, we assume that all the resource owners behave rationally and make choices that achieve more revenue.

Table 1. Parameters setting

Parameters	Values
P	[0, 1)
e	[3, 6] MUs
Ts	[4, 20] MUs
T_0	[4, 6] MUs
Amount of resources	[50, 100]

- Fixed Pricing: It is a fixed pricing method, in which we assume the average price of the resource owners as the fixed price.
- DP-IRG (Dynamic Pricing based on Infinite Repeated Game): It is the dynamic pricing scheme based on infinite repeated game, which is the method proposed in our paper.
- S-P scheme: A Strategy-Proof resource dynamic pricing scheme for federated cloud, which is proposed in [6]. In the federated cloud, all the users and resource providers are rational and maximize their own interest when consuming and contributing resources.

5.1   Comparison with Fixed

In this section, we evaluate the performance of DP-IRG on the revenues of a resource owner. As shown in Fig. 3, we can find that the resource owner can achieve a higher resource price if its completion rate is higher, which can encourage the resource owners make effort to complete the tasks. In addition, for the resource owners with higher disutility (e), there will be more bad influence on themselves if they ensure the completion rate. Therefore, if we want to encourage the resource owners with higher disutility complete the task actively, we should give them a higher resource price, which has been confirmed by Fig. 3.

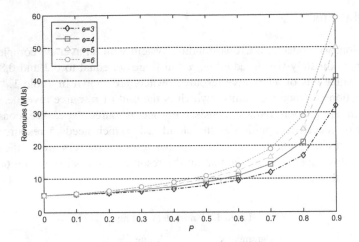

Fig. 3. Revenue vs completion rate

5.2 Comparison with Fixed Pricing

For the dynamic pricing scheme proposed in this paper (DP-IRG:), the revenue of a resource owner increases as their completion rate increases. Therefore, the resource owner can achieve higher revenue if they make efforts to improve their reputation. In addition, as shown in Fig. 4, the revenue grows increasingly faster as completion rate increases, which provides more encouragement to the resource owners. However, for Fixed Pricing, the revenues of resource owners are the same irrespective of their completion rate value. Therefore, Fixed Pricing cannot encourage resource owners to improve their reputations.

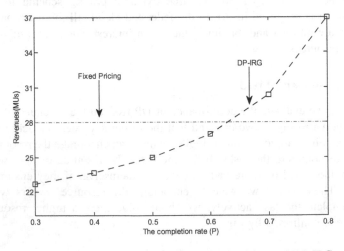

Fig. 4. Revenue comparison between Fixed Pricing and DP-IRG

For Fixed Pricing, all the resource owners in the Crowd-funding system will achieve the same revenues regardless of their service levels. Therefore, some rational resource owners may choose to not join the Crowd-funding system if the fixed price is not large enough. For DP-IRG, the price paid to resource owners can be adjusted dynamically according to different service levels (i.e., reputations). Therefore, the amount of cooperation with DP-IRG is always greater than the amount with Fixed Pricing, regardless of the number of resource owners, as shown in Fig. 5.

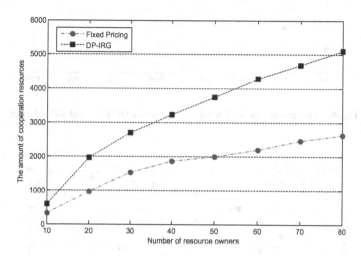

Fig. 5. Comparison of the amount of cooperation resources

5.3 Comparison with S-P Scheme

In this section, we evaluate the performance of our mechanism using total cost and number of successful completed requests as the performance measures. Total cost means all the expenses that broker needs to pay. The cloud broker' principle of allocating tasks is minimizing the cost on the premise of meeting user requests. Figure 6 shows the total cost when the user requests the task completion rate not less than 70%. The overall task completion rate means the ratio of the number of tasks completed to the number of requests. With the requirement of task completion rate not less than 70%, the total cost of DP-IRG are always less than the cost of S-P scheme no matter how many requests there are, as shown in Fig. 6. In addition, when there are same numbers of requests, the completed tasks with DP-IRG are always more than that with S-P scheme, as shown in Fig. 7.

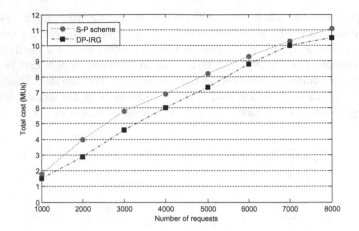

Fig. 6. The comparison of total cost between DP-IRG and I-P scheme

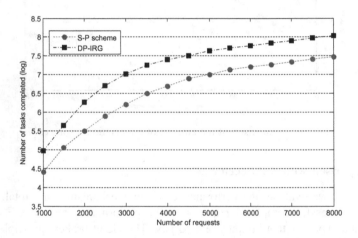

Fig. 7. The comparison of completed task number DP-IRG and I-P scheme

6   Conclusion

For a resource Crowd-funding system, the resource providers always have different service levels. Therefore, a dynamic pricing scheme, which is reasonable for different resource providers, is proposed in this paper based on an infinite repeated game between the Crowd-funding broker and resource providers. In addition, there are many additional issues worth studying, such as resource management and task scheduling, which will be the focus of our future work.

References

1. Portaluri, G., Giordano, S., Kliazovich, D., et al.: A power efficient genetic algorithm for resource allocation in cloud computing data centers. In: 2014 IEEE 3rd International Conference on Cloud Networking (CloudNet), pp. 58–63. IEEE (2014)
2. Teng, F., Magoules, F.: Resource pricing and equilibrium allocation policy in cloud computing. In: 2010 IEEE 10th International Conference on Computer and Information Technology (CIT), pp. 195–202. IEEE (2010)
3. Jøsang, A., Ismail, R., Boyd, C.: A survey of trust and reputation systems for online service provision. Decis. Support Syst. 43(2), 618–644 (2007)
4. Shneidman, J., Parkes, David C.: Rationality and self-interest in peer to peer networks. In: Kaashoek, M.F., Stoica, I. (eds.) IPTPS 2003. LNCS, vol. 2735, pp. 139–148. Springer, Heidelberg (2003). https://doi.org/10.1007/978-3-540-45172-3_13
5. Mouline, I.: Why assumptions about cloud performance can be dangerous to your business. Cloud Comput. J. 2(3), 24–28 (2009)
6. Mihailescu, M., Teo, Y.M.: Dynamic resource pricing on federated clouds. In: Proceedings of the 2010 10th IEEE/ACM International Conference on Cluster, Cloud and Grid Computing, pp. 513–517. IEEE Computer Society (2010)
7. Kaewpuang, R., Niyato, D., Wang, P., et al.: A framework for cooperative resource management in mobile cloud computing. IEEE J. Sel. Areas Commun. 31(12), 2685–2700 (2013)
8. Niyato, D., Vasilakos, A.V., Kun, Z.: Resource and revenue sharing with coalition formation of cloud providers: game theoretic approach. In: 2011 11th IEEE/ACM International Symposium on Cluster, Cloud and Grid Computing (CCGrid), pp. 215–224. IEEE (2011)
9. Zhang, N., Yang, X., Zhang, M., et al.: Crowd-funding: a new resource cooperation mode for mobile cloud computing. PLoS ONE 11(12), c0167657 (2016)
10. Androutsellis-Theotokis, S.: A survey of peer-to-peer file sharing technologies (2002)
11. Mehta, V., Shaikh, Z., Kaza, K., et al.: A crowd-cloud architecture for big data analytics. In: 2016 Twenty Second National Conference on Communication (NCC), pp. 1–6. IEEE (2016)
12. Song, C., Liu, M., Dai, X.: Remote cloud or local crowd: communicating and sharing the crowdsensing data. In: 2015 IEEE Fifth International Conference on Big Data and Cloud Computing (BDCloud), pp. 293–297. IEEE (2015)
13. Sharifi, L., et al.: Energy efficiency dilemma: P2P-cloud vs. data center. In: 2014 IEEE 6th International Conference on Cloud Computing Technology and Science (2014)
14. Teo, Y.M., Mihailescu, M.: A strategy-proof pricing scheme for multiple resource type allocations. In: Proceedings of the 38th International Conference on Parallel Processing, Vienna, Austria, pp. 172–179 (2009)

Multi-choice Virtual Machine Allocation with Time Windows in Cloud Computing

Jixian Zhang[1], Ning Xie[1], Xuejie Zhang[1], and Weidong Li[2(✉)]

[1] School of Information Science and Engineering, Yunnan University,
Kunming 650504, Yunnan, People's Republic of China
denonji@163.com, 493561346@qq.com, xjzhang@ynu.edu.cn
[2] School of Mathematics and Statistics, Yunnan University,
Kunming 650504, Yunnan, People's Republic of China
weidong@ynu.edu.cn

Abstract. Virtual machine allocation is a core problem in cloud computing. Most cloud computing platforms allow users to submit one requirement, which does not satisfy the diversity of user demands and also reduces the incomes of the platform. We propose a novel model, called multi-choice virtual machine allocation (MCVMA) with time windows, where the users can enter and leave the system at any time and submit multiple requirements. We design an optimal algorithm based on dynamic programming and a heuristic algorithm based on the resource scarcity and density for the MCVMA problem with time windows. We experimentally analyze both algorithms in terms of social welfare, execution time, resource utilization and users served.

Keywords: Cloud computing · Multiple requirements · Heuristic algorithm · Online · Virtual resource allocation

1 Introduction

In recent years, the demands for cloud computing resources from businesses and individuals have been continuously increasing. In infrastructure as a service (IaaS), cloud providers offer easily accessible, abstracted, virtualized, and dynamically scalable resources that are allocated to users. For example, Microsoft Azure and Amazon EC2 [1] assemble different types of cloud resources to create virtual machine (VM) instances that can be supplied to users. Essentially, the cloud provider supplies resources to the user, and for convenience, we also refer to the cloud provider as a resource provider.

One of the major problems in cloud resource allocation is how to effectively allocate VM instances. Zaman and Grosu [11] formulated the problem of VM allocation in clouds as a combinatorial auction problem and proposed two mechanisms to solve it. Nejad et al. [9] constructed an integer program for the VM allocation problem, which is equivalent to the multidimensional knapsack problem, and designed a polynomial-time approximation algorithm. When the number of resources is bounded, Mashayekhy et al. [8] designed a polynomial time approximation scheme mechanism for the VM allocation problem considered in [9]. Liu et al. [6] studied a novel model of

© Springer Nature Switzerland AG 2019
S. Li (Ed.): GPC 2018, LNCS 11204, pp. 182–195, 2019.
https://doi.org/10.1007/978-3-030-15093-8_13

VM allocation in heterogeneous clouds, which is a variant of the multidimensional multiple knapsack problem. Shi et al. [10] presented an online auction framework for dynamic VM allocation. Recently, Mashayekhy et al. [7] considered the multi-choice reward-based scheduling problem with time windows, and designed a polynomial-time approximation scheme.

The VM allocation problem is mostly related to the interval scheduling problem with a resource constraint. Darmann et al. [5] designed a deterministic $(1/2 - \epsilon)$-approximation algorithm for the interval scheduling problem with a resource constraint, where $\epsilon > 0$. Angelelli and Filippi [3] studied the complexity of interval scheduling with a resource constraint. Angelelli et al. [2] designed several algorithm for the interval scheduling problem with a resource constraint, including approximation algorithm, a column generation scheme for the exact solution, several greedy heuristics and a restricted enumeration heuristic.

Motivated by the models in [2, 7], we consider the multi-choice virtual machine allocation (MCVMA) with time windows, where each user can submit several alternative requirement as in [7], and each requirement has a time window and multiple resource requirements, generalizing the model in [2]. In Sect. 2, we describe the model of MCVMA with time windows. In Sect. 3, we design a dynamic programming algorithm. In Sect. 4, we propose a heuristic Algorithm based on resource scarcity and density. In Sect. 5, we give our experimental results. We give conclusion in the last section.

2 Preliminaries

Assume that a resource provider that has a total of m types of different resources, and the capacity of each resource is represented by the vector $\vec{C} = (c_1, c_2, \ldots, c_m)$; each resource unit cost is represented by the vector $\vec{V} = (v_1, v_2, \ldots, v_m)$. A total of n users arrive at the system to request resources at different times. Assume that user i submits his requirements at time t_i, leaving the system at d_i. Each user can submit multiple requirements and the corresponding execution time and value, but only one requirement can be satisfied. For resource providers, the usage of resources cannot exceed the total amount of resources at any time.

For example, for a job, the user may have a variety of considerations. One option is to use a high-performance VM over a short time period to complete the job, and another option is to use a general-performance VM over a long time period to complete the job. The resource allocation model consists of two aspects. The first is resource provider information. We denote a resource provider offering m types of resources $TP_1, TP_2, \ldots, TP_m$, and the resource capacities are represented by vector $\vec{C} = (c_1, c_2, \ldots, c_m)$. The unit costs of the resources is defined by vector $\vec{V} = (v_1, v_2, \ldots, v_m)$. The second part is user information. Let $U = \{1, 2, \ldots, n\}$ be the set of n users. The entire time period for resource allocation and usage is $[0, T]$. Assuming that user i submits their resource requirements at time t_i, the deadline for using these resources is d_i, and user i can submit k_i requirements, denoted by $K_i = \{1, 2, \ldots, k_i\}$. Thus, user i submits a bundle of requirements $S_i = [s_{i1}, s_{i2}, \ldots, s_{ik}]$, where only one of the k_i requirements can be satisfied. The requirement $s_{ik}, k \in K_i$ can be represented by the vector $(s_{ik1}, \ldots, s_{ikr}, \ldots, s_{ikm}, e_{ik}, b_{ik})'$, where $s_{ikr}, r \in R$ represents the amount of the r-th resource in the k-th requirement that is requested by user i, e_{ik} is the execution time of requirement k that must be used, and b_{ik} is the user valuation of the corresponding requirement. Because each requirement of user i has the same deadline d_i but has a different execution time e_{ik} for the requirement k of user i, the latest start time of resource allocation is $d_i - e_{ik}$, that is, the requirement should be allocated during the window period $[t_i, d_i - e_{ik})$; otherwise, it will not be allocated. The final submission information of user i is represented by vector $\theta_i = (t_i, d_i, S_i)$.

For example, in Table 1, user 1 submits requirement 1 and requirement 2 at time 2. Requirement 1 needs 2 units of TP_1 resource, 4 units of TP_2 resource, 5 units of TP_3 resource, and an execution time of 20 h, and the user is willing to pay 8 for requirement 1. At the same time, user 1 also submits requirement 2, which needs 4 units of TP_1 resource, 8 units of TP_2 resource, 5 units of TP_3 resource, and an execution time of 15 h, and the user is willing to pay 9 for requirement 2. The deadline of both requirements is 25. There are 2 requirements for user 1, but only one requirement will be allocated successfully. We denote $\theta_1 = (t_1, d_1, S_1)$, $t_1 = 2, d_1 = 25$, $S_1 = [s_{11}, s_{12}]$, $s_{11} = (2, 4, 5, 20, 8)', s_{12} = (4, 8, 5, 15, 9)'$.

Table 1. Submission requirements of user 1

Requirements	Arrival time	Execution time	Deadline	TP_1	TP_2	TP_3	Bid
Requirement 1	2	20	25	2	4	5	8
Requirement 2		15		4	8	5	9

For MCVMA with time windows, the provider pursues the maximization of social welfare, which can generate greater revenue. We introduce x_{ikt} to indicate whether the k-th requirement of user i is allocated at time t, where $x_{ikt} = 1$ indicates that the requirement is allocated at time t and $x_{ikt} = 0$ otherwise.

$$x_{ikt} = \begin{cases} 1, & \text{if the } k\text{-th requirement of user } i \text{ is allocated at } t, t \in [t_i, d_i - e_{ik}) \\ 0, & \text{otherwise} \end{cases} \tag{1}$$

We denote the total social welfare as V and formulate the MCVMA problem with time windows as an integer program as follows:

$$Objective : V = \max[\sum_{i \in U} \sum_{k=1}^{k_i} \sum_{t=1}^{T} (b_{ik} - \sum_{r=1}^{m} s_{ikr} \cdot v_r \cdot e_{ik}) \cdot x_{ikt}] \tag{2}$$

$$\text{Subject to}: \sum_{i \in U} \sum_{k=1}^{k_i} \sum_{\omega=t-e_{ik}}^{t-1} s_{ikr} \cdot x_{ik\omega} \leq c_r, \forall r \in R, \ \forall t = 0, 2, \ldots, T-1 \qquad (2a)$$

$$\sum_{r=1}^{m} s_{ikr} \cdot v_r \cdot e_{ik} \cdot \sum_{t=1}^{T} x_{ikt} \leq b_{ik}, \forall i \in U, \forall k \in K_i \qquad (2b)$$

$$\sum_{k=1}^{k_i} \sum_{t=1}^{T} x_{ikt} \leq 1, \ \forall i \in U \qquad (2c)$$

$$x_{ikt} \in \{0, 1\}, \forall i \in U, \forall k \in K_i, \forall t \in [0, T) \qquad (2d)$$

where V represents the total social welfare, which is the sum of all of the selected user requirement bids b_{ik} minus the corresponding cost $\sum_{r=1}^{m} s_{ikr} \cdot v_r$. Formula (2a) indicates that the resource allocation at any time t cannot exceed the capacity of any type of resource. Formula (2b) indicates that the cost of each requirement of user i must be no more than the corresponding valuation b_{ik}. Formula (2c) indicates that only one of the requirements submitted by user i can be allocated. Without loss of generality, we assume that each user has the same number of requirements, $k_i = K, i \in U$. A feasible solution of the problem is represented by a $n * KT$ matrix $\vec{X}$.

Table 2 and Fig. 1 show the results of the online multiple requirement allocation with a resource capacity of $\vec{C} = (5, 5, 5)$. The optimal solution is that the first user's requirement 1 starts at time 0, the second user's requirement 2 starts at time 1, the third user's requirement 1 starts at time 4, and the fourth user's requirement 2 starts at time 5. The final social welfare is $10 + 6 + 10 + 5 = 31$.

Table 2. Multiple requirement online allocation example

User	Requirements	Arrival time (t_i)	Execution time (e_{ik})	Deadline (d_i)	TP_1	TP_2	TP_3	Bid (b_{ik})
User 1	Requirement 1	0	3	6	3	2	1	10
	Requirement 2		3		2	4	3	9
User 2	Requirement 1	1	5	8	2	1	4	5
	Requirement 2		3		2	3	2	6
User 3	Requirement 1	3	4	15	1	3	2	10
	Requirement 2		7		2	1	3	9
User 4	Requirement 1	5	5	13	2	3	2	4
	Requirement 2		4		3	1	2	5

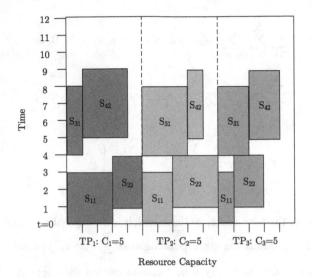

Fig. 1. Online multiple requirement allocation

3 Dynamic Programming Algorithm

Because the system cannot predict each user's arrival time, the optimal resource allocation of MCVMA with time windows must occur after collecting all user requirements. Thus, the optimal resource allocation of MCVMA with time windows is actually static (offline), which cannot satisfy the pay-as-you-go model of users. However, the optimal solution can be used as comparison data. We propose the optimal resource allocation algorithm DP_Optimal_A.

The DP_Optimal_A algorithm can be designed using dynamic programming (DP). However, the computational time increases exponentially with an increasing number of users, resource types, and user requirements. We use DP to obtain the optimal solution of the resource allocation problem as follows:

Initial value setting:

$$U(1, C') = \begin{cases} b_1, & \text{if } c'_{rj} = s_{1kr}, \ j \in [t_1, t_1 + e_{1k} - 1] \ , \ k \in K_i, \ r \in R \\ -\infty, & \text{otherwise} \end{cases} \text{, where,}$$

$$C' = \begin{bmatrix} c'_{11}, c'_{12}, \ldots, c'_{1T} \\ c'_{21}, c'_{22}, \ldots, c'_{2T} \\ \ldots \ldots \\ c'_{m1}, c'_{m2}, \ldots, c'_{mT} \end{bmatrix} \tag{3}$$

Recursive relationship:

$$U(i, C') = \max\{U(i-1, C'), \max_{k \in K_i,\, t \in [t_i, d_i - e_{ik})} (U(i-1, C' - C^{(ik)}) + b_{ik})\}, \text{ where,}$$

$$C^{(ik)} = \begin{bmatrix} 0 & \cdots & s_{ik1} & \cdots & s_{ik1} & \cdots & 0 \\ 0 & \cdots & s_{ik2} & \cdots & s_{ik2} & \cdots & 0 \\ 0 & \cdots & \cdots & \cdots & \cdots & \cdots & 0 \\ 0 & \cdots & s_{ikm} & \cdots & s_{ikm} & \cdots & 0 \end{bmatrix},\ t \in [t_i, d_i - e_{ik}) \tag{4}$$

$$1\ \cdots\ t\ \cdots\ t + e_{ik} - 1\ \cdots\ T - 1$$

$U(i, C')$ denotes the maximum social welfare of a user set $\{1, 2, \ldots i\}$ that has been allocated resources C' in period T. Because of the online resource allocation, the resource capacity C' is a matrix that ensures that any type of resource allocation at $[0, T-1]$ cannot exceed the amount. $C^{(ik)}$ indicates the resource request capacity s_{ik} and time period $[t, t + e_{ik} - 1]$ that must be allocated to the k-th requirement of user i. According to DP theory, the maximum social welfare in the case of the participation of user i can be obtained from the maximum of either $U(i-1, C')$ or $\max_{k \in K_i, t \in [t_i, d_i - e_{ik})} (U(i-1, C' - C^{(ik)}) + b_{ik})$.

Algorithm 1 shows the corresponding algorithm of DP_Optimal_A. In this algorithm, line 1 determines the initial social welfare and resources, and lines 2–10 implement a specific DP algorithm. The final results of the algorithm indicate that the optimal

---

Algorithm 1   Dynamic programming optimal resource allocation algorithm (DP_Optimal_A)

Input: All user requirements: Resource capacities: $\vec{\theta} = (\theta_1, \theta_2, \ldots, \theta_n)$; Resource capacities: $\vec{C} = (c_1, c_2, \ldots, c_m)$

Output: The optimal allocation includes which requirement should be allocated to the users: $\vec{X}$; Total social welfare: U_{sw}

1. Initial $U(1, C')$
2. for $i \leftarrow 1$ to n, $k \leftarrow 1$ to K, $t \leftarrow 1$ to T, $r \leftarrow 1$ to m do
3.   for $c'_{1t} \leftarrow 0$ to c_1, $c'_{2t} \leftarrow 0$ to $c_2 \ldots$, $c'_{mt} \leftarrow 0$ to c_m do
4.   if $s_{ik1} < c'_{1t}$ and $s_{ikr} < c'_{rt}, \ldots, s_{ikm} < c'_{mt}$ then
5. $U(i, C') = \max\{U(i-1, C'), \max_{k \in K_i}(U(i-1, C' - C^{(ik)}) + b_i)\}$
6.   else
7.   break from loop
8.   end if
9.   end for
10. end for
11. $U_{sw} \leftarrow U(i, C')$
12. find $\vec{X}$ by looking forward at $U(i, C')$
13. return $\vec{X}$, U_{sw}

---

social welfare is $U_{sw} \leftarrow U(i, C')$ and that the optimal allocation solution is X. The algorithm can find the optimal solution, but the time complexity of the algorithm is $O(kn(c_1 \ldots c_m)^T)$. A more efficient algorithm is needed for practical use.

4   A Heuristic Algorithm

Because the optimal algorithm DP_Optimal_A can not solve the online allocation, we propose a heuristic algorithm for MCVMA with time windows: the MCVMA_A algorithm. The MCVMA_A algorithm considers the allocation solution $\vec{X}^t$ and social welfare U_{swt} at time t. Because predicting the arrival time of each user is impossible, MCVMA_A is event driven and invoked whenever a new user submits his requirements or when a running user exits and releases his resources.

The MCVMA_A framework calculates the corresponding cost (lines 1–6) of all user requirements. If any of user i's requirements have a corresponding cost cp_{ik} that is higher than bid b_{ik}, then this requirement should be removed from his requirements (lines 3–5) to ensure the effectiveness of the subsequent allocation and payment algorithms.

Algorithm 2 Framework of MCVMA_A

Input: Current time: t ; The user requirements that in window period: $[t_i, d_i\text{-}e_{ik})$ but have not been allocated at current time t : $\vec{\theta}$; The users who have been allocated resources but whose jobs are not completed at the current time t : $\tilde{\theta}$; The unit price of each type of resource: $\vec{V} = (v_1\ v_2 \ldots v_m)$; The remaining capacity of each type of resource at current time t : $\vec{C^t} = (c_1^t, c_2^t, \ldots, c_m^t)$

Output: The allocation solution up to the current time t : $\vec{X}$; The social welfare up to the current time t : U_{swt}

1. for all $i \in \{i \mid \theta_i \in \vec{\theta}\}$ and $k \leftarrow 1$ to K

2. $cp_{ik} \leftarrow \sum_{r \in R} s_{ikr} \cdot v_r \cdot e_{ik}$ /* Calculate the cost of each user's requirements */

3.   if $cp_{ik} > b_{ik}$ then

4. $S_i \setminus \{s_{ik}\}$

5.   end if

6. end for

7. $U_{swt}, \overrightarrow{X^t} \leftarrow MCVMA\_A(\theta, \tilde{\theta}, C')$

8. $\vec{X} \leftarrow \vec{X} \cup \overrightarrow{X^t}$

9. $U_{sw} \leftarrow U_{sw} + U_{swt}$

10. return $\vec{X}, U_{sw}$

We use the concept of resource density sorting to obtain better results using the allocation algorithm. One of the most important parameters is the *resource scarcity* h_r, which represents the supply and demand of resource r at the current time t.

c_r^t represents the remaining amount of the r-type resource at time t. A higher value indicates a rarer resource. *Resource scarcity* is defined as follows:

$$\hat{h}_r = \frac{\sum\limits_{i \in U^*} \sum\limits_{k=1}^{k_i} s_{ikr}}{c_r^t}, \ \forall r \in R, \ U^* = \{i | \theta_i \in \vec{\theta}\} \tag{5}$$

$$h_r = \frac{2}{1 + e^{-\hat{h}_r}} - 1, \ h_r \in (0, 1) \tag{6}$$

We introduce the *resource density* f_{ik} as follows:

$$f_{ik} = \frac{b_{ik}}{\sum\limits_{r \in R} \left(\frac{s_{ikr}}{c_r} \cdot h_r \right) \cdot e_{ik}}, \ \forall i \in U, \ k \in K_i \tag{7}$$

where f_{ik} represents the resource density of the k-th requirement of user i. A higher *resource density* value indicates that the provider tends to allocate resources to this user.

Definition 1. *Reallocation strategy. In the allocation algorithm, the resource density will be recalculated after addressing a certain user's allocation to obtain more accurate results.*

The OMVA_A algorithm is a heuristic algorithm, and it is invoked when a new user arrives and submits his requirement or when an allocated user needs to release his resource.

Step 1. Recycle the released resource at the current time to $\overrightarrow{C^t} = (c_1^t, c_2^t, \ldots, c_m^t)$;

Step 2. According to the *resource density* f_{ik}, sort all requirements submitted at the current time t in non-increasing order to set D and allocate resources by the order. If any type of resource cannot satisfy the current allocation, then the entire allocation algorithm ends.

Step 3. During each allocation, after a certain number step of users is allocated, the *resource density* of the users who are not allocated needs to be recalculated and resorted according to the current resource capacities remaining for the subsequent allocation. This approach can guarantee that resource providers receive greater social welfare. The time complexity of the MCVMA_A algorithm is $O(n^2 k^2 m)$.

Theorem 1. *MCVMA_A is a feasible solution of formula (2).*

Proof. The MCVMA_A algorithm code (lines 7–11) indicates that any resource can be allocated before being consumed, which satisfies formula (2a). The code of Algorithm 2 (lines 1–6) ensures that every requirement cost for each user must be less than the valuation b_{ik}, which satisfies formula (2b). Line 18 indicates that each user has only one requirement to be allocated, which satisfies formula (2c). Because MCVMA_A is event driven, the solution $\overrightarrow{X^t}$ of each time t is satisfied by formulas (2a)–(2c). Therefore, in time period T, the total solution X is the union of the solutions $\overrightarrow{X^t}$, also satisfying formulas (2a)–(2c). Solution $\vec{X}$ is a feasible solution of formula (2).

Algorithm 3 MCVMA_A Algorithm

Input: The user requirements that in window period: $[t_i, d_i$-$e_{ik})$ but have not been allocated at current time t: $\vec{\theta}$; The users who have been allocated resources but whose jobs are not completed at the current time t: $\tilde{\theta}$; The remaining capacity of each type of resource at current time t: $\overrightarrow{C^t} = (c_1^t, c_2^t, ..., c_m^t)$

Output: The allocation solution up to the current time t: $\vec{X}$; The social welfare in this allocation process: U_{swt}

1. $\overrightarrow{C^t} \leftarrow recycle\_resources(\overrightarrow{C^t}, \tilde{\theta}, t)$
2. for all $i \in \{i \mid \theta_i \in \theta\}$ and $k \leftarrow 1$ to K do
3.   calculate the *resource density* f_{ik}
4. end for
5. $D \leftarrow descend\_sort(f_{ik})$
6. for all $i \in \{i \mid \theta_i \in \vec{\theta}\}$ and $k \leftarrow 1$ to K ,according to the non-increasing order
    in D do
7.   for all $r \leftarrow 1$ to m do
8.   if $(c_r^t - s_{ikr}) \leq 0$ or $\theta = \phi$ then
9.   return $\overrightarrow{X^t}$, U_{swt}
10.   end if
11.   end for
12.   for all $r \leftarrow 1$ to m do
13. $c_r^t \leftarrow c_r^t - s_{ikr}$
14.   end for
15. $x_{ikt} \leftarrow 1$
16. $X^t \leftarrow X^t \cup \{i\}$
17. $U_{swt} \leftarrow U_{swt} + b_{ik}$
18. $\vec{\theta} \leftarrow \vec{\theta} \setminus \{\theta_i\}$ /* indicates user i has already allocated */
19. $n \leftarrow n+1$
20.   if n mod $step = 0$ then /* reallocation */
21.   for all $i \in \theta, k \in K$ do
22. $D \leftarrow descend\_sort(f_{ik})$ /* resort f_{ik} */
23.   end for
24.   end if
25. end for
26. return $\overrightarrow{X^t}$, U_{swt}

Theorem 2. *The time complexity of the MCVMA_A algorithm is $O(n^2k^2m)$.*

Proof. According to the MCVMA_A algorithm, it is possible that k requirements with m resource types of n users could potentially be considered. Additionally, lines 20–24 also involve the reallocation strategy, and the worst case also needs to address n users' k requirements; thus, the complexity of the algorithm is $O(n^2k^2m)$.

5   Experimental Results

We rely on the well-studied and standardized workload DAS-2 [4] from Grid Work-loads as the test data for simulating user requirements. DAS-2 is provided by the Advanced School for Computing and Imaging (ASCI). The DAS-2 data set contains user job IDs and the corresponding resource requirement information. To ensure reasonable simulation data, we remove the jobs that have zero value from the data set. The experimental platform hardware is configured as follows: Pentium G630 CPU, 4 GB of memory, and 500 GB of storage. The experimental settings are as follows:

(1) We use every k jobs in the data set as one user's requirements. In each job, we use the CPU request, memory request, and storage request as three different resource types (TP_1, TP_2, TP_3) and the user arrival time, job start time, and job execution time to simulate the user requirements.

(2) We randomly generate a value from 1 to 10 to simulate users' requirement value b_{ik} and pre-set the capacity of various types of resources in $\vec{C}$ and the unit price of various types resources in $\vec{V}$.

(3) We set the reallocation *step* to 5, 10, 50, and 100 according to the users' number and the capacity of resources $\vec{C}$;

(4) We use IBM CPLEX12 to program the DP_Optimal_A algorithm to solve for the optimal solution of resource allocation.

(5) We use the C++ programming language to program the MCVMA_A algorithm to solve for a feasible solution of resource allocation.

The experimental cases are divided into small scale (cases 1–6), medium scale (cases 7–12) and large scale (cases 13–18) based on the number of users, which can reflect real environments. Table 3 shows the initial set of 18 cases as a prerequisite for all subsequent experiments. Each case of experiments verifies the MCVMA_A algorithm and DP_Optimal_A algorithm by setting a different number of requirements k and reallocation *step*. Without loss of generality, we use CPU, memory, and storage to represent multiple resources (TP_1, TP_2, TP_3), and each case has the same resource capacity (TP_1: 50, TP_2: 5,000, TP_3: 50,000). Figures 2, 3, 4 and 5 compare the social welfare, execution time, served users, and resource utilization of the two algorithms, respectively.

Figure 2 compares the MCVMA_A algorithm with the optimal algorithm DP_Optimal_A from the social welfare perspective. MCVMA_A-k or DP_Optimal_A-k represents the result obtained by the corresponding algorithm when the user submits k requirements. For a given capacity of resources, the number of requirements k of each user has a significant impact on social welfare. When the user submits 20 groups

(MCVMA_A-20, DP_Optimal-20), the resource provider will obtain a higher social welfare than when the user submits only 5 groups (MCVMA_A-5, DP_Optimal_A-5). The social welfare calculated by MCVMA_A has reached more than 90% of the optimal solution. The main difference between MCVMA_A and DP_Optimal_A is that the optimal algorithm uses the static allocation method to obtain all of the user requirements before allocation, but the optimal algorithm does not satisfy the online model. Figure 2 (10,000 users) illustrates that DP_Optimal_A cannot obtain the optimal solution when there is a sufficiently large number of users because the resource allocation problem is NP-hard.

Table 3. Experiment cases of multiple requirement online allocation

Case	User number (n)	Requirement number (k)	Reallocation ($step$)
1	100	5	5
2	100	5	10
3	100	5	None
4	100	20	5
5	100	20	10
6	100	20	None
7	1000	5	10
8	1000	5	50
9	1000	5	None
10	1000	20	10
11	1000	20	50
12	1000	20	None
13	10000	5	10
14	10000	5	100
15	10000	5	None
16	10000	20	10
17	10000	20	100
18	10000	20	None

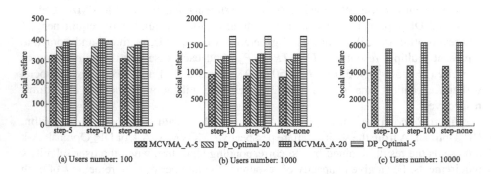

(a) Users number: 100   (b) Users number: 1000   (c) Users number: 10000

Fig. 2. MCVMA_A versus DP_Optimal_A: Social welfare

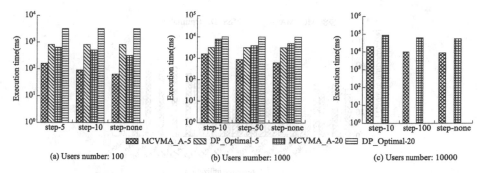

Fig. 3. MCVMA_A versus DP_Optimal_A: Execution time

Figure 3 compares the execution times of the MCVMA_A algorithm and the DP_Optimal_A algorithm, and it can be observed that the execution time of MCVMA_A is considerably faster than that of DP_Optimal_A. Even when the case has a large number of users (cases 13–18), the MCVMA_A algorithm is still able to calculate a feasible solution. For cases 13–18, we can observe that step changes do not have much impact on social welfare from Fig. 2, which shows that when the user number is large, we can improve the value of step to significantly reduce the execution time.

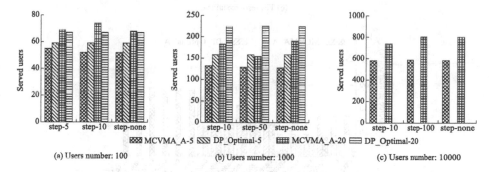

Fig. 4. MCVMA_A versus DP_Optimal_A: Served users

Figure 4 compares the served users of the MCVMA_A algorithm and the DP_Optimal_A algorithm. In Fig. 4(a), the served users of the two algorithms are more than 50%, whereas in Fig. 4(b), (c), the percentage of served users has decreased, mainly because in Fig. 4(b), (c), the resource usage time period does not increase with the number of users, and the competition for resources has become more intense; thus, the percentage of served users decreased.

Figure 5 compares the resource utilization of the MCVMA_A algorithm and the DP_Optimal_A algorithm. The utilization of resources under the online model is defined as (*The allocated resource usage time/Total resource availability time*). Because the online allocation is difficult and exhausts all the resources at the same time,

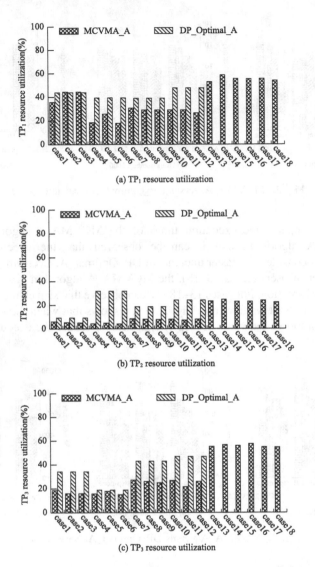

(a) TP₁ resource utilization

(b) TP₂ resource utilization

(c) TP₃ resource utilization

Fig. 5. MCVMA_A versus DP_Optimal_A: Resource utilization

the resource utilization is not high. However, we can also observe that MCVMA_A resource utilization can achieve approximately 60% of the optimal solution in most of the cases.

The experimental results show that the MCVMA_A algorithm can solve the problem of online multiple requirement, multiple resource allocation and pricing, and that it has advantages in terms of execution time and revenue.

6 Conclusion

We have presented two algorithms for the MCVMA problem with time windows. It is challenging and interesting to design a strategy-proof mechanism for the MCVMA problem with time windows as in [6–9].

Acknowledgments. The authors thank IBM for providing the full version of CPLEX12, which sets no limitations for solving for the optimal solution. This research is supported by the National Natural Science Foundation of China (61472345, 61662088 and 11663007), the Project of Natural Science Foundation of Yunnan Province of China (2014FA023, 2015FB115), and the Scientific Research Foundation of Yunnan Provincial Department of Education (2017ZZX228).

References

1. Amazon: Amazon ec2 instance types. https://amazonaws-china.com/cn/ec2/instance-types/
2. Angelelli, E., Bianchessi, N., Filippi, C.: Optimal interval scheduling with a resource constraint. Comput. Oper. Res. 51(3), 268–281 (2014)
3. Angelelli, E., Filippi, C.: On the complexity of interval scheduling with a resource constraint. Theoret. Comput. Sci. 412(29), 3650–3657 (2011)
4. ASCI: Grid workloads archives. http://gwa.ewi.tudelft.nl
5. Darmann, A., Pferschy, U., Schauer, J.: Resource allocation with time intervals. Theoret. Comput. Sci. 411(49), 4217–4234 (2010)
6. Liu, X., Li, W., Zhang, X.: Strategy-proof mechanism for provisioning and allocation virtual machines in heterogeneous clouds. IEEE Trans. Parallel Distrib. Syst. https://doi.org/10.1109/tpds.2017.2785815
7. Mashayekhy, L., Fisher, N., Grosu, D.: Truthful mechanisms for competitive reward-based scheduling. IEEE Trans. Comput. 65(7), 2299–2312 (2016)
8. Mashayekhy, L., Nejad, M.M., Grosu, D.: A PTAS mechanism for provisioning and allocation of heterogeneous cloud resources. IEEE Trans. Parallel Distrib. Syst. 26(9), 2386–2399 (2015)
9. Nejad, M.M., Mashayekhy, L., Grosu, D.: Truthful greedy mechanisms for dynamic virtual machine provisioning and allocation in clouds. IEEE Trans. Parallel Distrib. Syst. 26(2), 594–603 (2015)
10. Shi, W., Zhang, L., Wu, C., Li, Z., Lau, F.C.M.: An online auction framework for dynamic resource provisioning in cloud computing. IEEE/ACM Trans. Networking 24(4), 2060–2073 (2016)
11. Zaman, S., Grosu, D.: Combinatorial auction-based allocation of virtual machine instances in clouds. J. Parallel Distrib. Comput. 73(4), 495–508 (2013)

Fine-Gained Location Recommendation Based on User Textual Reviews in LBSNs

Yuanyi Chen[1(✉)], Zengwei Zheng[1], Lin Sun[1], Dan Chen[1],
and Minyi Guo[2]

[1] Hangzhou Key Laboratory for IoT Technology and Application,
Zhejiang University City College, Hangzhou, China
chenyuanyi@zucc.edu.cn
[2] Department of Computer Science and Engineering,
Shanghai Jiao Tong University, Shanghai, China

Abstract. As user-generated reviews from Location Based Social Networks (LBSNs) are becoming increasingly pervasive, exploiting sentiment analysis based on user's textual reviews for location recommendation has become a popular approach due to its explainable property and high prediction accuracy. However, the inherent limitations of existing methods make it difficult to discover what aspects that a user cared most about when visiting a location. In this study, we propose a fine-gained location recommendation model by jointly exploiting user's textual reviews and ratings from LBSNs, which considers not only the direct rating that a user would score on a location but also the compatibility between user's interested features and location's high-quality features. Specifically, the proposed recommendation model consists of three steps: (1) extracting feature-sentiment pairs from user's textual reviews; (2) learning to rank features using an Elo-based scheme; (3) making fine-gained location recommendation. Experiment results demonstrate that our proposed model can improve the recommendation performance compared with several state-of-the-art methods.

Keywords: Fine-gained location recommendation · User reviews ·
Sentiment analysis · LBSNs

1 Introduction

Numerous location recommendation models leveraging user's check-in records have been proposed over recent years with the rapid development of LBSNs. Among these recommendation models, collaborative filtering (CF) models and latent factor models are widely used, due to their good performance in providing personalized recommendation based on the wisdom of the crowds. However, existing CF models [9, 19] and latent factor models [13, 21] mainly focused on modelling the relations between user and location from user's check-in records, which cannot infer the actual rationale of the rating. For example, in the catering domain, a user may give a 5-star rating for taste, while another user may give the same rating but for the price. Since CF models and latent factor models lack such fine-grained analysis, they may fail to accurately model a target user's preference towards different aspects of a location.

© Springer Nature Switzerland AG 2019
S. Li (Ed.): GPC 2018, LNCS 11204, pp. 196–211, 2019.
https://doi.org/10.1007/978-3-030-15093-8_14

Recently, exploiting user's textual reviews for location recommendation has become a popular approach, since user's reviews offer the underlying reasons for the rating by discussing some specific aspects of the location. A few studies [2, 5–7, 15, 17, 23–25] have been proposed to enhance the interpretability of recommendation models by exploiting user-generated reviews. For example, the studies [7, 15, 17, 25] model users' rating behavior at the word level, which can uncover the latent topics employ in user's reviews. However, the recommendation process of these methods are non-transparent and the recommendation results are not explainable, as a topic may contain different aspects of a location and a user could express different opinions for various aspects in the same topic.

To improve the transparency of recommendation results, recent studies [2, 6, 23, 24] perform phrase-level sentiment analysis to discover user's opinion towards different aspects of a location. For instance, if a user visits a restaurant, and writes the review "the taste is perfect, but the price is a bit expensive!", the methods proposed in [2, 6, 23] can capture the feature-sentiment pairs like <*taste*; +1> and <*price*; −1>, which could provide finer-grained preference analysis of users, and make more accurate recommendation. Despite these methods achieve considerable performance improvement, they still suffer two limitations: (1) they cannot discover what aspects that a user cared most about when visiting a location; (2) they cannot distinguish what features of a location that are most valuable or least valuable to users, such kind of information is important for businesses to enhance user's experience.

In this paper, we propose a fine-gained location recommendation model by exploiting user's ratings along with textual reviews by the following three steps: (1) extracting aspect-sentiment pairs from user's textual reviews. Instead of modeling reviews at the word or topic level, we propose to model reviews in the aspect level with a few aspect-sentiment pairs; (2) learning to rank aspects using an Elo-based scheme. Given the aspect-sentiment pairs extracted from reviews, we utilize an Elo-based scheme to learn to rank user preferences over various aspects of a location; (3) making fine-gained location recommendation. We assume that a user's decision about whether or not to visit a location is based on several important aspects to him or her, rather than considering all hundreds of possible aspects.

The remainder of the paper is organized as follows. We first review related work in Sect. 2. Section 3 describes the overview of the proposed fine-gained recommendation model. Section 4 describes the proposed Elo-based method for learning to rank aspects. Then, we detail the fine-gained recommendation model in Sect. 5. Section 6 reports and discusses the experimental results. Finally, we present our conclusion and future work in Sect. 7.

2 Related Work

Existing studies on location recommendation based on sentiment analysis of textual reviews can be divided into word-level method and aspect-level method.

2.1 World-Level Method

Word-level method takes a review or a sentence as a whole, and analyses its sentiment directly. The work in [5] utilized a graph based recommendation framework to reconcile the tip and review information. In [17], the author proposed a hybrid preference model to unify user's preference by combining the preference extracted from user's check-ins and text-based tips. The work in [15] integrated location reviews into matrix factorization-based Bayesian personalized ranking for alleviating the cold-start problem in top-k location recommendation. For alleviating the sparsity problem, [26] utilized deep cooperative neural networks to learn location properties and user behaviors from online location reviews. In [25], the author proposed a bootstrapping approach to extract location adopters from review text. The work in [8] integrated ratings, reviews, user similarity and item similarity for location recommendation by combining matrix factorization with latent dirichlet allocation. TopicMF [1] jointly considered the ratings and accompanied review texts for location recommendation. The work in [7] integrated reviews into matrix factorization for location recommendation based Bayesian personalized ranking.

　　These methods mentioned above have made great efforts to fuse user's review information in location recommendation. However, these methods are lack of explainability by modeling user's sentiment in word-level, thus cannot obtain user's attitudes towards the specific aspect of a location.

2.2 Aspect-Level Method

Aspect-level method aims to discover user's opinion towards different aspects of a location. For example, EFM [24] first extracted explicit location aspects and user opinions by phrase-level sentiment analysis on user reviews, then made recommendation according to the specific location features to the user's interests and the hidden features learned. The work in [2] proposed a recommendation ranking strategy that combines similarity and sentiment to recommend locations according to user's opinion. In [6], the author modeled the user-location-aspect ternary relation as a heterogeneous tripartite graph, then regarded the recommendation task as a vertex ranking problem in the tripartite graph. The work in [23] incorporated textual reviews for recommendation through phrase-level sentiment analysis.

　　Through these aspect-level methods can generate _ne-grained location aspects extraction, the simple matrix factorization approach to optimize RMSE as a rating-based task fails to distinguish user's aspect-level preference of different locations. In addition, the complicated optimization algorithms to fuse the heterogeneous sources may cause greater errors.

　　Our proposed approach differs from the above-mentioned studies in the following two aspects: (1) we propose a simple and straightforward method to infer the aspects of a location that are most interesting to users and a location's high-quality aspects based on aspect-level sentiment analysis; (2) we make fine-gained location recommendation by jointly considering the direct rating that a user would score on a location and the compatibility between a user's interested aspects and a location's high-quality aspects.

3 Overview of Fine-Gained Location Recommendation

3.1 Preliminary

For ease of the following presentation, we define the key data structures and notations used in the proposed method.

Definition 1 (**Aspect**). An aspect a is an attribute or feature of a location, e.g., "service", "price", "environment" and "taste" for a restaurant.

Definition 2 (**Aspect-Sentiment Pair**). An aspect-sentiment pair is defined as a tuple $<a, o>$, where a is an aspect and o is the sentiment orientation towards the aspect. For example, for the piece of review 'the service is perfect, but the taste is terrible!', "the extracted aspect-sentiment phrases are $<service, +1>$ and $<taste, -1>$".

Definition 3 (**Location-Aspect Relation**). The relation strength between location p_j and feature a_k denotes as $w^{(jk)}$, which indicates the opinion of most users to aspect a_k of location p_j.

3.2 Recommendation Framework

Our proposed model produces top-k recommended locations by three phases: (1) extracting aspect-sentiment pairs from user's textual reviews; (2) learning to rank aspects based on aspect-sentiment Pairs; (3) making fine-gained location recommendation.

3.2.1 Extracting Aspect-Sentiment Pairs from User's Textual Reviews

As our focus is to learn to rank aspects for identifying what aspects users cared most about when they visit a location, we do not contribute to extract aspect-sentiment pairs from user's reviews, but instead exploit ASUM model [10] due to its high accuracy. For a given piece of review, we generate a set of aspect-sentiment pairs (A, O) using ASUM model to represent this review, where O is assigned as 1 or -1 according to the sentiment polarity that the user expressed on this aspect.

We collect a real-world dataset by crawling user's textual reviews about 2700 stores from Dianping site, which consists of multiple categories that matches with our research tasks (more details of this dataset are shown in Table 3). To verify the hypothesis that users usually care about different aspects for various locations, we select the most popular aspects in each category according to the frequencies they are mentioned in the textual reviews. In this dataset, we select top-10 most cared aspects for empirical analysis, as shown in Table 1. Based on simple observations we can find that:

- On pairwise level, all the six categories contain only two common aspects, and the others have at most four common aspects or no intersection at all.
- Among all these 36 mentioned aspects, only three appear in more than three categories.

Table 1. The top-10 most cared aspects of different categories.

Restaurant	Fashion	Kid store	Leisure	Education	Jewelery
'service'	'service'	'prices'	'massage'	'course'	'prices'
'taste'	'prices'	'style'	'service'	'Classroom'	'service'
'price'	'quality'	'clothes'	'price'	'prices'	'workmanship'
'dinner'	'clothes'	'discount'	'health'	'profession'	'style'
'clean'	'material'	'fabric'	'environment'	'environment'	'staff'
'Dessert'	'environment'	'branch'	'staff'	'service'	'environment'
'staff'	'collocation'	'service'	'decoration'	'discount'	'promotion'
'treat'	'staff'	'environment'	'promotion'	'atmosphere'	'discount'
'brunch'	'custom'	'packing'	'transportation'	'custom'	'after-sale'
'decoration'	'temperament'	'gift'	'skill'	'coffee'	'brand'

These observations imply that user's interests may vary with the categories, which is an important motivation for us to model user's preferences with the discrimination of different categories.

3.2.2 Learning to Rank Aspects Using Aspect-Sentiment Pairs
This step aims to discover the most interesting aspects of a location to most users and model a user's interests in different aspects of a location. Firstly, we extract pairwise preferences of aspects from the generated aspect-sentiment pairs. Then we utilize an Elo rating-based method [18] to learn to rank aspects. (Details of the method will be introduced in Sect. 4)

3.2.3 Making Fine-Gained Location Recommendation
After learning to rank aspects for users and locations, we combine both the direct rating and the ranked aspects to generate top-k recommendation. Specifically, we consider two issues when making recommendation: (1) the direct rating that a user would score on a location and (2) the matching degree between a user's interested aspects and a location's high-quality aspects. (Details of the method for making top-k recommendation will be introduced in Sect. 5)

4 Learning to Rank Aspects Using Aspect-Sentiment Pairs

In this section, we first present the problem statement of learning to rank aspects from user's textual reviews. Then we detail the proposed solution, an Elo rating-based method for ranking aspects of a location based on the extracted aspect-sentiment pairs. Ranking aspects for a user can be designed accordingly based on aspect-sentiment pairs.

Problem Statement. Let $A_s = \{a_1, a_2, \ldots a_{|As|}\}$ denote a finite aspect set of location p, (A_p, O_p) denote a set of aspect-sentiment pairs extracted from the textual reviews of p, the problem is ranking aspects according to the relation strength between aspect $a_i \in A_p$ and location p.

Problem Solution. Our solution for this problem consists of two phases:

(1) *Extract pairwise preference of aspects*: Given two aspects $a_i, a_j \in A_p$, a pairwise preference is extracted as a response from user's textual reviews. Either a_i is preferred to a_j (denoted $a_i \succ a_j$) or the other way around. Note pairwise preference labels may be non-transitive (due to irrationality or different personal preference), which means $a_i \prec a_k$ and $a_k \prec a_j$ cannot deduce $a_i \prec a_j$.

(2) *Estimate relation strength of aspects based on pairwise preferences*: We utilize a linear score function based on Elo rating-based scheme to estimate relation strength of aspects. More exactly, the Elo rating-based scheme is shown in Algorithm 1. First, as shown in Line 2, we calculate the winning expectation of aspects according to pairwise preference. Then, as depicted in Line 3, we update the Elo point of aspects according to their winning expectation. Finally, we utilize the ultimate Elo points as aspect's relation strength.

Algorithm 1. The Elo rating -based scheme for estimating relation strength

Require: 1) Location aspects $\{a_1, a_2, \ldots a_{|As|}\}$; 2) Pairwise preferences: $\Gamma = \{\ldots, a_i \prec a_j, \ldots\}$; 3) The starting Elo points of aspects: $\{R_1^0, \ldots, R_i^0, \ldots, R_{|A|}^0\}$; 4) Elo parameters: $\Sigma_E = \{\alpha_E, \beta_E, K_E\}$.

Ensure: Relation strength of aspects: $\{R_1, \ldots, R_i, \ldots, R_{|A|}\}$

1: **for** each pairwise preference $a_i \prec a_j \in \Gamma$ **do**

2:     Calculate winning expectation of a_i, a_j :
$$E_i = \frac{1}{1+\alpha_E^{(R_j-R_i)/\beta_E}}, E_j = \frac{1}{1+\alpha_E^{(R_i-R_j)/\beta_E}}$$

3:     Update the Elo point of a_i, a_j :
$$R_i \leftarrow R_i - K_E * E_i , R_j \leftarrow R_j - K_E * (1 - E_j)$$

4: **end for**

5: **return** The ultimate Elo rating: $\{R_1, \ldots, R_i, \ldots, R_{|A|}\}$.

---

5 Making Fine-Gained Location Recommendation

We make fine-gained location recommendation by jointly considering the direct rating that a user would score on a location and the matching degree between a user's interested aspects and a location's high-quality aspects.

5.1 Estimating the Direct User-Location Rating

Formally, let M be the number of users and N the number of locations, the rating database is denoted as a $N \times M$ rating matrix R, each row of R represents a user, each column of R represents a location, each element r_{ij} of R represents the rating of user i towards location j. Traditionally, the Regularized Singular Value Decomposition [14] model is employed to predict missing values. The basic idea is using low-rank matrix factorization approach seeks to approximate the rating matrix R by a multiplication of f-rank factors $R = U^T V$, where $U \in R^{f \times M}$ and $V \in R^{f \times N}$. The objective function is equivalent to minimize the sum of squared errors with quadratic regularization terms as follows:

$$L = \min_{U,V} \frac{1}{2} \sum_{i=1}^{M} \sum_{j=1}^{N} c_{ij} \left(r_{ij} - u_i^T v_j \right)^2 + \frac{\lambda_1}{2} \left(\|U\|_F^2 + \|V\|_F^2 \right) \tag{1}$$

where u_i and v_j are column vectors with f values, c_{ij} is the indicator function that is equal to 1 if user i rated location j and equal to 0 otherwise, λ_u, λ_v represent the regularization parameters, and $\|\cdot\|_F$ is the Frobenius norm of matrices.

However, matrix factorization methods are not capable of computing meaningful recommendations for entirely new users and locations. To tackle such cases, we utilize K-Nearest-Neighbor (KNN) mapping to estimate latent factors of entirely new users and locations based on additional review information. The general form of rating estimation by mapping from location aspects to factors is (the mapping method for users can be designed accordingly):

$$R_{ij}^{rating} = \sum_{h=1}^{f} u_{ih} \phi_h(A_j) \tag{2}$$

where $\phi_h: R^n \mapsto R$ denotes the function that maps the location aspects of j to the factor with index h.

We utilize kNN regression to map the aspect space to the factor space for each factor. Let N_k denotes the k nearest neighbors in terms of pearson coefficient of location features, then the factor can be estimated by:

$$\phi_h(A_j) = \frac{\sum_{j \in N_k(i)} sim(A_i, A_j) * v_{ij}}{\sum_{j \in N_k(i)} sim(A_i, A_j)} \tag{3}$$

where $sim(A_i, A_j)$ is the feature similarity of location and calculated by pearson coefficient:

$$sim(x, y) = \frac{<x - \bar{x}, y - \bar{y}>}{\|x - \bar{x}\| \|y - \bar{y}\|} \tag{4}$$

The intuition on how to map location aspects to factors can be explained by the following example:

Example 1: Suppose the rating matrix consists of three users and three locations, we train a matrix factorization model with $f = 2$ yields two matrices consisting of the user and location factor vectors, respectively:

$$U = \begin{bmatrix} 0.34 & 0.95 \\ 1.27 & 0.78 \\ 0.74 & 1.35 \end{bmatrix} V = \begin{bmatrix} ? & ? \\ 0.93 & 0.18 \\ 0.84 & 1.42 \\ 0.35 & 1.08 \end{bmatrix} \tag{5}$$

Table 2. The Elo points of location features

Feature	'Service'	'Price'	'Environment'	'Staff'
P1	845.7	974.1	1097.2	647.6
P2	945.9	942.8	1217.3	748.5
P3	1125.6	910.5	1457.5	1248.2
P4	1324.8	859.4	1218.6	1179.5

Every row in U corresponds to one user, each row in V corresponds to exactly one location. Suppose that the first location has not yet been rated by the three users, so row 1 in V does not contain any meaningful values. The Elo points based on aspect-sentiment pairs of the locations are shown in Table 2, then we can utilize KNN to estimate the factors of the first location (1-NN):

$$v_1 = \phi(A_1) = \left[\frac{0.93 * v_{21}}{0.93}, \frac{0.93 * v_{22}}{0.93}\right] = [0.93, 0.18] \tag{6}$$

5.2 Estimating the Aspect Rating

To evaluate the aspect rating by considering the consistency between a user's favored features and a location's aspects. We first estimate location's quality on different aspects. Let s_{ij} denotes the Elo score of location p_i's feature j based on aspect-sentiment pairs, c_{ij} denotes the number of times that aspect j mentioned in all the reviews of location p_i, then we estimate the quality of aspect j for location p_i by:

$$q_{ij} = \frac{1}{1 + e^{-s_{ij}c_{ij}}} \tag{7}$$

We assume that a user's decision about whether or not to visit a location is based on several important aspects to him or her, rather than considering all hundreds of possible aspects. For a given user-location pair (i, j), we could select u_i's favored aspects according to Eq. 7. Let $MAX_j = \{ind_1, ind_2, \ldots, ind_k\}$ denotes the k largest aspect's relation strength in A_j, we estimate a user's aspect rating for a location as

$$R_{ij}^{feature} = \sum_{l \in MAX_j} s_{jl} * q_{jl} \tag{8}$$

After estimating R_{ij}^{rating} and $R_{ij}^{feature}$, we calculate the final recommendation score of location j for user i by linear integration:

$$\hat{R}_{ij} = \alpha R_{ij}^{rating} + (1 - \alpha) R_{ij}^{feature} \tag{9}$$

where α is a scale parameter that controls the tradeoff between aspect-based score and direct user-location ratings. Note that we firstly normalize R_{ij}^{rating} and $R_{ij}^{feature}$ to be a value in (0, 1) in our experiment before calculate the final recommendation score.

6 Experiment Evaluation

In this section, we conduct extensive experiments to evaluate the performance of the proposed recommendation model. We first describe the settings of experiments including data sets, comparative methods and evaluation metric in Sect. 6.1. Then, we report and discuss the experimental results in Sect. 6.2

Table 3. Statistics of location's reviews (#Avg P, i.e., average reviews per location; *#Avg U*, i.e., average reviews per user)

	# of reviews	# of locations	# Avg P	# of users	# Avg U
Restaurant	508872	932	546	15321	33
Fashion	124369	763	163	8752	14
Kid store	72708	498	146	6619	11
Leisure	27244	278	98	3063	8
Education	11264	176	64	2105	5
Jewelry	2166	57	38	490	4

6.1 Experimental Settings

6.1.1 Datasets

Dianping[1] is China's largest location review site (similar to Yelp[2]) where users can freely rate locations and write their textual reviews. We crawl ratings and textual reviews about more than 2,700 locations from Dianping site. We pre-process these reviews by removing web URLs and non-Chinese words, and utilize ICTCLAS [20] for parsing and stemming. After pre-processing, this data set consists of 746,623 reviews and the density is 0.76%, as shown in Table 3. The sentiment seed words of ASUM model should not be aspect-specific evaluative words because they are assumed to be unknown. In this experiment, we use two sets of sentiment seed words to extract aspect-sentiment pairs of user's review: HowNet [3] and NTUSD [12].

6.1.2 Evaluated Techniques

We compare the proposed recommendation model with the following state-of-the-art competitors for evaluation:

---

[1] http://www.dianping.com/.

[2] https://www.yelp.com/.

- **Biased MF:** the method extends the basis matrix factorization (as shown in Eq. 1) by considering the biases and the objective function is as follows:

$$L = \min_{U,V} \frac{1}{2} \sum_{i=1}^{M} \sum_{j=1}^{N} c_{ij} (r_{ij} - u_i^T v_j)^2 + \frac{\lambda_1}{2} \left(\|U\|_F^2 + \|V\|_F^2 \right)$$
$$+ \frac{\lambda_2}{2} \left(\|U\|_{bF}^2 + \|V\|_{bF}^2 \right)$$

(10)

where U_b and V_b are the user bias and location bias respectively.

- **SVD++:** this method [11] extends two collaborating filtering models (e.g., latent factor models and neighborhood models) by exploiting both explicit and implicit feedback of the users.
- **ORec:** this method [22] generates location recommendation by two steps: (1) develop a supervised aspect-dependent approach to detect the polarity of a review, and (2) devise a unified recommendation framework to fuse review polarities with social links and geographical information.
- **AspectRec:** this method [16] generates location recommendation by two steps: (1) In offline phase, extracting the important aspects from each review along with their polarity weight to generate aspect summarization; (2) In online phase, using an user-based collaborative filtering technique to make recommendation.
- **TriRank:** this method [6] builds a user-item-aspect ternary relation by jointly considering users' ratings and affiliated reviews, and models the user-location-aspect ternary relation as a heterogeneous tripartite graph then casts the recommendation task as one of vertex ranking.
- **TopicMF:** this method [1] extends the basic matrix factorization model by simultaneously considering the ratings and accompanied review texts of users.

6.1.3   Evaluation Metric

Following the work in [4], we employ two standard metrics (i.e., precision and recall) in top-K recommendation for evaluation, which are defined as:

$$Precision@k = \frac{|P_{rec} \cap P_{actual}|}{K}$$
$$Recall@k = \frac{|P_{rec} \cap P_{actual}|}{|P_{actual}|}$$

(11)

where P_{rec} is a top-K recommendation list sorted in descending order of prediction values, P_{actual} are the locations a user has visited in the testing dataset.

We randomly divided the dataset into 80% for training and the rest 20% for testing. We set $\lambda_1 = 0.005$ and $\lambda_2 = 0.005$ for Biased MF. As for the hyper parameters of ASUM model, following existing work [10], we empirically set $\phi = 0$ for the negative seed words and 0.001 for all the other words. Similarly, for negative aspect-sentiment, we set $\phi = 0$ for the positive seed words and 0.001 for all the other words.

6.2   Experimental Results

We conduct two groups of experiments and report their performance with the well-tuned parameters. The first group is to evaluate the recommendation performance for all users while the second group is to evaluate the recommendation performance for cold-start users. For the two groups of experiments, we show only the performance where the number (K) of recommendation results is in the range [1...30], because a greater value of K is usually ignored for a top-k recommendation task.

6.2.1   Impact of Model Parameters

Tuning model parameters, such as the number of aspects per sentiment and the weighting factor α, are critical to the recommendation performance of jointly utilizing rating and reviews. We first fix the weighting factor $\alpha = 0.0$ and find that the optimal value for the number of aspects per sentiment (*#Aspects*) for each category of locations. Our goal is to investigate the performance of the proposed recommendation model when *#Aspects* increases from 10 to the maximum possible value 100. As for the weighting factor α, grid search and five-fold cross-validation are used for parameter tuning.

Table 4. *Precision@k* with different number of aspects per sentiment ($k = 10$, # of avgA denotes the number of aspects per sentiment

Category	# of avgA									
	10	20	30	40	50	60	70	80	90	100
Restaurant	0.108	0.127	0.144	0.178	0.186	0.197	**0.203**	0.19	0.183	0.176
Fashion	0.095	0.121	0.135	0.143	0.149	**0.153**	0.137	0.126	0.119	0.112
Kid store	0.107	0.122	0.147	0.162	**0.173**	0.163	0.157	0.143	0.135	0.121
Leisure	0.095	0.117	0.137	**0.148**	0.139	0.132	0.125	0.118	0.113	0.105
Education	0.103	0.128	0.142	**0.163**	0.154	0.146	0.139	0.132	0.125	0.118
Jewelry	0.089	0.113	**0.139**	0.132	0.128	0.124	0.117	0.112	0.105	0.093

Table 5. *Recall@k* with different number of aspects per sentiment ($k = 10$, # of avgA denotes the number of aspects per sentiment

Category	# of avgA									
	10	20	30	40	50	60	70	80	90	100
Restaurant	0.101	0.164	0.203	0.225	0.244	0.269	**0.281**	0.273	0.261	0.253
Fashion	0.088	0.137	0.173	0.208	0.241	**0.275**	0.243	0.232	0.221	0.202
Kid store	0.118	0.146	0.187	0.226	**0.253**	0.238	0.225	0.207	0.189	0.162
Leisure	0.103	0.156	0.178	**0.209**	0.192	0.184	0.175	0.169	0.162	0.153
Education	0.115	0.147	0.172	**0.194**	0.186	0.178	0.173	0.164	0.157	0.152
Jewelry	0.094	0.142	**0.183**	0.174	0.166	0.159	0.153	0.148	0.144	0.139

The results are shown in Tables 4 and 5 respectively, and larger *#Aspects* would lead to significant bias for all kinds of locations. From the two tables, we observe: (1) the best performance for different categories is achieved with different number of aspects per sentiment. In general, the optimal value of aspects per sentiment for different categories have a correlation with the number of reviews: the more the reviews, the larger the optimal value of aspects per sentiment. For instance, the optimal value of aspects per sentiment is 70 for Restaurant, while 40 for Education. These observations confirm our hypothesis in Sect. 5 that when making decisions, users usually care about several key aspects, and taking too many features into consideration could introduce noise into the models.

6.2.2 Recommendation Effectiveness for All Users

Figure 1a and b report the precision and recall of the recommendation algorithms for all users. we observe: (1) the *Precision@k* of all recommendation methods increase as the length of recommendation results increases, since increasing the number of recommendation results makes the data denser thus leading to better recommendation; (2) Our proposed method performs better than other baseline models (Biased MF, SVD ++, AspectRec, ORec, TopicMF and TriRank) significantly, showing the advantages of jointly considering user's ratings and textual reviews for learning user's fine-gained preference. For example, the *Recall@k* of our method is about 17.62% when $k = 10$, and the performance is improved by 38.6% and 25.62% compare with ORec and TriRank respectively; (3) the recommendation methods (TopicMF, TriRank and our method) jointly utilize user's ratings and reviews achieve better performance than merely utilizing ratings (Bised MF and SVD++) or textual reviews (AspectRec and ORec). For instance, the *Precision@10* of TriRank is about 22.6%, while 19.8% for ORec and 16.7% for Biased MF.

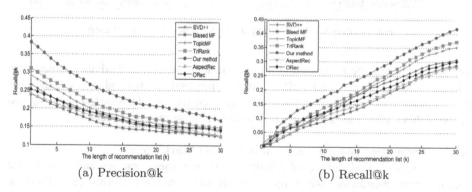

(a) Precision@k   (b) Recall@k

Fig. 1. Top-k recommendation performance for all users

To further investigate the recommendation performance for different kinds of locations, we report the *Precision@10* and *Recall@10* of all recommendation models for the six kinds of locations in Figs. 2 and 3. From the two figures, we observe the proposed method achieves the best recommendation performance in terms of all location categories, showing again the advantage of learning user's preference by

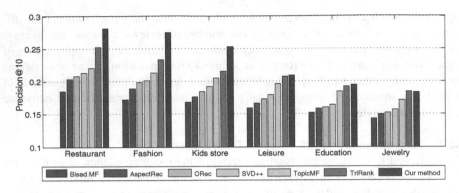

Fig. 2. *Precision@10* of different location categories for all users

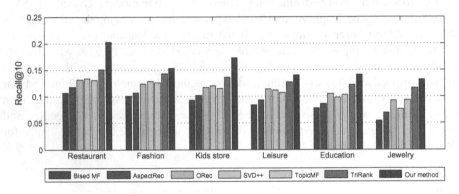

Fig. 3. *Recall@10* of different location categories for all users

jointly considering their ratings and reviews. In addition, our method also outperforms other methods (TopicMF and TriRank) that utilizes both ratings and reviews, showing users usually care about several key aspects when making decisions, and taking too many features into consideration could introduce noise into the models.

6.2.3 Recommendation Effectiveness for Cold-Start Users

To investigate the advantage of jointly considering user's ratings and reviews for location recommendation, we compare the recommendation performance of different algorithms for "cold-start" users in Fig. 4. In this experiment, we regard users whose location ratings or reviews are less than 5 as "cold-start" users. From Fig. 4a and b, we observe: (1) the performance of different recommendation algorithms for cold-start users degrades significantly compares to all users, showing data sparsity caused by few ratings or reviews bring serious challenge for location recommendation. For instance, the *Recall@10* of SVD++ for "cold-start" users drops 10.05% compare with all users; (2) the proposed method performs much better than baseline methods (i.e., Biased MF, SVD++, AspectRec, ORec, TopicMF and TriRank), showing the advantage of the proposed method by jointly considering user's ratings and reviews.

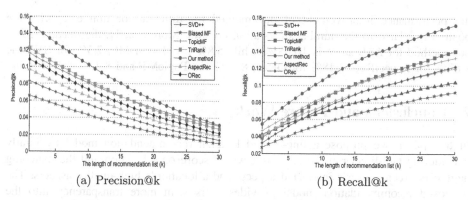

(a) Precision@k   (b) Recall@k

Fig. 4. Top-*k* recommendation performance for cold-start users

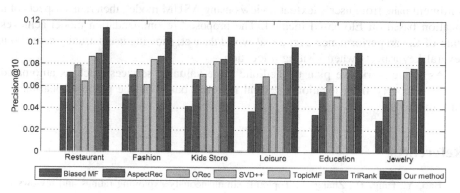

Fig. 5. *Precision@10* of different location categories for cold-start users

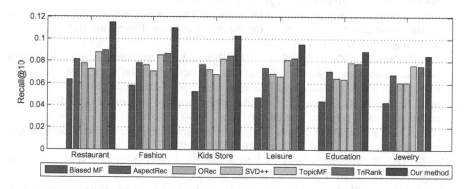

Fig. 6. *Recall@10* of different location categories for cold-start users

We also report the top-10 recommendation performance of different location categories for "cold-start" users in Figs. 5 and 6. The results suggest that, users usually care about several key aspects of locations when making decisions. We further observe the performance improvement of our method for different location categories are positively related to the number of location reviews. More exactly, the *Recall@10* of our method for *Kids store* is 10.08%, while 5.8% for Biased MF, 7.3% for SVD++, 7.87% for AspectRec, 7.48% for ORec, 8.37% for TopicMF and 8.42% for TriRank.

7   Conclusions and Future Work

In this paper, we propose a fine-gained location recommendation model by jointly considering the direct rating that a user would score on a location and the matching degree between a user's interested aspects and a location's high-quality aspects. The proposed recommendation model provides users with more transparency into the recommender system behavior and affords user interaction to further improve recommendations. More exactly, the fine-gained recommendation model first extracts aspect-sentiment pairs from user's textual reviews using ASUM model, then ranks aspects of a location based on Elo-based method. The proposed recommendation model achieves state-of-the-art performance over a real-world dataset, showing users usually care about several key aspects when visiting a location.

As future work, we plan to facilitate more valuable services (such as automated defect discovery from location's textual reviews) based on the proposed learning to rank methods.

References

1. Bao, Y., Fang, H., Zhang, J.: Topicmf: simultaneously exploiting ratings and reviews for recommendation. In: AAAI 2014, pp. 2–8 (2014)
2. Dong, R., O'Mahony, M.P., Schaal, M., McCarthy, K., Smyth, B.: Sentimental product recommendation. In: RecSys 2013, pp. 411–414 (2013)
3. Dong, Z., Dong, Q.: Hownet-a hybrid language and knowledge resource. In: Proceedings of the 2003 International Conference on Natural Language Processing and Knowledge Engineering, pp. 820–824. IEEE (2003)
4. Guo, Q., Sun, Z., Zhang, J., Chen, Q., Theng, Y.L.: Aspect-aware point-of-interest recommendation with geo-social influence. In: Adjunct Publication of the 25th Conference on User Modeling, Adaptation and Personalization, pp. 17–22. ACM (2017)
5. Gupta, S., Pathak, S., Mitra, B.: Complementary usage of tips and reviews for location recommendation in yelp. In: PAKDD 2015, pp. 720–731 (2015)
6. He, X., Chen, T., Kan, M., Chen, X.: Trirank: review-aware explainable recommendation by modeling aspects. In: CIKM 2015, pp. 1661–1670 (2015)
7. Hu, G., Dai, X.: Integrating reviews into personalized ranking for cold start recommendation. In: PAKDD 2017. pp. 708-720 (2017)
8. Huang, J., Zhong, N.: Leveraging item connections to improve social recommendations with ratings and reviews. In: WI 2016, pp. 185–191 (2016)
9. Jiang, S., Qian, X., Shen, J., Fu, Y., Mei, T.: Author topic model-based collaborative filtering for personalized poi recommendations. IEEE Trans. Multimedia **17**(6), 907–918 (2015)
10. Jo, Y., Oh, A.H.: Aspect and sentiment unification model for online review analysis. In: WSDM 2011. pp. 815–824 (2011)

11. Koren, Y.: Factorization meets the neighborhood: a multifaceted collaborative filtering model. In: SIGKDD 2008, pp. 426–434 (2008)
12. Ku, L.W., Liang, Y.T., Chen, H.H., et al.: Opinion extraction, summarization and tracking in news and blog corpora. In: AAAI Spring Symposium: Computational Approaches to Analyzing Weblogs, vol. 100107 (2006)
13. Liu, B., Xiong, H., Papadimitriou, S., Fu, Y., Yao, Z.: A general geographical probabilistic factor model for point of interest recommendation. IEEE Trans. Knowl. Data Eng. **27**(5), 1167–1179 (2015)
14. Paterek, A.: Improving regularized singular value decomposition for collaborative filtering. In: Proceedings of KDD Cup and Workshop, vol. 2007, pp. 5–8 (2007)
15. Song, T., Peng, Z., Wang, S., Fu, W., Hong, X., Yu, Philip S.: Review-Based Cross-Domain Recommendation Through Joint Tensor Factorization. In: Candan, S., Chen, L., Pedersen, T. B., Chang, L., Hua, W. (eds.) DASFAA 2017. LNCS, vol. 10177, pp. 525–540. Springer, Cham (2017). https://doi.org/10.1007/978-3-319-55753-3_33
16. Suresh, V., Roohi, S., Eirinaki, M.: Aspect-based opinion mining and recommendation system for restaurant reviews. In: RecSys 2014, pp. 361–362 (2014)
17. Yang, D., Zhang, D., Yu, Z., Wang, Z.: A Sentiment-enhanced personalized location recommendation system. In: Hypertext 2013, pp. 119–128. ACM (2013)
18. Yliniemi, L.M., Tumer, K.: Elo ratings for structural credit assignment in multiagent systems. In: AAAI (Late-Breaking Developments) (2013)
19. Zhang, C., Wang, K.: Poi recommendation through cross-region collaborative filtering. Knowl. Inf. Syst. **46**(2), 369–387 (2016)
20. Zhang, H.P., Yu, H.K., Xiong, D.Y., Liu, Q.: Hhmm-based chinese lexical analyzer ictclas. In: SIGHAN Workshop on Chinese Language Processing 2003, pp. 184–187 (2003)
21. Zhang, J.D., Chow, C.Y.: Crats: An lda-based model for jointly mining latent communities, regions, activities, topics, and sentiments from geosocial network data. IEEE Trans. Knowl. Data Eng. **28**(11), 2895–2909 (2016)
22. Zhang, J., Chow, C., Zheng, Y.: Orec: an opinion-based point-of-interest recommendation framework. In: CIKM 2015, pp. 1641–1650 (2015)
23. Zhang, Y.: Incorporating phrase-level sentiment analysis on textual reviews for personalized recommendation. In: WSDM 2015, pp. 435–440 (2015)
24. Zhang, Y., Lai, G., Zhang, M., Zhang, Y., Liu, Y., Ma, S.: Explicit factor models for explainable recommendation based on phrase-level sentiment analysis. In: SIGIR 2014. pp. 83–92 (2014)
25. Zhao, W.X., Wang, J., He, Y., Wen, J., Chang, E.Y., Li, X.: Mining product adopter information from online reviews for improving product recommendation. TKDD **10**(3), 29:1–29:23 (2016)
26. Zheng, L., Noroozi, V., Yu, P.S.: Joint deep modeling of users and items using reviews for recommendation. In: WSDM 2017, pp. 425–434 (2017)

Social and Urban Computing

Estimating Origin-Destination Flows Using Radio Frequency Identification Data

Chaoxiong Chen[1,2(✉)], Linjiang Zheng[1,2(✉)], Chen Cui[1,2],
and Weining Liu[1,2]

[1] Key Laboratory of Dependable Service Computing in Cyber Physical Society,
Chongqing University, Ministry of Education, Chongqing 400030, China
cxchen@cqu.edu.cn
[2] College of Computer Science, Chongqing University,
Chongqing 400030, China

Abstract. The origin-destination (OD) demand is a critical information source used in the traffic strategic planning and management. The Radio Frequency Identification (RFID) is an advanced technique to collect traffic data. In this paper, daily origin-destination trips were inferred from the RFID data. Locations of RFID readers are considered as the origins and destinations. However, the sparseness of RFID data leads uncertainty to the destination of trip. To handle this problem, an approach was proposed to estimate the OD matrix. At first, the driving time of trip-legs in all trajectories are calculated by the driving time of taxis, which can be distinguished from the RFID data. And then, the stay, the last pass-by RFID reader of a trip, is inferred based on the calculated driving time. Finally, we extracted daily origin-destination trips for all vehicles. Using the proposed method, a case study was developed employing the real-world data collected in Chongqing, China, which demonstrated the effectiveness of our proposed approach.

Keywords: RFID data · OD matrix · Trip generation · Data mining

1 Introduction

One of the critical input information used in the strategic planning and management of road network is the pattern of trip making of travelers in the city [1]. This trip pattern is typically represented by an origin-destination (OD) matrix where the individual cells indicate the number of trips between a given origin and destination that begin during a given time period. The OD matrix not only represents the demand for trips in the city, but also reflects the spatial distribution of traffic flow in the network. It can be used to plan traffic routes [2, 3], discovery regular commuting patterns [4], and analyze land use properties [5, 6]. An accurate calculation of OD matrices could help us understand the use of space and the mobility of the city.

The origin-destination estimation has been studied for many years. The recent development in the technologies of intelligent transportation systems (ITS) has created an opportunity for rectifying some of the problems in trip pattern identification. For example, the automatic vehicle identification (AVI) technologies, including license

S. Li (Ed.): GPC 2018, LNCS 11204, pp. 215–225, 2019.
https://doi.org/10.1007/978-3-030-15093-8_15

plate-based AVI and the radio frequency identification (RFID), have been studied extensively in recent years. The license plate-based AVI system recognizes and extracts the vehicle license plate from the video when vehicles drive through the detection point equipped with camera. It is convenient and efficient to obtain traffic data. But the influence of lightness and rainy weather limits the accuracy of the detected traffic information.

RFID uses electromagnetic fields to automatically identify and track tags attached to objects, and obtain information stored in tags [7]. It has been widely exploited in the field of intelligent transportation. A radio-frequency identification system usually consists of three parts: tags, readers and information center. Tag containing information such as license plate, vehicle type is attached to vehicle. When a RFID tag-attached vehicle passes by readers installed along the road links, the time and information of vehicle are collected by system. Therefore, data is not recorded continuously, and it is sparse in spatial and temporal scale. Compared with license plate-based AVI, the information from RFID systems is more accurate, and less influenced by the weather and environment. This new technique is deployed in some cities in China, like Chongqing, Nanjing, to collect traffic data. Therefore, the daily OD matrix can be derived by matching the vehicle information at origins, destinations.

RFID systems monitor the movements of almost all vehicles in a city. It can provide accurate traffic statistics, flow volume between each reader could be easy to calculate. While RFID data is not recorded continuously as traditional GPS data, the data is recorded only when vehicles pass through RFID readers. The activities of vehicles between readers are uncertain, and the stops of trips are unsure.

The focus of this paper is the use of data from RFID systems to help estimate daily OD matrix in an urban environment. In this regard, our major challenge is to deal with the uncertainties caused by the sparsity. The rest of this paper is organized as follows: Sect. 2 reviews related work of OD matrix estimation. Section 3 details the methodology to estimate OD matrix using RFID data. In Sect. 4, we give a description of the case study area and data, and provide discussion of results. Section 5 presents the conclusions and the further research.

2 Related Work

Because of the large number of vehicles and the complexity of urban traffic, directly measuring OD matrices is impossible. To solve this problem, researchers have devised lots of approaches and models for estimating these matrices. According to the methods of data collection, there are mainly four categories of methods to estimate OD matrix: estimate from the survey, estimate from road traffic counts, estimate from traffic trajectory data and estimate from AVI. Traditionally, transportation engineers use household questionnaires or census and road surveys to develop methodologies for OD matrix estimation [8]. The survey is labor intensive and time consuming. Data compilation is complicated and prone to error, and the obtained data is not reliable.

Methods based on traffic volumes estimate OD matrix using road traffic counts derived from sensors such as loop detectors and radar. Estimation techniques or models such as maximum likelihood, generalized least squares, maximal entropy and optimization are employed with traffic assignment matrix [9], which is a prior knowledge of a link choice proportion matrix. Their accuracies are, however, limited by the fact the traffic assignment matrix or a prior OD matrix is difficult to obtain.

With the wide deployment of GPS devices, digital cameras, smart cards and mobile phone, unprecedented footprint information can be collected and accessed. Trajectories of people and vehicles can be extracted from their digitized footprints. These trajectories are viewed as the representative of urban traffic, and serve as important data sources for calculating trip patterns in urban. Lu et al. [10] presented a visual system to explore OD patterns of interested regions based on taxi trajectories. Demissie et al. [11] used mobile phone data to estimate the origin-destination flows between districts of Senegal. Alexander et al. [12] estimated average daily origin-destination trips using call detail records (CDRs) data from millions of anonymized users. Alsger et al. [13] validated and improved the existing O-D estimation method using the unique smart card fare data of the South-East Queensland public transport network which includes data on both boarding stops and alighting stops. While these data only encompasses the movements of sampled vehicles in a city, just taxis, buses or floating cars.

The aforementioned Methods are generally based on the conventional data collection techniques. License-plate based AVI systems and RFID based systems are the new techniques for collecting traffic data. Dixon and Rilett [1] employed license-plate based AVI data to calculate the traffic assignment matrix and the prior OD matrix which were used to estimate freeway ramp-ramp OD flow. Feng et al. [14] extracted OD flow from the reconstructed trajectories which were constructed using license plate recognition data and traditional detection data with particle filter. Guo et al. [15] derived the preliminary OD matrix and dynamic assignment matrix from RFID data by matching the data collected in origins and destinations along the road links. But the uncertainties of the sparseness data are not considered in their work.

3 Methodology

Before introducing the methodology, some definitions are given.

Definition 1: Trajectory. A vehicle *trajectory* Tr_i is a sequence of time-ordered spatial readers, $Tr_i : (r_1, t_1) \rightarrow (r_2, t_2) \rightarrow \cdots \rightarrow (r_n, t_n)$, where v_i is the EID (unique identification of the vehicle) of vehicle, t_n is the time vehicle v_i passes by the reader r_n, each reader has a geospatial coordinate c.

Definition 2: Trip-leg. A *trip-leg* l_{mn} is a pair of readers (r_m, r_n) in *Trajectory,* having a $l_{mn}.dist$, where *Dist* is a function calculating the road network distance between two points. And denote the *trip-leg* l_{mn} in Tr_i as l_{mn}^i, the travel time t_{mn}^i is calculated based on Eq. 2.

$$l_{mn}.dist = Dist(r_m.c, r_n.c); \tag{1}$$

$$t_{mn}^i = t_n - t_m. \tag{2}$$

The idea of our method is very simple. Trip is a unidirectional movement from a reader of origin to a reader of destination. RFID readers are taken as origins and destinations of vehicles. The trajectory of a vehicle is comprised of trips (at least one trip). Because of the uncertainties caused by data, the stay of vehicles can not directly obtain. But it can be estimated by calculating the driving time of trip-legs. The proposed methodology for identifying the daily OD matrix is a three-step process: trip-leg driving time calculation, stay estimation and trip extraction.

3.1 Trip-Leg Driving Time Calculation

The driving time of a road segment is affected by many factors, such as weather, road conditions, traffic volume, and time in a day. Vehicles are driving slowly in the morning and evening rush hour, and traffic jams may even occur. Intuitively, taxis always roam around the city, looking for guests and providing services. Though the taxis on a road segment may be quite different from the entire set of vehicle, their travel speed could be similar [16]. Given this, we can infer the driving time based on the sampled traffic data.

To address this issue, we split the trajectories of all vehicles, and eventually obtain 138612 trip-legs. At the same time, we calculate the travel time for each trip-leg in the trajectory of each vehicle based on Eq. 2. Taxis are identified by the codes of vehicle (seen in Sect. 4.1), vehicle type = 'K33', plate type = '02' and property = 'D'. Then, we can calculate the driving time for each trip-leg covered by the trajectories of taxis. As shown in Fig. 1, two taxis v_1 and v_2 travel four trip-legs l_{12}, l_{23}, l_{34}, and l_{35}, passing by five readers, generating two trajectories Tr_1 and Tr_2. We can compute the average and variance of driving time of a trip-leg.

To distinguish rush hours in the morning and evening, we split one day into four time slots: 0:00 to 6:00, 6:00 to 12:00, 12:00 to 18:00 and 18:00 to 24:00. S_k^z is denoted as the time slot for the zth time slot of the kth day (totally 7 days), $z = 0, 1, 2, 3$, where 0th time slot represents the time span of 0:00 to 6:00. We calculate the mean driving time for each time slot and its variance as Eqs. 3 and 4.

$$\overline{t_{mn}} | S_k^z = \frac{\sum_i t_{mn}^i}{N}, z = 0, 1, 2, 3; k = 1, 2, \cdots 7 \tag{3}$$

$$d_t | S_k^z = var\left(\sum_i t_{mn}^i \right), z = 0, 1, 2, 3; k = 1, 2, \cdots 7 \tag{4}$$

For instance, the average driving time of l_{12} is $(t_{12}^1 + t_{12}^2)/2$. To ensure the quality of the calculated mean time, we require a trip-leg to be traveled by taxis for certain times, at least 2 times in a time slot. Otherwise, the trip-leg is considered absent of enough data. But it can be inferred by similar traffic conditions from other days. Traffic is generally similar on workdays, and similarly weekends. The average driving time and variance formulate a trip-leg's traffic conditions, which will be used in the stay extraction.

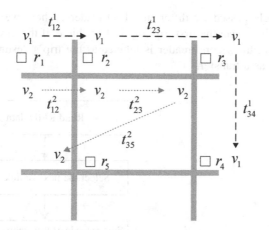

Fig. 1. Calculating the average driving time of a trip-leg

3.2 Stay Estimation

Furthermore, we wish to distinguish vehicles' stay locations (i.e., which reader is the last pass-by locations of trip). After calculating the average driving time and variance of each trip-leg from taxis, we define a time threshold of the trip-leg as:

$$T = \overline{t_{mn}}|S_k^z + 3 * (d_t|S_k^z), \qquad (5)$$

However, taxis are a small part of vehicle set, and the proportion is less than 1% (i.e. 15018 taxis in dataset). Moreover, there are still some trip-legs that are not traversed by taxis in the dataset. Targeting this issue, we set $T = 60$ min. In the existing literature [13, 17], T is regarded as an empirical constant, usually take one hour.

For each trip-leg l_{mn}^i in the trajectory Tr_i, if the travel time t_{mn}^i is greater than T of the corresponding time slot at the same day, the RFID reader r_m will be regarded as the stay location of the last trip. Therefore, reader r_m is the destination of vehicle in the last trip, and read r_n is the origin of the next strip. But if the travel time t_{mn}^i is less than T, the next *trip-leg* $l_{n(n+1)}^i$ in Tr_i will be considered, and dealt with in the same way until the last one.

3.3 Trip Extraction

In this step, we extracted origin-destination trips for all vehicles. The algorithm basically estimates the stay locations and chains the trip-legs wherever there is a stop among them to generate vehicle's O-D trips. To accelerate the process through all the vehicles, a search list containing the EIDs of all vehicles is created. As shown in Fig. 2, first choose a vehicle's EID from the database and then find all records about it for a given day. Sort records of the vehicle based on their timestamps, which represent the order that the vehicle passed by different RFID readers. Then, we obtained the trajectory of the vehicle. Select the first trip-leg of the vehicle, if the travel time is greater than the threshold T, the second reader is labeled as the trip's destination; otherwise, labeled the trip-leg as a transfer.

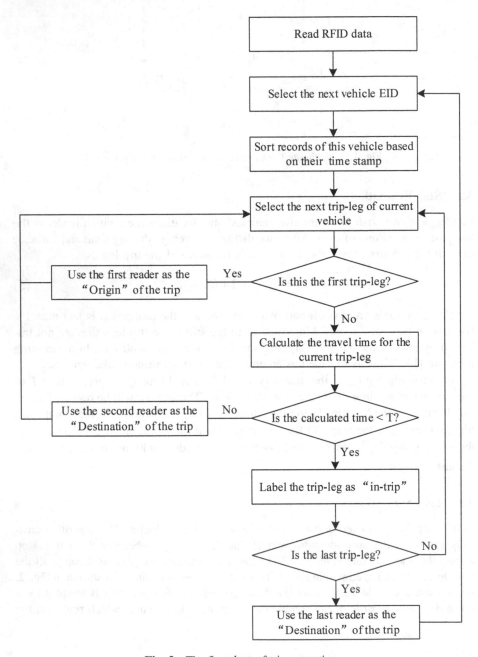

Fig. 2. The flowchart of trip extraction.

If the currently processed trip-leg is the last one of the vehicle's trajectory, the last reader is labeled as the trip's destination. The search process continues for all vehicles in the given day. A trip of each vehicle is denoted by $trip(v_i, o, d, t_o)$, where o and d are readers, and t_o is the starting time of the trip. Trips with the same origin and destination readers are grouped by day, and we obtain daily OD matrix.

4 Case Study

4.1 Data Description

The RFID data analyzed in this study were obtained from Chongqing, China. Almost vehicles in Chongqing have equipped with RFID tags. The tags contain electronically stored vehicle information, such as the EID, the code of the vehicle type, the code of the plate type, and the property of the vehicle. The EID is associated with the license plate, and it is encrypted to protect privacy. More than 1,000 RFID readers are deployed on major roads such as bridges and tunnels. There are 688 functioning RFID readers in the study area. Each reader contains information comprising: ID, name, direction, IP address, coordinates. The readers are installed aside the roads, read vehicle EID, reader ID and timestamp, and generate a record when a vehicle with an RFID tag passes by. The sample record data is shown as Table 1. The studied dataset contains about 80 million records from roughly 1.2 million vehicles in Chongqing main area over a period of a week, from February 29 to March 6 in 2016. The detailed statistics data are shown in Table 2.

Table 1. The sample record data of RFID system.

EID	Reader ID	Time stamp
1000003	R115	2016-03-02 08:53:24
1000005	R31	2016-02-29 09:13:06
1000008	R42	2016-02-29 09:40:46

Table 2. The detailed statistics of the dataset.

Date	Number of records	Number of vehicles
2016-02-29	11453809	778387
2016-03-01	11233288	763630
2016-03-02	11120763	758514
2016-03-03	11273198	760153
2016-03-04	11677016	794827
2016-03-05	11294460	743939
2016-03-06	10843602	739282
Total	78896136	1235655

4.2 Data Filtering

To improve the quality of the data and obtain a more accurate evaluation of the OD estimation, the data was filtered with some records excluded. For example, records with missing values and duplicate records are discarded. We assume that a vehicle will not pass by a RFID reader multiple times in a short time (set as 60 s). Therefore, the first of multiple records that a vehicle passing through the same reader was kept, and the rest were excluded. Vehicles with only a single record were also discarded, as the OD flow requires at least two records. The data cleaning process excludes 32248 records, accounting for less than 0.5% of the dataset. When calculating the travel time of trip-legs, we found that some of the travel times were abnormal. They are very small, some are equal to zero. We calculated the average travel speed of each trip-leg through Eq. 6. Trip-legs with zero travel time are discarded. And Trip-legs with an average travel speed of more than 150 km/h are also excluded.

$$\bar{v} = l_{mn}.dist/t_{mn}^i \tag{6}$$

4.3 Results and Validation

This section discusses the results of the proposed approach, and demonstrates the effectiveness of our method. After extracting OD-trips from all trajectories, we obtained trips from origin readers to destination readers. And then we derived the daily OD matrix, illustrated in Fig. 3. To have a close look at the results, there is high number of trips on workdays compared to the weekends. The minimum number of trips is on Sunday, March 6, 1388443 trips. Friday has the largest number of trips, there are 1536290 trips. Followed by Monday, 1522121 trips. The number of trips on Tuesdays and Thursdays is close, 1484821 trips and 1477289 trips, respectively. During week-days, the minimum number of trips is on Wednesday, having 1465496 trips. And Saturday, there are 1435460 trips.

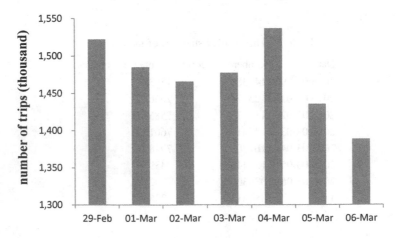

Fig. 3. Number of estimated OD trips during a week

To evaluate the proposed approach, we randomly select 200 vehicles in the dataset, and recruit five university students to extract OD trips from the vehicles' trajectories. All the participants are familiar with Chongqing urban area. Each trajectory is annotated by all of them, and the final OD trips of each trajectory, considered as 'ground truth', are determined by majority voting ($\geq$ three participants). The measures of effectiveness that we used in the analyses are the root-mean-squared error (RMSE) and the mean absolute error ratio (MAER) shown in Eqs. 7 and 8, respectively.

$$\text{RMSE} = \sqrt{\frac{\sum_i^n (x_i - \widehat{x_i})^2}{n}}, \tag{7}$$

$$\text{MAER} = \frac{1}{n}\sum_i^n \left(\frac{|x_i - \widehat{x_i}|}{\widehat{x_i}}\right), \tag{8}$$

where $n = 200$, 200 vehicles' trajectories being compared; x_i = the estimated number of trips for the ith trajectory; and $\widehat{x_i}$ = the number of trips for the ith trajectory in 'ground truth'. The RMSE of our algorithm is about 2.28, and the MAER is roughly 2.65, which demonstrated the effectiveness of our proposed approach.

5 Conclusion

With the increasing application of advanced traffic detection techniques, such as RFID based AVI systems, direct calculation of OD matrix is potentially possible. But the RFID data is sparse in time and space, leading uncertainty to the destination of a trip. Therefore, in this paper, we presented an algorithm to estimate daily OD matrix using the taxi trajectories in the dataset. Specifically, we first calculated the driving time for trip-legs in all trajectories, derived from the driving time of taxis. Second, we estimated stays (i.e., the last pass-by RFID reader of trips) based on the calculated driving times. Finally, we extracted daily origin-destination trips for all vehicles. Using the proposed method, a case study was developed employing the real-world data collected in Chongqing, China. The experimental results demonstrated the effectiveness of our proposed approach.

In the further, we plan to reconstruct the vehicles' trajectories from RFID data to improve the estimation of OD matrix. A more fine-grained trip pattern of different vehicles will also be studied. Besides, we would like to explore the OD patterns of interested region with visual analysis. Furthermore, it is interesting to identify unlicensed taxis with trip patterns derived from RFID data.

References

1. Dixon, M.P., Rilett, L.R.: Population origin-destination estimation using automatic vehicle identification and volume data. J. Transp. Eng. ASCE **131**(2), 75–82 (2005)
2. Chen, C., Zhang, D., Guo, B., Ma, X., Pan, G., Wu, Z.: TripPlanner: personalized trip planning leveraging heterogeneous crowdsourced digital footprints. IEEE Trans. Intell. Transp. Syst. **16**(3), 1259–1273 (2015)
3. Chen, C., Zhang, D., Li, N., Zhou, Z.H.: B-planner: planning bidirectional night bus routes using large-scale taxi GPS traces. IEEE Trans. Intell. Transp. Syst. **15**(4), 1451–1465 (2014)
4. Peng, C., Jin, X., Wong, K.C., Shi, M., Liò, P.: Collective human mobility pattern from taxi trips in urban area. PLoS ONE **7**(4), e34487 (2012)
5. Pan, G., Qi, G., Wu, Z., Zhang, D., Li, S.: Land-use classification using taxi GPS traces. IEEE Trans. Intell. Transp. Syst. **14**(1), 113–123 (2013)
6. Yuan, J., Zheng, Y., Xie, X.: Discovering regions of different functions in a city using human mobility and POIs. In: Proceedings of the 18th ACM SIGKDD International Conference on Knowledge Discovery and Data Mining - KDD 2012, p. 186. ACM Press, New York (2012)
7. Radio-frequency identification. https://en.wikipedia.org/wiki/Radio-frequency_identification, Accessed 16 Jan 2018
8. Calabrese, F., Di Lorenzo, G., Liang, L., Ratti, C.: Estimating origin-destination flows using mobile phone location data. Pervasive Comput. IEEE **10**(4), 36–44 (2011)
9. Ickowicz, A., Sparks, R.: Estimation of an origin/destination matrix: application to a ferry transport data. Public Transp. **7**(2), 235–258 (2015). https://doi.org/10.1007/s12469-015-0102-y
10. Lu, M., Liang, J., Wang, Z., Yuan, X.: Exploring OD patterns of interested region based on taxi trajectories. J. Vis. **19**(4), 811–821 (2016). https://doi.org/10.1007/s12650-016-0357-7
11. Demissie, M.G., Antunes, F., Bento, C., Phithakkitnukoon, S., Sukhvibul, T.: Inferring origin-destination flows using mobile phone data: a case study of Senegal. In: 2016 13th International Conference on Electrical Engineering/Electronics, Computer, Telecommunications and Information Technology, ECTI-CON (2016). https://doi.org/10.1109/ECTICon.2016.7561328
12. Alexander, L., Jiang, S., Murga, M., González, M.C.: Origin-destination trips by purpose and time of day inferred from mobile phone data. Transp. Res. C Emerg. Technol. **58**, 240–250 (2015). https://doi.org/10.1016/j.trc.2015.02.018
13. Alsger, A., Assemi, B., Mesbah, M., Ferreira, L.: Validating and improving public transport origin-destination estimation algorithm using smart card fare data. Transp. Res. C Emerg. Technol. **68**, 490–506 (2016)
14. Feng, Y., Sun, J., Chen, P.: Vehicle trajectory reconstruction using automatic vehicle identification and traffic count data. J. Adv. Transp. **49**(2), 174–194 (2015). https://doi.org/10.1002/atr.1260
15. Guo, J., Liu, Y., Li, X., Huang, W., Cao, J., Wei, Y.: Enhanced least square based dynamic OD matrix estimation using Radio Frequency Identification data. Math. Comput. Simul. (2017). https://doi.org/10.1016/j.matcom.2017.10.014

16. Shang, J., Zheng, Y., Tong, W., Chang, E., Yu, Y.: Inferring gas consumption and pollution emission of vehicles throughout a city. In: Proceedings of the 20th ACM SIGKDD International Conference on Knowledge Discovery and Data Mining - KDD 2014, pp. 1027–1036. ACM Press, New York (2014). https://doi.org/10.1145/2623330.2623653
17. Wang, Y., et al.: Unlicensed taxis detection service based on large-scale vehicles mobility data. In: Proceedings of the 2017 IEEE 24th International Conference on Web Services, ICWS 2017, pp. 857–861. Institute of Electrical and Electronics Engineers Inc. (2017). https://doi.org/10.1109/ICWS.2017.106

A Multi-task Decomposition and Reorganization Scheme for Collective Computing Using Extended Task-Tree

Zhenhua Zhang[1], Yunlong Zhao[1(✉)], Yang Li[2], Kun Zhu[1], and Ran Wang[1]

[1] Nanjing University of Aeronautics and Astronautics, Nanjing 211106, China
{zzh2016, zhaoyunlong}@nuaa.edu.cn
[2] Harbin Engineering University, Harbin 150001, China

Abstract. Task management has always been a key issue in collective computing, including task decomposition, distribution, execution and results integration, but there is little research on task decomposition. In order to improve multi-tasks execution efficiency and promote the full utilization of collective resources, a task decomposition model based on extended task-tree is proposed in this paper. Meanwhile, a series of pruning and reorganization algorithms are proposed, and the performance of the algorithms is analyzed and evaluated. Experiments verify that the proposed algorithms outperform traditional methods, and prove that the practicality and efficiency of the strategy.

Keywords: Task decomposition · Task reorganization · Collective computing

1 Introduction

The rapid advance in diverse computing devices, cloud technologies, and human-computer interaction leads to a new computing framework referred to as collective computing [1]. In collective computing, a large number of computing devices self-organizing and cooperating to complete various types of computing tasks. Comparing to the previous framework, the collective computing has some advantages over ubiquitous computing as: (1) more resources, extensible devices and the types of that; (2) more extensive tasks support without specialized infrastructure; (3) more intelligent since it involves human participants and (4) high scalability and spatiotemporal coverage. The collective computing could provide services for many kinds of works through the network of many computing resources integrated and collaborated.

The collective computing is efficient due to provide many devices for parallel performing different parts of a task. This paper decomposes tasks to take advantage of these devices to reduce the total completion time of the tasks. Due to the workload of some tasks is too large to be completed independently by a single device, the task can be successfully decomposed by multiple devices. After the task is decomposed, it can be successfully completed by multiple devices. This framework leads huge tasks into one platform, which makes lots of the same part of different tasks perform repetitively also wasting many resources. However, task decomposition avoids repetitive execution

© Springer Nature Switzerland AG 2019
S. Li (Ed.): GPC 2018, LNCS 11204, pp. 226–240, 2019.
https://doi.org/10.1007/978-3-030-15093-8_16

of the same part of different tasks. Furthermore, the task may occupy many unnecessary resources for this moment or deploy some resources frequently producing excessed consumptions like in communication if the decomposition of the task is coarse-grained or fine-grained excessively. However, most researches adopt the form of "AND/OR Tree" [10–14], the sequence relationship between tasks is not reflected, the decomposition and pruning operations are not thorough enough.

In this paper, we introduce a realistic strategy to decompose the task and assess its rationality. Our contribution is as follows, we present a universal task decomposition model based on task-tree for managing the association among subtasks. We use *Jaccard* similarity coefficient [15] of the keywords to prune the unnecessary tasks optimally. And a reorganization mechanism keeps equilibrium of the tasks, in order to avoid to abuse needless resources. There are three main stages to complete task decomposition including: task-tree initialization, tree pruning, and task reorganization. Finally, we prove the practicality and efficiency of the strategy by implementation and performance evaluation.

The rest of this paper is organized as follows. The related work is discussed in Sect. 2. We formulate the problem and introduce the task decomposition model in Sect. 3. Section 4 describes the detailed relationship of the tasks in task-tree construction. The process of task-tree optimization shows in Sect. 5. We present the system implementation and experimental result in Sect. 6. Finally, we conclude the paper in Sect. 7.

2 Related Work

Abowd proposed that the collective computing as the fourth generation of computing technology following ubiquitous computing [1]. The collective computing is an emerging research area with open issues and challenges. The collective computing is a computational model of human collectives and machine collectives collaboration, which deals with the complex tasks that existing computing technology cannot accomplish by integrating a large number of unknown users (human collectives) and computing resources (machine collectives) on the internet [2].

The research in collective computing mainly focuses on the field of task assignment, privacy protection, incentive mechanism and so on, but few researches on task decomposition. So this paper draws on a lot of task decomposition literature from the many fields such as multi-agent system, cloud computing etc.

A main task is decomposed into simpler subtasks, which will be processed separately, and the results of subtasks will be combined to obtain the final output [3]. The most typical method is sequential decomposition, which is characterized by the fact that the output of one subtask is the input of the next subtask. The completion of each phase requires three steps, including identification, filtering and extraction [4]. Hirth et al. put forward the process of data quality for each phase, collect all the subtasks from the first stage and assign them to the workers in the second stage for evaluation, and in the third stage, generate the final output. In [5], the model is trained first, the refined, and finally classified according to the quality of the result and submitted. In the literature [6], all the images of the food are first labeled. In the second phase, every image is identified by description, match and vote. In the final stage, estimate the number of each food item identified. Another typical approach is concurrent processing, which is a method that will execute a set of independent jobs concurrently to save the execution time. For example, SCRIBE, a system that provides instantaneous captions for the deaf in real time [7], in which the main task is decomposed into subtasks and assigned to different people for executed concurrently. When a worker starts recording, different subtitle workers try to capture different recording sections, which are merged and sent back to the user. Separate sequential decomposition or parallel decomposition is relative to a particular application, not a generic form.

The literature [8] explores the task decomposition between local resources and cloud from the perspective of non-cooperative game and proves the existence and uniqueness of Nash equilibrium. The literature [9] constructs a game model of task decomposition among cloud users by solving the Nash equilibrium solution. AND/OR tree is adopted to decompose the task [10–14], it is difficult to reflect the sequential relationship between tasks. When pruning the tree, only pruning the tasks of "or" association, and the pruned tasks are not reorganized, and the model performance is very limited.

3   Problem Formulation

In this section, we first present a task decomposition model, and then describe the problem formally.

3.1   Task Decomposition Model

The main reasons for the decomposition of tasks in a collective computing environment are:

(1) The workload of some tasks in collective computing is so large that a single entity cannot be completed independently within a limited period of time;
(2) After the task is decomposed, the degree of parallelism can be improved to shorten the tasks' completion time;
(3) After the task is decomposed into multiple subtasks, the task granularity is reduced, which is conducive to the further integration of the tasks so as to avoid repeated execution of the same tasks.

Based on the above reasons, this paper constructs a task decomposition model. As shown in Fig. 1, a task decomposition model with hierarchical structure adopting task-tree. Different from the previous works, the model contains three kinds of associations: "Serial", "Concurrent" and "Or".

In the process of constructing a task decomposition tree, each task is decomposed into the lower layer according to one of the three decomposition methods ("Serial", "Concurrent", "Or"), until the task is decomposed into irreducible atomic tasks.

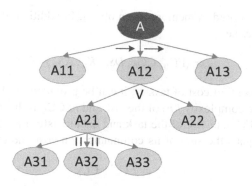

Fig. 1. Task-tree

According to the characteristics of collective computing and multi-task coordination, the following principles should be followed in task decomposition process:

(1) Independence. The subtasks obtained after decomposition should have certain independence so as to help each entity handle each task independently and reduce the cost of coordination and communication between each other;

(2) Hierarchy. A task can be decomposed into multiple subtasks; subtasks also can be decomposed into multiple lower layer subtasks. The tasks are decomposed from complex to simple, a complex task can be divided into multiple simple subtasks;

(3) Combination. A larger task can be completed through the proper combination of subtasks, another task can be completed by the appropriate transformation;

(4) Equilibrium. The overhead of each subtask should be balanced as much as possible to avoid the execution time of each subtask is greatly different and affect the overall system efficiency.

3.2 Problem Formulation

Due to the large number of tasks and the large differences in the size of tasks in the environment of collective computing, it is considered to decompose and reorganize several tasks to shorten the overhead of the tasks. Meanwhile as much as possible to avoiding over-decomposition, so as to reduce the number of tasks obtained by decomposition.

In this paper, we decompose and reorganize the tasks by the task-tree model, so the first goal is to find a unified and optimal construction method of task-tree. The reasonable construction of the task-tree facilitates the programming language description of the tasks and further facilitates the optimization of the task-tree, including operations such as pruning and composition. After getting a task-tree of a task decomposition process, the second goal is to prune the task-tree to get the best result.

Assuming that the main task set posted by task publisher is represented as $T = \{T_1, T_2, \ldots, T_n\}$. For all the main task T can also be completely decomposed into a set of subtasks $S = \{s_1, s_2, \ldots, s_m\}$. For the each task $t_k \in T \cup S (1 < k < n + m)$ in the set of system tasks is stored in memory as an object. In addition, the paper formalizes each task into 5-tuple, i.e.

$$t_k = \langle EC_{t_k}, CC_{t_k}, DS_{t_k}, K_{t_k}, AS_{t_k} \rangle \tag{1}$$

where EC_{t_k} is the execution cost of task t_k, it can be provided by the task publisher or calculated by the time complexity $f(n)$ of the task and n; CC_{t_k} is the communication cost that depends on the I/O data size of the task and data transfer rate; Moreover, the total cost of the task t_k equals the sum of its communication cost and execution cost, i.e.

$$Cost_{t_k} = EC_{t_k} + CC_{t_k} \tag{2}$$

$DS_{t_k}(DS_{t_k} \in S)$ represents a direct subtasks collection of task t_k after a single decomposition. Note that the direct subtasks collection DS_{t_k} can be empty, and called task t_k a leaf task, otherwise it is an internal task; K_{t_k} is the keywords collection of the publisher's description for task t_k; AS_{t_k} is association between direct subtasks of task t_k, including "Serial", "Concurrent", "Or".

4 Task-Tree Construction

In this section, we introduce the process of tasks decomposition and the task-tree construction.

In collective computing, a large number of tasks often emerge in a short period. Most of the tasks are heterogeneous and come from different publishers. However, these tasks are often related. Decomposing and reorganizing these tasks can reduce the total cost of performing all tasks to a certain extent. The typical task decomposition process is to use "AND/OR" tree in the form of hierarchical decomposition. However, this method of decomposition has some limitations, because it cannot well represent the order of execution of the task, and is not conducive to determine whether the task is over-decomposed. For these reasons, some problems need to be solved, such as determining the way of task-tree construction and determining the degree of task decomposition based on the task-tree so as to avoid excessive execution overhead and excessive number of tasks caused by excessive decomposition of the task.

In this paper, the structure of a task-tree is proposed, it can well represent the order relationship between tasks. Adopting this structure can effectively help us solve the mentioned problems above. From the above we can see that the construction of task-tree is the first step of task decomposition. As shown in Fig. 1, this kind of task-tree decomposes tasks according to three kinds of associations, namely "Serial", "Concurrent" and "Or", denoted "→", "‖" and "$\bigvee$" respectively:

(1) $t_i \rightarrow t_j$, means that t_i and t_j must be executed sequentially, and t_j must be executed after t_i has been executed. A parents task is decomposed into a series of serial subtasks, the input of the latter subtask depends on the output of the previous task. Moreover, the parent task is completed when all its subtasks are completed, and the completion time of the parents task depends on the sum of the completion times of all the subtasks.

(2) $t_i \| t_j$, means that t_i and t_j can be executed concurrently without affecting each other. A parents task is decomposed into a series of concurrent subtasks, there is no execution order restriction between them, so have better independence. Moreover, the parent task is completed when all its subtasks are completed, and the completion time of the parents task depends on the completion time of the largest subtask.

(3) $t_i \vee t_j$, means that one of t_i and t_j need be executed. If a parents task is decomposed into a series of "or" subtasks, both these subtasks and their parents task should have the same inputs and outputs, the difference is the process of handling the task. Therefore, the subtasks have the same communication costs but different execution costs. Moreover, the parent task is completed when one of its subtasks are completed, and the completion time of the parents task depends on the completion time of the minimum subtask.

Therefore, the task decomposition method of this paper is defined as: the tasks publishers decompose their main tasks according to the three kinds of associations mentioned above, and then the newly acquired subtasks are further decomposed into lower layers, and repeat this step until the tasks cannot be further divided. Eventually, the system will link up all the main tasks released by the task publishers over a period of time to form the root task of the task-tree. This paper has some definitions and conventions on the process of the tasks decomposition:

Definition 1 (Complete Decomposition). *The process of a task being completely decomposed into some irreducible atomic tasks, called Completely Decomposed.*

Definition 2 (Direct Subtasks). *The direct subtasks of a task is the set of all subtasks obtained after the task has been decomposed one time, excluding the subtasks that continue to be decomposed downwards. In the task-tree, a collection of all subtasks that appear to be adjacent to the task. Note that the number of direct subtasks for any task should be greater than 2 (i.e. $|DS_t| \geq 2$), otherwise the task is treated as has not been decomposed.*

Definition 3 (Leaf Tasks). *If a task has no subtasks, called a leaf task, which is the leaf node of the task-tree. In our system, all leaf tasks will be executed.*

Definition 4 (Internal Tasks). *If a task can continue to be decomposed into multiple subtasks, called an internal task, which is the internal node of the task-tree. Internal tasks are not executed directly by the system. The completion of an internal task depends on whether all leaf tasks in its subtasks are executed successfully.*

Convention 1. *All direct subtasks of each task in the task-tree must have only one kind of association.*

As shown in Fig. 2(a), there are multiple associations between all direct subtasks of a task, which should be transformed into the form shown in Fig. 2(b) when the task-tree is initialized.

Convention 2. *The association between the subtasks of the adjacent layers in the task-tree should be different.*

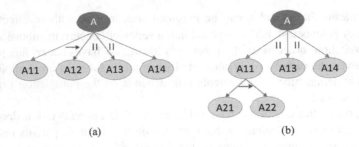

Fig. 2. (a) Multiple associations in the same layer. (b) The transformed result of (a).

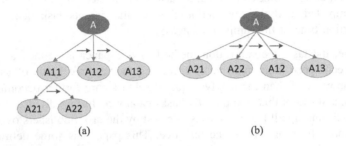

Fig. 3. (a) The same association between adjacent layers. (b) The transformed result of (a).

As shown in Fig. 3(a), the adjacent layers have the same association, which should be transformed into the form shown in Fig. 3(b) when the task-tree is initialized.

5 Task-Tree Optimization

In this section, we optimize the task decomposition process, in other words, prune and reorganize the task-tree properly. The entire process includes the initialization of the task-tree, similar tasks detection, task-tree pruning, and tasks reorganization and so on. We will introduce the core algorithms of each optimization process in detail. The optimization process greatly reduces the cost of tasks and the number of subtasks.

5.1 Task-Tree Initialization

As the first step in the optimization process, the initialization is the basis for all subsequent optimization processes. At this step, the system receives a large number of tasks from the publisher during a short period of time, as well as the preliminary decomposition results of their tasks. However, task decomposition provided by task publishers is often disorganized and does not meets system requirements. Therefore, it is necessary to reconstruct the respective task decomposition results provided by the publisher according to the Conventions in Sect. 4 to form a new task-tree that meets the system requirements.

Eventually, the system will link up all the main tasks released by the task publishers over a period of time to form the root task of the task-tree. Later in this section, we will minimize the cost of the root task of the task-tree and, on this basis, try to reduce the number of leaf tasks in the tree.

5.2 Similar Tasks Detection

As the collective computing system, it has a large number of heterogeneous tasks from different publishers. However, when the system receives a publisher's new task, it can try to detect whether a similar task has been executed successfully in history. In this paper, tasks similarity detection is not the focus of our attention, so we simply calculate the *Jaccard* similarity coefficient [15] according to the keywords set that the publisher describes for their tasks, so as to determine whether there is similarity between the new task and the historical tasks. The *Jaccard* similarity coefficient is denoted as:

$$SIM(K_{t_{new}}, K_{t_{hist}}) = \frac{|K_{t_{new}} \cap K_{t_{hist}}|}{|K_{t_{new}} \cup K_{t_{hist}}|} \tag{3}$$

where $K_{t_{new}}$ and $K_{t_{hist}}$ represent the keywords set of new task and the historical tasks respectively. If the *jaccard* similarity coefficient between a historical task and a new task is greater than a threshold θ_{SIM}, so the two tasks are considered highly similar and the historical task is pushed to the publisher to decide whether the two tasks are same tasks. If they are same tasks, the data of the historical task can be used directly as a result of the new task to avoid repeated execution of the same tasks to further improve the system efficiency. By decomposing the task, the granularity of task can be reduced, so that the detection rate of similar tasks can be improved. In the process of task decomposition, if we encounter the same task, we can prune it directly.

5.3 Task-Tree Pruning

In this section, we consider pruning the task-tree. The reason is that tasks may be redundant during the decomposition process. Unnecessary decomposition, not only doesn't reduce the tasks' cost, but also increase the number of tasks. However, in the collective computing system, the more the number of tasks, the more the number of the entities needed to carry out the tasks, which inevitably increases the burden of the system.

Post-decomposition Cost Calculation. We can get the cost of each task before it is decomposed according to Eq. (2). However, before pruning a task-tree, we also need to calculate the cost of each task after it has been completely decomposed. Here, we define the *Post-decomposition Cost* of the task:

Definition 5 (Post-decomposition Cost). *The cost of the task being decomposed is called Post-decomposition Cost, denoted as DC_t. It is equal to the sum of the post-decomposition communication cost and post-decomposition execution cost. The following represents it with **d-cost**.*

In fact, there is no necessary relation between the *Post-decomposition Cost* and the *Cost* of the task that have not been decomposed. For a leaf task that cannot be further decomposed, its *Post-decomposition Cost* equals *Cost*. For an internal task, its *Post-decomposition Cost* may be greater or less than *Cost* according to the different associations.

According to the analysis in Sect. 4, if a task can be decomposed into a series of serial subtasks, the *d-cost* of the task depends on sum of the *d-cost* of all the subtasks, and minus the cost of similar tasks, formulated as:

$$DC_t = \sum_{i=1}^{|DS_t|} DC_{\{DS_t^i - T_{SIM}\}} \tag{4}$$

Subject to:

$$T_{SIM} = \{t | SIM(K_t, K_{t_{hist}}) > \theta_{SIM}, t \in T \cup S\} \tag{5}$$

where $|DS_t|$ is the number of the direct subtasks of the task t; DS_t^i is the ith direct subtask of the task t; $\{DS_t^i - T_{SIM}\}$ represents the direct subtasks collection of task t without the tasks in similar tasks collection T_{SIM}.

If a task can be decomposed into a series of concurrent subtasks, the subtasks can be executed at the same time, but the task's completion time depends on the largest subtask, formulated as:

$$DC_t = max\{DC_{\{DS_t^i - T_{SIM}\}} | i = 1, 2, \ldots, |DS_t|\} \tag{6}$$

Subject to Eq. (5).

Similarly, if a task can be decomposed into a series of "or" subtasks, then all the subtasks can get the same output, so we should pick the task with the minimum *d-cost*, formulated as:

$$DC_t = \begin{cases} min\{DC_{DS_t^i} | i = 1, 2, \ldots, |DS_t|\} & \text{if } \forall DS_t^i \notin T_{SIM} \\ 0 & \text{if } \exists DS_t^i \in T_{SIM} \end{cases} \tag{7}$$

Subject to Eq. (5).

Note that, we should calculate the *d-cost* in the task-tree from the bottom up, starting at the leaf task.

"Serial" Tasks Pruning. For an internal task t_{in}, if its subtasks' association is "Serial", i.e. $AS_{t_{in}} = $ "*Serial*", and then comparing the *d-cost* and *Cost*, that is, to compare the cost of the task before and after decomposition. If $DC_{t_{in}} > Cost_{t_{in}}$, then all of its subtasks $S_{t_{in}}$ will be pruned out, i.e. $S_{t_{in}} = \emptyset$, otherwise, not pruning. While a task's serial subtasks are pruned out, the internal task becomes a leaf task, and its *d-cost* should be recalculated as $DC_{t_{in}} = Cost_{t_{in}}$. The pseudo code is presented in Algorithm 1. The algorithm complexity is $O(n)$. Here n represents the number of internal tasks.

After pruning of serial tasks, the minimum d-$cost$ of all the tasks in the task-tree is finally obtained, and the subsequent pruning operations will not reduce the tasks' total cost any more. Therefore, the serial tasks pruning operation began before other pruning operations is necessary. The "Or" tasks pruning will preserve the subtask with the minimum d-$cost$. The "Concurrent" tasks pruning will reduces the number of leaf tasks without reducing the total d-$cost$ of the task-tree.

"Or" Tasks Pruning. The "Or" subtasks represent different solutions to their parents task, so pruning the "Or" branch only requires that the branch with the minimum d-$cost$ be retained. After pruning, the parents task has only one direct subtask, so we need to replace the parents task with the subtask so that it satisfies Definition 2. After the replacement, if the situation shown in Fig. 3 appears, we should consider moving the subtasks further up to meet the Convention 2. The pseudo code is presented in Algorithm 2. The algorithm complexity is (n). Here n represents the number of internal tasks.

"Concurrent" Tasks Pruning. The reason for pruning concurrent subtasks is that the d-$cost$ of the parents task depends on the d-$cost$ of the largest subtask. Therefore, it is possible to reduce the degree of decomposition of the task appropriately and not even decompose the task if the d-$cost$ of other tasks is less than the maximum d-$cost$.

First, for a parents task t which $AS_t = $ "$Concurrent$", excluding its direct subtasks with the maximum d-$cost$ (max_{DS_t}), if there are some non-leaf subtasks in the remaining direct subtasks of task t that are less than the maximum d-$cost$ and can continue to be decomposed, performing the following operations on these subtasks.

Assuming that the ith direct subtask of the task t is DS_t^i, satisfying $DC_{DS_t^i} < max_{DS_t}$ and DS_t^i is an internal task. The first is to determine whether the task DS_t^i is less than the maximum d-$cost$ if it is has not been decomposed, i.e. $Cost_{DS_t^i} < max_{DS_t}$. If the result is true, the decomposition operation is stopped, the branch of the task DS_t^i is completely pruned out (i.e. $DS_t^i = \emptyset$), and the task's d-$cost$ is modified, i.e. $DC_{DS_t^i} = Cost_{DS_t^i}$; If the result is false, continue to decompose the task DS_t^i to the lower layer, and constantly update the task's d-$cost$, until the d-$cost$ is less than the maximum d-$cost$ (i.e. $DC_{DS_t^i} < max_{DS_t}$), and then prune the extra branches. Of course, every time the decomposition, the subtask DS_t^i and all of its subtasks require a bottom-up recalculation of the d-$cost$, according to Eqs. (4–7). The pseudo code is presented in Algorithm 3. The algorithm complexity is $O(n^2)$. Here n represents the number of internal tasks.

5.4 Tasks Reorganization

After pruning the task-tree, the tasks need to be reorganized to further reduce the number of leaf tasks. For a task t, if it can be decomposed into a series of concurrent direct subtasks DS_t and the maximum task d-$cost$ in the subtasks is max_{DS_t}, the remaining leaf tasks are combined to form new leaf tasks, and the new leaf tasks' cost must be less than max_{DS_t}. This can be considered as the minimum set covering problem (SCP). For example, if all four concurrent direct subtasks of a task t are $DS_t = \{t_1, t_2, t_3, t_4\}$, and $\{DC_{t_1} = 10, DC_{t_2} = 5, DC_{t_3} = 4, DC_{t_4} = 3\}$, so the maximum d-$cost$ is $max_{DS_t} = DC_{t_1} = 10$. So if all the remaining subtasks excluding task t_1, we can get three minimum set covering results: $r_1 = \{\{t_1\}, \{t_2, t_3\}, \{t_4\}\}$, $r_2 = \{\{t_1\}, \{t_2\}, \{t_3, t_4\}\}$, $r_3 = \{\{t_1\}, \{t_3\}, \{t_2, t_4\}\}$. Finally, taking into account the equilibrium, select the combination r_2 with the smallest variance $var_{r_2} = 4.22$. The pseudo code is presented in Algorithm 4. The algorithm complexity is $O(n \cdot 2^m)$. Here n represents the number of internal tasks, m represents the number of subtasks per internal tasks. Note that SCP was originally a NP-hard problem. But in our problem, a task is only decomposed into a limited number of subtasks, so the result of the minimal set covering can be traversed in polynomial time.

Algorithm 1: "Serial" Tasks Pruning Algorithm.	**Algorithm 2**: "Or" Tasks Pruning Algorithm.
Input: The initial task-tree with d-$cost$ calculated by Eq. (4~7). **Output**: The task-tree that has been pruned. 1. For each internal task t_{in}, run the following programs from the bottom up in the task-tree: 2. If $AS_{t_{in}} = "Serial"$ & $DC_{t_{in}} > Cost_{t_{in}}$: 3. (pruning:) 4. $DS_{t_{in}} = \emptyset$ 5. $DC_{t_{in}} = Cost_{t_{in}}$	**Input**: The output of algorithm 1. **Output**: The task-tree that has been pruned. 1. For each internal task t_{in}, run the following programs from the top to bottom in the task-tree: 2. If $AS_{t_{in}} = "Or"$: 3. Find the direct subtask t_{min} with minimum d-$cost$, and the rest of subtasks are pruned out. 4. Replace t_{in} with t_{min}, i.e. $t_{in} = t_{min}$. 5. If $AS_{t_{in}} = AS_{t_{in}'s \, parents \, task}$: 6. Replace t_{in} with $DS_{t_{in}}$: $t_{in} = DS_{t_{in}}$
Algorithm 3: "Concurrent" Tasks Pruning Algorithm.	**Algorithm 4**: Tasks Reorganization Algorithm.
Input: The output of algorithm 2. **Output**: The task-tree that has been pruned. 1. For each internal task t_{in}, run the following programs from the top to bottom in the task-tree: 2. If $AS_{t_{in}} = "Concurrent"$: 3. Get the maximum d-$cost$ in the subtasks, denoted as $max_{DS_{t_{in}}}$ and its corresponding task is t_{max}. 4. The other internal tasks t in $DS_{t_{in}}$, redecompose t from top to bottom, and update DC_t constantly. 5. Until $DC_t < max_{DS_{t_{in}}}$: 6. Prune the extra branches.	**Input**: The output of algorithm 3. **Output**: The task-tree that has been reorganized. 1. For each internal task t_{in}, run the following programs in the task-tree: 2. If $AS_{t_{in}} = "Concurrent"$: 3. Get the maximum d-$cost$ in the subtasks, denoted as $max_{DS_{t_{in}}}$ and its corresponding task is t_{max}. 4. Get all subsets, while subset $A_i \subset \{t | DS_{t_{in}} - t_{max}, t \, is \, a \, leaf \, task\} = B$, and $Cost_{A_i} < max_{DS_{t_{in}}}$. 5. Select several subsets as a combination, their union set equals B, and the intersection between any two subsets is $\emptyset$. 6. Calculate the variance of all the combinations obtained by step 5, preserving a combination of the minimum variance.

6 Evaluations

In this section, we evaluate the performance of our proposed algorithms. In order to verify the superiority of the algorithm in the number of subtasks and the overall cost of the system, we validate the results through the collective computing simulation experiments with multitask.

6.1 The Parameters of Simulation Experiments

This paper adopts the task-tree to describe the decomposition of the task. In fact, the initial task-tree should be provided by the tasks' publishers with the tasks' configuration files. Nevertheless, in our simulation environment, the task-tree is randomly constructed by our system according to the rules of Sect. 4. The values and descriptions of some of these parameters are shown in Table 1.

Table 1. Parameter settings.

Parameter	Description	Value
l_{max}	The maximum number of layers of task-tree	8
num	The number of subtasks that can be decomposed for each task	2–5
num_{main}	The number of main tasks published by the task publishers that the system received during a fixed period of time	2–10
$Cost_{in}$	The input cost of main task (unit: s)	5–20
$Cost_{out}$	The output cost of main task (unit: s).	5–20
$Cost_{exe}$	The execution cost of leaf task (unit: s)	5–20
p_{de}	The probability of a task being decomposed	0.5
p_{or}	The probability of a task being "Or" decomposed	0.05
p_{se}	The probability of a task being "Serially" decomposed	0.15
p_{co}	The probability of a task being "Concurrently" decomposed	0.3

Some other explanations for simulation experiments are as follows:

(4) We create a new virtual root task as the parent task to link all the main tasks published by the task publishers. Ultimately, the system gets the minimum *d-cost* of the root task, as well as the minimum number of leaf tasks.

(5) The I/O cost of each main task is randomly generated by the system according to the normal distribution. In simulation environment, the I/O costs of other tasks are randomly allocated by the main task. The communication cost of each task is equal to the sum of its input cost and output cost, in seconds. In fact, the communication cost equals the quotient of the size of communication and the data transfer rate.

(6) The execution cost of each leaf task are randomly generated by the system according to the normal distribution. In simulation environment, the execution costs of the other tasks are calculated by leaf tasks.

Our simulation environment fully considers the actual situation, and the parameter settings are reasonable and adjustable. Therefore, it has a great reference value for the actual situation.

6.2   Analysis of Result

In this simulation experiment, the system randomly generated 4 main tasks simulate the tasks publisher. First of all, the four tasks are independent before decomposed, so they can be executed concurrently. By linking these four tasks to the root task, the cost of root task depends on the largest tasks in the four tasks. Then the four tasks are completely decomposed to get the new cost of the root task. And then pruning and reorganizing the task-tree after decomposed completely, each step can get the new number of leaf tasks and the new cost of root task.

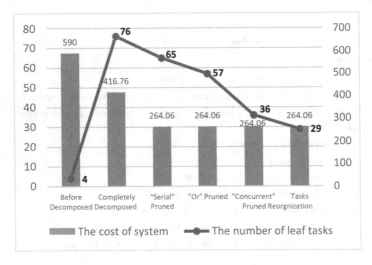

Fig. 4. The result of simulation

This simulation result is shown in Fig. 4. As we can seen from the figure, the total system cost is reduced by 29.4% after the main tasks are completely decomposed. After the task-tree was pruned and reorganized, the total system cost was further reduced by 36.6%. However, for the number of leaf tasks, the number of leaf tasks increased by 19 times after completely decomposed. But in collective computing, in order to reduce the total system cost as much as possible, it is bound to bring about an increase in the number of leaf tasks when decomposing a task, which is unavoidable. However, the number of leaf tasks decreased by 61.8% after pruning and reorganization.

It is worth nothing that the total system cost after does not decrease because the cost of the parents task for a series of "or" subtasks is stored according to the cost of the smallest subtasks from the beginning.

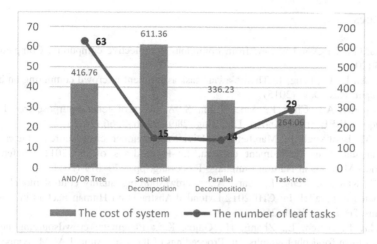

Fig. 5. The comparison between our model and other typical models.

The result of our model compared with the traditional method is shown in Fig. 5. The traditional method adopted a "AND/OR" tree model that decomposed the tasks completely, and only pruned the "or" tasks, so resulting in a great deal of redundancy. It is difficult for a "AND/OR" tree model to see the sequential relationship of tasks from the model. Other typical methods include sequential decomposition and parallel decomposition, but they are only a single form of decomposition and do not perform any pruning and reorganization operations. The task-tree model in this paper preserves the sequential relationship between tasks, which is helpful to pruning and reorganizing tasks. As we can see from Fig. 5, our model are superior to the traditional model in terms of both the number of leaf tasks and the cost of the system.

7 Conclusion

This paper puts forward not only the construction of our task-tree model based on hierarchical structure, but also some optimization methods such as pruning and reorganization of the task-tree. Compared with the traditional method of task decomposition, our model can extract the order of execution between tasks and further reduce the number of leaf tasks while keeping the total cost of the task to a minimum. The final experiment shows that our work has good performance. It can reduce the total system cost of 55%, compared with the traditional method, it can reduce the number of leaf tasks by 54%. However, it is necessary to further study the task decomposition methods in the collective computing environment, considering the heterogeneity of the execution entities and the incentive mechanism.

Acknowledgement. This research was supported by Defense Industrial Technology Development Program under Grant No. JCKY2016605B006, Six talent peaks project in Jiangsu Province under Grant No. XYDXXJS-031.

References

1. Abowd, G.D.: Beyond weiser: from ubiquitous to collective computing. Computer **49**(1), 17–23 (2016)
2. Zhang, X., Li, G., Feng, J.: Theme-aware task assignment in crowd computing on big data. J. Comput. Res. Dev. (2015)
3. Chittilappilly, A.I., Chen, L., Amer-Yahia, S.: A survey of general-purpose crowdsourcing techniques. IEEE Trans. Knowl. Data Eng. **28**(9), 2246–2266 (2016)
4. Negri, M., Bentivogli, L., Marchetti, A.: Divide and conquer: crowdsourcing the creation of cross-lingual textual entailment corpora. In: Proceedings of the 2011 Conference on Empirical Methods in Natural Language Processing, pp. 670–679 (2011)
5. Zhu, S., Kane, S., Feng, J., Sears, A.: A crowdsourcing quality control model for tasks distributed in parallel. In: CHI 2012 Extended Abstracts on Human Factors in Computing Systems, 2012, pp. 2501–2506
6. Noronha, J., Hysen, E., Zhang, H., Gajos, K.Z.: Platemate: crowdsourcing nutritional analysis from food photographs. In: Proceedings of the 24th Annual ACM Symposium on User Interface Software and Technology, pp. 1–12 (2011)
7. Lasecki, W., et al.: Real-time captioning by groups of non-experts. In: Proceedings of the 25th Annual ACM Symposium on User Interface Software and Technology, pp. 23–34 (2012)
8. Nahir, A., Orda, A., Raz, D.: Workload factoring with the cloud: a game-theoretic perspective. Proc. - IEEE INFOCOM **131**(5), 2566–2570 (2012)
9. Tao, J., Xiao-Hong, W.U., Yong-Gen, G.U.: Study of Cloud Computing Task Factoring Based on Game Theory. Sci. Technol. Eng. (2013)
10. Zhang, R.G., Liu, J., Zhang, J.F., et al.: Study on task decomposition and coordination modeling in product concurrent design based on multi-agent system. J. Taiyuan Heavy Mach. Inst. **23**(2), 166–169 (2002)
11. Song, J.P.: An improved algorithm to solve the task partition problem in MDOCEM. J. Hubei Univ. (2007)
12. Zeng, X.S., Song, M.Y., Xiao-Bo, Z.: Research of the algorithms for task cooperation execution based on multi-agent system. J. Comput. Appl. **26**(8), 1918–1922 (2006)
13. Xiao, Z.L., Yue, X.B., Zhou, H.: Multi-agent task decomposition algorithm based on and-or dependence graph. Comput. Eng. Des. **30**(2), 426–428 (2009)
14. Qing-shan, L.I., et al.: Collaboration strategy for software dynamic evolution of multi-agent system. J. Central South Univ. **22**(7), 2629–2637 (2015)
15. Niwattanakul, S., Singthongchai, J., Naenudorn, E., et al.: Using of Jaccard Coefficient for Keywords Similarity. Lecture Notes in Engineering & Computer Science, vol. 2202(1) (2013)

CompetitiveBike: Competitive Prediction of Bike-Sharing Apps Using Heterogeneous Crowdsourced Data

Yi Ouyang[1], Bin Guo[1]($\boxtimes$), Xinjiang Lu[1], Qi Han[2], Tong Guo[1], and Zhiwen Yu[1]

[1] Northwestern Polytechnical University, Xi'an, Shaanxi, China
guob@nwpu.edu.cn
[2] Colorado School of Mines, Golden, CO, USA

Abstract. In recent years, bike-sharing systems have been deployed in many cities, which provide an economical lifestyle. With the prevalence of bike-sharing systems, a lot of companies join the market, leading to increasingly fierce competition. To be competitive, bike-sharing companies and app developers need to make strategic decisions for mobile apps development. Therefore, it is significant to predict and compare the popularity of different bike-sharing apps. However, existing works mostly focus on predicting the popularity of a single app, the popularity contest among different apps has not been well explored yet. In this paper, we aim to forecast the popularity contest between Mobike and Ofo, two most popular bike-sharing apps in China. We develop CompetitiveBike, a system to predict the popularity contest among bike-sharing apps. Moreover, we conduct experiments on real-world datasets collected from 11 app stores and Sina Weibo, and the experiments demonstrate the effectiveness of our approach.

Keywords: Bike-sharing app · Mobile app · Competitive prediction · Popularity contest · Crowdsourced data

1 Introduction

In recent years, shared transportation has grown tremendously, which provides us an economical lifestyle. Among the various forms of shared transportation, public bike-sharing systems [1–3] have been widely deployed in many metropolitan areas (e.g. New York City in the US and Beijing in China). A bike-sharing system provides short-term bike rental service with many bicycle stations distributed in a city [4]. A user can rent a bike at a nearby bike station, and return it at another bike station near his/her destination. The worldwide prevalence of bike-sharing systems has inspired lots of active research, such as bike demand prediction [5–7], bike rebalancing optimization [8], and bike lanes planning [9].

More recently, station-less bicycle-sharing systems are becoming the mainstream in many big cities in China such as Beijing and Shanghai. Mobike[1] and

---

[1] https://en.wikipedia.org/wiki/Mobike.

© Springer Nature Switzerland AG 2019
S. Li (Ed.): GPC 2018, LNCS 11204, pp. 241–255, 2019.
https://doi.org/10.1007/978-3-030-15093-8_17

Ofo[2] are two most popular station-less bicycle-sharing systems. Unlike traditional public bike-sharing systems, station-less bike sharing systems aim to solve "the last one mile" issue for users. Using the Mobike/Ofo mobile app, users can search and unlock nearby bikes. When users arrive at their destinations, they do not have to return the bikes to the designated bike station. Instead, they can park the bicycles at a location more convenient for them. Therefore, it is easier for users to rent and return bikes than traditional bike-sharing systems.

As bike-sharing apps become increasingly popular, a lot of companies join the bike-sharing market, leading to fierce competition. To thrive in this competitive market, it is vital for bike-sharing companies and app developers to understand their competitors and then make strategic decisions accordingly [10] for mobile app development and evolution [11]. Therefore, it is significant and necessary to predict and compare the future popularity of different bike-sharing apps.

With the rapid development of mobile social media, heterogeneous crowd-sourced data [12,13] can bring multi-dimensional and rich information about bike-sharing apps. When users download and install a mobile app, they may submit user experience to the app store [14–16]. Specifically, users may upload their requirements (e.g. functional requirements), preferences (e.g. UI preferences) or sentiment (e.g. positive, negative) through reviews, as well as their satisfaction level through ratings. Online social media is another way to share the user experience of a mobile app. When users actually use the bike, they may share the ride experience on social media. Specifically, users may record the feeling of the ride, the advantages and disadvantages of the bike/system, or the comparison with other bikes/systems. Both users' online and offline experience will affect the popularity of the apps, thereby affecting their popularity contest outcome. Therefore, app store data and microblogging data are complementary, and can describe a mobile app from different perspectives. In this paper, we study the problem of competitive prediction of bike-sharing apps using heterogeneous app store data and microblogging data.

To the best of our knowledge, the problem of predicting the competitiveness of mobile apps has not been well investigated in the literature. There are several challenging questions to be answered. How to forecast the popularity contest outcomes of bike-sharing apps? How to extract effective features to characterize the competitiveness of bike-sharing apps from heterogeneous crowdsourced data?

To answer these questions, we propose CompetitiveBike, a system that predicts the outcomes of the popularity contest among bike-sharing apps leveraging heterogeneous app store data and microblogging data. We first obtain app descriptive statistics and sentiment information from app store data, and descriptive statistics and comparative information from microblogging data. Using these data, we extract both coarse-grained and fine-grained competitive features. Finally, we train a regression model to predict the outcomes of popularity contest. We make the following contributions.

---

[2] https://en.wikipedia.org/wiki/Ofo_(bike_sharing).

(1) This work is the first to study the problem of competitive prediction of bike-sharing apps. We use two indicators for the comparison: (i) competitive relationship to indicate which app is more popular; and (ii) competitive intensity to measure the popularity gap between the two apps/systems.

(2) To predict popularity contest between apps, we extract features from different perspectives including the descriptive information of apps, users' sentiment, and comparative opinions. Using the basic information, we further extract two novel features: coarse-grained and fine-grained competitive features, and choose Random Forest for prediction.

(3) To evaluate CompetitiveBike, we collect data about Mobike and Ofo from 11 app stores and Sina Weibo. With the data collected, we conduct extensive experiments from different perspectives. We find that the Random Forest model performs well on *competitive relationship* prediction (the Accuracy is 71.4%) as well as *competitive intensity* prediction (the RMSE is 0.1886). A combination of the coarse-grained and fine-grained competitive features improves performance in popularity contest prediction, and a combination of data from app store and microblogging also improves performance in popularity contest prediction. The results demonstrate the effectiveness of our approach.

2   Related Work

2.1   App Popularity Prediction

Recently, a significant effort has been spent on predicting popularity of mobile app [17–20]. Zhu et al. [17] proposed the Popularity-based Hidden Markov Model (PHMM) to model the popularity information of mobile apps. Wang et al. [18] proposed a hierarchical model to forecast the app downloads. Malmi [19] found that there existed connection between app popularity and the past popularity of other apps from the same publisher. Finkelstein et al. [20] found that there is a strong correlation between rating and the downloads.

Our work differs from and potentially outperforms the previous work in several aspects. First, we focus on the problem of competitive prediction of bike-sharing apps, instead of the prediction of a single app. Second, we predict the popularity contest leveraging heterogeneous crowdsourced data (i.e., app store data and microblogging data) that are often complementary and can reflect mobile app popularity contest from different perspectives.

2.2   Competitive Analysis

Competitive analysis involves the early identification of potential risks and opportunities to help managers making strategic decisions for an enterprise [10]. Jin et al. [21] selected subjective sentences from reviews which discuss common features of competing products. He et al. [22] analyzed the textual content on the social media of the three largest pizza chains, and the results revealed the business value of comparing social media content. Tkachenko and Lauw [23] proposed a generative model for comparative sentences, jointly modeling two levels of comparative relations: the level of sentences and the level of entity pairs. Zhang et al. [24] proposed to scan reviews to update a product comparison network.

These studies conduct competitive analysis simply via semantic analysis of users' opinion. In contrast, our work extracts features from different perspectives including the descriptive information of apps, user's sentiment, and comparative opinions. Using the basic information, we further extract coarse-grained and fine-grained competitive features, and train a model to predict popularity contest.

3   Data Acquisition and Analysis

3.1   App Store Data

We collected data from 11 mainstream Android app stores[3] in China, including: Wandoujia, Huawei, 360, Meizu, OPPO, VIVO, Yingyongbao, Xiaomi, Baidu, Lenovo and Anzhi market. An overview of app store data is listed in Table 1.

Table 1. Basic statistics of the app store data

Property	Statistics
App stores	11
Time span	04/22/2016–03/14/2017
Reviews of Mobike	69,228
Reviews of Ofo	13,928
Total downloads of Mobike	35,591,757
Total downloads of Ofo	30,423,077

We collected data between 04/22/2016 and 03/14/2017. At the beginning, these two apps were still relatively new and they are not as popular now, so there were not a lot of data. To ensure prediction accuracy, the actual time span of the app store data we use is from 06/20/2016 to 03/12/2017, exactly 38 weeks.

Figure 1 shows the downloads of the two apps over different lengths: daily, weekly, and monthly. We can observe that their downloads are all increasing, and for the recent months, Mobike and Ofo have more comparable downloads. The weekly records are more appropriate for competitive prediction, because the daily records fluctuate too much and the monthly records are too sparse. Therefore, we set the length of time window as one week.

---

[3] Data from Google Play is more sparse than these app stores as Mobike and Ofo users are mainly from China, so we did not collect data from Google Play.

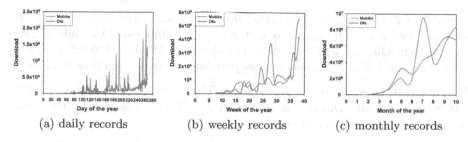

(a) daily records   (b) weekly records   (c) monthly records

Fig. 1. The daily, weekly and monthly downloads of Mobike and Ofo

3.2 Microblogging Data

We crawled three microblogging datasets from Sina Weibo[4], the most popular microblogging service in China. The first dataset was crawled by using a combination of the two keywords "Mobike" and "Ofo", we refer it as the "Mobike & Ofo". The second one was crawled by using the keyword "Mobike", we refer it as the "Mobike". The third one was crawled by using the keyword "Ofo", we refer it as the "Ofo". An overview of three datasets is listed in Table 2.

Table 2. Basic statistics of the microblogging data

Dataset	Time span	Microblogs	Users	Reposts	Comments	Likes
Mobike & Ofo	06/21/2016–03/14/2017	11,176	8,725	34,801	35,646	31,295
Mobike	04/22/2016–03/14/2017	52,718	40,187	151,126	207,926	181,560
Ofo	05/30/2016–03/14/2017	43,746	35,752	145,882	181,815	170,644

4 Problem Statement and System Framework

4.1 Problem Statement

The problem can be stated as follows: given the app store data and microblogging data about Mobike and Ofo, we want to predict which app will be more popular in the future.

Definition 1 *Popularity Contest.* Inspired by [25], the popularity of Mobike (or Ofo) can be measured by the downloads, and the popularity contest (PC) between Mobike and Ofo can be defined by the difference in their downloads D_m and D_o:

$$PC = \frac{D_m - D_o}{D_m + D_o} \tag{1}$$

---

[4] https://weibo.com/.

Definition 2 *Competitive Relationship.* The competitive relationship (CR) between Mobike and Ofo can be one of the two possibilities: (1) Mobike is more popular than Ofo, or (2) Ofo is more popular than Mobike. According to Formula (1), when $PC > 0$, Mobike is more popular; otherwise, Ofo is more popular.

Definition 3 *Competitive Intensity.* The competitive intensity (CI) between Mobike and Ofo is the absolute value of PC. The smaller the value, the higher the competitive intensity is.

Formally, we extract feature set X from app store data and microblogging data, then we want to predict the popularity contest Y. Let $X = \{x_1, ...x_N\}$ and $Y = \{y_1, ...y_N\}$, given $X^{(1:t+1)}(= \{X^{(1)}, ..., X^{(t+1)}\})$ and $Y^{(1:t)}(= \{Y^{(1)}, ..., Y^{(t)}\})$, our objective is to predict $Y^{(t+1)}$.

4.2   System Framework

The overview of the framework is illustrated in Fig. 2, which mainly consists of three layers: data preparation, feature extraction, and competitive prediction.

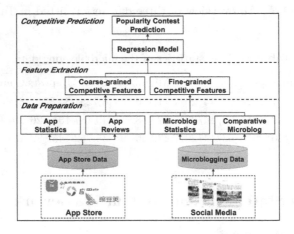

Fig. 2. System framework.

Data Preparation. We obtain app statistics and reviewers' sentiment from app store data, and microblogging statistics and comparative information from microblogging data.

Feature Extraction. To effectively extract and quantify the factors impacting mobile app popularity contest, we extract features from different perspectives including the inherent descriptive information of apps, users' sentiment, and comparative opinions. With this information, we further extract two novel sets of features: coarse-grained and fine-grained competitive features.

Competitive Prediction. With these two extracted feature sets, we train a model to predict the popularity contest between Mobike and Ofo.

5   Popularity Contest Prediction

In this section, we first analyze the factors impacting the popularity contest between Mobike and Ofo, then extract coarse-grained and fine-grained competitive features from these factors to characterize popularity contest. Finally, we train a model to predict popularity contest.

5.1   Coarse-Grained Competitive Features

Features from App Store. When users download and install a mobile app, they may submit reviews and ratings to the app store. For example, a user wrote: *"The Mobike app cannot launch today, it was still okay yesterday, what's the matter? It's terrible!"*. According to the review, we believe that app store data (e.g. reviews, ratings) can reflect users' online experience with the app. Typically, users may upload their requirements (e.g. functional requirements), preferences (e.g. UI preferences), or sentiment (e.g. positive, negative) through reviews, and they may also rate the app based on their overall satisfaction. Therefore, we extract features from reviews and ratings to characterize popularity contest.

App Statistics. Generally, the numerical statistics of reviews and ratings in each time window can reflect the popularity of the app. In other words, a bigger number of reviews and a higher rating score may indicate that the app is more popular. We use the difference between app's review number DN (and rating scores DS) to characterize popularity contest. A small value of DN (and DS) indicates that they have similar number of reviews (and rating score), thus their competition is more intense.

Sentiment Similarity. Besides numerical statistics, app reviews can express users' sentiment. We use a Chinese sentiment analyzer called SnowNLP[5] to analyze the sentiment of reviews. We calculate the sentiment value s_i of each review at time instant t_i, then we obtain the sentiment distribution vector $\mathbf{v_i} = (p_1, p_2, p_3)$ at time t_i, where p_1, p_2, p_3 is corresponds to negative, neutral and positive sentiment proportion respectively.

The extracted sentiment sequences are only for a single app, when we consider the competition between two apps, we compute sentiment similarity to capture the difference of users' sentiment about these apps, and the similarity can be measured by calculating the cosine similarity [26]. The higher similarity means that users' opinions about them are more similar, and the competition between them is more intense.

Features from Microblogging. When users ride the bike of different apps, they may share their riding experience on social media. An example of a microblog is like this: *"This is my first ride of Mobike, it is so cool!"*. We believe online social media is another way to express users' riding experience. Therefore,

---

[5] https://github.com/isnowfy/snownlp

we extract features from microblogging data to help understand the popularity contest of different apps.

Microblogging Statistics. In the "Mobike & Ofo" dataset, the number of microblogs, users, reposts, comments, and likes can reflect the attention about Mobike and Ofo on microblogging, the bigger value indicates more intense competition between Mobike and Ofo.

In the "Mobike" dataset, more microblogs that contain the keyword "Ofo" imply that Ofo is more frequently mentioned in the "Mobike" dataset. We use the ratio (R_{om}) of "Ofo" and "Mobike" to characterize the competition. Formally, $R_{om} = \frac{MN_o}{MN_m}$, where MN_o and MN_m represent the number of microblog that contains "Ofo" and "Mobike", respectively. Similarly, in the "Ofo" dataset, we use the ratio (R_{mo}) of "Mobike" and "Ofo" to characterize the competition. The higher ratios, the more intense competition.

Comparative Analysis. In addition to the numerical statistics, the textual information in microblog content is also valuable. The "Mobike & Ofo" dataset often contains the comparison between Mobike and Ofo. Let us consider a microblog: "*Mobike is too heavy, and it is uncomfortable to ride. It is also slightly expensive. Of course, there are some aspects where Mobike is better than Ofo, such as: Mobike is more solid than Ofo, and its bell is also better.*" According to this post, we observe that (1) there exists comparison between Mobike and Ofo; (2) a single microblog may compare the apps many times on different aspects (e.g. price, quality); (3) each comparison can discuss the advantages and disadvantages of the bike. Therefore, we need to address three issues in comparative analysis: (1) how to identify comparison between Mobike and Ofo; (2) how to calculate the comparison count; (3) how to determine the *comparison direction*, which means whether Mobike is better than Ofo, or Ofo is better than Mibike. We next describe our methods to address these issues.

First, the occurrences of comparative words such as "better" often indicate comparison and these comparative words are usually adjective or adverb. Therefore, to identify the comparison, we try to determine whether there exist comparative words in microblogs. Specifically, we use a Chinese lexical analyzer called Jieba[6] to annotate part of speech, and extract adjectives and adverbs to build a dictionary. We then determine whether there exist comparative words by querying the dictionary and filtering out microblogs without comparative words. After this, all the remaining microblogs contain comparison between Mobike and Ofo.

Next, when calculating the comparison count, we do not need to differentiate which aspects are in comparison. We can count the number of comparative word to determine the comparison count.

Last, the sentiment of the comparative words can be used to infer comparison direction. In the example above, "Mobike is more solid than Ofo" implies that Mobike is better than Ofo. We divide the dictionary into two sub-dictionaries: positive and negative. With a positive comparative word, 1 is added to its own score; with a negative comparative word, 1 is added to the score of the com-

---

[6] https://github.com/fxsjy/jieba

petitor. This way, we can obtain the comparison direction scores for Mobike and Ofo. We use the scores to characterize popularity contest.

5.2   Fine-Grained Competitive Features

Each coarse-grained competitive feature is a time series with time window of one week. In each time window, we extract the temporal dynamics of the coarse-grained competitive features as the fine-grained competitive features to characterize the trend of the sequence [27].

Overall Descriptive Statistics describe the basic properties of the coarse-grained competitive features from multiple aspects. We extract the mean, standard deviation, median, minimum and maximum as features.

Hopping Counts can effectively describe the "pulse" of sequence and is calculated as the number of elements whose values are greater than their next element. This feature is used to characterize the fluctuation of the sequences.

Lengths of Longest Monotonous Subsequences describe the size of gradient descent or ascent patterns in a sequence. We examine the longest monotone (including increasing and decreasing) subsequences, and use the lengths of these two subsequences to describe the tendency of the sequence.

5.3   Popularity Contest Prediction

With these two extracted feature sets, we want to predict the popularity contest in the future, we use regression-based methods. Since the extracted features are sequences, and the time window is one week, we treat successive several weeks as the training set, then compare the state-of-the-art regression models. Section 6 has the details on the models we compared and the one we eventually use.

6   Performance Evaluation

6.1   Experimental Setup

Comparison Settings. To demonstrate the effectiveness of different types of features, we divide the extracted features into two categories: (1) *coarse-grained competitive features* (CF); (2) *fine-grained competitive features* (FF).

To demonstrate the effectiveness of heterogeneous crowdsourced data, we divide the features into another two categories according to the data source: (1) *features from app store data* (AF); (2) *features from microblogging data* (MF).

Regarding algorithm comparison, in the phase of *competitive relationship* prediction, we evaluate three state-of-the-art classification algorithms: Decision Tree (DT), Adaboost and Random Forest (RF). In the phase of *competitive intensity* prediction, we evaluate two state-of-the-art regression algorithms: Support Vector Regression (SVR) and Random Forest (RF).

To conduct popularity contest prediction, we use the following setup: we use ten successive weeks as the training set and the next one week as the test set.

Baseline Algorithms. For popularity contest prediction, we use the following methods as the baselines:

- *Last_predcition*: it predicts the popularity contest using the last one week, i.e. $Y^{(t+1)} = Y^t$. We refer it as "Last".
- *CF*: it predicts the popularity contest using the coarse-grained competitive features alone.
- *FF*: it predicts the popularity contest using the fine-grained competitive features alone.
- *AF*: it predicts the popularity contest using the features from app store alone.
- *MF*: it predicts the popularity contest using the features from microblogging platform alone.

Evaluation Metrics. For popularity contest prediction, we measure the prediction performance using the following metrics:

- In the phase of *competitive relationship* prediction, we use *Accuracy, Precision, Recall, F-measure* as the evaluation metrics. Higher values of these metrics means the better performance in *competitive relationship* prediction.
- In the phase of *competitive intensity* prediction, we use *RMSE* as the evaluation metric. A smaller RMSE means the better performance in *competitive intensity* prediction.

6.2   Experimental Results

Comparison of Different Algorithms. We want to compare the effectiveness of different algorithms in popularity contest: *competitive relationship* and *competitive intensity*.

Regarding the *competitive relationship* prediction, Fig. 3 shows the Accuracy, Precision, Recall and F-measure of DT, Adaboost and RF. We observe that RF outperforms the other algorithms, with the Accuracy of 71.4%, and the state-of-the-art classification algorithms outperforms the baselines.

Regarding the *competitive intensity* prediction, Table 3 shows the RMSE of Last, SVR, and RF. We observe that RF again outperforms other algorithms, and the RMSE of the baseline is much larger than RF regression algorithm.

In summary, the state-of-the-art machine learning algorithms can train a better learning model by using the proposed features. RF performs well on *competitive relationship* prediction as well as *competitive intensity* prediction. Therefore, we choose Random Forest (RF) as the default predictor for predicting popularity contest.

Comparison of Different Features. We try to determine whether the combination of the coarse-grained and fine-grained competitive features can improve the performance of prediction. Therefore, we compare the CF, FF, and CF+FF, respectively.

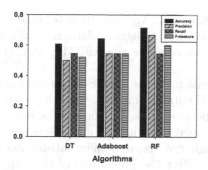

Table 3. RMSE of algorithms

Last	SVR	RF
0.2280	0.2185	0.1886

Fig. 3. Comparison of algorithms.

Figure 4 shows the Accuracy, Precision, Recall and F-measure of CF, FF and CF+FF. We observe that FF outperforms CF, with the Accuracy of 67.9%, while CF is 60.7%. This is because FF is generated based on CF, and it can reflect the fine-grained tendency of CF. Furthermore, the combination of the coarse-grained and fine-grained competitive features (CF+FF) improves the performance in *competitive relationship* prediction, compared with CF and FF alone.

Table 4 shows the RMSE of CF, FF and CF+FF. We can observe that FF outperforms CF, and can reflect the temporal dynamics of the CF. Furthermore, the combination of the coarse-grained and fine-grained competitive features (CF+FF) improves the performance in *competitive intensity* prediction, compared with CF and FF alone.

In summary, FF outperforms CF in both *competitive relationship* and *competitive intensity* prediction, and the combination of the coarse-grained and fine-grained competitive features (CF+FF) can further improve the performance in competition prediction.

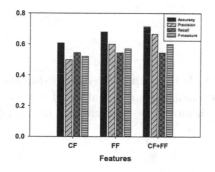

Table 4. RMSE of features

CF	FF	CF+FF
0.2059	0.1980	0.1886

Fig. 4. Comparison of features.

Comparison of Different Data Sources. We aim to determine whether the combination of app store data and microblogging data can improve the performance of prediction. Therefore, we compare the AF, MF, and AF+MF, respectively.

Figure 5 shows the Accuracy, Precision, Recall and F-measure of AF, MF and AF+MF. We can observe that AF outperforms MF, with the Accuracy of 64.3%, while MF is 60.7%. This is because that AF constitutes reviews and scores which can reflect users' online experience with the app. Users may report their sentiment or requirement through reviews, and their satisfaction degree through rating scores. It will directly affect the popularity of the app, therefore will affect the popularity contest. In contrast, MF reflects the popularity contest indirectly. Furthermore, the combination of features from app store and microblogging (AF+MF) improves the performance in *competitive relationship* prediction, compared with AF and MF alone.

Table 5 shows the RMSE of AF, MF and AF+MF. We can observe that AF outperforms MF, because AF will directly affect the popularity of the mobile app, while MF reflects the competition indirectly. Furthermore, the combination of features from app store and microblogging (AF+MF) improves the performance in *competitive intensity* prediction, compared with AF and MF alone.

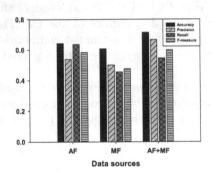

Table 5. RMSE of data sources

AF	MF	AF+MF
0.1965	0.2062	0.1886

Fig. 5. Comparison of data sources.

In summary, AF outperforms MF in both *competitive relationship* and *competitive intensity* prediction, and the combination of features from app store and microblogging (AF+MF) further improve the performance in competition prediction.

7 Conclusion

In this paper, we focus on the problem of competitive prediction over Mobike and Ofo. We propose CompetitiveBike to predict the popularity contest between Mobike and Ofo leveraging heterogeneous app store data and microblogging data. Specifically, we first extract features from different perspectives including the inherent descriptive information of apps, users' sentiment, and comparative opinions. With the basic information, we further extract two sets of novel features: coarse-grained and fine-grained competitive features. Finally, we choose the Random Forest algorithm to predict the popularity contest. Moreover, we collect data about two bike-sharing apps from 11 online mobile app stores and Sina Weibo, implement extensive experimental studies, and the results demonstrate the effectiveness of our approach.

In the future work, we will enrich our problem statement and system framework by learning from the classical economic theories on competitive analysis [28,29]. In the stage of data preparation, we will filter the fake and malicious user feedbacks [30], which may affect the prediction of popularity contest. In order to provide competitive analysis for mobile apps, we will view the mobile apps competition as a long-term event, and generate the event storyline [31] and present descriptive information regarding popularity contest to enrich the competitive analysis. Besides, we will improve the prediction model by analyzing the couplings [32,33] among features and determining their mutual influence. Morever, we will collect more categories of apps to enrich our datasets, not only focus on the popular apps, but also the new emerging apps, and extend the generality of our approach to other apps.

Acknowledgments. This work was partially supported by the National Key R&D Program of China (No. 2017YFB1001800), and the National Natural Science Foundation of China (No. 61332005, 61772428, 61725205).

References

1. DeMaio, P.: Bike-sharing: history, impacts, models of provision, and future. J. Public Transp. **12**(4), 3 (2009)
2. Shaheen, S., Guzman, S., Zhang, H.: Bikesharing in europe, the americas, and asia: past, present, and future. Transp. Res. Rec. J. Transp. Res. Board **2143**, 159–167 (2010)
3. Pucher, J., Dill, J., Handy, S.: Infrastructure, programs, and policies to increase bicycling: an international review. Prev. Med. **50**, S106–S125 (2010)
4. Liu, J., Sun, L., Chen, W., Xiong, H.: Rebalancing bike sharing systems: a multi-source data smart optimization. In: Proceedings of the 22nd ACM SIGKDD International Conference on Knowledge Discovery and Data Mining, pp. 1005–1014. ACM (2016)
5. Chen, L., et al.: Bike sharing station placement leveraging heterogeneous urban open data. In: Proceedings of the 2015 ACM International Joint Conference on Pervasive and Ubiquitous Computing, pp. 571–575. ACM (2015)
6. Chen, L., et al.: Dynamic cluster-based over-demand prediction in bike sharing systems. In: Proceedings of the 2016 ACM International Joint Conference on Pervasive and Ubiquitous Computing, pp. 841–852. ACM (2016)
7. Liu, J., Sun, L., Li, Q., Ming, J., Liu, Y., Xiong, H.: Functional zone based hierarchical demand prediction for bike system expansion. In: Proceedings of the 23rd ACM SIGKDD International Conference on Knowledge Discovery and Data Mining, pp. 957–966. ACM (2017)

8. Singla, A., Santoni, M., Bartók, G., Mukerji, P., Meenen, M., Krause, A.: Incentivizing users for balancing bike sharing systems. In: AAAI, pp. 723–729 (2015)
9. Bao, J., He, T., Ruan, S., Li, Y., Zheng, Y.: Planning bike lanes based on sharing-bikes' trajectories. In: Proceedings of the 23rd ACM SIGKDD International Conference on Knowledge Discovery and Data Mining, pp. 1377–1386. ACM (2017)
10. Xu, K., Liao, S.S., Li, J., Song, Y.: Mining comparative opinions from customer reviews for competitive intelligence. Decis. Support Syst. **50**(4), 743–754 (2011)
11. Di Sorbo, A., et al.: What would users change in my app? Summarizing app reviews for recommending software changes. In: Proceedings of the 2016 24th ACM SIGSOFT International Symposium on Foundations of Software Engineering, pp. 499–510. ACM (2016)
12. Guo, B., et al.: Mobile crowd sensing and computing: the review of an emerging human-powered sensing paradigm. ACM Comput. Surv. (CSUR) **48**(1), 7 (2015)
13. Guo, B., Chen, C., Zhang, D., Yu, Z., Chin, A.: Mobile crowd sensing and computing: when participatory sensing meets participatory social media. IEEE Commun. Mag. **54**(2), 131–137 (2016)
14. Martin, W., Sarro, F., Jia, Y., Zhang, Y., Harman, M.: A survey of app store analysis for software engineering. IEEE Trans. Software Eng. **43**(9), 817–847 (2017)
15. Fu, B., Lin, J., Li, L., Faloutsos, C., Hong, J., Sadeh, N.: Why people hate your app: making sense of user feedback in a mobile app store. In: Proceedings of the 19th ACM SIGKDD International Conference on Knowledge Discovery and Data Mining, pp. 1276–1284. ACM (2013)
16. Gu, X., Kim, S.: "What parts of your apps are loved by users?" (T). In: 2015 30th IEEE/ACM International Conference on Automated Software Engineering (ASE), pp. 760–770. IEEE (2015)
17. Zhu, H., Liu, C., Ge, Y., Xiong, H., Chen, E.: Popularity modeling for mobile apps: a sequential approach. IEEE Trans. Cybern. **45**(7), 1303–1314 (2015)
18. Wang, Y., Yuan, N.J., Sun, Y., Qin, C., Xie, X.: App download forecasting: an evolutionary hierarchical competition approach. In: Twenty-Sixth International Joint Conference on Artificial Intelligence, pp. 2978–2984 (2017)
19. Malmi, E.: Quality matters: usage-based app popularity prediction. In: Proceedings of the 2014 ACM International Joint Conference on Pervasive and Ubiquitous Computing: Adjunct Publication, pp. 391–396. ACM (2014)
20. Finkelstein, A., Harman, M., Jia, Y., Sarro, F., Zhang, Y.: Mining app stores: extracting technical, business and customer rating information for analysis and prediction. RN **13**, 21 (2013)
21. Jin, J., Ji, P., Gu, R.: Identifying comparative customer requirements from product online reviews for competitor analysis. Eng. Appl. Artif. Intell. **49**, 61–73 (2016)
22. He, W., Zha, S., Li, L.: Social media competitive analysis and text mining: a case study in the pizza industry. Int. J. Inf. Manage. **33**(3), 464–472 (2013)
23. Tkachenko, M., Lauw, H.: Comparative relation generative model. IEEE Trans. Knowl. Data Eng. **29**, 771–783 (2016)
24. Zhang, Z., Guo, C., Goes, P.: Product comparison networks for competitive analysis of online word-of-mouth. ACM Trans. Manag. Inf. Syst. (TMIS) **3**(4), 20 (2013)
25. DeSarbo, W.S., Grewal, R., Wind, J.: Who competes with whom? A demand-based perspective for identifying and representing asymmetric competition. Strateg. Manag. J. **27**(2), 101–129 (2006)
26. Ouyang, Y., Guo, B., Zhang, J., Yu, Z., Zhou, X.: SentiStory: multi-grained sentiment analysis and event summarization with crowdsourced social media data. Pers. Ubiquit. Comput. 1–15 (2016)

27. Lu, X., Yu, Z., Sun, L., Liu, C., Xiong, H., Guan, C.: Characterizing the life cycle of point of interests using human mobility patterns. In: Proceedings of the 2016 ACM International Joint Conference on Pervasive and Ubiquitous Computing, pp. 1052–1063. ACM (2016)

28. Bergen, M., Peteraf, M.A.: Competitor identification and competitor analysis: a broad-based managerial approach. Manag. Decis. Econ. **23**(4–5), 157–169 (2002)

29. Borodin, A., El-Yaniv, R.: Online Computation and Competitive Analysis. Cambridge University Press, Cambridge (2005)

30. Hovy, D.: The enemy in your own camp: how well can we detect statistically-generated fake reviews-an adversarial study. In: Proceedings of the 54th Annual Meeting of the Association for Computational Linguistics, vol. 2, pp. 351–356 (2016)

31. Guo, B., et al.: CrowdStory: fine-grained event storyline generation by fusion of multi-modal crowdsourced data. Proc. ACM Interact. Mobile Wearable Ubiquit. Technol. **1**(3), 55 (2017)

32. Cao, L., Ou, Y., Philip, S.Y.: Coupled behavior analysis with applications. IEEE Trans. Knowl. Data Eng. **24**(8), 1378–1392 (2012)

33. Cao, L., et al.: Behavior informatics: a new perspective. IEEE Intell. Syst. **29**(4), 62–80 (2014)

Dual World Network Model Based Social Information Competitive Dissemination

Ze-lin Zang[1,2], Jia-hui Li[1], Ling-yun Xu[1], and Xu-sheng Kang[1(✉)]

[1] School of Computer and Computing Science,
Zhejiang University City College, Hangzhou 310015, China
kangxs@zucc.edu.cn
[2] College of Computer Science and Technology,
Zhejiang University of Technology, Hangzhou 310027, China

Abstract. The study of the competitive dissemination of various social information is of great significance to product marketing, political competition, and public opinion. Based on the existing small-world network model, this paper establishes a dual world network model that combines geographical factors to describe the information dissemination in society from two aspects of human relations and geographical relations. In addition, in order to describe the competitive relationship of a variety of opinions, the Opinion Acceptance Rules (OAR) were designed and simulated in the MATLAB environment. Therefore, this paper proves a lot of communication phenomena such as information explosion, information balance, and information island.

Keywords: Competitive dissemination · Various social information ·
Dual world network

1 Introduction

Social information dissemination [1–4] refers to the process of proliferation and extinction of information in social networks. Mail, phone, TV, Facebook, or WeChat can be transmitted through text or multimedia as a medium for information dissemination. To a certain extent, social relations can be regarded as a kind of trust relationship on which information dissemination depends [5]. A set of social relationship (like Facebook "news feed") can be seen as a channel for information dissemination.

According to the competitive nature of information dissemination process, information dissemination research can be divided into two categories. One is focused on the study of a single viewpoint and the spread of a single event in the society, such as Karumanchi et al. [6], Ryan et al. [7] and others have put forward some effective research model in this field. The other is to consider more complex situations and study the competitive spread of multiple views in the same event. There is no doubt that the application of the second types of research is more extensive.

In addition, Studying the objective laws of opposing views dissemination can make the information spread more effectively in the community. It plays an significant role in the product marketing [8], political argument [9] and opinion guidance [10]. The study of the opposing views dissemination is to bring in interaction effect mechanism of opposing views based on the study of the law of information dissemination. It aims at figuring out how the information is disseminated in the social network. Specifically, it is to clarify that under what circumstances, people will adhere to their original views. In what circumstances, people will give up their original views and believe a new viewpoint.

The common mathematical models of information dissemination are SIS, SIR [11] model, ER random network model [12], small world network model (small world network model) [13–16] and so on. Among which, the small world network model has a wide range of attention and is considered to be more effective [17]. small world network model can be used to describe the natural network structure of [7] shortest path, clustering characteristics [14, 18]. In addition, the six-degree separation theory [19] is derived from small world network model.

Because the small world network model can effectively describe the relationship between objects, such as cooperation/competition among members of society, links between brain neurons, excitatory dissemination relationship, and social information dissemination. With the aid of the method of describing the relationship between the small world network, a large number of scholars have used the small world network to research the relationship above. For example, Sullivan et al. [20] combine the small world network model with the weaving learning theory to explore the relationship between the relationship intensity of the members and the learning ability of the company. Erkaymaz et al. [21] introduces the topology of the small world into the artificial neural network and uses an improved neural network to diagnose diabetes. A better diagnosis is achieved.

In all applications, the use of a small world network to solve the problem of information dissemination is a hot issue. In this field, all the studies are divided into three categories.

The first class of researchers used the small world network to apply to a particular field of communication and explored the spread and propagation of the real life. For example, Xu et al. [22] use small world network model to explore the spread of mobile social networking (MSN). Pin Luarn uses small world network model to explore the spread of mainstream social networking sites. Wang et al. [23] use a small world network model to discuss the spread of safety information on product quality. This kind of research has a strong application and can play a role in predicting and guiding information dissemination in a certain field.

The second types of researchers have the impact of the internal structure of the small world network on the communication of social information. For example, Lima et al. [24] study the containment strategy for bad information dissemination through the mobility of small world network nodes. Wei et al. [25] study the relationship between the intensity of the connections between the small world network nodes and the speed of the communication of social information.

There are many other scholars studying the key nodes in the network - the impact of super-communicators [26–28] on social communication.

The third types of researchers are trying to improve the small world network so that they are more satisfied with the requirements of social communication. Hodas et al. [28] studied the characteristics of social information dissemination on the Internet based on the small world network model by extracting a large amount of communication information from Twitter. On the basis of the node propagation model, Zhou et al. [29] added a lot of spreading factors, proposed a fine-grained online social network information dissemination model, and used KRI's information extraction method to gain a lot of information on the Weibo for model verification.

As a universal relational network model, the small world network is bound to be unable to describe a specific social structure very carefully. For example, when describing the social communication problem in a certain area, the net- work structure established by it is not different from the network structure in other regions and does not reflect the difference in different regions. In order to make the small world network more suitable for the request of the information dissemination problem. In this paper, a dual world network model is proposed.

Most importantly, the small world network model focuses on the network structure of social relationships [30], but information dissemination problem does not only conform to its social relations. In many ways of communication (such as people face to face communication, Mail, telephone, television and so on), geographical relations is also an important part of information dissemination. In addition, research shows that people are more likely to believe in friends who are geographically close to each other. So the impact of geographic factors on the problem of information dissemination is also worth introducing in the model.

In communication, The opposing view is the opposing viewpoint that every individual holds in same event [31]. Every individual can only choose to accept one in all opposing views. The acceptance of a personal view is influenced by many aspects, such as personal position, historical factor, and the heart of the crowd [32]. Specifically, people choose to accept or exclude a viewpoint that will be affected by the social environment. At present, the theory about the spread of opposing views in society is not very complete.

Based on the several points above, the main contributions of this paper are as follows: By developing the small world network model, the paper uses the impact of geographical factors on social information dissemination to first propose the establishment of dual world network model (the relationship world and the geographical world) and model for the information dissemination mechanism. The dual world network model also considers changing the nonweighted graph of social structure and geographical factors into a weighted graph. Based on the human characteristics like mass-following psychology and perspective inertia, the Opinion Accept Rules (OAR) is designed to describe each person for the acceptance of the view, thus discuss the laws of society as a whole.

2 Small World Network and Establishment of the Dual World Network

2.1 Small World Network

The small world network is between the regular network and stochastic network model. As one of the complex network model, the small world has both small world characteristics and clustering characteristics [33].

In reality, it was proved that most networks have small world characteristics, so the small world network can be used to depict actual real network. This paper introduces the establishment of the WS network method proposed by Watts to establish a small world network model [34].

The small world network model starts from the N nodes nearest neighbor network. 1 Put the adjacent coupled network into a ring. 2 Reconnect each side of the network with probability p. One endpoint is maintained, and the other endpoint is selected as a random node in the network.

And specify that there can be only one edge between any two different nodes. The coefficient P specifies the randomness of the network, as shown in Fig. 1:

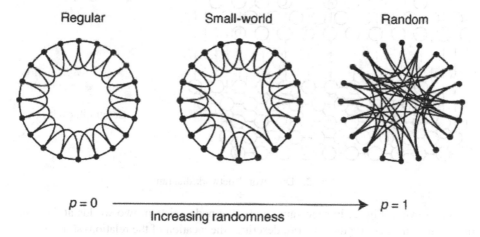

Fig. 1. Sketch map of small world network

In the small world network, $P = 0$ represents a fully defined network; $P = 1$ represents a completely random network. By adjusting the value of P, we can control the changes from the regular network to the random network, thus generated the small world network.

2.2 Conceptual Proposition and Mathematical Description of Dual World Network Model

In any dissemination of social information, a relationship network is not established entirely random. Its establishment is not only affected by the geographical environment, cultural factors, but also by the information media and the impact of randomness. So small world network model cannot description the social network accurately. The main reason is that the small social model does not take into account the media and geographical factors on social network.

On this paper, the dual world models are established based on the two constraints of geographical factors and relationship factors.

The diagram of the dual world network model is as shown in the Fig. 2.

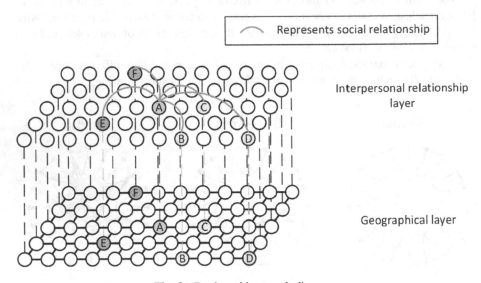

Fig. 2. Dual world network diagram

As shown in Fig. 2, For the same node, the node exists in two worlds at the same time, One is the geographical world, describes the location of the relation- ship between nodes. another is the relational world, describing the dissemination of information between nodes. For example, Node A has information exchange relationship with five other nodes (B, C, D, E) in information dissemination world, and the nodes are not geographically connected with them.

Dual world information dissemination model can be expressed as the following mathematical form:

In the undirected right connected graph $G(V, E)$, where V represents the range of unicom figure, E represents the link of unicom figure.

$$NET_t = NET_{t-1} \cup B_t \tag{1}$$

In (1), NET_t represents the state of acceptance of a certain message by the network at t. When $D(x, y)$ node receives this message, $NET_t = 1$, $NET_t = 0$ when not accepted, where:

$$B_t = \{v | v \in V, e_t^v > p\} \tag{2}$$

$$e_t^v = \frac{M_v^t}{d_v} \tag{3}$$

$$M_v^t = \{v' | (v', v) \in E, v' \in NET_{t-1}\} \tag{4}$$

In (2) and (3), B_t is the Transfer matrix of NET_t, and e_t^v represents the adoption of a problem in the scope of a node's friends. B represents the trust degree of this node's friend node. And p represents its trust threshold for a certain information.

2.3 The Opinion Accept Rules

In order to simulate the spread of information in society, this paper designs an Opinion Accept Rules (OAR) to simulate the interpersonal relationship through the dual world model.

When choosing the social viewpoint, social members will be influenced by other social members around them. Therefore, based on the psychology of conformity, this paper designs the Opinion Accept Rules. The mathematical description of OAR is shown as (5)

$$
\begin{aligned}
&\text{if } P(i) = 0 \text{ and } \frac{N_{in}}{N_a} > \alpha_i \\
&\quad T(i) = i \\
&\text{if } P(i) = 0 \text{ and } \frac{N_{in}}{N_a} \leq \alpha_i \\
&\quad T(i) = 0 \\
&\text{if } P(i) = 1 \text{ or } 2 \text{ or } \cdots \text{ or } n \text{ and } \frac{N_{in}}{N_a} > \alpha_i k_i \\
&\quad T(i) = i \\
&\text{if } P(i) = 1 \text{ or } 2 \text{ or } \cdots \text{ or } n \text{ and } \frac{N_{in}}{N_a} \leq \alpha_i k_i \\
&\quad T(i) = i
\end{aligned}
\tag{5}
$$

In (5), $P(i)$ represents the degree of the node believe in the information. $true(1)$ specifies that the information is accepted. Indicates the number of people who believe in the message in their friend circle, and represents the total friend number. The friend circle of Node i include all nodes in the dual world network which are related to node i. N_b represents the confidence coefficient, which is the coefficient of the model.

T_i represents the opinion accept transition matrix can describe the transition of people who do not believe in the message to accept the message. After all the nodes' confidence transfer matrix has been updated, the matrix P is updated by OR operation.

3   Model Simulation Results

3.1   Multi-viewpoint of the Simulation Results

In order to simplify the problem, this article defines the social area as a uniform population of square areas, select the model parameters in the Table 1.

Table 1. A single viewpoint of the variable name value comparison table

Variable name	N_f	L_f	$[W_c, W_k]$
Variable meaning friend number friend distance network size value			
Value	10	10	[500, 500]
Variable name	T_m	T_v	B_t
Variable meaning friend number friend distance network size value			
Value	10	10	[500, 500]

The simulation diagram of the propagation of opposing views is shown in the Fig. 3.

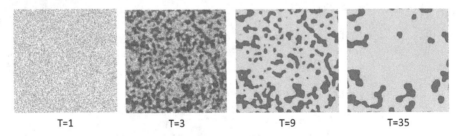

| T=1 | T=3 | T=9 | T=35 |

Fig. 3. Multi-view simulation of communication

Specific time dependent believers percentage of B and C shown as Table 2.

Table 2 shows the propagation status of the two viewpoints corresponding to the current situation in the figure and lists the proportions of beliefs of viewpoints B and C at different times. As shown above is the viewpoint B and the opposing view of C simulation of communication icon. In the figure, white indicates a node that does not believe any viewpoint of B and C, red indicates a node that believes viewpoint B, and blue indicates a node that believes viewpoint C.

Table 2. A time-varying table of people's persistence of opinions in a population

Time	T = 1	T = 2	T = 3	T = 4	T = 5	T = 7	T = 9	T = 11
PerB	0.12	0.24	0.54	0.65	0.68	0.70	0.71	0.72
PerC	0.11	0.18	0.33	0.29	0.26	0.24	0.23	0.22
Time	T = 15	T = 17	T = 19	T = 21	T = 23	T = 25	T = 27	T = 29
PerB	0.73	0.74	0.77	0.78	0.78	0.79	0.79	0.80
perC	0.21	0.20	0.17	0.16	0.15	0.14	0.14	0.14

The concrete manifestation is that the beliefs of viewpoint B and C increase at the same time until the saturation occurs. At $T = 5$, essentially all nodes believe in the B and viewpoint C and enter the saturation situation.

When $T > 5$, the two viewpoints are in a state of exponential growth. The increase in the number of people who believe in one viewpoint leads to another view that the reduction in the number of people is in a state of competitive communication. In the set parameters, the author presupposes that people in this dissemination have a tendency of accepting viewpoint B to be 0.01% more than viewpoint C.

However, in the final result of dissemination, viewpoint B occupies a great advantage, It can be seen that the ability of the message to spread has a great influence on the spread of the message.

Multi-perspective sensitive analysis shown in Fig. 4.

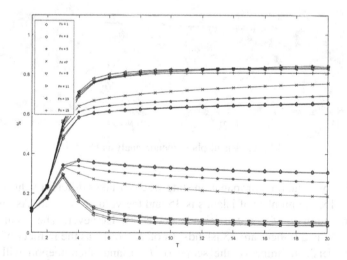

Fig. 4. Multi-perspective sensitive analysis

As shown in Fig. 4, the simulation result of changing the friendship range Nf is shown in the figure. The horizontal axis is the time scale and the vertical axis is the percentage of the number of people who believe it.

In the period of opinion communication $(T <= 5)$, no matter how the N_f changes, the proportion of beliefs in both viewpoint will increase rapidly. But the notion of B is growing faster than the viewpoint C, so at the end of the propagation phase, more nodes accept the viewpoint B.

In the next phase of the competition, the viewpoint B will eventually yield large-scale support because of its own dissemination advantages and the existing mass advantages. But at the end of the day, the idea that viewpoint B does not hold a 100% approval rate due to the information silos will be discussed later.

3.2   Discussion on the Phenomenon of Islands of Interest

In the situation of information's competitively dissemination, advantageous views will erode disadvantage views. It means that with the passage of time, due to the influence of the social network, there will be people who have weak point turn to believe the advantageous view. But often weak views will not be completely eroded since Information Island will appear in the social network. Similar to the concept of small world network, groups that have high rate information flow exist in the social network. Because the mutual friend relationship between nodes is very concentrated, in the premise of confidence in their internal rules, people tend to hold the same attitude to a question, is also easier to exclude outside in the opposite view. As the last figure in the last, the red spots being left are results of isolated island phenomenon.

Island phenomenon analysis chart shown in Fig. 5.

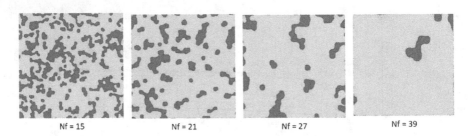

$Nf = 15$ $Nf = 21$ $Nf = 27$ $Nf = 39$

Fig. 5. Island phenomenon analysis chart

Figure 5 shows the final propagation state when N_f takes different values. When the N_f number is 15, the number of islands is 35 and the volume is small. As the N_f value increases, the number of isolated islands decreases. However, The volume will be larger. When N_f is 27, there are 13 islands. In other words, as the number of friends N_f increase, or friends to improve the scope of Lf, island phenomenon will gradually decline, that is, the society will be more and more tend to take the same view on a certain issue. In other words, with the escalation of communication means, the island phenomenon will be weakened.

4   Conclusions

This article establishes a dual-world model that is related to geographical factors to explore and simulate the process of the spread of social information. By MATLAB software to solve the following conclusions.

With the increase in the scope of making friends, there will be a leap in information dissemination speed. But after a threshold, the spread of information there will be a not smooth phenomenon. Nodes more friends, it will make the social relations more closely, which will bring about a corresponding increase in the speed of communication, but in the simulation results can also see the number of friends will have a roof effect, that is, when the number of friends After reaching a certain large value, the result of the spread of the sensitivity of the number of friends is not high.

In addition, under the circumstance of competition and dissemination, the advantage of occupying a small number of advantages expands its advantages at the initial stage of dissemination. The greater the social scope and the longer the stage of dissemination, the more obvious its advantages will be expanded.

There is an island phenomenon in society, that is, there is a small group insensitive to changes in external information. However, with the reduction of the scope of making friends, the island phenomenon will be weakened.

References

1. Del Vicario, M., et al.: The spreading of misinformation online. Proc. Natl. Acad. Sci. **113**(3), 554–559 (2016)
2. Lokot, T., Diakopoulos, N.: News bots: automating news and information dissemination on twitter. Digit. Journal. **4**(6), 682–699 (2016)
3. Zhou, X., Wu, B., Jin, Q.: User role identification based on social behavior and networking analysis for information dissemination. Futur. Gener. Comput. Syst. **117**, 107–117 (2017)
4. Househ, M.: Communicating ebola through social media and electronic news media outlets: a cross-sectional study. Health Inform. J. **22**(3), 470–478 (2016)
5. Yap, H.-Y., Lim, T.-M.: Trusted social node: evaluating the effect of trust and trust variance to maximize social influence in a multilevel social node influential diffusion model. In: Gervasi, O., et al. (eds.) ICCSA 2016. LNCS, vol. 9789, pp. 530–542. Springer, Cham (2016). https://doi.org/10.1007/978-3-319-42089-9_37
6. Karumanchi, G., Muralidharan, S., Prakash, R.: Information dissemination in partitionable mobile ad hoc networks. In: Proceedings of the 18th IEEE Symposium on Reliable Distributed Systems 1999, pp. 4–13. IEEE (1999)
7. Ryan, J.O.: Method and system for audio information dissemination using various modes of transmission, 4 June 1996. US Patent 5,524,051
8. Nascimento, A.M., Da Silveira, D.S.: A systematic mapping study on using social media for business process improvement. Comput. Hum. Behav. **73**, 670–675 (2017)
9. Cornand, C., Heinemann, F.: Optimal degree of public information dissemination. Econ. J. **118**(528), 718–742 (2008)
10. Shan, S., Liu, M., Xiaobo, X.: Analysis of the key influencing factors of haze information dissemination behavior and motivation in WeChat. Inf. Discov. Deliv. **45**(1), 21–29 (2017)
11. Shulgin, B., Stone, L., Agur, Z.: Pulse vaccination strategy in the sir epidemic model. Bull. Math. Biol. **60**(6), 1123–1148 (1998)

12. Ruths, D., Ruths, J.: Estimating the minimum control count of random network models. Sci. Rep. **6**, 19818 (2016)
13. Pennings, P.S., Holmes, S.P., Shafer, R.W.: HIV-1 transmission networks in a small world. J. Infect. Dis. **209**(2), 180 (2014)
14. Watts, D.J.: Small-World Networks in the Oxford Handbook on the Economics of Networks. Oxford University Press, Oxford (2016)
15. Bassett, D.S., Bullmore, E.T.: Small-world brain networks revisited. Neurosci. **23**(5), 499–516 (2017)
16. Boccaletti, S., Latora, V., Moreno, Y., Chavez, M., Hwang, D.-U.: Complex networks: structure and dynamics. Phys. Rep. **424**(4), 175–308 (2006)
17. Liu, M., Li, D., Qin, P., Liu, C., Wang, H., Wang, F.: Epidemics in interconnected small-world networks. PLoS ONE **10**(3), e0120701 (2015)
18. Watts, D.J., Strogatz, S.H.: Collective dynamics of 'small-world' networks. Nature **393** (6684), 440 (1998)
19. Shu, W., Chuang, Y.-H.: The perceived benefits of six-degree-separation social networks. Internet Res. **21**(1), 26–45 (2011)
20. Sullivan, B.N., Tang, Y., Marquis, C.: Persistently learning: how small-world network imprints affect subsequent firm learning. Strat. Organ. **12**(3), 180–199 (2014)
21. Erkaymaz, O., Ozer, M.: Impact of small-world network topology on the conventional artificial neural network for the diagnosis of diabetes. Chaos, Solitons Fractals **83**, 178–185 (2016)
22. Xu, Q., Su, Z., Zhang, K., Ren, P., Shen, X.S.: Epidemic information dissemination in mobile social networks with opportunistic links. IEEE Trans. Emerg. Top. Comput. **3**(3), 399–409 (2015)
23. Wang, L., Lei, C., Yingcheng, X., Yang, Y., Shan, S., Xiaobo, X.: An information dissemination model of product quality and safety based on scale- free networks. Inf. Technol. Manag. **15**(3), 211–221 (2014)
24. Lima, A., De Domenico, M., Pejovic, V., Musolesi, M.: Disease containment strategies based on mobility and information dissemination. Sci. Rep. **5**, 10650 (2015)
25. Wei, J., Bu, B., Guo, X., Gollagher, M.: The process of crisis information dissemination: impacts of the strength of ties in social networks. Kybernetes **43**(2), 178–191 (2014)
26. Pietrucha, F.J.: Supercommunicator: Explaining the Complicated So Anyone Can Understand. AMACOM Div American Mgmt Assn, New York City (2014)
27. Chowell, G., Viboud, C., Hyman, J.M., Simonsen, L.: The western africa ebola virus disease epidemic exhibits both global exponential and local polynomial growth rates. PLoS Curr. **7** (2015). https://www.ncbi.nlm.nih.gov/pmc/articles/PMC4322058/
28. Hodas, N.O., Lerman, K.: The simple rules of social contagion. Sci. Rep. **4**, 4343 (2014)
29. Zhou, D., Han, W.-B., Wang, Y.-J.: Social network information transmission model based on node and information characteristics. J. Integr. Plant Biol. **52**(1), 156–166 (2015)
30. Zhang, S., Luo, X., Xuan, J., Chen, X., Xu, W.: Discovering small-world in association link networks for association learning. World Wide Web **17**(2), 229–254 (2014)
31. Romeu, J.L.: Statistical methods for communication science (2006)
32. Li, J., Zhoua, Q., Zheng, X., Moga, L.M., Neculita, M.: Influencing factors of rural human consumption in china. Econ. Appl. Inform. **20**(3), 117–127 (2014). Annals of the University Dunarea de Jos of Galati: Fascicle: I
33. Stone, T.E., McKay, S.R.: Majority-vote model on a dynamic small-world network. Phys. A Stat. Mech. Appl. **419**, 437–443 (2015)
34. Jones, C., Hesterly, W.S., Borgatti, S.P.: A general theory of network governance: exchange conditions and social mechanisms. Acad. Manag. Rev. **22**(4), 911–945 (1997)

Parallel and Distribution Systems, Optimization

WarmCache: A Comprehensive Distributed Storage System Combining Replication, Erasure Codes and Buffer Cache

Brian A. Ignacio, Chentao Wu$^{(\boxtimes)}$, and Jie Li

Shanghai Key Laboratory of Scalable Computing and Systems,
Department of Computer Science and Engineering, Shanghai Jiao Tong University,
Shanghai, China
bignacio5@sjtu.edu.cn, {wuct,lijie}@cs.sjtu.edu.cn

Abstract. A tiered storage system uses replication method to provide both high reliability and availability, which stores three replicas over different nodes in the clusters. Erasure codes (EC) such as Reed-Solomon (RS) are increasingly utilized to further reduce the storage overhead while providing low I/O performance and availability. Existing solutions nowadays implement heterogeneous storage systems either using triple replication, erasure coding methods or a combination of both, although involves high performance gap between each data layer. To address this problem, in this paper, we introduce WarmCache, a new data layer for warm data by having one copy stored using erasure coding and the other copy in memory data layer. Using one copy in erasure coding data layer ensures data reliability, while the other copy in memory data layer provides fast I/O performance.

Keywords: Erasure codes · Storage overhead · I/O performance · Replication · Cache

1 Introduction

The upper limit of data processing is constantly growing each year which implies storage utilization is exponential increasing. Data centers have begun to analyze their data in order to optimize their clusters efficiency: (1) realizing that most existing data is almost never accessed (archival or cold data); (2) while a shorter proportion of data is more frequently accessed. Tiered storage systems [1] groups data based on the business Key Performance Indicator ("KPI") allocating each group on different storage devices such as SSD for frequently accessed data and HDD for less accessed data. The goal of tiered storage is to reduce storage cost while keeping system utilization optimized. Distributed File Systems (DFS) such as Hystor [2] and Hadoop [3] implement tiered storage in their architecture.

Typically Distributed File Systems (DFS) use triple replications to ensure data reliability and availability. Although it is simple and effective, replication

© Springer Nature Switzerland AG 2019
S. Li (Ed.): GPC 2018, LNCS 11204, pp. 269–283, 2019.
https://doi.org/10.1007/978-3-030-15093-8_19

have a 3× storage cost in the default case. While some previous work [4,5] tries to provide adaptive number of replicas based on access demand to optimize storage cost, the Quality of Service (QoS) cannot be guaranteed. Another approach used in DFS to provide reliability is the implementation of Erasure Codes ("EC"), which divides data in smaller chunks with parity information resulting in less storage overhead. As shown in previous publications [6], we can observe there exist many kind of codes for storage systems based on RAID [7]. A major category of erasure codes are the Maximum Distance Separable ("MDS") codes, such as Reed-Solomon ("RS") [8] and STAR codes [9]. Another type are the non-MDS codes, such as Locally Repairable Codes ("LRC") [10], Local Reconstruction Codes ("LRC") [11], GRID Codes [12] and HoVER Codes [13].

Although Erasure Coding (EC) gives less storage overhead, it has higher I/O cost than replication since encoding/decoding tasks are involved in typical I/O operations such as read, write and block reconstruction. Recent work [14–17] have shown different ways of combining replication and erasure coding in tiered storage system, mostly using replication for hot data and erasure coding for cold data with a data monitoring mechanism that transform data from a type to another.

However, [18] shows that it exists a huge gap between the replication and erasure coding, since there is a lot of encoding bandwidth and computation cost to move a data from a layer to another, which decreases the system performance. Reducing the gap between replication and erasure coding is crucial since the latter involves high system resources utilization as shown in [18], where at least a median of 95,500 blocks (180 TB) is transferred between racks every day in a Facebook's warehouse cluster due to block recovery operations inherited of erasure coding methodologies. Degraded system utilization can affect clusters normal operations colliding with regular business operations causing extra failures. Furthermore, the data transformation gap between EC and replication doesn't take into account warm data.

To address this problem, we propose WarmCache, a new warm data layer that uses a 2.5× storage cost by allocating one replica in memory and the other replica encoded as erasure coding. It allows one replica for fast I/O operations and the other copy encoded by erasure codes to ensure reliability. The proposed approach can improve the I/O performance of overall distributed file systems by improving the I/O performance on erasure coding layer with lazy persisted replica in memory.

We make the following contributions:

– We propose a new warm data layer, which uses one replica in memory and the other encoded using a erasure code;
– To demonstrate the effectiveness of the warm data layer, we conduct several experiments. The results show that, by adding warm data layer into distributed file systems, we can achieve higher I/O performance with low additional storage cost (Fig. 1).

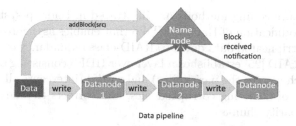

Data pipeline

Fig. 1. Writing data with triple replication in Hadoop [3]. As shown in this figure, a typical Hadoop system consists of four nodes, one namenode and three data nodes. The Namenode provides the management of namespace and datanodes. The Datanodes store the digital data. By default, Hadoop write three replicas in different nodes sequentially, sending an acknowledgement to the Namenode after writing a replica in a Datanode, which is responsible to keep data consistency.

The rest of the paper is organized as follows. Section 2 contains a detailed discussion of the performance of triple replication systems, erasure code systems and combination approaches as well as their drawbacks. Section 3 shows in depth the design of our new warm data layer. Section 4 shows the evaluation results. Lastly, Sect. 5 gives a summary the contributions of our paper.

2   Related Work

2.1   Triple Replication

For distributed storage systems, it is essential to ensure reliability since they are built by interconnected cheap commodity hardware. The simplest way is to use replication, store copies of data over multiple locations to tolerate node failures and provide high reliability of the system. Distributed File Systems (DFS) such as the Google Filesystem (GFS) [19] and Hadoop HDFS [3] implement triple copies over available nodes as standard behavior. Load balancing is achieved by distributing utilization across the replicas. In most cases, triple replications work well in terms of reliability and performance, but the high monetary cost on storage devices motivates the search of new methods. Approaches such as *CDRM* [4] and *Scarlett* [5] optimizes the number of replicas based on data access patterns and system capacities.

2.2   Erasure Codes

The implementation of erasure codes [20] in DFS can further reduce the storage overhead replacing the traditional triple replication. Theoretically it can reduce the storage cost at least 50% in practical systems [21].

Erasure codes [20] separates data into k chunks. Another m chunks of coding information are calculated from original k chunks, the parity chunks combining them into n chunks $(n = k+m)$, which are stored in n disks, respectively [22,23].

Various erasure coding methods are illustrated in Hadoop systems. DiskReduce [24] is a modification of Hadoop HDFS that enables asynchronous compression of initially triplicated data down to RAID-class redundancy overheads. Facebook's HDFS-RAID [25] establishes a layer over HDFS consisting of a *RaidNode* (a daemon which creates the parity files) and *raidfs* (intercepts all calls to HDFS client in pursue of detecting corrupted files, and corrects the files based on the corresponding parity chunks).

2.3 Combination of Triple Replication and Erasure Codes

The advantages of triple replication are high I/O performance and fast data recovery. The advantages of erasure coding is that the storage cost is typically $1.5\times$ compared to the $3\times$ storage cost of triple replication. Compared to triple replication, EC has lower I/O performance and slower data reconstruction. Therefore replication is typically used for hot data while erasure coding is used for cold data. Different solutions vary the transformation method between these data layers.

ERMS [14] takes advantages of a Complex Event Processing engine to distinguish real-time data and providing an elastic replication policy, incrementing replicas for hot data using a replication manager and erasure coding for cold data, while reducing the number of replicas of cool data. Another solution is encoding-aware replication (EAR) [15] which carefully determines where to place the replicas in order to reduce cross-rack replica download traffic while applying encoding operations.

MICS [16] organizes each data in two parts, a master node (MN) for full copy and a Erasure Coded Chain (ECC) for $k + m$ chunks with modified read and write protocols. CAROM [17] stores data using (m, k) EC meanwhile it creates a replica from EC for read/write operations as a cache version for frequently accessed files.

Table 1. Advantages and disadvantages of existing approaches

Redundancy approaches	Advantages	Disadvantages
Replication	I/O performance fast data recovery	Storage overhead
Erasure coding	Storage overhead	I/O performance slow data recovery
Replication & EC combination	Fast I/O performance low storage overhead	Data transition complexity [18]
WarmCache	Fast I/O performance low storage overhead Low transformation complexity	Memory overhead

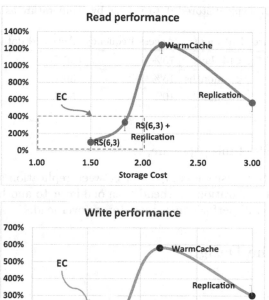

Fig. 2. Performance of WarmCache vs traditional implementations. RS(6,3)+ 3x is the combination of triple replication and RS(6,3) erasure code. Note that different EC implementations requires various storage costs (typically between 100% and 200%), and here we show the results for RS(6,3).

2.4 Motivation of Our Idea

Based on previous literatures, we conclude that both replication and erasure code redundancy offers advantages for hot and cold data, respectively. But they inquire high processing and monitoring resources to maintain each data in the corresponding data layer such as encoding latency and traffic cost, number of replicas to ensure performance and availability as well as location awareness. WarmCache add a data layer for warm data in combination with triple replication for hot data and erasure code for cold data to improve read and write operations without higher storage cost. Our results are summarized in Table 1 and Fig. 2, where we take the read and write performance of RS(6,3) erasure coding DFS as baseline and compared it with replication, the combination of erasure coding and replication and our proposed solution. The results shows that our solution can provide increased performance at low additional storage overhead.

Table 2. Typical stored data volume by temperature as in [1]

Temperature	Data age	Access frequency	Accessed data volume
HOT	<14 days	75%	20%
WARM	<3 months	15%	30%
COLD	>3 months	10%	50%

The design goals of this paper are,

- Smooth the data transition complexity between replication and EC [18];
- Reduce the data migration overhead from one layer to another;
- Increase the overall performance over storage overhead.

3   WarmCache Design

3.1   Key Idea

The hot data layer is based on triple replication, the cold data layer is implemented using erasure coding. The warm data layer uses one replica stored in memory data layer and the other replica encoded and stored in erasure coding. The architecture of WarmCache is shown in Fig. 3. We conducted our tests using Alluxio [26] as in-memory filesystem with Hadoop [3] as under filesystem. The hot data layer uses Hadoop triple replication, the warm data layer keeps a in-memory replica in Alluxio and a encoded replica in Hadoop using Reed-Solomon (RS) erasure coding with $k = 6$ data blocks and $m = 3$ parity blocks and the cold data layer keep only an encoded copy using the previous encoding method.

Using in-memory replica for the warm data results in better IO performance for archival data that shows an increased access in the later analyzed period, an smoother transition from replication to a erasure coding approach due to the fact that a in-memory replica provides high IO performance meanwhile encoding operation can be tasked in parallel. Data can be lazily moved in erasure coding due to memory blocks eviction when other files require better access time.

We classify the data by its *temperature*, which is based on the data block access frequency [27]. A typical data volume and access frequency is observed in Table 2.

3.2   Identify Data Temperatures and Thresholds

We define λ_{WARM} and λ_{COLD} as the access frequency threshold for 'warm' and 'cold' data, respectively. We use the function $S_{block}(f)$ to define the current state of the data block based on the access frequency f, which is calculated by observing the number of times a block is accessed within a period of time defined by the application needs (hourly, daily, ...). The overall access time uses the

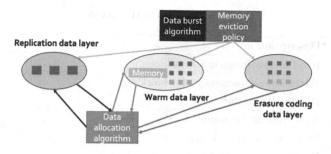

Fig. 3. Architecture of WarmCache. We implement a hot data layer using triple replication for frequently accessed data, a warm layer made by one encoded replica using Reed-Solomon erasure coding and other replica stored in memory data layer, and a cold layer using Reed-Solomon erasure coding for rarely accessed data.

bandwidth (BW) and seek time (T_{seek_m}) analysis as defined in [28] to establish the preference to put data blocks in a specific data layer.

$$T_{READ_m}(i) = \frac{Size(i)}{BW_{READ_m}} + T_{seek_m} \tag{1}$$

$$T_{WRITE_m}(i) = \frac{Size(i)}{BW_{WRITE_m}} + T_{seek_m} \tag{2}$$

$$C_m(i) = T_{READ_m}(i) * f_{read}(i) + T_{WRITE_m}(i) * f_{write}(i) \tag{3}$$

$$S_{block}(i) = \begin{cases} HOT & \lambda_{WARM} < C_m(i) \\ WARM & \lambda_{COLD} < C_m(i) < \lambda_{WARM} \\ COLD & C_m(i) < \lambda_{COLD} \end{cases}$$

From Eqs. (1) and (2) [28], we observe that using a replica in memory reduces $T_{READ_m}(i)$ and $T_{WRITE_m}(i)$ due to faster bandwidth and slower T_{seek_m} for the warm data layer.

3.3 Data Reallocation Policies

Data block B_i initially is located in the 'hot' state. The data allocation algorithm changes B_i based on its current overall access time $C_m(i)$. The stored metadata of B_i is used to review the latest three states of the block. If the block states have a circular variation (i.e. $hot \rightarrow warm \rightarrow hot$) and the difference between the threshold λ and $C_m(i)$ is smaller than $\Delta_{circular}$ we skip the reallocation of the file for a next execution of the algorithm. If the difference is bigger than $\Delta_{circular}$ we schedule the reallocation to the corresponding layer. This helps avoid excesive reallocation of data block with $C_m(i)$ close to thresholds. The algorithm takes into account the workload of each datanode to decide in which node to allocate the replica by considering the node's bandwidth and the access time of other replicas already in memory or in the hot data layer for the cases where capacity is almost reached (Table 3).

Table 3. Symbols in this paper

Symbol	Description
$f_{read}(i)$	Read access frequency of a block i
$f_{write}(i)$	Write access frequency of a block i
BW_m	Bandwidth of datanode m
T_{seek_m}	Seek time of datanode m
$T_{READ_m}(i)$	Read time of data block i
$T_{Write_m}(i)$	Write time of data block i
$C_m(i)$	Overall access time of a block i
$S_{block}(i)$	Temperature state of the block (hot, warm or cold)
λ_{HOT}	Access frequency threshold for hot data
λ_{WARM}	Access frequency threshold for warm data
λ_{COLD}	Access frequency threshold for cold data
$\Delta_{circular}$	Minimum acceptable difference $|C_m(i) - \lambda_{state}|$ for data reallocation
Δ_{burst}	Minimum increment of $C_m(i)$ to cache a replica for IO performance
f_{Rep}	Number of replicas of a block

If data increases $C_m(i)$ over an upper state layer's threshold for a short period of time (depending on the applications requirements) and go back to the previous average access time, a temporarily cache of a block's B_i replicas is created depending of its current state in a memory data storage layer, such as [29]. Δ_{burst}, λ_{burst} and $C_{ref}(i)$ are custom defined by the application as the minimum time window to allow data burst caching, the data burst increment percentage over the threshold and the closest access time reference for a base comparison (i.e. a day average of $C_m(i)$), respectively.

3.4   Memory Eviction Policies

Since memory data layer devices possess limited storage capacity compared with other storage devices, a mechanism to preserve most "important" blocks is required. In our experiments we used simple Least Recently Used (LRU), but as shown by [30] different workloads perform better under a different eviction policy, which are easily pluggable in modern distributed filesystems such as Alluxio [26].

4   Evaluation

To demonstrate the effectiveness of WarmCache, in this section, we implement WarmCache into Hadoop and Alluxio and conduct several experiments by running different I/O workloads.

4.1 Methodology

We compare the following solutions in this evaluation,

1. Triple replication method [3], which is referred to as "HOT" in the evaluation using a storage cost of 3 replicas.
2. Erasure Coding approach. Typically, RS (6,3) erasure code (referred to as "COLD") is a popular choice in distributed storage systems [24,25] using a storage cost of 1.5 replicas.
3. Combination scheme of replication and erasure coding [27], which is referred to "HC" in the evaluation using a storage cost of 1.83 replicas.
4. The approach presented in this paper (WarmCache), which is the combination of replication, erasure coding and the warm data layer, one replica in a memory layer and the other encoded using erasure coding as explained in Sect. 3.1; hereby referred as "HWC" using a storage cost of 2.17 replicas. This value comes from the average overhead shown in our tests by implementing the previous explained design.

We implement the previous approaches into a real Hadoop system, and use TestDFSIO to evaluate the I/O performance. TestDFSIO is a benchmark tool included with Hadoop's distributions to measure read and write I/O performance, which is helpful to test HDFS network and hardware bottlenecks without involving Map-Reduce operations. Statistical Workload Injector for MapReduce (SWIM) [31] includes workload repositories from real life production systems, workload synthesis tools to generate representative test workloads by sampling historical cluster traces and replay tools to execute these workloads. Our experiments use Facebook's 2009 first 50 Jobs repository which includes an array of jobs as explained in [32].

Table 4. Hardware setup

#	Server model	Cores	RAM	Storage units
1	Dell PowerEdge R370	(24) Xeon(R) CPU E5-2620	16 GB	8 1TB HDD RAID-1
2	Dell PowerEdge R370	(24) Xeon(R) CPU E5-2620	16 GB	8 1TB HDD (4× HDD 1× per VM)
3	Dell PowerEdge R370	(24) Xeon(R) CPU E5-2620	16 GB	16 250GB SSD (4× per VM)
4	Dell PowerEdge R370	(24) Xeon(R) CPU E5-2620	16 GB	14 250GB SSD (3× per VM)

The hardware configuration of our Alluxio-Hadoop system is shown in Table 4 and Alluxio version 1.6.0 for all nodes.

Table 5. Configuration

#	Hadoop version	Virtual machine	RAM	Storage
1	Hadoop 3.0.0 -alpha4	None	16 GB	RAID1
2	Hadoop 3.0.0 -alpha4	3 VM	4 GB each	4 HDD 1TB 1 HDD Spare
3	Hadoop 3.0.0 -alpha4	3 VM	4 GB each	4 250GB SSD each
4	Hadoop 3.0.0 -alpha4	3 VM	4 GB each	3 250GB SSD each

We use the following metrics in our evaluation (Table 5),

1. Throughput (megabytes/seconds) $= \frac{\sum_{i=0}^{N} filesize_i}{\sum_{i=0}^{N} time_i}$: The overall amount of megabytes processed in the TestDFSIO job duration.
2. Average I/O Rate (megabytes/seconds) $= \frac{\sum_{i=0}^{N} \frac{filesize_i}{time_i}}{N}$: The average of the megabytes processed in TestDFSIO duration of all map jobs.

For SWIM tests we measure the metrics as below,

1. Total time spent by all maps in occupied slots (seconds): The total time taken to execute the map tasks.
2. Total time spent by all reduce in occupied slots (seconds): The total time taken to execute reduce tasks.
3. Jobs duration (milliseconds): Duration of each job.

For the figures, we take erasure coding layer (referred here as "COLD") as basis of comparison, and we show others solutions as multiples of the previous. To provide fair comparison, we compared our proposed solution with others solution using each metrics over storage cost in Table 6 as follows:

$$metric = \frac{metric_{HWC} - metric_{otherSolution}}{storageCost_{otherSolution}} \tag{4}$$

4.2 Experimental Results

In this subsection, we present the experimental results of WarmCache compared to other popular approaches.

I/O Performance. The results of I/O performance are shown in Fig. 4 throught Fig. 5. In Fig. 4, we observe that HWC perform 6.7× COLD, 4.4× HC and 3.19× below than HOT in terms of write throughput. For the read throughput in Fig. 6, HWC is 16× COLD, 12.9× HC and 10.8× HOT.

We summarize the improvements of WarmCache versus other approaches in terms of I/O performance over the storage cost in Table 6 (performance results in Fig. 4 through Fig. 5 divided by the storage cost in X-axis). From this table, we observe HWC gives up to 3.6× better performance over other solutions. In the case of read tests, HWC executes up to 3.16× over other solutions. The average I/O rate over storage cost for write tests is up to 3.02× and up to 7.6× in read tests (Fig. 7).

Table 6. Improvements of WarmCache over other approaches in terms of read/write throughput and average I/O rate over storage Cost

	Solutions	Throughput	Average I/O rate
Write	HWC/HOT	164.06%	151.14%
	HWC/HC	167.31%	145.24%
	HWC/COLD	363.09%	302.32%
Read	HWC/HOT	315.64%	204.40%
	HWC/HC	328.21%	216.21%
	HWC/COLD	1015.19%	762.02%

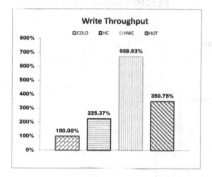

Fig. 4. Write throughput

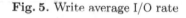

Fig. 5. Write average I/O rate

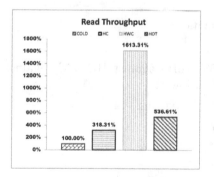

Fig. 6. Read throughput

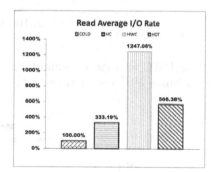

Fig. 7. Read average I/O rate

SWIM Evaluation. The evaluation on Facebook's workloads are shown in Figs. 8, 9, 10. We observe in Fig. 8 that HWC take 29.24% less time in map tasks compared to COLD, 17.16% less than HC and 15.20% more than HOT. Similar Fig. 9 shows that the average total time in reduce tasks is up to 7.47% over COLD, 20.35% faster than HC and 9.64% less time than HOT. Analyzing Fig. 10 jobs duration average shows that HWC is 17.63% faster than COLD, 36.16% than HC and 5.54% than HOT.

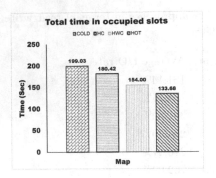

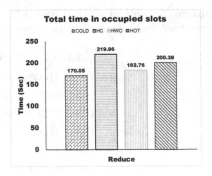

Fig. 8. Map time   **Fig. 9.** Reduce time

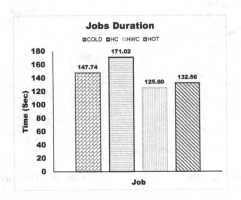

Fig. 10. Jobs duration

The HWC storage consumption is 8.33% worse against HC, 25% against COLD but is 50% better than HOT as seen in Fig. 11.

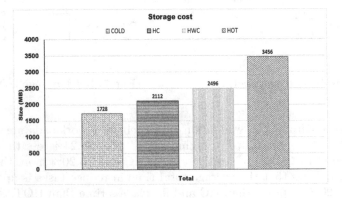

Fig. 11. Storage cost of WarmCache vs various data redundancy approaches (HOT, COLD, HC & HWC). Original file size for each file is 128 MB.

4.3 Analysis

From the results in previous subsection, WarmCache shows great advantages in terms of the performance/reliability over storage cost. There are many reasons to achieve these gains. First, the use of one replica in erasure coding data layer and the other replica in memory data layer being processed in parallel achieves high I/O performance, due to a preference to consume the replica in memory over the one stored using EC. Since memory replica is allocated by considering data node workload, imbalanced datanodes are minimized. Second, the warm data layer provides faster data recovery compared to erasure coding methods due to existence of in-memory replica, since we can either recover from the in-memory replica (faster recovery) or other EC chunks. Third, compared to replication method, the storage cost of WarmCache is much lower.

Job average durations are slightly reduced due to fewer written output bytes (HWC overall data volume is reduced compared to other solutions), which translate in more CPU and I/O cost in the map reduce job intermediate files used in the reduce phase.

In our design we take Table 2 as typical data volume reference. DFS file access distribution analysis provide correct data volume and access frequency temperature percentages. In our experiments we found that 50% or more storage volume for cold data, 20% or less for the hot data and the remaining percentage for the warm data achieve the higher I/O performance per storage unit.

5 Conclusion

This work introduces a novel approach called WarmCache for tiered storage systems, which conceptualizes a new data layer for warm data. It implements one copy in memory data layer and the other in a erasure coding layer, providing both high data availability and reliability. To demonstrate the effectiveness of Warm-Cache, we implement WarmCache into a real Alluxio-Hadoop system. And in our experiments, compared to other popular solutions for tiered storage, Warm-Cache shows (1) higher read/write performance over storage cost; (2) lower job duration over storage cost for SWIM jobs due to better chunk access frequency classification and (3) faster recovery process than erasure coding method.

References

1. Teradata: The impact of data temperature on the data warehouse, August 2012. http://www.teradata.com/Resources/White-Papers/The-Impact-of-Data-Temperature-on-the-Data-Warehouse/
2. Chen, F., Koufaty, D.A., Zhang, X.: Hystor: making the best use of solid state drives in high performance storage systems. In: Proceedings on Supercomputing, pp. 22–32. ACM (2011)
3. Shvachko, K., Kuang, H., Radia, S., Chansler, R.: The hadoop distributed file system. In: 2010 IEEE 26th Symposium on Mass Storage Systems and Technologies (MSST), pp. 1–10. IEEE (2010)

4. Wei, Q., Veeravalli, B., Gong, B., Zeng, L., Feng, D.: CDRM: a cost-effective dynamic replication management scheme for cloud storage cluster. In: 2010 IEEE International Conference on Cluster Computing (CLUSTER), pp. 188–196. IEEE (2010)
5. Ananthanarayanan, G., et al.: Scarlett: coping with skewed content popularity in mapreduce clusters. In: Proceedings of the Sixth Conference on Computer Systems, pp. 287–300. ACM (2011)
6. Plank, J.S.: Erasure codes for storage systems: a brief primer. Usenix Mag. **38**(6), 44–50 (2013)
7. Patterson, D.A., Gibson, G., Katz, R.H.: A case for redundant arrays of inexpensive disks (RAID), vol. 17. ACM, New York (1988)
8. Plank, J.S., et al.: A tutorial on reed-solomon coding for fault-tolerance in raid-like systems. Softw. Pract. Exp. **27**(9), 995–1012 (1997)
9. Huang, C., Xu, L.: STAR: an efficient coding scheme for correcting triple storage node failures. IEEE Trans. Comput. **57**(7), 889–901 (2008)
10. Sathiamoorthy, M., et al.: Xoring elephants: novel erasure codes for big data. In: Proceedings of the 39th International Conference on Very Large Data Bases, PVLDB 2013, pp. 325–336. VLDB Endowment (2013)
11. Huang, C., et al.: Erasure coding in windows azure storage. In: Proceedings of the 2012 USENIX Conference on Annual Technical Conference, USENIX ATC 2012, Berkeley, CA, USA, p. 2. USENIX Association (2012)
12. Li, M., Shu, J., Zheng, W.: Grid codes: Strip-based erasure codes with high fault tolerance for storage systems. ACM Trans. Storage (TOS) **4**(4), 15 (2009)
13. Hafner, J.L.: Hover erasure codes for disk arrays. In: 2006 International Conference on Dependable Systems and Networks, DSN 2006, pp. 217–226. IEEE (2006)
14. Cheng, Z., et al.: ERMS: an elastic replication management system for HDFS. In: 2012 IEEE International Conference on Cluster Computing Workshops, Cluster Workshops, pp. 32–40. IEEE (2012)
15. Li, R., Hu, Y., Lee, P.P.: Enabling efficient and reliable transition from replication to erasure coding for clustered file systems. In: 2015 45th Annual IEEE/IFIP International Conference on Dependable Systems and Networks (DSN), pp. 148–159. IEEE (2015)
16. Tang, Y., et al.: MICS: mingling chained storage combining replication and erasure coding. In: 2015 IEEE 34th Symposium on Reliable Distributed Systems, SRDS, pp. 192–201. IEEE (2015)
17. Ma, Y., Nandagopal, T., Puttaswamy, K.P., Banerjee, S.: An ensemble of replication and erasure codes for cloud file systems. In: 2013 Proceedings IEEE INFOCOM, pp. 1276–1284. IEEE (2013)
18. Rashmi, K.V., et al.: A solution to the network challenges of data recovery in erasure-coded distributed storage systems: a study on the facebook warehouse cluster. In: Proceedings of the 5th USENIX Conference on Hot Topics in Storage and File Systems, Berkeley, CA, USA, pp. 3–8 (2013)
19. Ghemawat, S., Gobioff, H., Leung, S.T.: The google file system. In: ACM SIGOPS Operating Systems Review, vol. 37, pp. 29–43. ACM (2003)
20. Plank, J.S.: T1: erasure codes for storage applications. In: Proceedings of the 4th USENIX Conference on File and Storage Technologies, pp. 1–74 (2005)
21. Weatherspoon, H., Kubiatowicz, J.D.: Erasure coding vs. replication: a quantitative comparison. In: Druschel, P., Kaashoek, F., Rowstron, A. (eds.) IPTPS 2002. LNCS, vol. 2429, pp. 328–337. Springer, Heidelberg (2002). https://doi.org/10.1007/3-540-45748-8_31

22. Plank, J.S., Luo, J., Schuman, C.D., Xu, L., Wilcox-O'Hearn, Z.: A performance evaluation and examination of open-source erasure coding libraries for storage. In: Proceedings of the 7th Conference on File and Storage Technologies, Berkeley, CA, USA, pp. 253–265 (2009)

23. Dimakis, A.G., Godfrey, P.B., Wu, Y., Wainwright, M.J., Ramchandran, K.: Network coding for distributed storage systems. IEEE Trans. Inf. Theory **56**(9), 4539–4551 (2010)

24. Fan, B., Tantisiriroj, W., Xiao, L., Gibson, G.: DiskReduce: RAID for data-intensive scalable computing. In: Proceedings of the 4th Annual Workshop on Petascale Data Storage, pp. 6–10. ACM (2009)

25. Facebook: Erasure coded HDFS, November 2011. https://github.com/facebook/hadoop-20

26. Alluxio Open Foundation: Alluxio (2012). http://www.alluxio.org/

27. Subramanyam, R.: HDFS heterogeneous storage resource management based on data temperature. In: 2015 International Conference on Cloud and Autonomic Computing, ICCAC, pp. 232–235. IEEE (2015)

28. Zhou, W., Feng, D., Tan, Z., Zheng, Y.: PAHDFS: preference-aware hdfs for hybrid storage. In: Wang, G., Zomaya, A., Perez, G.M., Li, K. (eds.) ICA3PP 2015. LNCS, vol. 9529, pp. 3–17. Springer, Cham (2015). https://doi.org/10.1007/978-3-319-27122-4_1

29. Wang, T., et al.: BurstMem:: a high-performance burst buffer system for scientific applications. In: 2014 IEEE International Conference on Big Data, Big Data, pp. 71–79. IEEE (2014)

30. Shu, P., Gu, R., Dong, Q., Yuan, C., Huang, Y.: Accelerating big data applications on tiered storage system with various eviction policies. In: 2016 IEEE Trustcom/BigDataSE/ SPA, pp. 1350–1357. IEEE (2016)

31. Chen, Y., Ganapathi, A., Griffith, R., Katz, R.: The case for evaluating mapreduce performance using workload suites. In: 2011 IEEE 19th International Symposium on Modeling, Analysis & Simulation of Computer and Telecommunication Systems, MASCOTS, pp. 390–399. IEEE (2011)

32. Chen, Y., Alspaugh, S., Katz, R.: Interactive analytical processing in big data systems: a cross-industry study of mapreduce workloads. Proc. VLDB Endowment **5**(12), 1802–1813 (2012)

imBBO: An Improved Biogeography-Based Optimization Algorithm

Kai Shi[1,2], Huiqun Yu[1(✉)], Guisheng Fan[1(✉)], Xingguang Yang[1],
and Zheng Song[3]

[1] Department of Computer Science and Engineering,
East China University of Science and Technology, Shanghai 200237, China
BH4AWS@163.com, xingguang2955@163.com,
{yhq,gsfan}@ecust.edu.cn
[2] Shanghai Key Laboratory of Computer Software Evaluating and Testing,
Shanghai 201112, China
[3] The Third Research Institute of the Ministry of Public Security,
Shanghai, China
songzheng@stars.org.cn

Abstract. Biogeography based Optimization (BBO) is a new evolutionary optimization algorithm based on the science of biogeography for global optimization. However, its direct-copying-based migration and random mutation operators make it easily possess local exploitation ability. To enhance the performance of BBO, we propose an improved BBO algorithm called imBBO. A hybrid migration operation is designed to further improve the population diversity and enhance the algorithm exploration ability. Empirical results demonstrate that our imBBO effectively gains the high optimization performance by comparing with the original BBO and three BBO variants for 23 out of 30 CEC'2017 benchmarks. Moreover, our imBBO presents a faster convergence speed.

Keywords: Hybrid migration · Biogeography-based Optimization · Global optimization

1 Introduction

Optimization problems are of increasing importance in modern science and engineering fields, especially for the global continuous optimization problem. During the past decade, they have turned to be more complicated and diversified commensurate with the unceasing progress of science and technology [41]. The major challenge of the global continuous optimization is that the problems to be optimized may have many local optima. This issue is particularly challenging when the dimension is high [15]. Thus, numerous optimization techniques have been advanced to handle these problems [2, 20]. Currently, the most popular method is meta-heuristics, such as Genetic Algorithms (GAs) [6], Evolution Strategy (ES) [42], Particle Swarm Optimization (PSO) [21], Differential Evolution (DE) [40], Ant Colony Optimization (ACO) [9] and Biogeography-based Optimization (BBO) [36].

© Springer Nature Switzerland AG 2019
S. Li (Ed.): GPC 2018, LNCS 11204, pp. 284–297, 2019.
https://doi.org/10.1007/978-3-030-15093-8_20

Biogeography-based Optimization (BBO) proposed by Dan Simon in 2008 is a novel Evolutionary Algorithm (EA) for global optimization based on the equilibrium theory of island biogeography. Different from other population-based algorithms, in BBO, poor solutions can improve their qualities by accepting new features from good ones [19]. Concretely, a habitat in BBO algorithm is called an island and a group of habitats construct the ecosystem. The habitat suitability index (HSI) presents the habitability of an island, which can be analogized as the fitness values to evaluate the problems. The HSI is always determined by a series of independent decision variables called Suitability Index Variable (SIV), such as the features of temperature, disease and earthquake in real world. Specially, the most important characteristics in BBO are migration and mutation operation. The migration, including immigration and emigration process, is designed to conditionally share the SIV information among habitats in an ecosystem. The mutation is a probabilistic operation that can randomly modify the habitat SIVs based on the a priori probability of the habitat.

Similar to other EAs, the BBO algorithm has also some certain weaknesses. The probabilistic migration can make the population share different information among solutions to guide good exploitation ability [14]. However, *its directcopying-based migration and random mutation operators make BBO lack enough exploration ability and cannot improve the diversity of population* [15, 41]. Although its convergence speed is relatively fast at the beginning of the evolutionary process, it easily falls into local optima. To mitigate these weakness, an improved BBO variant (called imBBO) is proposed in this paper.

Our contributions of this paper are summarized as follows:

- An imBBO algorithm is proposed to mitigate part weakness of BBO for global optimization problems, which is composed of a hybrid migration operation and a scalable direction mutation operation.
- A hybrid migration operation is designed based on the combination of the DE theory and a scalable method, which helps our imBBO to improve the population diversity and enhance the algorithm exploration ability.
- We conduct an optimization performance comparison among our imBBO, the original BBO and three BBO variants. Empirical results demonstrate that our algorithm effectively outperforms the competitors for 23 out of 30 CEC'2017 benchmarks. Moreover, our imBBO presents a faster convergence speed.

The rest of this paper is organized as follows. Related works are discussed in Sect. 2. Section 3 defines the problem formulation. Section 4 proposes our imBBO algorithm. We describe our experimental setup in Sect. 5 and present the results in Sect. 6. Section 7 discusses our experimental results by answering two research questions. Section 8 concludes the paper.

2 Related Work

For global optimization problem, many references have showed that meta-heuristic algorithms [30], including the categories of nature-inspired meta-heuristic algorithms, physics-based algorithms and swarm-based methods, become much more popular to solve these problems or various engineering applications [28]. Part of typical algorithms include the first category of Genetic Algorithms (GAs) [6], Genetic Programming (GP) [29] and Biogeography-based Optimization (BBO) [36], the second category of Simulated Annealing (SA) [5], Gravitational Search Algorithm (GSA) [33] and Artificial Chemical Reaction Optimization Algorithm (ACROA) [1] and the third category of Particle Swarm Optimization Algorithm (PSO) [21], Ant Colony Optimization (ACO) [9] and Artificial Bee Colony Algorithm (ABC) [18].

This paper focuses on the Biogeography-based Optimization [36], which shows the excellent performance on various unconstrained or constrained benchmarks [7]. During the last decade, more BBO variants are proposed and are available in literature [8, 24, 37]. On the one hand, Mehmet et al. proposed an oppositional Biogeography-based Optimization [11] which is composed of the opposition based learning and migration rates of original BBO in 2009. To enhance the mutation operation, a real coded BBO algorithm [15] was proposed in 2010. Similarly, there are some additional BBO variants that the migration operation is also modified, such as the literatures [13, 17, 19, 20, 22, 41]. Moreover, Haiping Ma analyzed the equilibrium of migration models [23]. Haiping Ma and Dan Simon discussed migration models using Markov theory [25] and the blended BBO [26] for constrained optimization. Considering the random initial generation of population, Simon et al. focused on the re-initialization and local search in Linearized BBO [38].

On the other hand, Wenyin et al. proposed a combination DE/BBO [14] that combines the DE algorithm with BBO to improve the searching capability. Moreover, a hybrid BBO [27] was proposed in 2014, which combine the various EAs with BBO in different ways. That is, two types of hybridization named as iteration-level hybridization and algorithm-level hybridization are used.

Naturally, BBO has been widely applied to solve the real-world and engineering problems, such as the sensor selection [36], power system optimization [32], economic load dispatch [4] and antenna design [12, 39].

3 Problem Definition

For the different areas of engineering or scientific application, the optimization problems should be solved to achieve approximate optimal solutions. However, different problems may have the special constraints and conflicting objectives. So an effective method is to design a global search algorithm to find these optimal or near-optimal solutions.

Without loss of generality, the mathematical expression presents that the unconstrained continuous global minimization problem can formulize as a pair (S, f), where the $S \subseteq R^D$ is a bounded set based on R^D and $f : S \rightarrow R$ is a Ddimensional real-valued function. Finally, the purpose of these problems is to find a point $X^* \in S$ [15] and the

X^* belongs to a D-dimensional ($D \in \{1, 2, 3, \cdots\}$) vector. Thus, the $f(X^*)$ value is the global minimum on $S[1]$. More specifically, it is required to find an $X^* \in S$, as the formula 1:

$$\forall X \in S : f(X^*) \leq f(X) \tag{1}$$

Note that the function f does not need to be continuous but it must be bounded. Moreover, the different variables contain the different bound in realworld constraint optimization problems. We only consider the unconstrained and continuous functions optimization in this paper.

As mentioned in above, the major challenge in global continuous optimization problem is that the optimization problems to be solved may easily lead EAs to trap into local optima. These issues are particularly challenging when the problem has the high dimension. So, one of the effective methods is that different EAs adopt the special search and modification method for different optimization problems.

4 Our Approach

In this section, we firstly introduce our algorithm implementation in Algorithm 1. Then, a hybrid migration operation is discussed in detail.

Algorithm 1. The main algorithm structure of imBBO
Input: objectives, constraint condition, maximum iterations and migration rate, population size NP, different algorithm parameters, etc.
Output: the optimal objective fitness, solutions and iteration optimum curves
1: **Begin**
2:　　　Generating the initial population
3:　　　Setting relevant algorithm Parameters
4:　　　Evaluating fitness values for each individual in NP
5:　　　**while** *the halting criterion is not satisfied* **do**
6:　　　　　Elites inheritance operation
7:　　　　　**for** *each individual* **do**
8:　　　　　　　modifying the number of species
9:　　　　　**end**
10:　　　　Calculating migration rate λ_i and μ_i for each habitat X_i
11:　　　　Modifying the population size with a hybrid migration operation shown in **Section 4.2**
12:　　　　Enhancing exploration with the mutation operation referring from [34]
13:　　　　Other mechanisms, such as data validation and boundary checking
14:　　　　Evaluating the fitness for each individual in NP
15:　　　　Elites selection operation
16:　　　**end**
17: **end**

4.1   The Structure of imBBO

Our algorithm structure is showed in Algorithm 1, which is mainly composed of a hybrid migration operation introduced in Sect. 4.2 and a scalable direction mutation operation derived from our previous work [34].

Concretely, the hybrid migration operation combines the DE theory with a scalable method to further improve the population diversity, exploit the population information and enhance the algorithm exploration ability. Moreover, considering the excellent algorithm performance of iCPBBOCO [34], we still try to adopt the same mutation operation into our imBBO algorithm.

To make a fair comparison, all parameters values of our algorithm are set by referring to [14, 34, 36], which is shown in Table 1. The modification of migration and mutation operation are marked from Line 11 to Line 12 in Algorithm 1.

4.2   Hybrid Migration Operation

A hybrid migration operation is deployed in our imBBO algorithm, which combines the DE theory [14, 31] with the component of migration operation from our previous iCPBBOCO algorithm [34]. The core idea of this proposed method is that good solutions would be less destroyed, while poor solutions can accept a lot of new features from good solutions. Furthermore, the implementation is listed in formula 2, which is composed of the relevant DE theory and a scalable method in formula 3 and formula 4, respectively. To enhance the diversity, we define a parameter ξ to further guide the specific migration operation.

$$Migraiton \begin{cases} Formual & 3 & rand \leq \xi \\ Formual & 4 & rand > \xi \end{cases} \tag{2}$$

For operations of the DE theory, the main mathematical principle is illustrated in formula 3. Recently, the DE algorithm is used for global permutation based combinatorial optimization problems [14], successfully. It is good at exploring the search space and locating the region of global minimum. It uses the distance and direction information from the current population and the characteristics of problem to guide the further search. So, we try to adopt it to our migration operation.

$$\begin{aligned} U_i(j) &= \begin{cases} U_i(j) + f_1 * d_1 + f_2 * d_2 & rand \leq \kappa \\ U_i(j) - f_1 * d_3 - f_2 * d_2 & rand > \kappa \end{cases} \\ d_1 &= U_1(j) - U_{r_1}(j), d_2 = U_{r_2}(j) - U_{r_3}(j) \\ d_3 &= U_{NP}(j) - U_{r_1}(j) \end{aligned} \tag{3}$$

To inherit the excellent performance, we define the standard parameters setting of DE theory in our algorithm. In formula 3, the variable $U_i(j)$ presents the $j - th$ decision variable in $i - th$ solution. According to the principle of DE theory [14, 16, 31], we randomly select three additional population r_1, r_2 and r_3, where $r_1 \neq r_2 \neq r_3$. A series of variables, such as d_1, d_2 and d_3, are the difference values of corresponding populations.

Then, two scale factors of f_1 and f_2 are related with the lower/upper bound for immigration probability per individual, which indicates that how much amount of differential variation [10] will influence on target. Similarly, we also define the guide parameter κ.

$$U_i(j) = \begin{cases} \partial * U_i(j) + (1 - \partial) * U_j(j) & rand \leq \varphi \\ (1 - \partial) * U_i(j) + \partial * U_j(j) & rand > \varphi \end{cases} \tag{4}$$

In description of formula 4, it is an additional method to perform the migration operation. The new offspring solution comes from a different combination of the source solution of U_i and a target solution of U_j. We define a scalable factor ∂ to decide the migration size of individuals. To maintain the stability, we exchanges the scalable coefficient ∂ among each other. That is, the purpose is to apply for asymmetrical migration, further enhance potential population diversity and exploit the population information.

5   Experimental Setup

In this section, we describe settings of our conducted experiments to evaluate ouralgorithm performance. Concretely, we detail the benchmarks, the performance criteria and algorithms setting.

5.1   Benchmark Functions

To evaluate the performance of the proposed algorithm, 30 benchmarks from CEC'2017 [3] which is the latest set of benchmarks are employed in our experiments. These benchmarks are divided into four categories, including unimodal functions (F01-F03), simple multimodal functions (F04-F10), hybrid functions (F11-F20) and composition functions (F21-F30). *The more complex benchmarks in evaluation process, the better performance superiority of competition will be shown.*

Currently, these benchmarks from CEC'2017 [3] are related with the real-parameter single objective optimization without making use of the exact equations of the test functions. Some benchmarks are developing with novel features such as new basic problems, composing test problems by extracting features dimension-wise from several problems, graded level of linkages, rotated trap problems, and so on.

5.2   Performance Mertrics

We evaluate the performance of our algorithm in terms of two aspects [14, 34], which are described as follows in detail.

- **Error:** The error value of a solution X is defined as $f(X) - f(X^*)$, where X^* is the standard, global optimization of the objective. The minimum average error is recorded when the maximum number of fitness function evaluations ($maxFEs$) reached in 30 independent runnings. Moreover, we also calculate the average median values.

- **Convergence graphs:** We present the convergence speed of our algorithm compared to the competitors. In order to observably demonstrate differences, we recalculate all of error values by the logarithm (log) function.

5.3 Algorithm Settings

Table 1 lists the values of key parameters of imBBO. To enable a fair performance comparison to the competitors, we use the same settings as the ones reported by [14, 15, 19, 20, 34, 36]. All of algorithms are developed by Matlab. We generate the initial population by uniform random initialization within the search space. According to [3], we set the problem dimension $D = 100$. That is, the search rang is $[-100, 100]$. Moreover, we define the *maxFEs* is 10000 * 10. All algorithms need to be terminated when reaching *maxFEs* or the error value is smaller than 10^{-08}.

Table 1. Overview of parameters setting of imBBO and BBO variants.

Parameter (imBBO and Variant BBOs)	Default
Population size: NP	100
Habitat modification probability	1.0
Mutation probability	0.005
Maximum migration rate: $I\&E$	1.0 & 1.0
Number of elites	2
Migration scaling factor: ∂	0.3
Migration guide parameter of ξ, κ, φ	0.5

The competitors in our experiments include original BBO [36] and three BBO variants. They are the MOBBO [20], the PBBO [19] and the RCBBO [15] algorithms.

5.4 Measurement Settings

All measurements of each algorithm are performed on Windows 7 (64bit) Machine with Intel Core i5-4690 K CPU 3.5 GHz and 16 GB RAM. Each algorithm variant is compute-bound and not memory intensive.

To reduce measurement fluctuations caused by randomness (e.g., the randomness of performing migration and mutation), we independently execute each algorithm 30 times. We take both the median and mean values of the measurements for analysis.

6 Results

In this section, we report the experiment results in detail. we aim at answering the following two research questions.

RQ1: What is the performance of our imBBO, compared to the BBO and three BBO variants? (Section 6.1)

RQ2: How fast is the convergence speed of our imBBO? (Section 6.2).

6.1 Performance Results

Table 2 records the experimental results of the comparison between our imBBO algorithm and competitors when applied to 30 CEC'2017 benchmarks in reaching to *maxFEs*. The columns BBO, imBBO, MOBBO, PBBO and RCBBO list the measured results of each algorithm. We report the median and mean values for 30 executions of each algorithm. The rows of the table record the measured details for each benchmark. Moreover, we highlight the median and mean values in bold, which are the best for each benchmark.

Our experimental results reveal that our imBBO outperforms the competitors for 23 out of 30 CEC'2017 benchmarks, except for F1, F3, F4, F7, F12, F17 and F24. Analyzing the experimental results, our algorithm achieve the better algorithm performance of the hybrid and composition functions by comparing with the unimodal and simple multimodal functions. Hence, we conjecture that the components of our imBBO make an effect to be the complex searching.

6.2 Convergence Speed

It is interesting to understand the convergence speed of our approach compared to the others. The convergence is an important metric to illustrate whether or not a algorithm has reached to the steady state.

Our experimental results demonstrate that the convergence speed of imBBO is much faster than other competitors, that part of representative curves are shown in Fig. 1 (blue line).

Table 2. Experimental results of the best imBBO algorithm and four competitors (BBO, MOBBO, PBBO and RCBBO) for CEC'2017 continuous optimization benchmarks. The number in gray indicates a better case than others.

Fun	BBO		imBBO		MOBBO		PBBO		RCBBO	
	Mean	Median	Mean	Median	Mean	Median	Mean	Median	Mean	Median
F1	9.785465559e+05	9.785465597e+05	4.41871763e+03	7.1507775e+03	2.53159472e+03	2.53170605e+03	4.1612408e+03	4.1613762e+03	1.71890871e+03	1.71896613e+03
F2	2.000000000e+02	2.00000000e+02	2.00000000e+02	2.00000000e+02	2.00000000e+02	2.00000000e+02	2.00000000e+02	2.00000000e+02	2.00000000e+02	2.00000000e+02
F3	3.02100739e+02	3.03155855e+02	3.01816730e+02	3.03558234e+02	3.00001702e+02	3.00006563e+02	3.00000178e+02	3.00000796e+02	3.00000027e+02	3.00000226e+02
F4	4.00346418e+02	9.33701140e+02	4.036373026e+02	5.43359428e+02	4.00071176e+02	2.76890951e+02	4.000407536e+02	7.48551891e+02	4.00041288e+02	4.00041288e+02
F5	5.06478575e+02	5.32419313e+02	5.03958244e+02	5.91300152e+02	5.27195472e+02	6.17464876e+02	5.07296361e+02	5.24968718e+02	5.29039986e+02	5.29039986e+02
F6	6.00770684e+02	6.00770684e+02	6.000317436e+02	6.00055788e+02	6.00000019e+02	6.00000028e+02	6.0000179e+02	6.00000324e+02	6.00000148e+02	6.00000212e+02
F7	7.17307487e+02	7.65299560e+02	7.14000767e+02	8.36989979e+02	7.39455246e+02	8.37840225e+02	7.131157346e+02	7.51249471e+02	7.15929047e+02	7.64095823e+02
F8	8.07056264e+02	8.30904774e+02	8.04645032e+02	8.92635597e+02	8.14924371e+02	8.75000612e+02	8.09617935e+02	8.28257554e+02	8.05638100e+02	8.22949725e+02
F9	9.00230815e+02	9.002308315e+02	9.00006111e+02	9.00042873e+02	9.00000002e+02	9.00000010e+02	9.00000001e+02	9.00000003e+02	9.00000002e+02	9.00000002e+02
F10	1.14053976e+03	1.69256700e+03	1.09599842e+03	3.87124069e+03	1.43290929e+03	3.44390925e+03	1.125363026e+03	1.5309751e+03	1.27499313e+03	1.60924493e+03
F11	1.10434240e+03	1.10488514e+03	1.103285366e+03	1.10418539e+03	1.10670450e+03	1.10673822e+03	1.10329167e+03	1.1037043e+03	1.10382540e+03	1.10427200e+03
F12	1.48175677e+05	1.52795682e+05	1.00555983e+04	1.01425263e+04	1.13226031e+04	1.14057230e+04	1.18658035e+04	1.19145648e+04	8.424520200e+03	8.45423220e+03
F13	6.557960900e+04	8.42129974e+04	1.41701869e+03	5.24908604e+03	5.45733377e+03	5.45767596e+03	7.48492112e+03	7.48649670e+03	5.91490499e+03	5.91603604e+03
F14	1.00071584e+04	1.00071584e+04	1.41241283e+03	1.41286831e+03	4.72304439e+03	4.72361057e+03	5.60220131e+03	5.60246475e+03	2.43395772e+03	2.43395772e+03
F15	1.50526300e+04	1.74918763e+04	1.508564600e+03	1.50920856e+03	2.07423301e+03	2.09189423e+03	7.63392195e+03	7.64775851e+03	7.63849994e+03	7.71640052e+03
F16	1.69425497e+03	2.20331472e+03	1.600517530e+03	2.37628643e+03	1.75793246e+03	3.07561961e+03	1.70538053e+03	2.14143763e+03	1.70278290e+03	2.25736477e+03
F17	1.70881113e+03	1.70884909e+03	1.71149583e+03	1.71282842e+03	1.71774219e+03	1.71830341e+03	1.70233694e+03	1.70234369e+03	1.70490187e+03	1.70690651e+03
F18	3.97270951e+04	5.79282919e+04	1.940850180e+03	2.28397015e+04	1.96776333e+04	2.07415248e+04	7.935789127e+03	8.24699658e+03	9.15867099e+03	9.64824598e+03
F19	1.44886269e+04	1.48099770e+04	1.90225673e+03	1.90271468e+03	1.10238448e+04	1.18148600e+04	7.11491750e+03	7.18647289e+03	7.43230272e+03	7.50711877e+03
F20	2.00924831e+03	2.04420085e+03	2.000035726e+03	2.39494215e+03	2.00826263e+03	2.50251636e+03	2.00304019e+03	2.02695457e+03	2.00489042e+03	2.03422884e+03
F21	2.30685508e+03	2.34346330e+03	2.268158426e+03	2.38344694e+03	2.28357371e+03	2.54850391e+03	2.30426053e+03	2.33671683e+03	2.30707922e+03	2.33465071e+03
F22	2.30198510e+03	2.46215658e+03	2.279929706e+03	2.41139681e+03	2.99967460e+03	3.45666747e+03	2.30055774e+03	2.48318293e+03	2.30255758e+03	2.46838853e+03
F23	2.61084751e+03	2.64257156e+03	2.605999576e+03	2.69254276e+03	2.61590866e+03	3.08815452e+03	2.61131148e+03	2.63693551e+03	2.61064042e+03	2.63504022e+03
F24	2.73755609e+03	2.79252244e+03	2.73783730e+03	2.82845379e+03	2.74936662e+03	3.04478493e+03	2.737225866e+03	2.77454706e+03	2.73671628e+03	2.78472691e+03
F25	2.93955468e+03	3.22498366e+03	2.929250046e+03	3.06733172e+03	2.93207429e+03	3.85377698e+03	2.93407338e+03	3.20763851e+03	2.93769720e+03	3.18967729e+03
F26	2.91266154e+03	3.41706205e+03	2.896800506e+03	3.20103977e+03	2.93612574e+03	4.83753566e+03	2.93812322e+03	3.36356031e+03	2.93472024e+03	3.38539297e+03
F27	3.09152760e+03	3.14130899e+03	3.089187256e+03	3.12398320e+03	3.09817376e+03	3.76244369e+03	3.09398997e+03	3.13456653e+03	3.09287695e+03	3.13275115e+03
F28	3.23216686e+03	3.51546787e+03	3.11000900e+03	3.38377055e+03	3.32476041e+03	4.22682020e+03	3.32606876e+03	3.50122091e+03	3.29696272e+03	3.49274291e+03
F29	3.15241031e+03	5.62143058e+03	3.139521046e+03	4.83003860e+03	3.15871975e+03	3.66231126e+03	3.15836831e+03	1.10467175e+05	3.15406069e+03	3.98577471e+04
F30	7.24472571e+03	7.40423622e+03	5.167328906e+03	5.50406827e+03	1.46830694e+04	1.92864281e+04	6.24876854e+03	8.50512952e+03	7.73866767e+03	9.27910406e+03

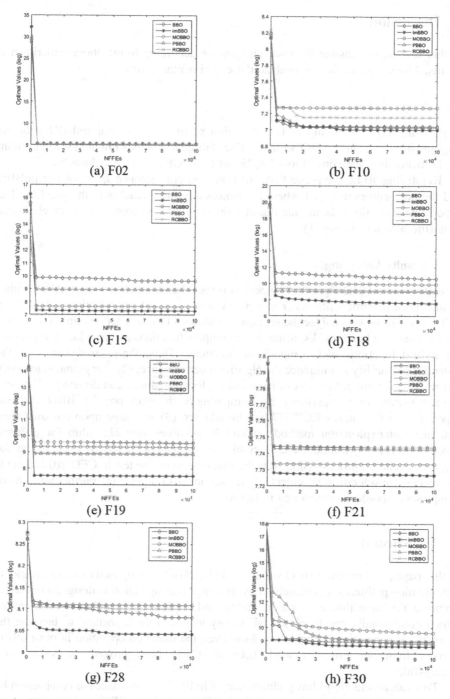

Fig. 1. The convergence curves of imBBO algorithm and four competitors (BBO, MOBBO, PBBO and RCBBO) for F02, F10, F15, F18, F19, F21, F28 and F30 benchmarks. The X-axis shows the number of iteration (NFFEs). The Y-axis presents the algorithm values of each iteration (log-error). (Color figure online)

7 Discussion

In this section, we answer the research question according to the above experimental results. Moreover, we further analyze the experimental results.

7.1 Research Questions

Regarding RQ1, we compare the algorithm performance of our imBBO with the original BBO and three BBO variants. The statistical results show that our algorithm works better than the competitors for 23 out of 30 CEC'2017 benchmarks.

Regarding RQ2, we present part of functions convergence speed of our imBBO and the competitors in Fig. 1 when the evolution running reaches to the *maxFEs*. The experimental results indicate that our algorithm quick converges to a relatively stable state (the blue line in Fig. 1).

7.2 Results Analyzing

Although our imBBO algorithm outperforms the other competitors, the final algorithm results present some differences compared with standard optimization values from CEC'2017. Thus, we analyze the reasons as follows: (1) The functions from CEC'2017 are the latest test benchmarks. Much more complex functions are introduced, especially for hybrid functions and composition functions. Thus, these conditions take the potential probability to influence the algorithm performance. (2) Our global is to verify our imBBO for mitigating part of issues of exploration ability and diversity. We focus on the performance superiority by comparing with other popular BBO variants, especially for the latest CEC'2017 benchmarks set. (3) We insist upon our own view that different exploration methods should be involved into algorithm for different objectives. That is, different characteristics of objectives should be analyzed at the begin of evolution process. Since it is the first time to do the test in CEC'2017, there is no special consideration of objectives in our imBBO algorithm. Furthermore, more components should be developed in future.

8 Conclusion

In this paper, an improved BBO variant called imBBO, is proposed to solve the global optimization problems. Concretely, a hybrid migration operation is designed to further improve the algorithm exploration ability and exploit the population information, which conditionally combines the DE theory with a scalable method to increase the diversity of population in formula 4. Moreover, the mutation operation in our imBBO derives from our previous work because it has been proved its performance successfully.

To evaluate the algorithm performance of imBBO, we conduct the comparison by evaluating our algorithm to the original BBO and three BBO variants based on 30 CEC'2017 benchmarks [3] with different characteristics. Empirical experimental results demonstrate that our algorithm effectively outperforms the competitors for 23 out of 30 benchmarks.

In future work, the influence of population size, other parameters tuning and the problem dimension will be further studied. Additional, this research just focus on the unconstrained global numerical optimization problems. Another work will extend our imBBO to address some constrained, real-world optimization problems, such as virtual machine consolidation problems [35, 43].

Acknowledgments. This work is partially supported by the NSF of China under grants No. 61772200, 61702334 and No. 61472139, Shanghai Pujiang Talent Program under grants No. 17PJ1401900, Shanghai Municipal Natural Science Foundation under Grants No. 17ZR1406900 and 17ZR1429700, Educational Research Fund of ECUST under Grant No. ZH1726108, the Collaborative Innovation Foundation of Shanghai Institute of Technology under Grants No. XTCX2016-20, the Opening Project of Key Lab of Information Network Security of Ministry of Public Security Under No. C17604, Key Lab of Information Network Security of Ministry of Public Security Under No. C17604.

References

1. Alatas, B.: ACROA: artificial chemical reaction optimization algorithm for global optimization. Expert Syst. Appl. **38**(10), 13170–13180 (2011)
2. Arora, J.S.: Jan A. Snyman, practical mathematical optimization: an introduction to basic optimization theory and classical and new gradient-based algorithms. Struct. Multi. Optim. **31**(3), 249–249 (2006)
3. Awad, N., Ali, M., Liang, B., Qu, B., Suganthan, P.: Problem definitions and evaluation criteria for the CEC 2017 special session and competition on single objective bound constrained real-parameter numerical optimization. Technical report (2016). http://www.ntu. edu.sg/home/EPNSugan/index_files/CEC2017
4. Bhattacharya, A., Chattopadhyay, P.K.: Hybrid differential evolution with biogeography-based optimization algorithm for solution of economic emission load dispatch problems. Expert Syst. Appl. **38**(11), 14001–14010 (2011)
5. Černý, V.: Thermodynamical approach to the traveling salesman problem: AN efficient simulation algorithm. J. Optim. Theory Appl. **45**(1), 41–51 (1985)
6. Deep, K., Thakur, M.: A new mutation operator for real coded genetic algorithms. Appl. Math. Comput. **193**(1), 211–230 (2007)
7. Du, D., Simon, D., Ergezer, M.: Biogeography-based optimization combined with evolutionary strategy and immigration refusal. In: Proceedings of International Conference on Systems, Man and Cybernetics, San Antonio, USA, pp. 997–1002 (2009)
8. Ekta, M.K.: Biogeography based optimization: a review. In: International Conference on Computing for Sustainable Global Development (2015)
9. Ellabib, I., Calamai, P.H., Basir, O.A.: Exchange strategies for multiple Ant Colony System. Inf. Sci. **177**(5), 1248–1264 (2007)
10. Engelbrecht, A.P.: Computational Intelligence - An Introduction, 2nd edn. Wiley, Hoboken (2007)
11. Ergezer, M., Simon, D., Du, D.: Oppositional biogeography-based optimization. In: Proceedings of the IEEE International Conference on Systems, Manand Cybernetics, San Antonio, USA. pp. 1009–1014 (2009)

12. Feng, S.L., Zhu, Q.X., Gong, X.J., Zhong, S.: Hybridizing biogeography-based optimization with differential evolution for motif discovery problem. Appl. Mech. Mater. **457–458**(4), 309–312 (2014)
13. Garg, V., Deep, K.: Performance of Laplacian biogeography-based optimization algorithm on CEC 2014 continuous optimization benchmarks and camera calibration problem. Swarm Evol. Comput. **27**, 132–144 (2016)
14. Gong, W., Cai, Z., Ling, C.X.: DE/BBO: a hybrid differential evolution with biogeography-based optimization for global numerical optimization. Soft. Comput. **15**(4), 645–665 (2010)
15. Gong, W., Cai, Z., Ling, C.X., Li, H.: A real-coded biogeography-based optimization with mutation. Appl. Math. Comput. **216**(9), 2749–2758 (2010)
16. Jadon, S.S., Tiwari, R., Sharma, H., Bansal, J.C.: Hybrid artificial bee colony algorithm with differential evolution. Appl. Soft Comput. **58**, 11–24 (2017)
17. Kanoongo, S., Jain, P.: Blended biogeography based optimization for different economic load dispatch problem. In: Proceedings of the 25th International Conference on Electrical and Computer Engineering (CCECE), Montreal, QC, Canada, pp. 1–5 (2012)
18. Karaboga, D., Basturk, B.: Artificial bee colony (ABC) optimization algorithm for solving constrained optimization problems. In: Melin, P., Castillo, O., Aguilar, L.T., Kacprzyk, J., Pedrycz, W. (eds.) IFSA 2007. LNCS (LNAI), vol. 4529, pp. 789–798. Springer, Heidelberg (2007). https://doi.org/10.1007/978-3-540-72950-1_77
19. Li, X., Wang, J., Zhou, J., Yin, M.: A perturb biogeography based optimization with mutation for global numerical optimization. Appl. Math. Comput. **218**(2), 598–609 (2011)
20. Li, X., Yin, M.: Multi-operator based biogeography based optimization with mutation for global numerical optimization. Comput. Math Appl. **64**(9), 2833–2844 (2012)
21. Liang, J.J., Qin, A.K., Suganthan, P.N., Baskar, S.: Comprehensive learning particle swarm optimizer for global optimization of multimodal functions. IEEE Trans. Evol. Comput. **10**(3), 281–295 (2006)
22. Lohokare, M.R., Panigrahi, B.K., Pattnaik, S.S., Devi, S., Mohapatra, A.: Neighborhood search-driven accelerated biogeography-based optimization for optimal load dispatch. IEEE Trans. Syst. Man Cybern. Part C **42**(5), 641–652 (2012)
23. Ma, H.: An analysis of the equilibrium of migration models for biogeography-based optimization. Inf. Sci. **180**(18), 3444–3464 (2010)
24. Ma, H., Simon, D.: Biogeography-based optimization with blended migration for constrained optimization problems. In: Proceedings of the International Conference on Genetic and Evolutionary Computation Conference (GECCO), Portland, Oregon, USA, pp. 417–418 (2010)
25. Ma, H., Simon, D.: Analysis of migration models of biogeography-based optimization using markov theory. Eng. Appl. AI **24**(6), 1052–1060 (2011)
26. Ma, H., Simon, D.: Blended biogeography-based optimization for constrained optimization. Eng. Appl. AI **24**(3), 517–525 (2011)
27. Ma, H., Simon, D., Fei, M., Shu, X., Chen, Z.: Hybrid biogeography-based evolutionary algorithms. Eng. Appl. AI **30**, 213–224 (2014)
28. Mirjalili, S., Lewis, A.: The whale optimization algorithm. Adv. Eng. Softw. **95**, 51–67 (2016)
29. O'Reilly, U.: Genetic programming II: automatic discovery of reusable programs. Artif. Life **1**(4), 439–441 (1994)
30. Pholdee, N., Bureerat, S.: Comparative performance of meta-heuristic algorithms for mass minimisation of trusses with dynamic constraints. Adv. Eng. Softw. **75**, 1–13 (2014)
31. Price, K., Storn, R.M., Lampinen, J.A.: Differential Evolution: A Practical Approach to Global Optimization. Springer, Heidelberg (2006). https://doi.org/10.1007/3-540-31306-0

32. Rarick, R.A., Simon, D., Villaseca, F.E., Vyakaranam, B.: Biogeography-based optimization and the solution of the power flow problem. In: Proceedings of the IEEE International Conference on Systems, Man and Cybernetics, San Antonio, TX, USA, pp. 1003–1008 (2009)

33. Rashedi, E., Nezamabadi-pour, H., Saryazdi, S.: GSA: a gravitational search algorithm. Inf. Sci. **179**(13), 2232–2248 (2009)

34. Shi, K., Yu, H., Fan, G., Luo, F.: iCPBBOCO: a combination evaluation algorithm based on the extensional BBO. In: Proceedings of International Conference on Internet of Things (iThings) and IEEE Green Computing and Communications (GreenCom) and IEEE Cyber, Physical and Social Computing (CPSCom) and IEEE Smart Data (SmartData), Chengdu, China, pp. 717–723 (2016)

35. Shi, K., Yu, H., Luo, F., Fan, G.: Multi-objective biogeography-based method to optimize virtual machine consolidation. In: Proceedings of 28th International Conference on Software Engineering and Knowledge Engineering (SEKE), Redwood City, San Francisco Bay, USA, pp. 225–230 (2016)

36. Simon, D.: Biogeography-based optimization. IEEE Trans. Evol. Comput. **12**(6), 702–713 (2008)

37. Simon, D.: A dynamic system model of biogeography-based optimization. Appl. Soft Comput. **11**(8), 5652–5661 (2011)

38. Simon, D., Omran, M.G.H., Clerc, M.: Linearized biogeography-based optimization with re-initialization and local search. Inf. Sci. **267**, 140–157 (2014)

39. Singh, U., Singh, D., Kaur, C.: Hybrid differential evolution with biogeography based optimization for Yagi-Uda antenna design. In: Proceedings of the International Conference on Circuit, Power and Computing Technologies, pp. 1163–1167 (2015)

40. Storn, R., Price, K.V.: Differential evolution - A simple and efficient heuristic for global optimization over continuous spaces. J. Glob. Optim. **11**(4), 341–359 (1997)

41. Xiong, G., Shi, D., Duan, X.: Enhancing the performance of biogeography-based optimization using polyphyletic migration operator and orthogonal learning. Comput. OR **41**, 125–139 (2014)

42. Yao, X., Liu, Y.: Fast evolution strategies. In: Angeline, P.J., Reynolds, R.G., McDonnell, J.R., Eberhart, R. (eds.) EP 1997. LNCS, vol. 1213, pp. 149–161. Springer, Heidelberg (1997). https://doi.org/10.1007/BFb0014808

43. Zheng, Q., et al.: Virtual machine consolidated placement based on multi-objective biogeography-based optimization. Futur. Gener. Comput. Syst. **54**, 95–122 (2016)

An Efficient Consensus Protocol for Real-Time Permissioned Blockchains Under Non-Byzantine Conditions

Gengrui Zhang[1,2]([⊠]) and Chengzhong Xu[1,2]([⊠])

[1] Shenzhen Institutes of Advanced Technology,
Chinese Academy of Sciences, Beijing, China
{gr.zhang, cz.xu}@siat.ac.cn
[2] Shenzhen College of Advanced Technology,
University of Chinese Academy of Sciences, Beijing, China

Abstract. Blockchains are increasingly used in the collaboration between business as a trusted distributed ledger. Coping with massive data transactions raises the requirement of real-time safety of blockchains. The celebrated Raft protocol has limitations of being a consensus protocol for permissioned blockchains where a strong consistency is needed between clients and servers. In this work, we propose a new consensus protocol called Dynasty which ensures the real-time safety and the liveness under all non-Byzantine conditions. We design and implement a three-layer permissioned blockchain framework which tolerates f failures with $2f + 1$ correct servers based on Dynasty. We demonstrate the blockchain as a service in an application of used-vehicle trading management and evaluate the performance of the blockchain framework in terms of throughput and latency. Experimental results show that while the latency in different scales of the system increases as expected, the number of committed transactions per second stabilizes at a point within less than 8% difference after a warming-up period.

1 Introduction

Since Bitcoin [18] has been widely used to exchange for currencies, products and services, the blockchain technology where the transactions are stored in a distributed ledger accomplishes the solution to validate transactions without the presence of any trusted authority. At present, the applications based on blockchains have been extended from digital currency to other financial applications such as Ripple [20], Dash [8], BlackCoin [13] and Hyperledger Fabric [5]. Using blockchains makes it unnecessary to use a third party to provide credit support for the transactions. Each node that joins the system maintains the same chain at any verifiable time and adds the same block when the system reaches a consensus. Depending on the applicable scenarios, the blockchains are divided into two categories, namely, permissionless and permissioned.

Acting as a real-time trusted system is becoming an urgent requirement of blockchains. The real-time safety requests that the blockchain systems are able to provide the real-time consistency of every committed transaction at any time since it has committed. The permissionless blockchains use Proof-of-X [2, 13, 18] based protocols to provide consensus services with no management of the node identity. For example, the Proof-of-Work protocol, firstly used in Bitcoin, guarantees the correctness when no party can control over 51% of the computing power and offers a good scalability of nodes and clients but a poor performance in throughput and latency on a widely-open ledger. Due to the high latency and the low throughput, permissionless blockchains can not meet the requirement in building a real-time service blockchain driven by a high volume of transactions. In addition, the huge energy consumption of Proof-of-Work based protocols in solving cryptographic puzzles reduces the value of the commercial use. On the contrary, the IDs of all other servers are known in the permissioned blockchains since the nodes added in the system are already licensed by the node management. The consensus protocols such as Paxos [15], PBFT [6] and Raft [19] manage the fault-tolerance based on the theory of the state-machine replication, which makes a greatly improved performance in throughput and latency but brings a limited scalability [21]. The above protocols make the permissioned blockchains are well suited to commercial applications such as cross-border transactions between banks and smart contrast between business cooperation. Raft is used by R3 Corda [4], QuorumChain, Hyperledger Kafka [5] to provide a fast blockchain consensus service with high throughput and low latency of transactions. Raft guarantees the tolerance in any $f < n/2$ crashes on n nodes in total.

However, the Raft protocol can not ensure the real-time safety in terms of committing transactions. A majority of cluster does not commit the transaction when the client receives a notify from a Raft-based blockchain system. Although Raft can guarantee the ultimate consistency of the commitment, the real-time consistency also needs to be considered in building permissioned blockchains especially when the transactions have sequential dependencies.

In this paper, we present an efficient consensus protocol called Dynasty which ensures the real-time safety and the liveness in building permissioned blockchains. Furthermore, Dynasty performances a good throughput and scalability with $O(N)$ messages in each round. We design and implement a three-layer permissioned blockchain framework to serve real-time requests of clients based on Dynasty protocol. The blockchain could tolerate f failures with $2f + 1$ correct servers under all non Byzantine conditions such as crash, network delays, packet omission and record tampering. We believe that the trading business model based on this blockchain framework is superior to those centralized, both in data protection and transaction transparency.

In summary, we made the following contributions:

1. *Dynasty consensus protocol.* The new protocol is designed for building real-time permissioned blockchains. Its properties guarantee the real-time safety and the liveness under all non-Byzantine conditions. It handles the consensus with a message complexity of $O(N)$ on broadcasts.
2. *Application in used-vehicle trading.* The implemented trading model for managing the trades of used vehicles is a distributed ledger based on the blockchain framework. It makes the trading transactions recorded chronologically and irreversible. The evaluation suggests that the blockchain based on Dynasty is more practicable in terms of throughput and latency.

The remainder of this paper discusses the related work of consensus problems and blockchain applications (Sect. 2), describes the design and analysis of the Dynasty consensus protocol (Sect. 3), demonstrates the blockchain framework and its three layers (Sect. 4), and presents the case study with a used-vehicle trading application and the evaluation (Sect. 5).

2 Related Work

2.1 Consensus Protocols in Blockchains

The blockchains, whether permissionless or permissioned, are distributed ledgers that rely on consensus protocols to achieve agreements. The consensus problem is to get a group of n processes in a distributed system to agree on a value. A consensus protocol is an algorithm that produces such an agreement. Correct consensus protocols must satisfy the following three conditions [1]: agreement, termination and validity, which means that all processes that decide choose the same value, all non-faulty processes eventually decide, and the common output value is an input value of some process.

Consensus Protocols for Permissionless Blockchains. Bitcoin [18] uses the Proof-of-Work protocol to avoid charging real money by requiring senders to demonstrate that they have expended processing time in solving a cryptographic puzzle. The work-based certification protocols require a lot of consumption to record transactions so that fraudsters pay much more than the reward when they modify the history record. In permissonless blockchains, there are also some other protocols such as Bitcoin-NG [10], Proof-of-Stake [13] and Proof-of-Activity [2] accomplish the consensus even more efficient and applicable in some additional requirements than the PoW protocol.

Consensus Protocols for Permissioned Blockchains. Unlike the permissionless blockchain that opens to the whole network, the nodes that join the system are not required a permission, permissioned blockchain is more suitable for specific cooperative transactions as it faces the users with a certain permission. Some consensus protocols such as Paxos [7, 15, 16], PBFT [6], XFT [17] and Raft [19] could handle the consensus between servers in permissioned blockchains. Although the celebrated PBFT protocol achieves the Byzantine agreement [6, 11, 14], it brings a lot of problems such as the low efficiency of message delivery, the heavy capacity of network and the additional costs of the strong requirement of Byzantine broadcast. PBFT tolerates f Byzantine faults with the cost of $O(N^2)$ by broadcasting signed messages. However, the Byzantine problem rarely happens in such a system where the nodes are licensed to join in. When using PBFT as the consensus protocol of permissioned blockchains, the additional costs in terms of resources and message delivery cause a great loss on the throughput and scalability.

Some protocols such as Paxos and Raft have a better efficiency than PBFT under non-Byzantine conditions. Paxos ensures that a single value among the proposed values is chosen [3, 15]. Although the Basic Paxos meets the requirements of safety and liveness by its two phases of operating actions in accepting proposed values, the livelock could happen when a next propose phase always comes before the previous accept phase, and such disadvantages of Basic Paxos make it not practical enough to build a real-time service blockchain.

Raft guarantees the election safety, leader append-only, log matching, leader completeness and state machine safety. Those proofs are given in its paper in Section 5 [19]. The features of log replication and leader election make it superior to other consensus algorithms, both for understanding and as a foundation for implementation of blockchains.

Although the Raft protocol has so many advantages of being a consensus protocol, it does not meet the requirements of the real-time safety of permissioned blockchains driven by transactions. We will discuss the limitations of Raft in building real-time blockchains in the next Section.

2.2 Blockchain Applications

Blockchain technology benefits shared economy applications as a financial technology (FinTech) in providing a wide variety of services spanning areas such as payment, financial management, insurance business and asset registration. [12] points out that contracts, transactions, and the records of them are among the defining structures in our economic, legal, and political systems. Blockchain adoption has the power to transition new and existing models of insurance, including P2P insurance, parametric insurance and microinsurance, into a new digital age.

The blockchain-based smart contracts have been rapidly developed in the financial economy. These contracts can be partially or fully executed or enforced without human interactions. A smart contract could be enabled by scalable program-instruction that initializes and executes an agreement. These automate payments and the transfer of currency or other assets as negotiated conditions are met. Firms could be built on contracts, from incorporation to buyer-supplier relationships to employee relations [12]. In Hyperledger fabric [5] and Ethereum [22], smart contracts are mostly used more specifically in the sense of general purpose computation that takes place on a blockchain.

3 Design and Analysis of Dynasty Consensus Protocol

In this section, we present the Dynasty protocol in detail. First, we discuss the limitations of Raft in being a consensus protocol for permissioned blockchains. Then we demonstrate the Dynasty protocol and its features in terms of real-time safety, liveness and fault-tolerance.

3.1 Raft and Its Limitations

Raft protocol manages the agreement of producing the same outputs from the identical sequences of commands of the replicated state machine. In Raft, each server runs one state of leader, follower, or candidate at any given time. In normal operations there is exactly one leader and all of the other servers are followers. The system time of Raft is divided into terms. Terms are numbered with consecutive integers, and a single leader manages the cluster until the end of the term.

Leader Election. The servers begin with the initial state as a follower. From Fig. 1(a) we can see that when a timeout happens on a specific follower, it becomes a candidate and starts an election by asking other servers to vote in its term. If the candidate is elected by a majority, it must has received more than half servers' tickets, then it becomes the new leader of the current term and begins to take the authority of leading the consensus progress with other servers.

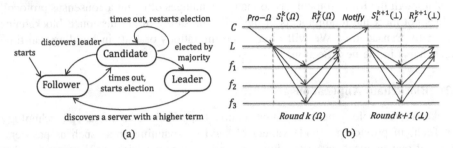

Fig. 1. The three states of server (a) and normal case operations of log replication (b) of Raft

Log Replication. Raft uses remote procedure calls (RPCs) to communicate between servers. During log replications, the leader appends commands to its log as a new entry, then issues AppendEntries RPCs in parallel to each of the other servers to replicate the entry. The arguments in AppendEntries RPC are: term, *leaderId*, *prevLogIndex*, *prevLogTerm*, *entries[]*, *leaderCommit*. The *term* is the leader's current term; the *leaderId* is used for redirecting clients to a new leader when the crash happens on the elder leader; the *prevLogIndex* is the index of log entry immediately preceding new ones; the *prevLogTerm* is the term of *PrevLogIndex* entry; the *entries[]* is log entries to store commands, and it could be empty for heartbeat; the *leaderCommit* is the leader's local commit index. Formally, we denote S_l^k with $\left\{ term, L_{id}, Index_{log}^{prev}, term_{log}^{prev}, Index_{cmt}^l \right\}$ to represent the arguments in AppendEntries RPC in a round k. Thus $S_l^k(\Omega)$ stands for the *entries[]* in AppendEntries RPC is a record Ω. The $S_l^k(\bot)$ represents the heartbeat that contains an empty *entries[]*.

Figure 2(b) shows the normal operations in a 4-node Raft cluster in round k and $k + 1$. In round k, Ω is proposed to the leader by a client. Then, the leader broadcasts $S_l^k(\Omega)$ to followers: f_1, f_2, f_3. Followers log Ω and send back $R_f^k(\Omega)$ to leader. When the leader receives a majority replies of $R_f^k(\Omega)$, the consensus is reached for committing Ω. By this time, leader increases $Index_{cmt}^l$ and notifies the proposed client that Ω has been committed. In Raft, if a majority of the cluster has responded the return value, the commit is made, otherwise the commit is refused.

The Safety Limitations. In a Raft-based blockchain, there is no majority server has committed Ω before the proposed clients are notified by leader. Since followers are only passive to receive the message and give the return value, the followers always increase their $Index_{cmt}^{f}$ by synchronizing the $Index_{cmt}^{l}$ in the next round. As a result, Ω cannot be committed on leader and followers in the same round.

In Raft, the *unsafety condition* happens when the proposed client is notified in round k until a majority cluster commits in round $k + 1$. In log replication, followers never know the consensus result in round k, which results in that the $Index_{cmt}^{f}$ won't be increased until followers synchronize the $Index_{cmt}^{l}$ in $S_{l}^{k+1}(*)$ (where * represents either Ω or $\perp$) in round $k + 1$ even the leader has notified the proposed client. What's worse, the unsafety time could be much longer in the case that the leader l crashes after it notified the proposed client. During a period between l has crashed and a new leader l^{*} with a higher term $t^{*}(t^{*} > t \text{ in } l)$ is elected, the unsafety time gets much longer since there is no $S_{l}^{k+1}(*)$ for followers to synchronize their $Index_{cmt}^{f}$ from $Index_{cmt}^{l}$.

Although the followers could eventually commit Ω according to the properties of Raft, the *unsafety condition* still cannot be tolerated when we use the real-time blockchains as a service. For example, for smart contracts, we assume two contracts A and B where A could be paid individually but B depends on A has been paid first. In above conditions, when the notice that A was accomplished is sent to the proposed client, the contract B could be proposed to the system. At this time, actually, there is no cluster has committed A besides the current leader but B is proposed by client. If the current leader crashes, until a new leader l^{*} is elected, the log replication stops but the transactions are blocking in the committing queue. We do not allow those transactions that have dependency relationship block in the same committing queue. Since the new leader issues $S_{l}^{*}(*)$ in parallel to each server to replicate entries, the contract B could be accomplished before A on a majority server. As a result, the dependence relationship of $A \rightarrow B$ changes to $B \rightarrow A$.

We need a confirmation that a majority of the cluster has stored the record and provides an interface for querying requests before the leader notifies external clients. The Dynasty protocol perfectly solves the *unsafety condition* when we design real-time service blockchains.

3.2 Dynasty Consensus Protocol

Dynasty is a consensus protocol which ensures the real-time safety and the liveness. When designing Dynasty, we refer to the basic idea of strong leader and leader election features of Raft. The Dynasty protocol is totally different in the process of log replication from Raft, we present a new method called two-step commit to ensure the safety property to meet the requirements of building a real-time permissioned blockchain.

Overview. The Dynasty protocol uses a log phase and a commit phase to replicate a proposed record. In the log phase, we denote $S_{l_{log}}^{k}$ with Sym_{log}, $term$, L_{id}, $Index_{log}^{prev}$, $term_{log}^{prev}$, $Index_{cmt}^{l}$, where Sym_{log} is a symbol character to distinguish the RPC type. When a client proposes a record Ω to leader, the leader broadcasts $S_{l_{log}}^{k}$ to other servers to start a consensus process. In the commit phase, the symbol character in $S_{l_{cmt}}^{k}(\Omega)$

changes to Sym_{cmt}. We also $S^k_{l_{cmt}}(\perp)$ where the symbol is $Sym_{\perp}$ to stand for the heartbeat in round k to maintain the authority of the leader.

In Dynasty, heartbeat is only used for resetting followers' timer. Followers never synchronize the $Index^l_{cmt}$ through $S^k_l(\perp)$ and the arguments of $S^k_l(\perp)$ change to $Sym_{\perp}$, term, L_{id}, $Index^{prev}_{log}$, $term^{prev}_{log}$. The $S^k_l(\perp)$ is sent between a fixed interval $Time_{\perp}$ from the leader. To maintain the liveness and stability of the system, we set $Time_f \gg Time_{\perp}$ to tolerate some network delays and guarantee the timeout never happens on followers before the $S^k_l(\perp)$ comes in normal cases.

The *Legality Check (LC)* for receivers in round k includes message-type validity and leader state validity. The follower resets its timer when, $S^k_{l_{cmt}}(\Omega)$ or $S^k_l(\perp)$ passes the *LC*. The requirements are as following:

- *term* of receiver $\leq$ *term* in S^k_l. (both for *log, cmt, $\perp$*)
- $\exists$ term of Ω at $Index^{prev}_{log}$ matches $term^{prev}_{log}$.

Two-step Commit. Dynasty uses two-step commit to ensure that leader always commits a record after a majority cluster has done. In Fig. 2 we illustrate a process of committing Ω in round k. The two-step includes two phases: *Log Phase* and *Commit Phase*, following shows the details.

First Step

- A client (*C*) sends a propose request with record Ω to leader (*L*).
- *Log Phase*:
  - *L* broadcasts $S^k_{l_{log}}(\Omega)$ to other servers.
  - If the received $S^k_{l_{log}}(\Omega)$ passes the *LC*, followers send back $R^k_{f_{log}}(\Omega)$ to *L* indicates that Ω has logged and agrees to commit.

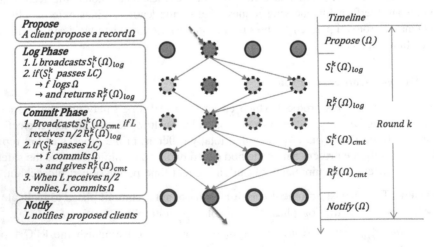

Fig. 2. Two-step commit ensures that the leader always commits after a majority cluster has committed. Timeline shows the happens-before process.

Second Step

- *Commit Phase*:
  - L broadcasts commit requests $S_{l_{cmt}}^k(\Omega)$ to followers when the number of received $R_{f_{log}}^k(\Omega)$ is more than half.
  - If the received $S_{l_{cmt}}^k(\Omega)$ passes the LC, the follower: commits Ω; increases $Index_{cmt}^f$; sends back $R_{f_{cmt}}^k(\Omega)$ to L.
  - Once the number of received $R_{f_{cmt}}^k(\Omega)$ more than $n/2$, the leader L: commits Ω; increases $Index_{cmt}^l$.
- The leader L notifies the proposed client that Ω has been safely committed.

During the normal case, a committed record involves $4N$ messages in total as in each step, the leader only broadcasts $O(N)$ messages and the followers only passively sends back to the leader with $O(1)$ message. Thus, the Dynasty protocol reduces the message complexity from exponential level to a constant degree than PBFT.

View-Change. The current leader in the blockchain may crash down or have a high network latency, which may result in a time that the blockchain system could neither have any communication to external clients nor proceed any consensus process since the followers only passively response to the leader. The view-change begins on these conditions, a new leader needs to be elected to ensures the liveness of the system so as to the blockchain never goes to a termination. Referring to the idea of the leader election in Raft, Dynasty protocol also defines three states of each server: leader, follower, or candidate. A follower increases its term t to $t+1$, starts an election and becomes to a candidate when its timer has a timeout. The new leader comes from the candidate which has received a majority votes in the whole cluster must have the highest term value. We do not modify the election part (Section 5.2 in [19] of Raft since it does not have a bad influence in building the real-time blockchain in this work. The blockchain based on Dynasty cannot provide any service during the election time.

Analysis of Dynasty Protocol

The Dynasty protocol ensures the safety and the liveness properties in building real-time service permissioned blockchains. The two-step commit provides a real-time safety promise in building blockchains than Raft. This helps the Dynasty protocol reduces the conflicts between the interactions with clients and makes the transaction driven more practicable.

Property 1. (Safety) Two-step commit ensures that a majority of cluster has committed before the leader sends a notice to the proposed client.

Proof. In Dynasty, the *unsafety condition* never happens. In normal case, as shown in Fig. 2, the timeline shows that the leader commits a record Ω when it has received majority $R_{f_{cmt}}^k(\Omega)$ replies, so that the notify won't be sent before two-step commit has been accomplished. In the case of leader crashes, the two-step commit also ensures the real-time safety. In the log phase, if the leader crashes, the consensus fails and a new leader l^* continues to broadcast the $S_{l^*}^{k+1}(\perp)$ after being elected, the system aborts the second step of committing Ω. There is no follower commits the logged Ω and the safety is ensured. In the commit phase, if the leader crashes and a new leader must be elected according to the election method. The elected leader must the one that has committed Ω since in the *Legality Check*, it guarantees the leader's $Index_{log}^{prev}$ is more up-to-date than a majority server in the system. The new leader l^* broadcasts $S_{l_{log}^*}^{k+1}(\Omega)$ to form a new process of committing Ω. Above all, the leader always commits after a majority cluster has committed despite of any conditions. □

The liveness is promised by non-stop elections until a new leader is elected in the view-change. A new election must start when a follower's or a candidate's timer triggers a timeout. Assume that the current leader l crashes, after at most $\min\{Timeout_{f_i}\}$ time passes, a candidate begins to start an election with a higher term t^* ($t^* > t$). When multiple candidates get the same number of votes, the election in term t^* fails but a new election must be start after the $\min\{Timeout_{Candi_i}\}$. Since the term of the system never decreases on any conditions, the system never goes back to a previous time. In order to avoid a long period of election, each server's timer is initialized by random times.

When we use Dynasty to implement a real-time permissioned blockchain, after the consensus process, the following works step to store the record from memory to disk and provide an interface for clients to handle query quests.

4   Design of Blockchains Based on Dynasty Protocol

In this section, we demonstrate the structure of the blockchain in detail. In the first subsection, we present the basic idea of the three-layer design, the detailed block structure and the query method. In the second subsection, we give the fault-tolerance discussions of the blockchain.

4.1   Three-Layer Design

The designed blockchain framework is divided into two parts: cluster and clients. The servers with states of leader and follower make up the cluster where the records could be stored after being committed. The client communicates with the internal system by sending propose-request and query-request to the current leader.

The three-layer design is shown in Fig. 3, in the consensus layer, the leader is not only a bridge that communicates with the clients in terms of receiving requests and sending notices, but also manages the consensus with other servers in the blockchain. The storage layer packs the committed records to blocks and stores them on local server once the consensus has been reached. The query layer responds to the query requests from the clients and gives back the result according to the *QMethod*.

Consensus Layer. The protocol in consensus layer is the designed Dynasty in Sect. 3. Consensus layer handles the proposals from clients and maintains the consistency between servers in the blockchain.

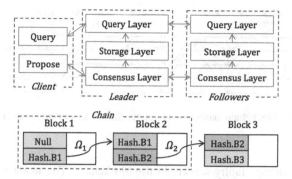

Fig. 3. The three-layer design. The storage layer saves the value passed from the consensus layer. The query layer provides an interface between clients and servers.

Storage Layer. The principle of the block structure in the storage layer is packing a committed record into single block concatenated with a unique *BlockId*. We use the hash of the block's address to act as the unique Id. Each new block holds its own hash and pervious block's hash, and links to a previous one by the previous block's hash up to the first. Figure 3 shows a chain that owns three blocks with one being built. The block 3 is waiting to pack the next record from the consensus layer. Block 2 links to Block 1 and these two are both available to the query layer. In this way, a unique chain is formed on different servers in the whole system.

Once a record is committed, it has been stored in a distributed ledger on different servers. This single record block structure makes it quicker to add a new record into a chain, and the *BlockId* makes it more efficient to the query service to locate a certain block.

Query Layer. In the query layer, the blockchain believes that the correct result is the most consistent result. When the current leader notifies the proposed client, according to Property 1, majority servers must have stored the record and the blocks are accessible for query layer to obtain. We use the *QMethod* strategy that chooses a value among a group of different values according to the majority principle to return the final result. Thus, the *QMethod* gives the correct result when the correct nodes are not less than $f + 1$ when f records are faulty.

4.2   Fault Tolerance

With above designs, this blockchain can tolerant f faults when the correct nodes are not less than $2f + 1$. For example, in a 5-node blockchain of Fig. 4, $S0$ is the leader and the other 4 are followers. $S3$ crashes and $S1$ gets a storage fault. In each step of the consensus process, the leader still gets enough correct replies ($R_{f_{log}}^k$ and $R_{f_{cmt}}^k$) from other servers. Thus, the consensus still could be reached. In terms of querying process, the *QMethod* also gives the correct result since correct servers are more than faulty ones.

In our Dynasty based blockchain, if f faults all happens on the consensus stage, the blockchain needs $f + 1$ well-behaved servers to ensure the enough replies. Similarly, the *QMethod* also need $f + 1$ correct backups to guarantee the correctness of final result

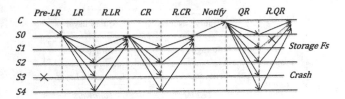

Fig. 4. A five-node blockchain based on Dynasty. Server *S3* crashes, server *S1* has a storage fault.

when there are *f* storage-faulty servers in the blockchain. The worst case is that those storage faulty nodes are non-crash servers so that the safety must be guaranteed at least *2f + 1* correct servers.

5   Case Study

We have implemented the blockchain with roughly 2000 lines of java code, not including tests, comments, or blank lines. We use multi-thread to send and receive packets, distinguish message types, countdown fixed time and other functions. The source code is freely available on GitHub [23]. The reminder of this section we present the application of used-vehicle trading model based on our blockchain framework and the evaluation of throughput and latency.

5.1   Used-Vehicle Trading Model

The traditional business management in used-vehicle trading is centralized so that transactions between buyers and sellers must get through a third party (agency). The buyers can only obtain vehicles' information through those agencies, which may breed business fraud cases such as the concealment of the vehicle's damage, the fake years of used and the missing information of transfers.

In a blockchain based trading management, the transactions and the vehicle's insurance information are recorded on a open ledger. No one can change the transaction records in the blockchain unless he can control more than half the nodes in the system. Each participating node proposes the information of vehicles, each record is stored on a majority of cluster after the consensus approach. Sellers and buyers are able to read the records by proposing query-request, and we do not need a third party to provide proofs of vehicles' information anymore.

5.2   Evaluation and Analysis

We measured the throughput and the latency of the designed blockchain on a cluster of 4, 8, 12 and 16 nodes. Each server has an E5-2630 2.40 GHz CPU, 64 GB RAM, 2 TB hard drive, running on Ubuntu 14.04.1, and connected to the other servers via 1 GB switch. In out tests, we do not consider any case of node failure, but allow the loss of packet. All the results are averaged over 5 independent runs.

Throughput. We measured the throughput as the number of successful transactions per second (TPS). The client sends continuous requests in order to saturate the network capacity to achieve a maximum throughput. We noticed that a warm-up period always appears at the beginning of our tests. During the warm-up period, since the transactions are in the order requested by the leader in turn sent to the servers in the cluster, the network capacity does not reach the peak at the beginning of the test. After the network has been saturated to the maximum, the leader and other servers maintain a smooth throughput that stabilizes at a point within tiny fluctuations.

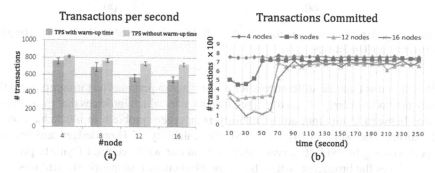

Fig. 5. The throughput due to the blockchain based on Dynasty.

From Fig. 5(a) we can see that along with the increasing scale of the cluster's size, either TPS with or without the warm-up period shows a downward tendency due to the requirement of more replies to reach the consensus. In our 240 s tests, the 4-node blockchain commits and stores averaged 1.8×10^8 transactions in total and the TPS averaged at 764 which is the maximum in all test cases. When the cluster's size increases to 16, the number of committed transactions falls to 1.3×10^5 and the TPS decreases to an average of 543.

We have analyzed the message complexity of Dynasty protocol in Sect. 3. During normal cases, a committed record involves $4N$ messages in the consensus layer as the two-step commit method takes $O(N)$ message in each step. The constant degree of the message complexity could be verified from Fig. 5(b), the average TPS has been measured in an interval of 10 s, and the throughput reduces less than 8% in each increasing 4 nodes of the cluster's size after the warm-up period. The peak performance is measured as the highest TPS per 40 s as shown in Fig. 6(a) that the increased cluster size is the major factor that causes the degradation during the warm-up period (first 120 s). After that, the cluster size alleviates the influence and the performance gap between different sizes of clusters gets smaller.

Latency. We measured the latency as the response time of each successful transactions. According to Dynasty protocol, each transaction needs two round $O(N)$ broadcasts, which makes the latency affected more by network delay. Figure 6(b) shows that the latency gets marked higher as expected along with the increasing sizes of the scale.

The above evaluations show that while the latency in different scales of the system increases as expected, the number of committed transactions per second stabilizes at a

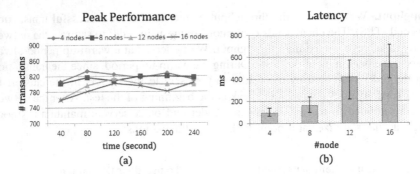

Fig. 6. The peak performance and the latency of the blockchain based on Dynasty. (a) The peak performance. (b) The latency.

point within less than 8% difference after a warming-up period. The throughput and the latency are both severely affected by the cluster's size, which indicates that measuring a balance between throughput and scalability is an urgent problem in blockchain-based applications. The work in [9] measured the scalability of Hyperledger and the test stopped working beyond 16 servers. Although in our work we used Dynasty protocol which reduces the broadcast cost to build the blockchain framework, we still observed that the performance reduces when the cluster size becomes larger.

6   Conclusion

Blockchains benefit commercial applications as a distributed ledger where transactions could be committed without the presence of any trusted authority. The permissioned blockchain works in reaching consensus, replicating state, broadcasting requests and storing transactions on server.

In this paper, we discussed the limitations of the real-time safety of Raft in terms of being a consensus protocol in blockchains. We designed and analyzed a new consensus protocol called Dynasty with its two-step commit method to ensure the real-time safety. We demonstrated a three-layer blockchain framework based on Dynasty, which requires $2f + 1$ correct servers to tolerate f failures under non-Byzantine conditions. We implemented the framework in a used-vehicle trading application and measured the evaluations of the blockchain on several clusters with different size. We observed that in our approach, the size of the blockchain for each 4-node increases, the throughput only takes 8 percentage reduction even the latency nearly doubly increases.

Acknowledgment. The authors would like to thank the anonymous reviewers for their insightful comments and valuable suggestions. This work was supported by the China National Basic Research Program (973 Program, No. 2015CB352400), NSFC under grant U1401258, Science and Technology Planning Project of Guangdong Province (2015B010129011, 2015A030310326), the Basic Research Program of Shenzhen (JCYJ20150630114942313).

References

1. Aspnes, J.: Randomized protocols for asynchronous consensus. Distrib. Comput. **16**(2), 165–175 (2003)
2. Bentov, I., Lee, C., Mizrahi, A., Rosenfeld, M.: Proof of activity: Extending bitcoin's proof of work via proof of stake (extended abstract). ACM SIGMETRICS Perform. Eval. Rev. **42** (3), 34–37 (2014)
3. Bolosky, W.J., Bradshaw, D., Haagens, R.B., Kusters, N.P., Li, P.: Paxos replicated state machines as the basis of a high-performance data store. In: Symposium on Networked Systems Design and Implementation (NSDI), pp. 141–154 (2011)
4. Brown, R.G.: Introducing R3 Corda: a distributed ledger designed for finanial services, 2016 (2017)
5. Cachin, C.: Architecture of the hyperledger blockchain fabric. In: Workshop on Distributed Cryptocurrencies and Consensus Ledgers (2016)
6. Castro, M., Liskov, B., et al.: Practical byzantine fault tolerance. In: OSDI, vol. 99, pp. 173–186 (1999)
7. Chandra, T.D., Griesemer, R., Redstone, J.: Paxos made live: an engineering perspective. In: Proceedings of the Twenty-Sixth Annual ACM Symposium on Principles of Distributed Computing, pp. 398–407. ACM (2007)
8. Dash. https://www.dash.org/
9. Dinh, T.T.A., Wang, J., Chen, G., Liu, R., Ooi, B.C., Tan, K.L.: Blockbench: a framework for analyzing private blockchains. In: Proceedings of the 2017 ACM International Conference on Management of Data, pp. 1085–1100. ACM (2017)
10. Eyal, I., Gencer, A.E., Sirer, E.G., Van Renesse, R.: Bitcoin-NG: a scalable blockchain protocol. In: NSDI, pp. 45–59 (2016)
11. Fischer, M.J., Lynch, N.A., Paterson, M.S.: Impossibility of distributed consensus with one faulty process. J. ACM (JACM) **32**(2), 374–382 (1985)
12. Iansiti, M., Lakhani, K.R.: The truth about blockchain. Harv. Bus. Rev. **95**(1), 118–127 (2017)
13. King, S., Nadal, S.: PPCoin: peer-to-peer crypto-currency with proof-of-stake. Self-published paper, 19 August 2012
14. Lamport, L., Shostak, R., Pease, M.: The byzantine generals problem. ACM Trans. Program. Lang. Syst. (TOPLAS) **4**(3), 382–401 (1982)
15. Lamport, L., et al.: Paxos made simple. ACM SIGACT News **32**(4), 18–25 (2001)
16. Lampson, B.: The ABCD's of paxos. In: PODC, vol. 1, p. 13 (2001)
17. Liu, S., Viotti, P., Cachin, C., Quema, V., Vukolic, M.: XFT: practical fault tolerance beyond crashes. In: OSDI, pp. 485–500 (2016)
18. Nakamoto, S.: Bitcoin: a peer-to-peer electronic cash system (2008)
19. Ongaro, D., Ousterhout, J.K.: In search of an understandable consensus algorithm. In: USENIX Annual Technical Conference, pp. 305–319 (2014)
20. Schwartz, D., Youngs, N., Britto, A.: The ripple protocol consensus algorithm. Ripple Labs Inc White Paper 5 (2014)
21. Vukolić, M.: The quest for scalable blockchain fabric: proof-of-work vs. BFT replication. In: Camenisch, J., Kesdoğan, D. (eds.) iNetSec 2015. LNCS, vol. 9591, pp. 112–125. Springer, Cham (2016). https://doi.org/10.1007/978-3-319-39028-4_9
22. Wood, G.: Ethereum: a secure decentralised generalised transaction ledger. Ethereum Project Yellow Paper 151 (2014)
23. Zhang, G.: Transaction based blockchains. https://github.com/thatisedward/Transaction-Based-Blockchain

EDF-Based Mixed-Criticality Systems with Weakly-Hard Timing Constraints

Hao Wu[1], Zonghua Gu[1(✉)], Hong Li[1], and Nenggan Zheng[2]

[1] College of Computer Science, Zhejiang University,
Hangzhou 310027, Zhejiang, China
zonghua@gmail.com
[2] Qiushi Institute of Advanced Studies, Zhejiang University,
Hangzhou 310027, Zhejiang, China
zng@cs.zju.edu.cn

Abstract. Safety-critical embedded systems are often subject to multiple certification requirements from different certification authorities, giving rise to the concept of Mixed-Criticality Systems. In the classical Mixed-Criticality Scheduling task model, all low-criticality tasks are dropped in high-criticality mode. This approach may not be very practical in reality, since it may cause serious degradation of Quality-of-Service (QoS) for low-criticality tasks. In this paper, we present *EDF with Virtual Deadlines-Weakly Hard (EDF-VD-WH)*, where a number of consecutive jobs of LO-crit tasks may be skipped in high-criticality mode, in order to provide a certain level of QoS for low-criticality tasks in high-criticality mode. We present schedulability analysis of EDF-VD-WH based on Demand Bound Functions, and perform experimental evaluation of schedulability acceptance ratios compared to the original EDF-VD.

Keywords: Mixed-Criticality Systems · Weakly-hard · EDF-VD

1 Introduction and Relate Work

Today's complex safety-critical embedded systems often need to integrate diverse subsystems with varying levels of *criticality* on a shared hardware platform. For example, in the automotive certification standard ISO 26262, there are 4 criticality levels Automotive Safety-Integrity Level (ASIL) A, B, C and D, with D being the highest criticality level. Furthermore, a system may be required to meet multiple certification requirements from different certification authorities. For such *Mixed-Criticality Systems*, the topic of *Mixed-Criticality Scheduling (MCS)* [1, 2] has been studied intensively in recent years, with the goal of simultaneously achieving strong temporal protection for high-criticality applications and efficient utilization of hardware resources.

For Fixed-Priority scheduling on a uniprocessor, a well-known MCS algorithm is *Adaptive Mixed Criticality (AMC)* [3], with two schedulability analysis algorithms: the more efficient but pessimistic AMC-rtb, and the more accurate but more computationally-expensive AMC-max. Zhao et al. presented integration of Preemption Threshold Scheduling with MCS [4, 5] to reduce system stack size, and resource synchronization protocols [6, 7] to protect shared variables while guaranteeing schedulability. For Earliest

© Springer Nature Switzerland AG 2019
S. Li (Ed.): GPC 2018, LNCS 11204, pp. 312–322, 2019.
https://doi.org/10.1007/978-3-030-15093-8_22

Deadline First (EDF) scheduling on a uniprocessor, Baruah et al. [8] presented EDF-VD (Earliest Deadline First with Virtual Deadlines), where each high-criticality (HI-crit) task is given two relative deadlines: its deadline in HI-crit mode is the 'real' deadline, and its deadline in low-criticality (LO-crit) mode that is smaller than its deadline in HI-crit mode. All HI-crit tasks have their deadlines reduced by the same reduction factor in LO-crit mode. Ekberg and Yi [9] presented a more general EDF-VD model, where each HI-crit task may have a different deadline reduction factor, and schedulability analysis based on *Demand-Bound Functions*. Zhang et al. [10] provided a tighter schedulability analysis for EDF-VD based on *unified Demand Bound Functions* that considers system behavior crossing multiple criticality levels.

In the classical MCS model, all LO-crit tasks are dropped immediately when the system switches from LO-crit to HI-crit mode. This approach may not be very practical in reality, since it can cause serious degradation of Quality-of-Service (QoS) for LO-crit tasks. Instead of dropping all LO-crit tasks, many authors have proposed techniques to provide reduced QoS for LO-crit tasks in HI-crit mode. Among the many different approaches reviewed in the survey paper [2], one effective approach is based on the concept of *elastic scheduling* [11], where a task's period may change dynamically at runtime depending on runtime workload and system mode. Su et al. presented Elastic Mixed-Critical (E-MC) task model [12]. The basic idea is to allow each LO-crit task to have a maximum period that defines its minimum service level, and its schedulability is guaranteed even in HI-crit mode. Su et al. [13] presented a *Dual-Rate Mixed-Criticality (DR-MC)* task model based on EDF-VD, where each LO-crit task is assigned a pair of periods to represent its service requirements in the LO and HI running modes, respectively. Its period in HI-crit mode is larger than that in LO-crit mode, hence it has degraded QoS in HI-crit mode compared to LO-crit mode. Su et al. [14] presented *Mode-Switch Fixed-Priority (MS-FP)*, where each LO-crit task may change both its period and priority when the system switches to HI-crit mode.

Another approach is based on *weakly-hard*, also called *(m, k)-firm* timing constraints. Gettings et al. [15] present *Adaptive Mixed Criticality-Weakly Hard (AMC-WH)* for Fixed-Priority scheduling. Instead of being dropped completely HI-crit mode, a number of *consecutive* jobs of LO-crit tasks may be skipped, characterized by the parameters (s_i, M_i): for a LO-crit task executing in HI-crit mode, at most consecutive s_i jobs can be skipped within M_i consecutive jobs, with $1 \leq s_i < M_i$. Figure 1 shows an example. This task model reduces system workload in HI-crit mode, freeing up capacity for HI-crit tasks while also providing a (degraded) QoS for LO-crit tasks. The assumption of allowing only consecutive job skips in a cycle is motivated by the fact that consecutive skips cause more severe performance degradation than non-consecutive skips in either control systems or multimedia systems, hence it is important to prevent a large number of consecutive job skips. It also simplifies the schedulability analysis considerably.

In this paper, we present *EDF with Virtual Deadlines-Weakly Hard (EDF-VD-WH)*, which is based on a similar weakly-hard task model to that in [15], but in the context of EDF scheduling instead of Fixed-Priority scheduling.

The remainder of the paper is organized as follows. The system models and preliminaries are discussed in Sect. 2. Section 3 reviews the unified Demand bound

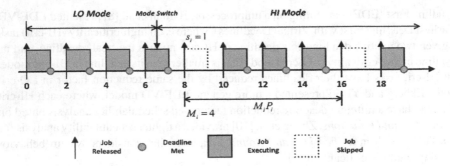

Fig. 1. Weakly-Hard task model as adopted in [15] and in this paper, with parameters $(s_i, M_i) = (1, 4)$.

analysis approach of EDF-VD and introduces our algorithm. Section 4 presents performance evaluation results, and Sect. 5 concludes the paper.

2 System Models and Preliminaries

2.1 Task Model

A software task is characterized by different estimates of its *Worst Case Execution Time (WCET)* [16] used for schedulability analysis at different criticality levels, e.g., one optimistic WCET estimate by the system designer, and another more pessimistic and safe WCET estimate by the safety certification authority. We consider a dual-criticality system, with 2 criticality levels High (HI) or Low (LO). Formally, a HI-crit task τ_i is characterized by the tuple $\left(C_i^{LO}, C_i^{HI}, D_i^{LO}, D_i^{HI}, P_i, \xi_i = HI\right)$. C_i^{HI} and C_i^{LO} denote the Worst-Case Execution Time (WCET) of task τ_i in HI-crit mode and LO-crit mode, respectively. D_i^{HI} and D_i^{LO} denote the relative deadlines of task τ_i in HI-crit mode and in LO-crit mode, respectively. We assume the implicit deadline model $(D_i^{HI} = P_i)$ in HI-crit mode, and $D_i^{LO} \leq P_i$ in LO-crit mode [5]. P_i is the period, or the minimal inter-arrival time of the sporadic task. A LO-crit task τ_i is characterized by the tuple $(C_i, D_i, P_i, \xi_i = LO, s_i, M_i)$, with the same WCET, relative deadline, and period in LO-crit or HI-crit mode. We assume the implicit deadline model $(D_i = P_i)$. For a LO-crit task executing in HI-crit mode, at most consecutive s_i jobs can be skipped within M_i consecutive jobs, with $1 \leq s_i < M_i$.

2.2 Notations

For a HI-crit task τ_i, its CPU utilizations in LO-crit and HI-crit modes are defined as:

$$u_{i_{HI}}^{LO} = \frac{C_i^{LO}}{P_i}, u_{i_{HI}}^{HI} = \frac{C_i^{HI}}{P_i}$$

For a LO-crit task τ_i, its CPU utilizations in LO-crit and HI-crit modes are defined as:

$$u_{i_{LO}}^{LO} = \frac{C_i}{P_i}, u_{i_{LO}}^{HI} = \frac{C_i(M_i - s_i)}{M_i P_i}$$

The system utilizations are further defined as:

$$U_{HI}^{LO} = \sum_{\tau_i \in \tau_{HI}} u_{i_{HI}}^{LO}, U_{HI}^{HI} = \sum_{\tau_i \in \tau_{HI}} u_{i_{HI}}^{HI}, U_{LO}^{LO} = \sum_{\tau_i \in \tau_{LO}} u_{i_{LO}}^{LO}, U_{LO}^{HI} = \sum_{\tau_i \in \tau_{LO}} u_{i_{LO}}^{HI}$$

$$U^{LO} = U_{HI}^{LO} + U_{LO}^{LO}, U^{HI} = U_{HI}^{HI} + U_{LO}^{HI}$$

3 Unified Demand Bound Analysis

3.1 Demand Bound Analysis in LO-Crit Mode

For an HI-crit task τ_i, its unified DBF in an interval of length ℓ in LO-crit mode is [10]:

$$dbf_{HI}^{LO}(\tau_i, \ell) = \left(\left\lfloor \frac{\ell - D_i^{LO}}{P_i} \right\rfloor + 1 \right) \cdot C_i^{LO} \tag{1}$$

For a LO-crit task τ_i, its unified DBF in an interval of length ℓ in LO-crit Mode is:

$$dbf_{LO}^{LO}(\tau_i, \ell) = \left(\left\lfloor \frac{\ell}{P_i} \right\rfloor \right) \cdot C_i \tag{2}$$

Theorem 1. A mixed-criticality weakly-hard taskset is schedulable in LO-crit mode if:

$$\forall \ell \geq 0: \sum_{\tau_i \in \tau_{LO}} dbf_{LO}^{LO}(\tau_i, \ell) + \sum_{\tau_i \in \tau_{HI}} dbf_{HI}^{LO}(\tau_i, \ell) \leq \ell \tag{3}$$

3.2 Demand Bound Analysis in HI-Crit Mode

We now derive the unified DBF for a HI-crit task τ_i in HI-crit mode, based on [10], but with weakly-hard timing constraints for LO-crit tasks.

In HI-crit mode, two types of jobs need to be considered:

Definition 1. (*Normal job*): a normal job has its release time after the criticality mode switch point.

Definition 2. (*Crossing job*): a crossing job has its release time before and absolute deadline after the criticality mode switch point, respectively.

Consider the time interval $[t^o, t^d]$ with length $\ell = t^o - t^d$. Suppose that the running mode switches at time t^* with $t^o \leq t^* \leq t^d$. We further define $x = t^d - t^*$. The maximum number of jobs of τ_i with both release times and absolute deadlines in a time interval of length ℓ is:

$$n_i(\ell) = max\left(\left\lfloor \left|\frac{\ell - D_i^{HI}}{P_i}\right| \right\rfloor + 1, 0\right) \tag{4}$$

The number of jobs of τ_i that have not finished their execution by time t^* whether its crossing job should be counted is:

$$m_i(\ell, x) = min(p_i(\ell, x), q_i(\ell, x)) \tag{5}$$

where

$$p_i(\ell, x) = \begin{cases} n_i(x) & x \bmod P_i < S_i \\ min(n_i(x) + 1, n_i(\ell)) & otherwise \end{cases} \tag{6}$$

$$q_i(\ell, x) = max\left(min\left(\left\lceil \frac{x - S^*}{P_i}\right\rceil, n_i(\ell)\right), n_i(x)\right) \tag{7}$$

$$S^* = min\{\forall \tau_i \in \tau_{HI} \,|\, D_i^{HI} - D_i^{LO}\} \tag{8}$$

Hence, the total workload of a HI-crit task τ_i in $[t^o, t^d]$ is calculated by:

$$dbf_{HI}^{HI}(\tau_i, \ell, x) = (n_i(\ell) - m_i(\ell, x)) \cdot C_i^{LO} + m_i(\ell, x) \cdot C_i^{HI} \tag{9}$$

The number of jobs of a skippable LO-crit task τ_i in LO-crit mode that have both release times and deadlines within the interval $[t^o, t^d]$ of length $\mathscr{y} = \ell - x$ is bounded by:

$$j_i(\ell, x) = \left\lceil \frac{\ell - x}{P_i} \right\rceil \tag{10}$$

The number of jobs of a skippable LO-crit task τ_i in Hi-Crit mode can be expressed as the number of jobs of τ_i assuming no skips, minus the number of skipped jobs in each cycle. We note that the maximum amount of execution occurs when the phasing of consecutive skips is at the end a cycle i.e. $N = 1, 2, \cdots, s_i$, shown in Fig. 2.

Therefore, after the mode switch time t^*, the number of jobs of τ_i executed within the interval $[t^*, t^d]$ of length x (including one crossing job and normal jobs, minus skipped jobs) is bounded by:

$$k_i(\ell, x) = \left\lfloor \frac{x}{P_i} \right\rfloor - \sum_{N=1}^{s_i} \left(\left\lfloor \frac{x - (M_i - N + 1)P_i}{M_i P_i} \right\rfloor\right) \tag{11}$$

Moreover, to rule out the case where the crossing job is released before t^o, it is further bounded by:

$$k_i^*(\ell, x) = min\left(k_i(\ell, x), \left\lfloor \frac{\ell}{P_i} \right\rfloor\right) \tag{12}$$

Hence, the total workload of a LO-crit task τ_i in $\left[t^o, t^d\right]$ is calculated by:

$$dbf_{LO}^{HI}(\tau_i, \ell, x) = \left(j_i(\ell, x) + k_i(\ell, x)\right) \cdot C_i \tag{13}$$

Theorem 2. A dual-criticality weakly-hard taskset is schedulable in HI-crit mode if it holds:

$$\forall \ell \geq x \geq 0 : \sum_{\tau_i \in \tau_{LO}} dbf_{LO}^{HI}(\tau_i, \ell, x) + \sum_{\tau_i \in \tau_{HI}} dbf_{HI}^{HI}(\tau_i, \ell, x) \leq \ell \tag{14}$$

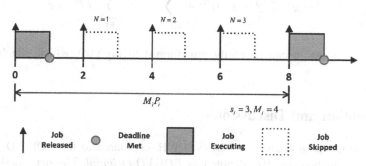

Fig. 2. Maximum workload in each cycle

3.3 Test Region

We present Theorem 3, which gives the test region of (ℓ, x) that needs to be checked for the schedulability condition in Theorem 2. (The theorem and its proof is similar to Theorem 4 in [13].)

Theorem 3. Let $x + y = \ell$. For schedulability test in Hi-Crit mode, only the values of (ℓ, x) that satisfies the following condition need to be considered.

$$\alpha x + \beta(\ell - x) < \gamma$$

where

$$\alpha = 1 - U^{HI}$$
$$\beta = 1 - U^{LO}$$
$$\gamma = \sum_{\tau_i \in \tau_{LO}} C_i + \sum_{\tau_i \in \tau_{HI}} C_i^{HI}$$

Proof. From the definitions of dbf_{LO}^{HI} and dbf_{HI}^{HI}, we have

$$dbf_{LO}^{HI}(\tau_i, \ell, x) \leq (\ell - x) \cdot u_{i_{LO}}^{LO} + x \cdot u_{i_{LO}}^{HI} + C_i$$
$$dbf_{HI}^{HI}(\tau_i, \ell, x) \leq (\ell - x) \cdot u_{i_{HI}}^{LO} + x \cdot u_{i_{HI}}^{HI} + C_i^{HI}$$

Sum up these equations for all task, we have:

$$\sum_{\tau_i \in \tau_{LO}} dbf_{LO}^{HI}(\tau_i, \ell, x) + \sum_{\tau_i \in \tau_{HI}} dbf_{HI}^{HI}(\tau_i, \ell, x)$$

$$\leq (\ell - x) \cdot U^{LO} + x \cdot U^{HI} + \sum_{\tau_i \in \tau_{LO}} C_i + \sum_{\tau_i \in \tau_{HI}} C_i^{HI}$$

Since the schedulability condition in Hi-Crit mode is violated, there exist a pair of (ℓ, x) such that:

$$(\ell - x) \cdot U^{LO} + x \cdot U^{HI} + \sum_{\tau_i \in \tau_{LO}} C_i + \sum_{\tau_i \in \tau_{HI}} C_i^{HI} > \ell$$

The above equation can be easily transformed to Eq. (14), which concludes the proof.   ∎

4   Evaluation and Discussions

We evaluate the performance of EDF-VD-WH, compared to the EDF-VD with the schedulability analysis in [10], denoted as *EDF-VD-Original*. The main performance metric is the ratio of schedulable tasksets. The experiment platform is AMD Athlon(tm) II X4 640 Processor with 7.8 GB main memory, running -64bit Ubuntu Linux release 12.04.

We use UUnifast algorithm [17] to generate synthetic tasksets. Taskset utilization is varied from 0.05 to 0.95 in steps of 0.05. For each utilization level, 1000 tasksets are randomly generated. Task periods are set according to a log-uniform distribution within the interval [10, 200]. D_i^{HI} values of HI-crit tasks or D_i values of LO-crit tasks are implicit, i.e. same as its period. D_i^{LO} values of HI-crit tasks are calculated by $D_i^{LO} = \lambda \cdot D_i^{HI}$, where $\lambda = U_{HI}^{LO} / (1 - U_{LO}^{LO})$. C_i^{LO} values of HI-crit tasks or C_i values of LO-crit tasks are set based on utilization and period, $C_i^{LO}(or C_i) = U_i \cdot P_i$. C_i^{HI} values are assigned by multiplying the C_i^{LO} by a Criticality Factor (CF). A Criticality Probability (CP) denoted the probability that a task would be designated as a HI-crit task. By default, each taskset contains 10 tasks with CP = 0.5 and CF = 2.

In the following, for experiments 1 to 3, the weakly-hard constraints (s_i, M_i) for all LO-crit tasks range from (1,5) to (4,5), respectively; for experiments 2 to 5, weighted schedulability is used to flatten the data from 3D to 2D as in [15], where higher utilization tasksets are more heavily-weighted than lower utilization tasksets.

Figure 3 shows the acceptance ratios as a function of the total utilization. As expected, EDF-VD-Original dominates EDF-VD-WH since all LO-crit tasks are dropped in HI-crit mode, but the differences are not very large. For EDF-VD-WH, more skippable jobs means higher acceptance ratio.

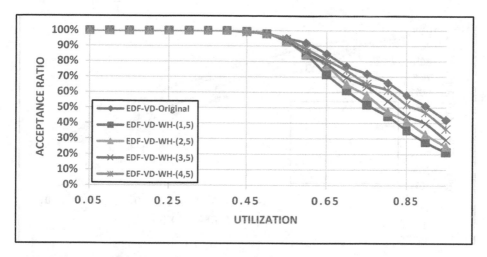

Fig. 3. Exp 1 – Acceptance ratio vs. different utilization levels, CP = 0.5 & CF = 2

Figures 4 and 5 illustrate the influences of HI-crit tasks. More HI-crit tasks in a taskset, or more pessimistic WCET estimates in HI-crit mode, lead to lower acceptance ratio.

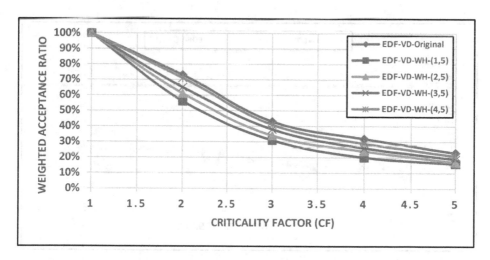

Fig. 4. Exp 2 – Weighted acceptance ratio vs. different Criticality Factors, with CP = 0.5

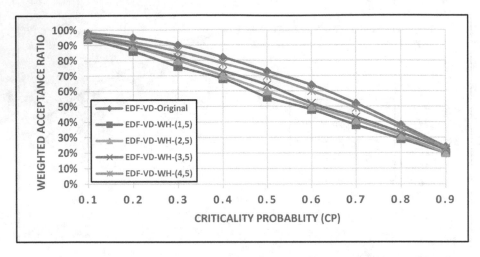

Fig. 5. Exp 3 – Weighted acceptance ratio vs. different Criticality Probabilities, with CF = 2

Figures 6 and 7 show the impact of weakly-hard constraint parameters s_i and M_i. Increasing s_i or decreasing M_i both lead to increased acceptance ratio.

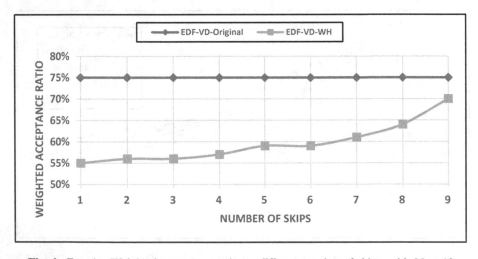

Fig. 6. Exp 4 – Weighted acceptance ratio vs. different number of skips, with $M_i = 10$

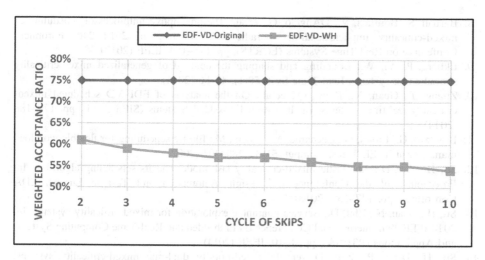

Fig. 7. Exp 5 – Weighted acceptance ratio vs. different cycle of skips, with $s_i = 1$

5 Conclusions

In this paper, we have presented schedulability analysis of weakly-hard mixed criticality system on a uniprocessor based on EDF-VD, and shown its advantages in improving service guarantees for LO-crit tasks. Performance evaluation indicates that it is possible to maintain a certain level of QoS for LO-crit tasks in HI-crit mode without affecting HI-crit tasks, instead of dropping all LO-crit tasks in HI-crit mode.

Acknowledgements. This work is partially supported by NSFC Project # 61672454; Zhejiang Provincial Natural Science Foundation Project # LY16F020007.

References

1. Vestal, S.: Preemptive scheduling of multi-criticality systems with varying degrees of execution time assurance. In: 28th IEEE International on Real-Time Systems Symposium, RTSS 2007, pp. 239–243. IEEE (2007)
2. Burns, A., Davis, R.I.: A survey of research into mixed criticality systems. ACM Comput. Surv. (CSUR) **50**(6), 82 (2017)
3. Baruah, S.K., Burns, A., Davis, R.I.: Response-time analysis for mixed criticality systems. In: 2011 IEEE 32nd Real-Time Systems Symposium (RTSS), pp. 34–43. IEEE (2011)
4. Zhao, Q., Gu, Z., Zeng, H., et al.: Schedulability analysis and stack size minimization with preemption thresholds and mixed-criticality scheduling. J. Syst. Arch. **83**, 57–74 (2018)
5. Zhao, Q., Gu, Z., Zeng, H.: Design optimization for AUTOSAR models with preemption thresholds and mixed-criticality scheduling. J. Syst. Arch. **72**, 61–68 (2017)
6. Zhao, Q., Gu, Z., Zeng, H.: Resource synchronization and preemption thresholds within mixed-criticality scheduling. ACM Trans. Embed. Comput. Syst. (TECS) **14**(4), 81 (2015)
7. Zhao, Q., Gu, Z., Zeng, H.: HLC-PCP: a resource synchronization protocol for certifiable mixed criticality scheduling. Embed. Syst. Lett. **6**(1), 8–11 (2014)

8. Baruah, S., Bonifaci, V., DAngelo, G., et al.: The preemptive uniprocessor scheduling of mixed-criticality implicit-deadline sporadic task systems. In: 2012 24th Euromicro Conference on Real-Time Systems (ECRTS), pp. 145–154. IEEE (2012)

9. Ekberg, P., Yi, W.: Bounding and shaping the demand of generalized mixed-criticality sporadic task systems. Real-Time Syst. 50(1), 48–86 (2014)

10. Zhang, T., Guan, N., Deng, Q., et al.: On the analysis of EDF-VD scheduled mixed-criticality real-time systems. In: Industrial Embedded Systems (SIES 2014), pp. 179–188 (2014)

11. Buttazzo, G., Lipari, G., Caccamo, M., Abeni, M.: Elastic scheduling for flexible workload management. IEEE Trans. Comput. 51(3), 289–302 (2002)

12. Su, H., Zhu, D.: An elastic mixed-criticality task model and its scheduling algorithm. In: Proceedings of the Conference on Design, Automation and Test in Europe. EDA Consortium, pp. 147–152 (2013)

13. Su, H., Guan, N., Zhu, D.: Service guarantee exploration for mixed-criticality systems. In: 2014 IEEE 20th International Conference on Embedded and Real-Time Computing Systems and Applications (RTCSA), pp. 1–10. IEEE (2014)

14. Su, H., Deng, P., Zhu, D., et al.: Fixed-priority dual-rate mixed-criticality systems: schedulability analysis and performance optimization. In: Embedded and Real-Time Computing Systems and Applications (RTCSA 2016), pp. 59–68 (2016)

15. Gettings, O., Quinton, S., Davis, R.I.: Mixed criticality systems with weakly-hard constraints. In: Proceedings of the 23rd International Conference on Real Time and Networks Systems, pp. 237–246. ACM (2015)

16. Lv, M., Gu, Z., Guan, N., et al.: Performance comparison of techniques on static path analysis of WCET. In: IEEE/IFIP International Conference on Embedded and Ubiquitous Computing, EUC 2008, vol. 1, pp. 104–111. IEEE (2008)

17. Bini, E., Buttazzo, G.C.: Measuring the performance of schedulability tests. Real-Time Syst. 30(1–2), 129–154 (2005)

GA-Based Mapping and Scheduling of HSDF Graphs on Multiprocessor Platforms

Hao Wu[1], Nenggan Zheng[2], Hong Li[1], and Zonghua Gu[1(✉)]

[1] College of Computer Science, Zhejiang University,
Hangzhou 310027, Zhejiang, China
zonghua@gmail.com
[2] Qiushi Academy for Advanced Studies, Zhejiang University,
Hangzhou 310027, Zhejiang, China
zng@cs.zju.edu.cn

Abstract. Synchronous Dataflow (SDF) is a widely-used model-of-computation for signal processing and multimedia applications. We address the problem of mapping a Homogeneous Synchronous Dataflow (HSDF) graph onto a multiprocessor platform with the objective of maximizing system throughput. Since the problem is a NP-hard combinatorial optimization problem, it computationally infeasible to use exhaustive search to obtain optimal solutions for large applications. In this paper, we apply Genetic Algorithms to search the design space of all possible actor-to-processor mappings and static-order schedules on each processor to find a near-optimal solution, and compare the performance and scalability of GA with the exact solution technique based on SAT solving.

Keywords: Synchronous Dataflow (SDF) · Genetic Algorithms ·
Multiprocessor systems

1 Introduction

1.1 SDF Graphs

In the dataflow paradigm, a program is represented as a directed graph, where nodes, called *actors*, represent computational modules, and directed edges represent communication channels between the modules. There are many variants of dataflow models, including Synchronous Dataflow (SDF), Cyclo-Static Dataflow (CSDF), Multi-Dimensional SDF (MDSDF), Boolean Dataflow (BDF), and others. Synchronous Dataflow (SDF) is a type of dataflow model where each actor invocation consumes and produces a constant number of data tokens. A SDF graph can be statically scheduled offline because of its static nature. SDF is used in a broad range of signal processing applications, including modems, multi-rate filter banks, and satellite receiver systems.

Formally, a SDF graph is defined as G = (V, E). It is a directed graph with a set of nodes V = $\{v_i \mid 1 \leq i \leq n\}$ which representing the actors, combined with a set of directed edges E = $\{e_{ij} \mid i, j \in [1, ..., n]\}$ from source actor v_i to sink actor v_j. When an actor v_i fires, the number of tokens it consumes (produces) on each input (output) edge

© Springer Nature Switzerland AG 2019
S. Li (Ed.): GPC 2018, LNCS 11204, pp. 323–333, 2019.
https://doi.org/10.1007/978-3-030-15093-8_23

e_{ij} is fixed and known at compile time, denoted as $cons(e_{ij})$ ($prod(e_{ij})$). Each edge e_{ij} has a known buffer size $buff_{ij}$, representing the maximum number of tokens that can be stored on it. Each edge e_{ij} may contain a number of initial tokens, also called delays, denoted as d_e.

As an example, Fig. 1(a) shows a simple SDF graph. Assume that buffer sizes on edges e_{AB} and e_{BA} are both 4. Each firing of actor A consumes 2 tokens on edge e_{BA}, and produces 2 tokens on edge e_{AB}; each firing of actor B consumes 3 tokens on edge e_{AB}, and produces 3 tokens on edge e_{BA}. One feasible static schedule is AABAB. Figure 1(b) shows how the numbers of tokens on the two edges evolve as each actor is fired in the sequence of AABAB. At the end of the schedule, the SDF graph goes back to the initial state shown in the left subfigure. Therefore, we can execute this sequence of actor firings repeatedly without any deadlock or buffer overflow conditions.

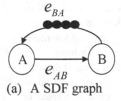

	init	A	A	B	A	B
e_{BA}	4	2	0	3	1	4
e_{AB}	0	2	4	1	3	0

(a) A SDF graph

(b) A   feasible   static   schedule   of   "AABAB".

Fig. 1. A SDF graph and a feasible static schedule.

Homogeneous Synchronous Dataflow (HSDF) is a special type of SDF with $cons(e_{ij}) = prod(e_{ij}) = 1$ for all edges. Every SDF graph can be converted to an equivalent HSDF graph, with possible large increase in model size. A *simple cycle* is a cycle where no actor appears more than once. The cycle mean of a simple cycle c in a HSDF graph is defined as $\mu_c = \sum_{i \in N(c)} (WCET(v_i)) / \sum_{e \in E(c)} d_e$, where N(c) is the set of all actors in cycle c; E(c) is the set of all edges in cycle c; $WCET(v_i)$ is the Worst-Case Execution Time of actor v_i. The *Maximum Cycle Mean (MCM)* of a HSDF graph G is defined as:

$$\mu(G) = \max_{c \in C(G)} \left(\sum_{i \in N(c)} (WCET(v_i)) / \sum_{e \in E(c)} d_e \right)$$

where C(G) is the set of simple cycles in graph G; N(c) is the set of all nodes traversed by cycle c, and E(c) is the set of all edges traversed by cycle c. A cycle is called the *critical cycle* if it has the largest MCM among all simple cycles in graph G. There may be multiple critical cycles which have the same MCM in a graph G. The maximum throughput of a HSDF graph is the inverse of its MCM, hence maximizing throughput of a HSDF graph is equivalent to minimizing its MCM. There are efficient polynomial-time algorithms for computing the MCM of a HSDF graph [9]. A *deadlock cycle* is a cycle with no tokens on any edge, which should be prevented during design space exploration.

1.2 Problem Formulation

We consider a homogeneous multiprocessor platform, where all processors are identical in terms of processing speed, hence each actor v_i's WCET(v_i) is independent of the specific processor that it is mapped to. The problem formulation is as follows:

Find an actor-to-processor mapping and static-order schedule on each processor with the objective of maximizing system throughput (minimizing MCM).

There are three steps in solving the optimization problem:

1. Map each actor to a processor
2. Find a static-order schedule for actors mapped onto the same processor
3. Perform deadlock detection, and calculate the MCM for a given deadlock-free configuration.

We apply Genetic Algorithms (GA) to solve this optimization problem. GA is a stochastic optimization algorithm based on the principles of biological evolution. It uses recombination of parent individuals, and gene mutations in each individual to generate new individuals. GA tries to identify good genes and keep them in the population gene pool, e.g., if an individual has a smaller MCM than another individual, then it is likely to contain some good genes that should be preserved in the gene pool. Individuals with good genes have a better chance of mating with other individuals and pass their good genes to the next generation. After multiple generations of evolution, it is expected that more and more good genes are accumulated in the population, which helps in the search for a good solution (but no optimality guarantees can be made).

2 Related Work

Task graphs can be viewed as a special case of HSDF graphs without any delay tokens, hence they should not contain any cycles, since any cycle will lead to a deadlock. In contrast, cycles are allowed in HSDF as long as there are non-zero initial tokens in the cycle, which means that actor firing in HSDF does not have to follow the edge precedence order. For scheduling of either task graphs or SDF graphs, there are two types of multiprocessor schedules: non-overlapped schedules, where consecutive iterations of application execution cannot overlap in time, and overlapped (pipelined) schedules, where consecutive iterations of application execution can overlap in time, which can achieve higher throughput by exploiting inter-iteration concurrency. Satish et al. [7] and Metzner et al. [8] considered mapping and non-overlapped scheduling of tasks graphs on multiprocessor systems, with the objective of minimizing the overall schedule length (makespan). In this paper, we consider SDF/HSDF graphs and overlapped schedules.

Stuijk et al. [9] presented an efficient state space exploration technique for calculating the Pareto space of throughput and storage trade-offs, which can be used to determine the minimal buffer space needed to execute a SDF graph under a given throughput constraint. They adopted the assumption that the number of processors is unlimited, hence the optimal schedule of an SDF graph with maximum throughput can be computed in polynomial time. However, when the number of available processors is less than the application's maximum degree of parallelism (the maximum number of actors that can fire in parallel), we need to find a scheduling strategy to order the firing of those actors that share the same processors, then the problem of mapping and scheduling of an SDF Graph on a multiprocessor platform becomes NP-hard. Stuijk et al. [10] addressed mapping of multiple SDF graphs on a multiprocessor platform based on Network-on-Chip (NoC). TDMA scheduling is used to provide performance isolation among applications, and static-order scheduling is used to schedule the actors within the same application. The optimization objective is to reduce the resource usage of each application (number of time slices it occupies in the TDMA schedule) in order to maximize the number of applications that can run in parallel. List scheduling, an efficient heuristic scheduling algorithm, is used to construct the static-order schedule.

Since finding the mapping and scheduling with the minimum MCM is a NP-hard combinatorial optimization problem, it is computationally infeasible to obtain optimal solutions for realistic large-size applications with any exact solution technique. With a small number of processors and/or tasks, exhaustive search may be feasible by listing all the possible mapping and scheduling strategies, but the search space grows exponentially with problem size. However, for certain critical applications, it may be important to search all possible scheduling and mapping choices to find an optimal solution. Liu et al. [1] presented an exact solution technique based on SAT solving for mapping a HSDF graph onto a multiprocessor platform with the objective of maximizing system throughput. Based on branch-and-bound and SAT-solving techniques, two optimization approaches are presented: Logic-Based Benders Decomposition (LBBD) approach, and the integrated approach. The integrated approach integrates branch-and-bound search into the SAT engine to achieve effective search tree pruning and better scalability. Although it returns exact optimal solutions to the mapping and scheduling problem and scales better than the LBBD approach, the integrated approach still has limited scalability: when the number of processor and the size of the HSDF graph are relatively large, the running time of the algorithm can be extremely long. Liu et al. [2–5] presented a series of follow-on work that adopted different assumptions (task graphs vs. SDF graphs), and used different optimization techniques (SAT modulo theories vs. SAT...).

3   Constructing Architecture-Aware Models

We consider a homogenous multiprocessor platform consisting of multiple identical processors connected with a communication substrate with guaranteed latency, which may be based on bus or NoC. For a given hardware platform and a given application HSDF graph, different actor-to-processor mappings and static-order scheduling on each processor will result in different system throughput. For each possible mapping and scheduling alternative, a Platform-Specific Model (PSM), or an *architecture-aware HSDF graph*, is generated, with additional actors and edges to model platform constraints such as limited buffer size, network latency, etc.

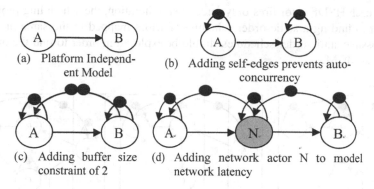

(a) Platform Independent Model

(b) Adding self-edges prevents auto-concurrency

(c) Adding buffer size constraint of 2

(d) Adding network actor N to model network latency

Fig. 2. Incremental addition of platform constraints to obtain an architecture–aware model.

As an example, Fig. 2 shows a series of transformations that gradually add platform constraints to the Platform-Independent model to obtain the final architecture-aware model, with the effect of gradually increasing MCM and decreasing maximum throughput. Figure 2(a) shows an HSDF graph without any platform constraint. Its MCM is undefined. Since it does not contain any cycles, maximum throughput is unlimited, and can grow without bound with increasing number of available processors. Figure 2(b) shows addition of self-edges with one token on each actor. This prevents *auto-concurrency*, so that an actor must finish one firing before starting the next one. Assuming WCET(A) = 10, WCET(B) = 6. MCM of the model in Fig. 2(b) is maximum of the cycles means of the two cycles "AA" and "BB": max(WCET (A)/1, WCET (B)/1) = 10. Figure 2(c) shows addition of buffer-size constraint of 2 tokens on edge BA. The MCM is still 10, since the newly-added cycle "BAB" has a Cycle Mean of $(10 + 6)/2 = 8$, which is smaller than 10. Hence it is not a critical cycle, and the buffer-size constraint does not affect the maximum system throughput. Figure 2(d) shows addition of a network actor N between actors A and B to model network latency between the two processors that actors A and B are mapped to. The self-edge on actor N means that only one token can be transmitted at any given time, and concurrent token transmission is not permitted, i.e., if WCET (N) = 4, transmission of one token between A and B consumes 4 time units, and transmission of 2 tokens consumes 8 time units. The MCM of the new HSDF graph is now the Cycle Mean of the cycle "NAN", which is $(10 + 4)/1 = 14$.

Once actor-to-processor mapping has been determined, there are 3 possible actor scheduling algorithms on each processor [6]:

1. *Fully-static*: in compile time, the exact firing time of each actor is determined in a TDMA schedule. It is typically used for scheduling between different applications.
2. *Self-timed*: Each actor fires immediately when its firing precedence is satisfied.
3. *Static-order*: The order of actor firing on the same processor is statically-determined at compile time. We adopt this approach in this paper.

Since each HSDF actor fires only once at each iteration, the scheduling problem is equivalent to finding a static-order schedule of actors mapped to the same processor, and all possible static-order schedules should be explored in order to find the one with the smallest MCM.

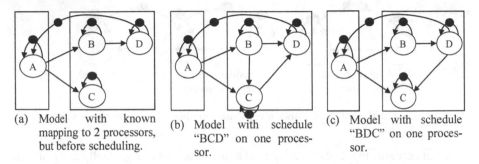

(a) Model with known mapping to 2 processors, but before scheduling.

(b) Model with schedule "BCD" on one processor.

(c) Model with schedule "BDC" on one processor.

Fig. 3. Model before scheduling, and 2 architecture-aware models with different schedules.

Considering the model in Fig. 3(a), with actor A mapped to one processor, and actors B, C and D mapped to another processor. For simplicity, we assume network latency between the 2 processors is 0, hence network actors are not necessary. There is no scheduling constraints among actors B, C and D on the same processor. Assume WCET of each actor is the same and equal to 5. The MCM is 15, which is the cycle mean of the critical cycle "ABDA". Figure 3(b) shows an architecture-aware model with schedule "BCD" on one processor, enforced by addition of two new edges B → C and C → D. A new cycle "ABCDA" is formed, and the MCM is 20, which is the cycle mean of the critical cycle "ABCDA". Figure 3(c) shows an architecture-aware model with schedule "BDC" on one processor, enforced by addition of one new edge D → C. The MCM is 15, which is the cycle mean of the critical cycle "ABDA" (same as Fig. 3(a)).

The following set of pre-defined static-order schedule constraints are always present regardless of the actual mapping and scheduling {A → B, B → D, A → C}. A full set of precedence constraints can be obtained from computing the transitive closure of the directed graph, e.g., A → B can be derived from A → B & B → D. Such pre-defined precedence constraints can be exploited to reduce the search space by setting the corresponding a_{ij} value in the order encoding matrix to 11, and holding these bits constant during the GA procedure.

4   GA-Based Optimization Algorithm

Given an HSDF graph, an initial population of individuals are generated. For each individual, the platform-aware model that encodes a specific mapping and scheduling alternative is generated. If there is no deadlock in the model, its MCM is calculated, and the minimum value seen so far is kept track of, and returned at the end of the algorithm, along with the corresponding architecture-aware model.

We use $|V| * \lceil \ln|P| \rceil$ bits to encode each actor-to-processor mapping alternative, where $|V|$ is the number of actors, and $|P|$ is the number of processors. $\lceil \ln|P| \rceil$ bits are needed to assign a unique binary string to represent each processor. Each actor can be mapped to any one of the $|P|$ processors, hence a total of $|V| * \lceil \ln|P| \rceil$ bits are needed to model all possible mappings of $|V|$ actors. For example, consider a 4-processor system ($|P| = 4$). We use strings 00, 01, 10, 11 to represent each of the 4 processors p0, p1, p2, p3, respectively. Consider an HSDF graph with 6 actors (v0 to v5). If actor-to-processor mapping is v0 on p0; v1 and v3 on p1; v2 on p2; v4 and v5 on p3, then this mapping is encoded as the binary string 000110011111.

For each pair of actors V_i and V_j, we use a string a_{ij} to encode the static-order schedule constraint between them:

1. $a_{ij} = 00$: V_i precedes V_j.
2. $a_{ij} = 01$: V_j precedes V_i.
3. $a_{ij} = 10$: V_i and V_j are mapped on two different processors, hence there is no precedence constraint between them.
4. $a_{ij} = 11$: the precedence constraint between V_i and V_j is fixed in the Platform-Independent Model, and should not be changed during the GA procedure.

For example, if 3 actors V_1, V_2 and V_3 are mapped on the same processor, with the static-order scheduling constraints: V_1 precedes V_3, and V_3 precedes V_2, then this mapping can be encoded with 3 binary strings concatenated together: $a_{12} = 00; a_{13} = 00; a_{23} = 01$. The resulting string is in turn concatenated with the binary string that encodes each actor-to-processor mapping alternative to form a *gene* used in the GA algorithm.

Algorithm 1 is used to generate an architecture-aware HSDF graph. Given an architecture-aware HSDF graph constructed by Algorithm 1, Algorithm 2 is used to check for deadlocks, and compute the MCM with the procedure *eval_mcm()*, a well-known algorithm for computing MCM of an HSDF graph [11]. If the architecture-aware HSDF contains a deadlock, then its MCM is set to be a large number (say 1000000), which should be larger than the maximum possible MCM, so that this individual will be eliminated as a feasible candidate solution.

Algorithm 1 Generate Architecture-aware HSDF Graph

HSDFgraph *GenerateNewGraph (map,comm,ord, HSDFgraph *g)
{
    addedge(map,ord,g); /* add edges to encode the static-order schedule of actors on the same processor*/
    addnode(map,comm,ord,g); /* add a communication actor between two connected actors mapped to two different processors*/
    addnewedge(g);   /* add edges between communication actors and application actors*/
    addselfedge(g);   /*add a self-edge on each node to disable auto-concurrency*/
    return(g);
}

Algorithm 2 MCM calculation

```
double eval_mcm(individual *ind)  /* for each individual ind, calculate mcm*/
{
   newG = GenerateNewGraph (ind,originalG); /*generate PSM HSDF :newG*/
   if deadlockfree(newG) then /*check for deadlocks*/
   return eval_mcm (newG); /*calculate MCM of newG*/
   else {
   print 'deadlocked';
   return 1000000;
   }
}
```

Algorithm 3 is used to perform recombination of individuals to generate new children individuals. The crossover point is chosen on the level of mapping. This is because the scheduling of actors on processors is generated depending on the mapping strategy. With mapping strategy changes, the static-order schedule on each processor should be generated accordingly.

Algorithm 3 One point crossover

```
void one_point_crossover(individual *ind1, individual *ind2)
/* Performs "one point crossover". Takes pointers to two individuals as arguments and
replaces them by the children (i.e. the children are stored in ind1 and ind2). */
{ int position, i;
  int *map_string_ind2;
  map_string_ind2 = (int *) malloc(sizeof(int) * ind2->num_T);
  for(i = 0; i < ind2->num_T; i++)
     map_string_ind2[i] = ind2->map_string[i];
  position = irand(ind2->num_T); /*randomly pick a crossover point*/
  for(i = 0; i < position; i++) {
     ind2->map_string[i] = ind1->map_string[i];
     ind1->map_string[i] = map_string_ind2[i];
  }
  free(map_string_ind2);
  re_sch(ind1);   /*change the static-order schedule*/
  re_sch(ind2);
  /* evaluate new inds */
  ind1->mcm = eval_mcm(ind1);
  ind2->mcm = eval_mcm(ind2);
}
```

Algorithms 4 and 5 show two types of gene mutation for each individual. In Algorithm 4, 1 bit in the individual's gene is selected randomly and set to 0 or 1 randomly; In Algorithm 5, all bits in the individual's gene are set randomly with probability *pro*, a hyper-parameter that controls the aggressiveness of mutation operations.

Algorithm 4 One bit mutation

```
void one_bit_mutation(individual *ind)
/* Randomly set a bit at a random position in the gene ind. */
{   int position;
    position = irand(ind->num_T);
    ind->map_string[position] = irand(ind->num_P);
    /* evaluate new ind */
    re_sch(ind);
    ind->mcm = eval_mcm(ind);
}
```

Algorithm 5 Independent bit mutation

```
void indep_bit_mutation(individual *ind, pro)
/* Randomly set every bit in the gene ind with probability 'pro'.*/
{   int i;
    for(i=0;i<ind->num_T;i++) {
      if (drand(1) <= pro)
        ind->map_string[position] = irand(ind->num_P);
    }
    /* evaluate new individual */
    re_sch(ind);
    ind->mcm = eval_mcm(ind);
}
```

---

5 Experimental Evaluation

The experiments are run on a Linux workstation with an Intel dual-core 2.4 GHZ 64-bit processor and 4 GB of main memory. We use the software tool SDF3 [12] to generate random HSDF graphs. The WCET of each application actor ranges from 10 to 50. The network actors have WCET set to 5. Buffer size of each edge ranges from 1–5 tokens; the probability of an edge have initial tokens is 0.2. We use the PISA software package to implement the GA. Following are several important hyper-parameters used in our implementation of GA:

1. Size of the initial population: set to 100. This parameter affects the maximum number of parents and children in GA and its running time.
2. Maximum number of generations: set to 20. (It should be set to a larger value for larger systems.)
3. Mutation probability: set to 50%.
4. Recombination probability: set to 80%.

Figure 4 compares the performance and scalability of GA-based algorithms with SAT-based exact solution technique. We use Minisat [13] as the SAT solver. The number of actors ranges from 2 to 20, and the number of processors is 2. Figure 4(a) shows that when the number of actors is relatively small (<10), GA can return the optimal solution since the search space is small. With a larger number of actors, GA can no longer find the optimal solution. If the search space is very large and the number

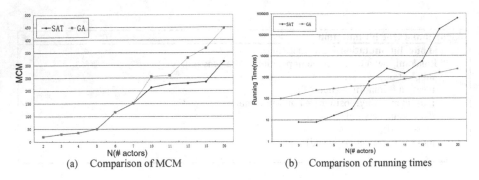

(a)   Comparison of MCM   (b)   Comparison of running times

Fig. 4. Experimental results.

of feasible solutions is extremely small (e.g., when most evolved individuals contain deadlocks that make them infeasible), GA may degenerate to blind random search that may not find a good solution. Figure 4(b) shows that the SAT solver running time increases at an exponential rate with the number of actors, making the SAT-solving approach infeasible for large applications.

6   Conclusions

In this paper, we address the problem of mapping an HSDF on a multiprocessor platform, with the object of maximizing total throughput. We use Genetic Algorithms to explore the search space: generate feasible actor-to-processor mapping and static-order scheduling of tasks on processor, then use graph theoretic techniques to calculate application throughput for each mapping and scheduling strategy in the quest for the optimal solution with maximum throughput. Even though no optimality guarantees can be made, experiment results indicate that GA can provide high-quality solutions when the search space is too large to handle with SAT solving.

Acknowledgements. This work is partially supported by NSFC Project # 61672454 and # 61602404; Zhejiang Provincial Natural Science Foundation Project # LY16F020007.

References

1. Liu, W., Yuan, M., He, X., Gu, Z.: Efficient SAT-based mapping and scheduling of homogeneous synchronous dataflow graphs for throughput optimization. In: Real-Time Systems Symposium (RTSS), pp. 492–504 (2008)
2. Liu, W., Gu, Z., Xu, J.: Efficient software synthesis for dynamic single appearance scheduling of synchronous dataflow. IEEE Embed. Syst. Lett. 1(3), 69–72 (2009)
3. Liu, W., Gu, Z., Xu, J., et al.: An efficient technique for analysis of minimal buffer requirements of synchronous dataflow graphs with model checking. In: IEEE/ACM International Conference on Hardware/Software Codesign and System Synthesis (CODES-ISSS), pp. 61–70 (2009)

4. Liu, W., Gu, Z., Xu, J., et al.: Satisfiability modulo graph theory for task mapping and scheduling on multiprocessor systems. IEEE Trans. Parallel Distrib. Syst. **22**(8), 1382–1389 (2011)

5. Liu, W., Gu, Z., Ye, Y.: Efficient SAT-based application mapping and scheduling on multiprocessor systems for throughput maximization. In: Compilers, Architecture and Synthesis for Embedded Systems (CASES), pp. 127–136 (2015)

6. Parhi, K.K., Messerschmitt, D.G.: Static rate-optimal scheduling of iterative data-flow programs via optimum unfolding. IEEE Trans. Comput. **40**(2), 178–195 (1991)

7. Satish, N., Ravindran, K., Keutzer, K.: A decomposition-based constraint optimization approach for statically scheduling task graphs with communication delays to multiprocessors. In: DATE 2007, pp. 57–62 (2007)

8. Metzner, A., Herde, C.: RTSAT–an optimal and efficient approach to the task allocation problem in distributed architectures. In: RTSS 2006, pp. 147–158 (2006)

9. Stuijk, S., Geilen, M., Basten, T.: Exploring tradeoffs in buffer requirements and throughput constraints for synchronous dataflow graphs. In: DAC 2006, pp. 899–904 (2006)

10. Stuijk, S., Basten, T., Geilen, M., Corporaal, H.: Multiprocessor resource allocation for throughput-constrained synchronous dataflow graphs. In: DAC 2007, pp. 777–782 (2007)

11. Dasdan, A., Gupta, R.K.: Faster maximum and minimum mean cycle algorithms for system-performance analysis. IEEE Trans. CAD Integr. Circ. Syst. **17**(10), 889–899 (1998)

12. Stuijk, S., Geilen, M., Basten, T.: SDF3: SDF for free. In: ACSD 2006, pp. 276–278 (2006)

13. Sorensson, N., Een, N.: Minisat v1. 13-a sat solver with conflict-clause minimization. In: SAT 2005, pp. 1–2 (2005)

Integration and Evaluation of a Contract-Based Flexible Real-Time Scheduling Framework in AUTOSAR OS

Ming Zhang[1], Nenggan Zheng[2(✉)], and Hong Li[1]

[1] College of Computer Science, Zhejiang University,
Hangzhou 310027, Zhejiang, China
[2] Qiushi Academy for Advanced Studies, Zhejiang University,
Hangzhou 310027, Zhejiang, China
zng@cs.zju.edu.cn

Abstract. FRESCOR (Framework for Real-time Embedded Systems based on COntRacts) is a flexible real-time scheduling architecture for real-time systems. This paper describes our implementation experience of integrating the FRESCOR framework in an AUTOSAR-compliant Real-Time Operating System. Performance evaluation shows that the performance overheads of integrating FRESCOR is acceptable on an embedded microcontroller.

Keywords: AUTOSAR · Real-Time Operating Systems · Contract-based

1 Introduction and Related Work

A contract is the interface between the application layer and the OS layer. The contract model is a high-level abstraction between applications and OS kernel to keep the application independent of specific scheduling algorithms [1]. It allows the developers to use a high level of abstraction to specify the application requirements instead of mapping the application requirements into scheduling parameters like priorities. FRESCOR [1, 2] is a real-time contract-based framework that supports multiple resources: processors, networks, memory, energy and other resources. It is a middleware framework that runs on top of an RTOS kernel, and aims to achieve flexible trade-offs between performance predictability and efficiency in resource utilization. FRESCOR can provide high-level Quality-of-Service (QoS) guarantees by satisfying the static and dynamic requirements through service contract model, enforcing online or offline schedulable acceptance test by means of contract negotiation and renegotiation, guaranteeing the minimum requirements by server-based resource reservation with contracts; distributing any spare capacity to maximize resource utilization; handling overload in a safe way by dynamic adaptation, etc.

AUTOSAR [3] is an open standardized automotive software architecture, jointly developed by automobile manufacturers, suppliers and tool developers. It aims at paving the way for innovative electronic systems that further improve performance, safety and environmental friendliness. AUTOSAR OS is a Real-Time Operating System (RTOS) standard within AUTOSAR. The FRESCOR framework was originally developed and

S. Li (Ed.): GPC 2018, LNCS 11204, pp. 334–341, 2019.
https://doi.org/10.1007/978-3-030-15093-8_24

evaluated on top of Linux OS and relatively-powerful x86 CPUs, but it has not been evaluated on more resource-constrained microcontrollers used in automotive control systems. In this paper, we describe implementation and performance evaluation of integrating FRESCOR with SmartOSEK OS, an AUTOSAR-compliant RTOS developed at Zhejiang University [4], on a typical low-cost microcontroller commonly used in in-vehicle embedded systems.

The remainder of the paper is organized as follows. In Sect. 2, we introduce the key modules of FRESCOR; in Sect. 3, we describe integration of FRESCOR with SmartOSEK OS; in Sect. 4, we present performance evaluation results; Sect. 5 presents conclusions and future work.

2 Key Modules of FRESCOR

Figure 1 shows the main modules in the FRESCOR framework. We implemented the following modules, highlighted with red boxes in the figure.

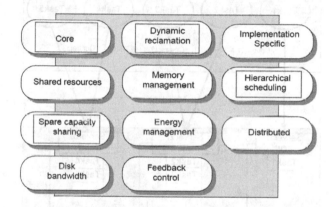

Fig. 1. Modules in the FRESCOR framework [1] (Color figure online)

2.1 Core Module

Core module specifies contract attributes and operations. Contract attributes are the parameters of the operations required to create contracts, negotiate contracts, bind tasks to server, and resource reservation mechanism that allows RTOS to guarantee the minimum resources requirements.

2.2 Dynamic Reclamation and Spare Capacity Module

Dynamic reclamation module is the functional module that carries out dynamic reclamation when there is any available capacity, while spare capacity module is responsible for distributing pending capacity that has not spent by servers. If there is any extra capacity left at runtime due to tasks' abnormal termination or completion ahead of schedule, it will be distributed among the different servers which have showed their desire or ability of using additional capacity according to the method specified in contracts.

2.3   Hierarchical Scheduling Module

In hierarchical scheduling, the global scheduler is used to determine which server should have the access to the uniprocessor, while a local scheduler is used in server to determine which task should execute when the server is active. Figure 2 shows the hierarchical scheduling framework of FRESCOR. The global scheduler is Sporadic Server (SS) on Fixed-Priority scheduling [5]. Sporadic Server is a bandwidth-preserving scheduling server, which preserves its budget/capacity until the next budget replenishment time if there are no ready tasks for the server. There are three possible local scheduling algorithms, e.g., Earliest Deadline First (EDF) for Server1, Fixed Priority (FP) for Server2, and First in First out (FIFO) for Server3 in Fig. 2.

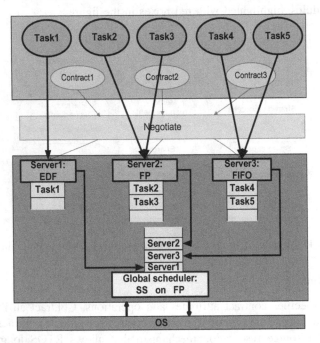

Fig. 2. Hierarchical scheduling framework of FRESCOR

When a contract has been specified, it should be negotiated, i.e., checked to see if the system is schedulable based on the specified contract. The system will check if it has enough resources to guarantee the minimum requirements of the new contract while keeping guarantees on all the previously accepted contracts. If the negotiation is successful, the system will reserve enough budgets to guarantee the new contract's minimum resources, and a server is created for that contract. The server is a special software task that is the runtime representation of the contract and it stores all the information of the contract. The server manages execution of multiple tasks that are bound to it.

3   Implementation Details

Figure 3 shows the SmartSAR software platform [4] with integrated FRESCOR modules. The following functionalities are implemented: resource reservation to guarantee the minimum requirements; negotiation and renegotiation to ensure schedulability; sharing spare capacity to have efficient maximum resource utilization.

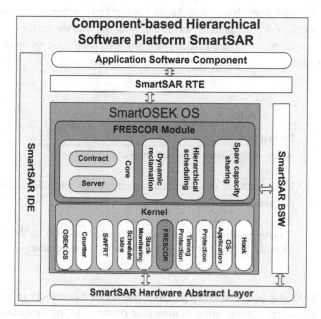

Fig. 3. SmartSAR software platform with integrated FRESCOR modules.

Tables 1, 2 and 3 show the main data structures of Sporadic Server.

Table 1. Attributes of the contract data structure.

Name	Description
max_period	Maximum server period
min_budget	Minimum budget (execution capacity) per server period
deadline	Deadline of the server
D = T	Whether deadline is equal to period
budget_overran	Current running task's budget has been exhausted
deadline_missed	The current task experienced a deadline miss
granularity	Indicates desire to use extra capacity, and how to use it: continuous or discrete
min_period	Minimum server period
max_budget	Maximum budget
utilization_set	Set of pairs {budget, period}
importance	A fixed priority used to distribute extra capacity in a continuous way
quality	A relative proportion used to distribute extra capacity among servers of the same importance in a continuous way
scheduling policy	Choice of the local scheduler: FP, EDF, FIFO

Table 2. Additional attributes of server structure.

Name	Description
priority	Parameter for scheduling
vres_id	Id for FRESCOR API
Task	Tasks bound to the server
task_num	Number of the bound tasks
task_finished_num	Number of finished bound tasks within one period
sporadic server	Sporadic server that is assigned to the server
spare_capacity	Spare capacity that has been distributed by system
used_capacity	Capacity that server has already used

Table 3. Structure of sporadic server.

Name	Description
replenishment queue	Array storing replenishment time and capacity
Alarm	Replenishment of consumed capacity occurs when the alarm expires
remaining budget	Accumulated budget not yet spent in the server
budget start	When sporadic server task goes into active status from the idle status; the capacity is being consumed while tasks are executing

Figure 4 shows the processing steps from the service contracts in the application layer, through the server representing contracts at runtime, to the hierarchical scheduler in the underlying implementation. First, the contract attributes are defined (Table 1). Then contracts need to be negotiated to preserve the minimum required resource budget for servers. The original FRESCOR framework uses Worse-Case Response Time (WCRT) analysis for schedulability test and online admission control [1]. If WCRT of a server exceeds its deadline, the server is not schedulable. However, the WCRT analysis equation has pseudo-polynomial time complexity [5], so its computation overhead may be too large to be used at runtime, especially on a low-cost microcontroller. Instead, we use the more efficient utilization bound tests for schedulability test, i.e., the contract is acceptable if total utilization of all accepted contracts does not exceed a pre-determined schedulable utilization bound, e.g., the Liu-and-Layland bound for fixed-priority scheduling [5].

When a contract is accepted, a scheduling server is created for it. Server is the control unit managing the minimum budget of resources, the amount of already consumed resources and obtained spare capacity, the replenishment method, and the local scheduling algorithm. Once tasks are assigned to a server, they will start execution and consume server's budget according to the contract established hierarchical scheduling method. When all tasks bound to the same server have finished execution in a server period, the unused capacity of that server is allocated to other servers that have expressed their need for more execution time. Tasks that leave or enter the system dynamically trigger renegotiation to redistribute the available spare capacity among servers. Renegotiation policy should comply with the contract definition.

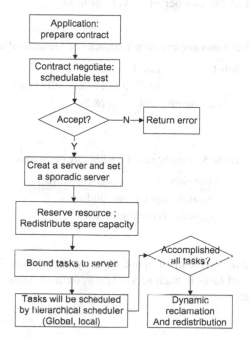

Fig. 4. Contract negotiation and resource reservation in FRESCOR.

4 Performance Evaluation

The framework has been implemented on a Freescale HCS12 development board with a 16 MHz microcontroller. We use the CodeWarrior Integrated Development Environment, which includes a simulator for performance evaluation. We performed measurements of various overheads, including: context-switch time of the hierarchical scheduling framework, overhead for negotiation and dynamic reclamation, and overhead for spare capacity distribution.

The maximum context switch time includes time for making a scheduling decision of selecting the highest-priority task, saving the context of old task, and loading the context of the new task. When there are three servers, each server has three tasks and different scheduling algorithms in the OS. Table 4 shows the maximum context switch latencies for different local scheduling algorithms (EDF, FP and FIFO). (Note that the context switching latencies are higher than those in [2], since we use a low-cost 16 MHz microcontroller, while [2] uses a 300 MHz x86 Pentium processor.)

Since AUTOSAR only supports FP scheduling, not EDF, and FIFO is not suitable for hard real-time systems, we adopt FP as the local scheduling algorithm in the next experiments. Table 5 shows average overhead of negotiation and dynamic reclamation, which is independent of the number of servers or tasks.

Table 4. Maximum context-switch latencies of hierarchical scheduler.

Global	Local		
	EDF (μs)	FP (μs)	FIFO (μs)
SS based on FP	70.4	68.2	65.9

Table 5. Overheads of FRESCOR operations.

Operation	Overhead (μs)
Dynamic negotiation	20.1
Dynamic reclamation	2.4

Figure 5 shows that overhead of spare capacity distribution increases monotonically with the number of tasks in each server (the system contains 3 servers, with equal number of tasks in each server).

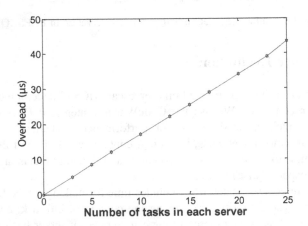

Fig. 5. Overhead of spare capacity distribution.

5   Conclusions and Future Work

In this paper, we present integration of the FRESCOR framework into an AUTOSAR-compliant RTOS to evaluate its applicability in severely resource-constrained embedded systems. Performance evaluation shows that the runtime overhead is acceptable even on a low-cost 16 MHz microcontroller. This brings many benefits of the FRESCOR framework into the domain of automotive embedded systems, including: the contract model allows the designer to specify complex timing requirements at a high-level of abstraction than manual priority assignment; schedulability and resource efficiency can be guaranteed by resource reservation and contract renegotiation/negotiation; resource utilization can be optimized by reclaiming spare processing capacity and distributing them among tasks. Future work includes integration with mixed-criticality scheduling [6–9], where the system consists of tasks with different levels of criticality or importance, and it is important to achieve performance isolation between tasks with different criticalities while achieving resource efficiency.

Acknowledgements. This work is partially supported by NSFC Project # 61672454 and # 61602404; Zhejiang Provincial Natural Science Foundation Project # LY16F020007.

References

1. Harbour, M.G., de Esteban M.T.: Architecture and contract model for integrated resources II. Deliverable of the FRESCOR project (DAC2v2) (2008)
2. Aldea, M., Bernat, G., Broster, I., et al.: FSF: a real-time scheduling architecture framework. In: Proceedings of the 12th IEEE Real-Time and Embedded Technology and Applications Symposium, pp. 113–124. IEEE (2006)
3. AUTOSAR Specification. http://www.autosar.org
4. Zhao, M., Wu, Z., Yang, G., et al.: SmartOSEK: a dependable platform for automobile electronics. In: The First International Conference on Embedded Software and System (2004). ISSN 0302-9743
5. Liu, J.W.S.: Real-Time Systems. Prentice Hall, Upper Saddle River (2000)
6. Zhao, Q., Gu, Z., Zeng, H., et al.: Schedulability analysis and stack size minimization with preemption thresholds and mixed-criticality scheduling. J. Syst. Archit. **83**, 57–74 (2018)
7. Zhao, Q., Gu, Z., Zeng, H.: Design optimization for AUTOSAR models with preemption thresholds and mixed-criticality scheduling. J. Syst. Archit. **72**, 61–68 (2017)
8. Zhao, Q., Gu, Z., Yao, M., et al.: HLC-PCP: a resource synchronization protocol for certifiable mixed criticality scheduling. J. Syst. Archit. **66**, 84–99 (2016)
9. Zhao, Q., Gu, Z., Zeng, H.: Resource synchronization and preemption thresholds within mixed-criticality scheduling. ACM Trans. Embed. Comput. Syst. (TECS) **14**(4), 81 (2015)

Pervasive Application

A Low-Cost Service Node Selection Method in Crowdsensing Based on Region-Characteristics

Zhenlong Peng[1,3], Jian An[2(✉)], Xiaolin Gui[1,4], Dong Liao[1,4], and RuoWei Gui[1,4]

[1] School of Electronics and Information Engineering, Xi'an Jiaotong University, No. 28, Xianning West Road, Xi'an 710049, People's Republic of China
{jxndpzl, liaod, r.w.gui}@stu.xjtu.edu.cn,
xlgui@mail.xjtu.edu.cn
[2] Xi'an Jiaotong University Shenzhen Research School, High-Tech Zone, Shenzhen 518057, People's Republic of China
anjian@mail.xjtu.edu.cn
[3] TSL Business School, Quanzhou Normal University, Donghai Streat, Quanzhou 362000, People's Republic of China
[4] Shaanxi Province Key Laboratory of Computer Network, No. 28. Xianning West Road, Xi'an 710049, People's Republic of China

Abstract. Crowdsensing is a human-centred perception model. Through the cooperation of multiple nodes, an entire sensing task is completed. To improve the efficiency of accomplishing sensing missions, a proper and cost-effective set of service nodes is needed to perform tasks. In this paper, we propose a low-cost service node selection method based on region features, which builds on relationships between task requirements and geographical locations. The method uses DBSCAN to cluster service nodes and calculate the centre point of each cluster. The region then is divided into regions according to rules of Voronoi diagram. Local feature vectors are constructed according to the historical records in each divided region. When a particular perception task arrives, Analytic Hierarchy Process (AHP) is used to match the feature vector of each region to mission requirements to get a certain number of service nodes satisfying the characteristics. To get a lower cost output, a revised Greedy Algorithm is designed to filter the exported service nodes to get the required low-cost service nodes. Experimental results suggest that the proposed method shows promise in improving service node selection accuracy and the timeliness of finishing tasks.

Keywords: Crowdsensing · Service node selection · Local feature vector

1 Introduction

With the continuous development of mobile Internet and sensor technology, people increasingly use mobile intelligent devices to facilitate the necessities in their lives. In this trend, Crowdsensing is gradually becoming the core of the mobile computing stage. Crowdsensing presents a new sensing paradigm based on the capacities of mobile

© Springer Nature Switzerland AG 2019
S. Li (Ed.): GPC 2018, LNCS 11204, pp. 345–356, 2019.
https://doi.org/10.1007/978-3-030-15093-8_25

devices and the interactions of a person or a group [1]. Crowdsensing has become a promising paradigm for cross-space and large-scale sensing [2], and it is pervasively leveraged to get urban conditions such as urban noise map [3], Navigating the last mile [4]. However, the selection of service nodes (e.g. task participants) has become the key to the success of crowdsensing services, because some solvers may not have the necessary abilities or may just want to get the reward without carefully performing tasks [5]. The previous researches have taken the nodes selection into account from a variety of aspects. Yu et al. [6] proposed a node prediction model based on the Markov chain (O2MM) and then introduced a prediction improvement based on social relations (SMLP). Daly et al. [7] exploited a novel social-based forwarding algorithm, that utilizes the real human mobility traces to enhance delivery performance. Ma, et al. [8] regarded the node selection as a multi attribute decision making problem, and then proposed a novel routing method for load-balancing, which can not only reduce the load of the whole network, but also balance the load of each participating node.

However, these algorithms focus more on node movements and the prediction of nodes. They only consider the relationship between position and node movement and make no reference to the connection between positions of nodes and characteristics of sensing tasks. Moreover, attributes and features of locations or regions are closely related to sensing tasks in many cases and have significant reference value for the accomplishment of the sensing tasks. A commonly approach called Grid Division [9] is pervasively leveraged in the prior literature. However, grids are often based on uniform partitioning of physical locations and do not take into account the characteristics of the location itself. In addition, many methods do not take the constraints (e.g. costs) into account in the process of completing tasks.

In this paper, we propose a low-cost service node selection method based on region features for sensing tasks. Firstly, the sensing region is divided according to the Voronoi diagram partition rules and each calculated center point becomes the center point of a corresponding Voronoi diagram. This partition rule has broad applications in many fields, such as visual imaging of invisible hazardous substances and uncertain data analysis. Secondly, the importance of each factor in the task is analyzed by AHP [10] method, and we can obtain the local feature vectors of each partitioned Voronoi diagram. We can obtain the matching degrees between the task and the local feature vector of each region and then obtain a sequence of candidate services nodes, and then the greedy algorithm is used to calculate the costs incurred by each selected node. Finally, some appropriate nodes in the region are selected to perform the sensing task. The whole process improves the accuracy, timeliness, and efficiency of node selection for completing the task. The main contributions of this paper are as follows.

- We propose a service node selection paradigm that combines task requirements with region features. Unlike most previous approaches that chose a node based only on task properties, our approach takes into account the relationship between task requirements and region characteristics, and the most matched nodes will be preferentially selected to fulfilled the tasks.
- We propose a region partitioning method based on the Voronoi diagram. This method aggregates sensed regions based on the attributes of the task, rather than just meshing them evenly across physical locations.

The rest of this paper is organized as follows. The second section presents the implementation process and the algorithmic of the method in detail. The third section applies the method to a real data set to evaluate the accuracy and speed by experimental comparisons. The final section summarizes this work.

2 ReLSNS Method

To overcome the shortcomings of the algorithms discussed above, this paper proposes a low-cost service node selection method (ReLSNS) based on local feature vectors. We define the whole sensing region as A, and the scale of A is determined by the sensing task requestors according to the task type and task granularity. The participant set that is used in a particular sensing task is defined as U. The number of participants is sufficient, otherwise, the quality of coverage may be degraded.

2.1 Frameworks of ReLSNS

As shown in Fig. 1, the specific steps of this method include three parts: region partitioning based on the Voronoi diagram [11], region matching based on feature vectors and the selection of nodes by the Greedy algorithm. In the different parts, there

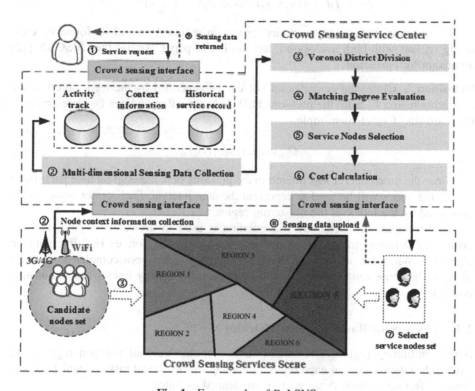

Fig. 1. Frameworks of ReLSNS

are some separate support modules, as shown in Fig. 1. The selection method proposed in this paper will be described in detail below.

2.2 Voronoi Diagram Partition

Compared with the traditional Grid Division method, Voronoi diagram-based region partitioning uses the characteristic of local similarity of the geographic region feature to divide the whole sensing region into multiple regions with irregular shape but different characteristics. Since Crowdsensing is a sensing mode that is centered on human beings, the regional characteristics are determined.

Definition 1. Track Point S_n is a trajectory point that represents the location of the mobile node $u(u \in U)$ at a time t, and is represented by a four-element tuple.

$$S_n = \{u, x, y, t\} \tag{1}$$

where t denotes time, x denotes the latitude of the node u at t, and y denotes the longitude of the node.

Definition 2. Track Point Set Td consists of all track point data of mobile nodes in U in region A, which can be represented as follows.

$$Td = \{(u, x, y, t) | \forall u \in U, (x, y) \in A\} \tag{2}$$

To discover clusters, high-activity regions in region A should be found by clustering regions with high track-point density. In this paper, we use the DBSCAN [12] algorithm for clustering.

Definition 3. Clustering center point O_{c_i} can be generated by the track points in the clustering region C_i from the previous DBSCAN algorithm and can be represented by the following four-element tuple.

$$O_{c_i} = \{i, x, y, [t_1, t_2]\}. \tag{3}$$

Since the time interval was filtered in the previous step, the original time parameters of sensing nodes will be used to represent the time interval. Each center point can be numbered by the index of the clustering region.

In this paper, the Delaunay triangulation is used to divide the region, and the process of Delaunay triangulation can be completed by the Convex Hull Interpolation algorithm. When the process of Delaunay triangulation has been completed, there are likely to be some center points that are located on boundary lines when delineating boundaries.

2.3 Eigenvector-Based Regions Matching

After partitioning region A to obtain some irregular Voronoi polygon regions, we extract the eigenvectors of each region according to a historical task set, which we call regional feature vectors; these vectors are defined as follows.

Definition 4. Task Feature Vector (*TFV*). A *TFV* is a vector of elements that describe the characteristics of a region. Each element of the vector reflects one characteristic of the region. In this paper, we also abstracted out four task-related regional characteristics.

Location information (*Li*). Local information, such as restaurants, hotels, shopping malls, banks, stadiums, and libraries, can be obtained by comparing maps from applications such as Amap and Google Maps.

Timeliness (*Tl*). This characteristic can be calculated by learning from historical data, namely, by calculating the average time spent by the nodes receiving tasks in the region from receiving tasks to uploading data.

Liveliness (*Ln*). Through the study of historical data, the total amount of effective task data that has been uploaded by nodes in an region can be obtained, which we call Liveliness.

Wi-Fi coverage information (*Wi*). By learning historical data, we can determine whether nodes upload historical task data through Wi-Fi, mobile 3G/4G network or other means in an region. Then, the ratio of Wi-Fi uploading to total uploading can be calculated as *Wi*.

For each Voronoi polygon region $a_i(a_i \in A)$, the task feature vector is denoted as *TFV* (a_i). *TFV* (a_i) can add new elements according to the specific sensing scenarios or task requirements. Next, we take these four factors as an example to explain how *TFV* (a_i) can be expressed in more detail as follows.

$$TFV(a_i) = (Li(a_i), Tl(a_i), Ln(a_i), Wi(a_i)) \qquad (4)$$

Where *Li* can be obtained by querying electronic map matching, and the resting three characteristics are obtained by analyzing historical effective task data. The method of calculating the four features in *TFV* (a_i) is illustrated.

Definition 5. Regional Functional Requirement set L_{ref}. This set is used to describe the collection of region functional attributes, such as dining region, culture and art region. L_{ref} can contain one or more regional functional attribute parameters.

The four elements in *TFV* (a_i), namely, position, length of time, number of times and proportion, have different meanings. To facilitate the calculation, the four elements of *TFV* (a_i) should be normalized to between 0 and 1. *Tl*, *Ln* and *Wi* in *TFV* (a_i) are represented by numbers, which are normalized to map eigenvalues to numbers between 0 and 1; *Li* in *TFV* (a_i) is obtained directly by matching *Li* (a_i) with L_{ref}, which is given by the task requestors. The normalization method is shown in Table 1.

After calculating an eigenvector for each region, we need to evaluate the preferences for each factor based on the assigned task and find a specific vector to represent the characteristics of the corresponding task. Therefore, two definitions are introduced here.

Table 1. Normalization method

Region feature	Normalization method
Location information (Li)	For every region a_i in region A, match $Li(a_i)$ with L_{ref}. If $Li(a_i)$ is contained in L_{ref}, set $Li'(a_i)$ as 1. Otherwise, set $Li'(a_i)$ as 0
Timeliness (Tl)	Find the smallest Tl in region A, which is denoted as a_{min}. Set $Tl'(a_{min})$ as 1, and compare the value of $Tl(a_{min})$ with the $Tl(a_i)$ values of all the regions in region A. Let $Tl'(a_i)$ denote the resulting ratio
Liveliness (Ln)	Find the largest Ln in region A, which is denoted as a_{max}. Set $Ln'(a_{max})$ as 1, and compare the value of $Ln(a_{max})$ to the $Ln(a_i)$ values of all the regions in region A. Let $Ln'(a_i)$ denote the resulting ratio
Wi-Fi coverage information (Wi)	Find the largest Wi in region A, which is denoted as a'_{max}. Set $Wi'(a'_{max})$ as 1, and compare the value of $Wi(a'_{max})$ with the $Wi(a_{max})$ values of all the regions in region A. Let $Wi'(a_{max})$ denote the resulting ratio

2.4   Selection of Service Nodes Based on Greedy Algorithm

After obtaining Region Matching Vector ζ, all regions are sorted in descending order according to the values of the elements in the vector. That is, the regions with high matching degrees are arranged in the front of a priority queue. The number of nodes that are required to perform a task is determined by the task requestors when the task is published, which is denoted as Q. Therefore, to improve the performance of the proposed algorithm, a certain number of regions are selected, which contain nearly $2Q$ nodes. We denote the number of selected regions as N and the number of selected nodes as K. These regions possess characteristics that match task characteristics. However, the distance cost is not considered. Therefore, further processing is required after selection.

Definition 6. Node Score Vector δ_k. This vector is used to measure the degree of matching of regional features and the cost-effectiveness of a node. δ_k is defined as follows.

$$\delta_k = \left(u_i, Score_i(M_{ref}), Score_k(Cost) \right) \tag{5}$$

Where u_i represents the number given to a specific region. $Score_i(M_{ref})$ represents the score obtained from the process of evaluating the matching degree of each region, and the cost score $Score_k(Cost)$ represents the score obtained from the calculated distance cost. In the following algorithms, service nodes can be selected according to these two values.

We can compute the scores in the node score vector δ_k according to the following rules. The values of $Score_i(M_{ref})$ can be scores from N to 1 in descending order, based on the order obtained by using the method that was described in the previous section, with each score decremented by one. The values of $Score_k(Cost)$ in the region score vector δ_k can be calculated according to the corresponding task latitude x_T and longitude y_T, using Algorithm 1.

Algorithm 1. Score the regions based on labor costs

Input: Task-featured Region Dataset $C = \{C_1, C_2, C_3, ..., C_N\}$, and Location of the specific task $O_T = (x_T, y_T)$

Output: Scores of labor cost of each region

1:    create a distance vector $D = \{D_1, D_2, D_3, ..., D_N\}$

2:    **for each** $C_i \in C$ **do**

3:      **while** not all nodes in C_i have been visited **do**

4:        select one unvisited node from C_i

5:        set it as visited

6:        calculate Manhattan distance from this node to O_T

7:        save the calculating result in D_k with the corresponding node identified number

8:      **end while**

9:    **end for**

10: **for each** $D_k \in D$ **do**

11: compare the value of D_k with other elements in the vector and rank

12: **end for**

13: Based on the order in D, score each region with a rule that a shorter calculated distance can have a higher score.

In Algorithm 1, the distance of each node is calculated and the corresponding Manhattan distance is saved in a distance vector D. Therefore, by using Algorithm 2, the distance costs of all the service nodes of each region can be considered. Finally, we can obtain the corresponding $Score_k (Cost)$ by using the rule that a node with a shorter distance to the target place has a higher cost score. After completing the above steps, the service nodes can be selected by using the following revised Greedy Algorithm.

3 Experiment

Experimental data is from the Wireless Topology Discovery (WTD) [13]. The WTD dataset contains data from 275 PDA users on the campus of the University of California, San Diego, for 11 weeks. Because the WTD dataset is too large, we choose data (1,106,308 records) from one week from six colleges of the University of California, San Diego campus center to simulate the above algorithms and method. There are 208 mobile nodes in the data set.

The experimental results of this method are compared with the social relation-based prediction model (SMLP) and the second-order Markov model (O2MM). Since this method is task-centered and selects nodes by region matching, it is very different from other node selection algorithms. Therefore, direct comparison with other algorithms is not convincing. Considering that SMLP and O2MM are trace prediction models of participatory Crowdsensing perception, this paper simulates the node selection algorithm based on their trajectory predictions to carry out the comparison with the method of this paper.

3.1 Region Partition Based on Voronoi Diagram

The trajectory of the 1,106,308 data instances is analyzed, and the region covered by all the nodes is partitioned. According to our algorithm, the region is divided into 79 regions. We set the clustering radius of the selected dataset to 100 m and the minimum number of nodes in each cluster to 5000. The sparse region of human flow in the original region is evenly divided into the Voronoi polygons where the nearest center points are located. The results are shown in following Fig. 2.

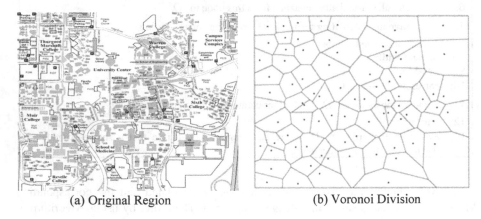

(a) Original Region   (b) Voronoi Division

Fig. 2. Voronoi diagram of experimental data

3.2 Evaluation of Service Node Selection Accuracy

The numbers of service nodes of tasks that were required for the simulation are 10, 20, 30, 40, 50 and 60. For each requirement, we perform four tests, and we count the numbers of nodes that satisfy the requirements of the task. Finally, the average numbers of nodes are calculated, which are regarded as accuracy rates of selected service nodes under different required numbers of nodes. Then, we calculate the accuracy rates between ReLSNS and Division Methods, which are shown in Fig. 3. It is obvious that using ReLSNS can provide task requestors with relatively stable service node selection results, while the results of traditional Grid Division are not guaranteed to be stable. The fluctuation is undesirable in the process of service node selection in practice.

In Fig. 4, ReLSNS is also compared with SMLP and O2MM. The accuracy rates of selected service nodes decrease as the required number of selected nodes increases. Because ReLSNS always chooses the service node with the highest matching degree, the more nodes that are selected, the lower the average matching degree of the selected service nodes, which results in a decrease in the accuracy of the selected set of nodes.

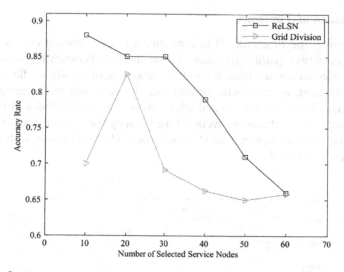

Fig. 3. Accuracy rate comparisons between ReLSNS and Division Methods

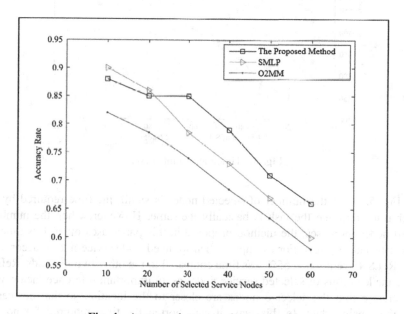

Fig. 4. Accuracy between relevant algorithms

According to the figure, when the number of service nodes in the selected node set is less than 20, the performance of ReLSNS in this paper is slightly worse than that of SMLP, which nonetheless still has nearly 90% accuracy. When the number of selected service nodes is more than 30, the accuracy of the proposed method is obviously higher than those of the other algorithms.

3.3   Evaluation of Timeliness

In Crowdsensing, the timeliness of task execution is an important criterion for measuring task execution quality. By obtaining the distance between the location of the node in the selected service node set and the nearest position where effective sensing data can be obtained, we can obtain the time taken by each node to successfully acquire the sensing task in the simulation and afterward obtain the timeliness of the selected node set. Then, we can obtain the average time taken by the selected nodes to complete the task under different required numbers of nodes. The comparison with other algorithms is shown in Fig. 5.

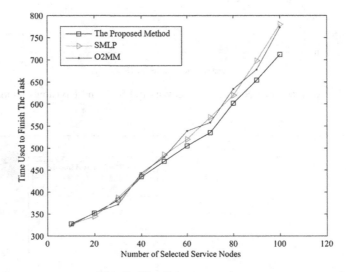

Fig. 5. Timeliness comparisons

In Fig. 5, when the number of selected nodes is small, the time required by each algorithm to complete the task is basically the same. However, when the number of selected nodes increases, the method proposed in this paper uses much less time than the other two algorithms. For example, when we need 100 service nodes, according to Fig. 5, ReLSNS requires 9.55% less time compared to the other two methods. ReLSNS can use the locations of selected service nodes as an important reference factor, which can guarantee that the selected nodes are closer to the location so that devices can obtain the sensing data. In this way, it can shorten the time required for nodes to perform the task and ensure that the set of selected service nodes can meet the requirement of high timeliness.

3.4   Generalization of Experimental Results

Based on the simulation tests of selected nodes' accuracy rate and the average time required for task execution, we can arrive at the following conclusions. First, when the number of tasks is small, the performance of the method based on region feature proposed in this paper is basically the same as the social relation models. But with the increasing of the number of selected service nodes, the method in this paper is superior to other two algorithms regarding accuracy and timeliness.

4 Conclusion

In this paper, a low-cost service node selection method based on region feature vector is proposed to solve the problem that the existing service node selection algorithms do not consider the association between sensing task and region. The method uses Voronoi polygons as the basis and constructs an eigenvector of each Voronoi polygon region. When the system releases a sensing task, this method can help task requestors to compute matching degrees between the task and region task feature vectors and select regions according to the overall score of each region. And finally, we can get regions with a high matching degree of task characteristics can then continue the progress of service nodes selection in these regions.

Acknowledgments. This work was supported partly by the NSFC Grant No 61502380, partly by Science and Technology Program of Shenzhen (JCYJ20170816100939373).

References

1. An, J., et al.: A crowdsourcing assignment model based on mobile crowd sensing in the internet of things. IEEE Internet Things 2(5), 358–369 (2015)
2. Guo, B., Yu, Z.W., Zhang, D.Q., et al.: From participatory sensing to mobile crowd sensing. In: Proceeding of the 12th IEEE International Conference on PerCom 2014, Budapest, Hungary, March 2014, pp. 593–598 (2014)
3. Qin, Z., Zhu, Y.: NoiseSense: a crowd sensing system for urban noise mapping service. In: Proceedings of ICPADS (2017)
4. Fan, X., Liu, J., Wang, Z., et al.: Navigating the last mile with crowdsourced driving information. In: Proceedings of INFOCOM, pp. 346–351 (2016)
5. Fiore, M., Nordio, A., Chiasserini, C.F.: Driving factors toward accurate mobile opportunistic sensing in Urban environments. IEEE Trans. Mob. Comput. 15(10), 2480–2493 (2016)
6. Yu, R.Y., Xia, X.Y., et al.: A location prediction algorithm with daily routines in location-based participatory sensing systems. Int. J. Distrub. Sens. N. 11(10), 1–12 (2015)
7. Pan, H., Crowcroft, J., Yoneki, E.: BUBBLE Rap: social-based forwarding in delay tolerant networks. IEEE Trans. Mob. Comput. 10(11), 1576–1589 (2011)
8. Ma, H., et al.: A multi attribute decision routing for load-balancing in crowd sensing network. Wireless Netw. 10, 1–16 (2017)
9. Guidi, F., Anna, G., Davide, D.: Personal mobile radars with millimeterwave massive arrays for indoor mapping. IEEE Trans. Mob. Comput. 15(6), 1471–1484 (2016)
10. Jin, J.W., Rothrock, L., McDermott, P.L., et al.: Using the analytic hierarchy process to examine judgment consistency in a complex multiattribute task. IEEE Trans. Syst. Man Cybern. C Appl. Rev. 40(5), 1105–1115 (2010)

11. Kao, B., Lee, S.D., Lee, F.K.F., et al.: Clustering uncertain data using voronoi diagrams and R-tree index. IEEE Trans. Knowl. Data Eng. **22**(9), 1219–1233 (2010)
12. Xu, X., et al.: DBSCAN revisited, revisited: why and how you should (Still) Use DBSCAN. ACM Transactions on Database Systems **42**(3), 19 (2017)
13. Wu, C.H., Lee, K.C., Chung, Y.C.: A Delaunay triangulation based method for wireless sensor network deployment. Comput. Commun. **30**(14), 2744–2752 (2007)

Electric Load Forecasting Based on Sparse Representation Model

Fangwan Huang[1,2], Xiangping Zheng[1,2], Zhiyong Yu[1,2,3(✉)],
Guanyi Yang[1,2], and Wenzhong Guo[1,2,3]

[1] College of Mathematics and Computer Science,
Fuzhou University, Fuzhou, China
yuzhiyong@fzu.edu.cn
[2] Fujian Provincial Key Laboratory of Network Computing and Intelligent
Information Processing, Fuzhou University, Fuzhou, China
[3] Key Laboratory of Spatial Data Mining and Information Sharing,
Ministry of Education, Fuzhou, China

Abstract. Accurate electric load forecasting can prevent the waste of power resources and plays a crucial role in smart grid. The time series of electric load collected by smart meters are non-linear and non-stationary, which poses a great challenge to the traditional forecasting methods. In this paper, sparse representation model (SRM) is proposed as a novel approach to tackle this challenge. The main idea of SRM is to obtain sparse representation coefficients by the training set and the part of over-complete dictionary, and the rest part of over-complete dictionary multiplied with sparse representation coefficients can be used to predict the future load value. Experimental results demonstrate that SRM is capable of forecasting the complex electric load time series effectively. It outperforms some popular machine learning methods such as Neural Network, SVM, and Random Forest.

Keywords: Electric load forecasting · Smart grid · Sparse representation

1 Introduction

As an important part of smart city, smart grid makes use of many intelligent acquisition equipment and advanced information management system, which can provide more efficient, reliable and environmentally friendly power services than traditional power grid [1]. Accurate electric load forecasting is one of the important functions of smart grid, because it plays an indispensable role in the planning and operation of the whole system. However, it is still a difficult problem, because there are many external factors influencing the change of load, such as weather conditions, temperature, type of date, seasonal factors, social activities and economic development [2].

Owing to the significance of electric load forecasting, there is a large and growing literature focused on it. In these studies, not only statistical methods but also machine learning methods were employed. Although extensive studies have been done, accurate electric load forecasting still remains the challenges in smart grid such as over-fitting and sensitivity to noise. In this paper, sparse representation model (SRM) is proposed

© Springer Nature Switzerland AG 2019
S. Li (Ed.): GPC 2018, LNCS 11204, pp. 357–369, 2019.
https://doi.org/10.1007/978-3-030-15093-8_26

as a novel approach to tackle this challenge. In the last decade, there has been an increasing interest in sparse representation, which has been successfully applied in the fields of signal processing, image analysis, pattern recognition and machine learning [3]. The beauty of sparse representation is that it allows us to capture important information from signals into a small number of components. Inspired by this, SRM is expected to be an effective method to deal with complex power load time series. The contributions of this paper are as follows:

- We employ sparse representation model as a novel approach to forecast electric load more accurately compared with some popular machine learning methods such as Neural Network, SVM, and Random Forest for two data sets.
- We utilize the analytic approach to construct the basic dictionary of SRM, which has the advantage of parameter-free adjustment. This makes SRM more general and convenient for electric load forecasting under different application backgrounds.
- We improve the performance of SRM further by constructing the over-complete dictionary with the combination of basic dictionary and external factors which had been proved to be correlated with load demand strongly.

The rest of the paper is organized as follows: Sect. 2 reviews the related work of electric load forecasting. The theory of sparse representation is introduced in Sect. 3. Sections 4 and 5 present the proposed technique and the results of experiments. Finally in Sect. 6, the conclusion and future work are stated.

2   Related Works

According to the duration of the forecasting, normally electric load forecasting can be classified into three categories: 1 h to 1 day or 1 week ahead for short-term, 1 month to 1 year ahead for medium-term, and more than 1 year ahead for long-term forecasting [4]. At present, there are many methods for electric load forecasting, including statistical methods commonly used in time series analysis and modern methods represented by artificial intelligence.

The statistical methods are mainly derived from the classical methods of time series analysis, including Exponential Smoothing (ES), Auto Regressive Integrated Moving Average (ARIMA), Generalized Auto Regressive Conditional Heteroskedasticity (GARCH) and a number of their variants [5]. However, these methods are only effective under the assumption that the historical values of time series are highly correlated with the predicted values, which makes them poor performance in full modeling of complex electric load time series and often results in lower accuracy.

Due to the limitation of statistical methods, with the rapid development of artificial intelligence, many modern methods have been applied for accurately forecast electric load in recent years such as fuzzy logic, expert systems, neural network, support vector machine and so on [6]. In addition, as Hinton et al. proved that better performance can be obtained through deep neural network than that of shallow network [7], the application of deep learning in electric load forecasting has also attracted extensive attention [8, 9]. These works successfully demonstrate the advantages of deep learning in electric load forecasting.

Finally, considering that each method has its own advantages and disadvantages in such aspects as prediction accuracy, algorithm complexity, parameter sensitivity and different forecasting horizon, the application of hybrid model is also a competitive competitor of the above methods. The main idea of hybrid model is to improve the overall prediction performance by maximizing the advantages of single model. It can be conclude that hybridization of two or more techniques shows better results than the individual models for load forecasting [10].

3 Preliminaries

From the perspective of mathematics, solving the sparse representation of a signal means to seek the sparsest linear combination of basis vectors from an over-complete dictionary to approximate it. An over-complete dictionary is a matrix with more columns than rows, and each of its columns is a basis vector that can be called an atom. The over-completeness of the dictionary provides more flexibility for the approximate representation of signal, which enables sparse representation to be robust to additive noise and occlusion by finding the sparse transformation domain [11].

Let us assume that a signal y can be represented in terms of a linear superposition of m basis vector $d_1, d_2, \ldots, d_m$ multiplied by their corresponding coefficients $\alpha_1, \alpha_2, \ldots, \alpha_m$ as follow:

$$y = \sum_{i=1}^{m} d_i \alpha_i \tag{1}$$

For convenience, the following linear systems can be obtained by using matrix-vector notation:

$$y = D\alpha \tag{2}$$

where the signal $y \in R^n$, the over-complete $D \in R^{n \times m}$ (n < m) and $\alpha \in R^m$ is the coefficient vector. When the majority of coefficients in α are zero, it is said that signal y has a sparse representation with respect to the dictionary D. Note that Eq. 2 is an underdetermined linear system having infinite number of solutions because D is a rectangular matrix having more columns than rows. To narrow down the choice to one well-defined solution, additional constraints are required. A familiar way to do this is to introduce an objective function and define a general optimization problem as follow:

$$\min_{\alpha} \|\alpha\|_0 \quad s.t. \ y = D\alpha \tag{3}$$

where $\|.\|_0$ represents the number of nonzero elements in the vector, which can also be regarded as the degree of sparsity. On account for the noise present in the real data, the constraint is often relaxed using a quadratic penalty function $\|y - D\alpha\|_2^2$ and following error-tolerant versions of Eq. 3 is solved.

$$\min_{\alpha} \|\alpha\|_0 \quad s.t. \ \|y - D\alpha\|_2^2 \le \delta \tag{4}$$

where δ is a small positive constant that can be considered as the reconstruction error. In addition, we also consider the variant that flip the objective and constraint:

$$\min_{\alpha} \|y - D\alpha\|_2^2 \quad s.t. \ \|\alpha\|_0 \le L \tag{5}$$

Furthermore, according to the Lagrange multiplier theorem, Eqs. 4, 5 are equivalent to the following unconstrained minimization problem with a proper value of λ.

$$\min_{\alpha} \frac{1}{2} \|y - D\alpha\|_2^2 + \lambda \|\alpha\|_0 \tag{6}$$

where λ refers to the Lagrange multiplier.

Unfortunately, sparse representation with l_0-norm minimization has been proved to be a NP-hard problem, which is difficult to find the global optimal solution in reasonable time [12]. In recent years, many algorithms have been proposed to solve approximate solutions. One of the algorithms is Matching Pursuit (MP) based on greedy strategy [13]. MP selects one column of dictionary at a time then approximating the solution step by step. At each step, MP looks for the atom that minimizes the reconstruction error. MP has some obvious problems. If the dictionary contains some similar atoms, it tends to match the residual over and over, that leads to converge slowly. A modification to the Matching Pursuit is called as Orthogonal Matching Pursuit (OMP), which accelerates convergence by modifying the formula of the reconstruction error [14]. More specifically, after selecting the atom at each step using the same greedy strategy as BP, OMP utilizes the standard least squares technique to calculate the best approximation over all the selected atoms and finally obtains the corresponding reconstruction error. This process can ensure that the reconstruction error of each step is completely orthogonal to the selected atoms [15].

Apart from the greedy strategy, another way is constrained optimization strategy which replaces the highly discontinuous l_0-norm with its closest convex approximation i.e. the l_1-norm. Recent literature has demonstrated that the solution using l_1-norm minimization with sufficient sparsity can be equivalent to the solution obtained by l_0-norm minimization with full probability [16]. Moreover, the l_1-norm optimization problem has an analytical solution and can be solved in polynomial time. Thus, we can transform Eqs. 3–6 to:

$$\min_{\alpha} \|\alpha\|_1 \quad s.t. \ y = D\alpha \tag{7}$$

$$\min_{\alpha} \|\alpha\|_1 \quad s.t. \ \|y - D\alpha\|_2^2 \le \delta \tag{8}$$

$$\min_{\alpha} \|y - D\alpha\|_2^2 \quad s.t. \ \|\alpha\|_1 \leq \kappa \tag{9}$$

$$\min_{\alpha} \frac{1}{2} \|y - D\alpha\|_2^2 + \lambda \|\alpha\|_1 \tag{10}$$

where $\|.\|_1$ is defined as the sum of absolute value of each entry in the vector. l_1-norm minimization can be solved by standard optimization tools like linear programming. This approach is known as the Basis Pursuit (BP) [17]. Although BP and its variants can obtain the global optimal solution of l_1-norm minimization, it is worth noting that there is a possibility of inconsistency with that of l_0-norm minimization, because l_0-norm fairly punishes all non-zero coefficients while l_1-norm punishes larger coefficients more.

4   SRM for Electric Load Forecasting

In this section, sparse representation model (SRM) is proposed as a novel approach to forecast electric load. The main idea of SRM is to obtain sparse representation coefficients by the training set and the part of over-complete dictionary, and the rest part of over-complete dictionary multiplied with sparse representation coefficients can be used to predict the future power load value. The framework of sparse representation model (SRM) for electric load forecasting is shown in Fig. 1.

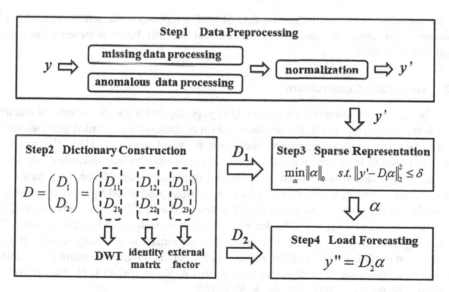

Fig. 1. The framework of sparse representation model.

4.1   Data Preprocessing

In order to establish a good foundation for the prediction model and generate a better output results, it is necessary to preprocess input data including the missing data recovery, anomalous data processing and normalization. Let us assume that $x(d, t)$ is the load value of d days at time t. It can be normally repaired by the data of the two moments before and after t at the same day and the data of the two days before and after d at the same moment. So the missing data recovery can be performed using the following formula:

$$x(d,t) = [x(d,t-1) + x(d,t+1) + x(d-1,t) + x(d+1,t)]/4 \qquad (11)$$

Moreover, the influence of random factors will produce some outliers which could reduce the predictive accuracy of the model. To deal with the anomalous data, quartile method is proposed to find the abnormal value which will be replaced by the reasonable value with the same method of filling the missing data. In the end, the normalization process can simplify the data calculation, accelerate the convergence of the program operation, and get the optimal solution easily. In standard normalization process, each input data point transform between interval 0 and 1. Data normalization can be performed using the following formula:

$$y = \frac{x - MinValue}{MaxValue - MinValue} \qquad (12)$$

where y represents the normalization data of load, x represents the actual value of load, *Minvalue* represents the minimum value of load and *MaxValue* represents the maximum value of load.

4.2   Dictionary Construction

The choice of over-complete dictionary D is very important for the success of electric load forecasting. In general, the implementation of dictionary is divided into two main approaches i.e., analytic approach and learning based approach [18]. The analytic approach need to formulate a mathematical model in order to lead towards dictionaries that are highly structured and numerically fast implementation. These dictionaries are also referred as implicit dictionaries which were usually achieved by transform domain methods including Wavelets based [19], Curvelets based [20], Contourlets based [21], Complex Wavelets [22] etc. The learning based approach need to train or infer the dictionary from a given set of examples. The dictionary is normally represented in terms of an explicit matrix and the matrix coefficients is required to adapt by machine learning algorithm such as Principal Component Analysis (PCA) [23], the method of optimal directions (MOD) [24], the K-SVD [25].

In this paper, we used the analytical dictionaries based on discrete wavelet transform (DWT) which has performed well for localization of singular points in many natural signals and images as the first part of over-complete dictionary D. The second part of over-complete dictionary D is the identity matrix which can be generated by delta function in order to avoid over-fitting. The above two parts constitute the basic dictionary. Moreover, in order to improve the precision of load forecasting effectively, the third part of over-complete dictionary D is designed with the combination of three external factors including average daily temperature, dates of holidays, day of week which had been proved to be correlated with load demand strongly. Over-complete dictionary D can be defined as follow:

$$D = \begin{pmatrix} D_1 \\ D_2 \end{pmatrix} = DWT + identity\,matrix + external\,factors\,matrix \qquad (13)$$

where D_1 is used to compute sparse representation coefficients α of the historical load data, D_2 is used to predict the future electric load.

4.3   Sparse Representation for Load Forecasting

Firstly, we need to obtain sparse representation coefficients α by approximating the constraint problem as follow:

$$\min_{\alpha} \|\alpha\|_0 \quad s.t. \ \|y' - D_1\alpha\|_2^2 \leq \delta \qquad (14)$$

where the training sample $y' \in R^m$ is the normalized data of the actual load data y, $D_1 \in R^{m \times [2(m+n)+3]}$ refers to the first m rows of the over-complete dictionary D constructed with DWT, identity matrix and three vectors of external factors. The OMP algorithm is used to solve the problem which is defined as Eq. 14.

Since the sparse representation coefficients $\alpha \in R^{[2(m+n)+3]}$ can be thought of as features extracted from the load data, the load value of the next n day $y'' \in R^n$ can be formulated as:

$$y'' = D_2\alpha \qquad (15)$$

where $D_2 \in R^{n \times [2(m+n)+3]}$ refers to the rest n rows of the over-complete dictionary D. The main steps of SRM-based Electricity Load Forecasting Algorithm have been summarized in Algorithm 1.

Algorithm 1 SRM-based Algorithm FOR Load Forecasting
Step 1: data preprocessing for input data
Step 2: Dictionary Construction

$$D = \begin{pmatrix} D_1 \\ D_2 \end{pmatrix} = DWT + identity\ matrix + external\ factors\ matrix$$

Step 3: Approximate the constraint problem by OMP:

$$\min_{\alpha} \|\alpha\|_0 \quad s.t. \|y' - D_1\alpha\|_2^2 \leq \delta$$

Initialization: $t = 0$, $r_0 = y'$, $D_1^{(0)} = \phi$

While $\|r_t\| > \delta$ do

    1) Find the best matching sample, i.e. the biggest inner product between r_t and $d_i \in D_1 - D_1^t$

    2) Update $D_1^{t+1} = [D_1^t,\ d_i]$

    3) Compute the sparse coefficients by using the least square algorithm $\alpha = \arg\min \|y' - D_1^{t+1}\alpha\|_2^2$

    4) Update the representation residual $r_{t+1} = y' - D_1^{t+1}\alpha$

    5) $t = t + 1$

    End

Step4   Load Forecasting by $y'' = D_2\alpha$

5 Experiments and Results

In this section, we present our experiments on applying sparse representation model (SRM) for electric load forecasting. In order to verify the influence of external factors on electric load forecasting, we constructed SRM I which chose the basic dictionary as overcomplete dictionary and SRM II which combined external factors with the basic dictionary respectively. We compared SRM I and SRM II with some popular machine learning methods such as Neural Network, SVM, and Random Forest for two data sets. Three evaluation criterions were used as performance metric: Mean Absolute Error (MAE), Root Mean Square Error (RMSE) and Mean Absolute Percentage Error (MAPE) defined as follows:

- Mean Absolute Error (MAE):

$$MAE = \frac{1}{n}\sum_{i=1}^{n} |y_i - \tilde{y}_i| \tag{16}$$

- Root Mean Squared Error (RMSE):

$$RMSE = \sqrt{\frac{1}{n}\sum_{i=1}^{n} (y_i - \tilde{y}_i)^2} \tag{17}$$

- Mean Absolute Percentage Error (MAPE):

$$MAPE = \frac{1}{n}\sum\nolimits_{i=1}^{n} \frac{|y_i - \tilde{y}_i|}{y_i} \tag{18}$$

where y_i and $\tilde{y}_i$ are the real and the predicted value of the ith day, n is the total number of data used for performance evaluation and comparison.

5.1   Experiment Results with the First Data Set

The first data set is provided by EUNITE network which organized a world-wide competition on electric load forecasting in 2001. Given information included electricity load demand recorded every half hour, average daily temperature, dates of holidays, day of week from January 1997 to January 1999. Without loss of generality, we only focused on the maximum values. In this paper, the data from 1997 to 1998 was used to train the model, and the data of January 1999 was used as the testing set.

It can be seen from Fig. 2 that all the five methods followed the trend of the original time series. Figure 3 shows more details of the forecasting results. We can clearly see that SRM II had superior performance over all the other methods. This is also verified from Table 1 which shows that SRM II achieved better forecasting performance with the smallest forecasting errors. The experimental results show that the dictionary with external factors is better than the basic dictionary in predicting accuracy.

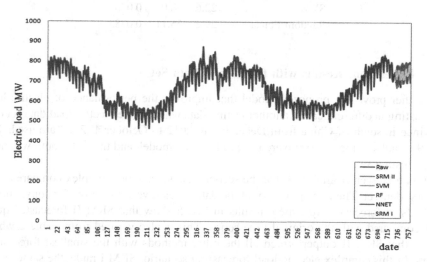

Fig. 2. In-sample and out-of-sample comparison for the first data set.

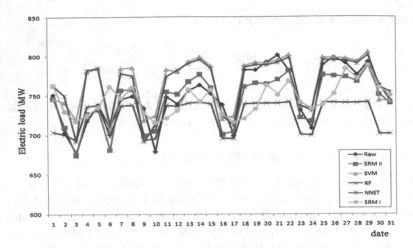

Fig. 3. Forecasting comparison for the first data set.

Table 1. Prediction results for the first data set.

Forecasting methods	MAE	RMSE	MAPE
SRM II	**13.8**	**17.01**	**0.0184**
SRM I	20.9	27.69	0.0294
Neural Network	20.4	26.08	0.0278
SVM	22.6	28.02	0.0309
Random Forest	29.2	35.11	0.0383

5.2   Experiment Results with the Second Data Set

To further prove the proposed model that improves the performance of electric load forecasting in different cases, another testing data which is the electric load data set of a province in south of China from December 8, 2012 to October 4, 2013 are used. The first 9 months of the data set were used to train the model, and the last month was used as the testing set.

It can be seen from Fig. 4 that the second data set is more complex compared with the first data set. The non-stationary of the data poses severe challenge for conventional methods to forecast. Experiment results in Fig. 5 show that SRM II forecasted quite well compared with the original time series. This is also verified from Table 2 which shows that SRM II outperformed all the other methods with the smallest forecasting errors. In this complex electric load forecasting scenario, SRM I ranks the same as the first dataset in terms of performance while Neural Network did not work although it performed quite well for the first data set.

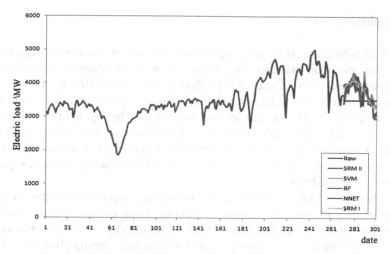

Fig. 4. In-sample and out-of-sample comparison for the second data set.

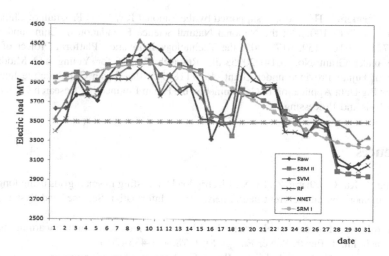

Fig. 5. Forecasting comparison for the second data set.

Table 2. Prediction results for the second data set.

Forecasting methods	MAE	RMSE	MAPE
SRM II	**138.8**	**177.1**	**0.0375**
SRM I	161.1	203.4	0.0448
Neural Network	293.5	393.5	0.0843
SVM	172.1	221.7	0.0473
Random Forest	167.6	205.0	0.0438

6 Conclusion

Electric load forecasting is of great significance for efficient energy management and better system planning of smart grid, because it not only improves the reliability of power system, but also reduces the production cost. Motivated by the successful application of sparse representation in other fields, this paper proposes a sparse representation model (SRM) for electric load forecasting. The highlight of this model is that three external factors influencing load change are extended on the basis of the analytical dictionary to achieve better prediction. Experiment results show that SRM can effectively improve the accuracy of electric load forecasting, which is superior to other forecasting methods.

For future work, we are interested in investigating other uses of sparse representation on smart grid, such as anomaly and change point detection, and clustering of households by consumption patterns. Finally we are also interested in evaluating how consumers adjust their consumption patterns based on these consumption forecasts, and building better tools and applications based on these improved load forecasts.

Acknowledgement. This work is supported by the National Key R&D Program of China under Grant No. 2017YFB1002000; the National Natural Science Foundation of China under Grant No. 61772136, 61672159, 61772005; the Technology Innovation Platform Project of Fujian Province under Grant No. 2014H2005; the Research Project for Young and Middle-aged Teachers of Fujian Province under Grant No. JT180045; the Fujian Collaborative Innovation Center for Big Data Application in Governments; the Fujian Engineering Research Center of Big Data Analysis and Processing.

References

1. Zheng, J., Xu, C., Zhang, Z., Li, X.: Electric load forecasting in smart grids using long-short-term-memory based recurrent neural network. In: Information Sciences and Systems. IEEE (2017)
2. Ertugrul, Ö.F.: Forecasting electricity load by a novel recurrent extreme learning machines approach. Int. J. Electr. Power Energy Syst. **78**, 429–435 (2016)
3. Zhang, Z., Xu, Y., Yang, J., Li, X., Zhang, D.: A survey of sparse representation: algorithms and applications. IEEE Access **3**, 490–530 (2015)
4. Alfares, H., Nazeeruddin, M.: Electric load forecasting: literature survey and classification of methods. Int. J. Syst. Sci. **33**(1), 23–34 (2002)
5. Weron, R.: Modeling and Forecasting Electricity Loads and Prices: A Statistical Approach, vol. 403. Wiley, Hoboken (2007)
6. Raza, M.Q., Khosravi, A.: A review on artificial intelligence based load demand forecasting techniques for smart grid and buildings. Renew. Sustain. Energy Rev. **50**, 1352–1372 (2015)
7. Hinton, G.E., Osindero, S., Teh, Y.W.: A fast learning algorithm for deep belief nets. Neural Comput. **18**(7), 1527–1554 (2006)
8. Kuremoto, T., Kimura, S., Kobayashi, K., Obayashi, M.: Time series forecasting using restricted Boltzmann machine. In: Huang, D.S., Gupta, P., Zhang, X., Premaratne, P. (eds.) ICIC 2012. CCIS, vol. 304, pp. 17–22. Springer, Heidelberg (2012). https://doi.org/10.1007/978-3-642-31837-5_3

9. Busseti, E., Osband, I., Wong, S.: Deep learning for time series modeling. Technical report, Stanford University (2012)
10. Xiao, L., Wang, J., Yang, X., Xiao, L.: A hybrid model based on data preprocessing for electrical power forecasting. Int. J. Electr. Power Energy Syst. **64**(64), 311–327 (2015)
11. Lewicki, M.S., Sejnowski, T.J.: Learning overcomplete representations. Neural Comput. **12** (2), 337–365 (2000)
12. Natarajan, B.K.: Sparse approximate solutions to linear systems. SIAM J. Comput. **24**(2), 227–234 (1995)
13. Mallat, S.G., Zhang, Z.: Matching pursuits with time-frequency dictionaries. IEEE Trans. Signal Process. **41**(12), 3397–3415 (1993)
14. Pati, Y.C., Rezaiifar, R., Krishnaprasad, P.S.: Orthogonal matching pursuit: recursive function approximation with applications to wavelet decomposition. In: 1993 Conference Record of the Twenty-Seventh Asilomar Conference on Signals, Systems and Computers, pp. 40–44. IEEE (1993)
15. Tropp, J.A., Gilbert, A.C.: Signal recovery from random measurements via orthogonal matching pursuit. IEEE Trans. Inf. Theory **53**(12), 4655–4666 (2007)
16. Donoho, D.L.: For most large underdetermined systems of linear equations the minimal ℓ1-norm solution is also the sparsest solution. Commun. Pure Appl. Math. **59**(6), 797–829 (2006)
17. Chen, S.S., Donoho, D.L., Saunders, M.A.: Atomic decomposition by basis pursuit. SIAM Rev. **43**(1), 129–159 (2001)
18. Qayyum, A., et al.: Designing of overcomplete dictionaries based on DCT and DWT. In: Biomedical Engineering & Sciences, pp. 134–139. IEEE (2016)
19. Daubechies, I.: The wavelet transform, time-frequency localization and signal analysis. IEEE Trans. Inf. Theory **36**(5), 961–1005 (1990)
20. Do, M.N., Vetterli, M.: The contourlet transform: an efficient directional multiresolution image representation. IEEE Trans. Image Process. **14**(12), 2091–2106 (2005)
21. Ma, J., Plonka, G.: The curvelet transform. IEEE Signal Process. Mag. **27**(2), 118–133 (2010)
22. Lewis, J.J., O'Callaghan, R.J., Nikolov, S.G., Bull, D.R., Canagarajah, N.: Pixel- and region-based image fusion with complex wavelets. Inf. Fusion **8**(2), 119–130 (2007)
23. Abdi, H., Williams, L.J.: Principal component analysis. Wiley Interdisc. Rev. Comput. Stat. **2**(4), 433–459 (2010)
24. Engan, K., Aase, S.O., Husoy, J.H.: Method of optimal directions for frame design. In: Proceedings of the 1999 IEEE International Conference on Acoustics, Speech, and Signal Processing, vol. 5, pp. 2443–2446. IEEE (1999)
25. Aharon, M., Elad, M., Bruckstein, A.: k-SVD: an algorithm for designing overcomplete dictionaries for sparse representation. IEEE Trans. Signal Process. **54**(11), 4311–4322 (2006)

Sensing Urban Structures and Crowd Dynamics with Mobility Big Data

Yan Liu[1], Longbiao Chen[1(✉)], Linjin Liu[1], Xiaoliang Fan[1],
Sheng Wu[3], Cheng Wang[1], and Jonathan Li[1,2]

[1] Fujian Key Laboratory of Sensing and Computing for Smart Cities,
Xiamen University, Xiamen, China
longbiaochen@xmu.edu.cn
[2] WatMos Lab, University of Waterloo, Waterloo, Canada
[3] Spatial Information Research Center of Fujian, Fuzhou University,
Fuzhou, China

Abstract. To facilitate efficient and effective city management, it is important for urban authorities to understand the regular functionalities of urban areas and the irregular crowd dynamics moving around the city. However, existing methods relying on manual surveys and statistics usually cost substantial time and labor, hindering the fine-grain characterization of urban structures and the in-depth understanding of crowd dynamics. In this paper, we leverage large-scale mobility data collected from vehicle GPS devices to analyze the dynamics of crowd movement in different urban areas in a low-cost and automatic manner. We extract the regular crowd movement patterns in different areas, detect the abnormal crowd movement flow peaks, and then interpret the influences of different types of urban events. More specifically, we first divide the city into fine-grained geographic regions and cluster them according to the similarity of crowd movement characteristics. Second, we detect anomaly traffic flow for each cluster area, interpret urban events for each abnormal flow point, and correlate urban events to the interpretation results. Finally, we determine the scope of urban events and use visualization techniques to demonstrate the impact of different types of urban events. We leverage the large-scale real-world datasets from Xiamen City for evaluation. Experimental results validate the effectiveness of our method, and several case studies in Xiamen are conducted.

Keywords: Crowdsensing · Mobility big data · Urban computing

1 Introduction

In order to facilitate efficient and effective city management, urban authorities usually need to understand the regular functionalities of urban areas and the irregular crowd dynamics moving around the city [15]. On one hand, urban planning, construction, and development have led to regular crowd movement patterns and structures to urban areas [20], such as central business districts (CBDs), residential areas, transit hubs. Meanwhile, the occurrences of certain urban events may break the regular crowd movement patterns in different areas of the city. For example, holding a concert in a stadium may lead to abnormal human flow peaks around the stadium and the city's

S. Li (Ed.): GPC 2018, LNCS 11204, pp. 370–389, 2019.
https://doi.org/10.1007/978-3-030-15093-8_27

transit hubs. Due to the lack of understanding of the urban structures and the dynamics of urban events, urban authorities cannot evaluate the impacts of urban events effectively in the city, which hinders the short-term event management and long-term urban planning [21]. Therefore, there is an urgent need for urban authorities to have a clear picture of urban structures, and to be able to analyze the crowd dynamics in urban areas caused by events.

Most existed studies consider urban regional structure and crowd dynamics separately. Many studies of urban regional structure have been conducted. For example, Yuan et al. [1] analyzed the structure of urban areas by dividing urban roads into different blocks with image processing technologies. Chen et al. [3] used mobile data to analyze the urban spatial structure. They employed the movement trajectory of cellphone users in urban area, and divided the urban area structure using the distribution of cell phone signal hotspots. On the other hand, extensive researches on crowd dynamics have been conducted based on the detection of urban events and its impact. Zhang et al. [12] proposed a method that can not only discover the occurrence time and venue of events, but also measure the scale of events. Chen et al. [11] used bike sharing data to check the occurrence of urban events. Besides, Li et al. [13] extracted events from social media, and discussed the annotation challenges and released a benchmark for the research community.

However, crowd dynamics can be used to depict the urban structure, as well as reflect the impacts in the structure caused by urban events. Regular flows of crowds reflect the structure, and breaks of the regular flows is probably caused by urban events. For example, when a particular region hold a large-scale trade show, traffic managers would take odd or even number restrictions of the vehicle license to prevent traffic jams. As a consequence, the traffic in other regions could be influenced.

Currently, there are types of data could sense the movements of urban crowds, such as social network check-in data, call detail records, monitoring video data and so on. But these kinds of data could be constrained to user's preference, violate user's privacy, or there is time delay over the occurrence and record. On the contrary, data from taxi GPS trajectories and social media network are out of these problems. Moreover, with the increasing number of taxis with GPS equipment in the city, real-time locational information is available for the study of the movement of urban crowds. And also, with the development of the social media network, detailed information of urban events could be reached to do the analytics. So we choose taxi GPS trajectories and social media network as the sources of data for analysis.

In this paper, we propose a two-phase framework to explore the impact of urban events on crowd dynamic. In the *urban structure portrait* phase, first, the urban area is meshed by the features of urban geographical area, then the taxi GPS trajectory data is mapped into the corresponding area grid. Secondly, we extract the spatial-temporal characteristics of the daily flow in a single grid, the distance-constrained clustering algorithm (DCCA) [11] is used to cluster the regional grids according to their structural similarity and then the flow characteristics of the clustered areas are extracted. In the *crowd dynamic characterization* phase, we detect anomaly traffic flow for each cluster area, interpret urban events for each abnormal flow point, and correlate urban events to the interpretation results. Finally, we determine the scope of urban events and use visualization techniques to demonstrate the impact of different types of urban events.

The main contributions of this paper are summarized as follows:

(1) We conduct in-depth analytics of the urban structures and crowd dynamics, this may benefit the management and deployment for urban authorities in the future.
(2) A novel two-phase framework to model urban structures and crowd dynamics. In the first phase, we first divide the city into fine-grained geographic regions and cluster them according to the similarity of crowd movement characteristics. In the second phase, we detect anomaly traffic flow for each cluster area, interpret urban events for each abnormal flow point, and correlate urban events to the interpretation results. We determine the scope of urban events and use visualization techniques to demonstrate the impact of different types of urban events.
(3) We evaluate our method on real-world taxi GPS dataset collected from Xiamen, China. Results demonstrate that our method effectively characterizes the urban structure and crowd dynamics, and outperforms baseline methods.

The rest of the paper is organized as follows: Sect. 2 reviews the related work. Section 3 is dataset description. We propose a two-phase framework is in Sect. 4. Sections 5 and 6 present the proposed technique. Section 7 is the results of experiments. Finally in Sect. 8, the conclusion and future work are stated.

2 Related Work

This paper investigates two aspects of research work: (1) analysis of urban regional structure, and (2) research on the impact of urban events on crowd dynamic.

2.1 Analysis of Urban Regional Structure

Yuan et al. [1] used image processing technology to extract information from remote sensing images of urban roads, and divided different areas by roads. The method does not consider the structural characteristics of regional population movement. Esch et al. [2] estimated population distribution through the construction density and then identified different urban structures. This method ignores the problem that the density of buildings cannot represent the density of the population. Chen et al. [3] reflected the crowd dynamics and urban spatial structure through the conversation of mobile phone users. However, the information of mobile phone users has strict privacy and is not easy to obtain. Chen et al. [11] utilized the usage of base station traffic data to divide the urban area into different structures. Besides, grid-based areas segmentation was extensively used in geospatial related analysis [4–6]. This paper divided the urban structure by the similarity of the daily traffic characteristics of the regional grid, and brings together the similar-shaped grids to ensure that each area has its own unique structural characteristics.

2.2 Research on the Impact of Urban Events on Crowd Dynamic

Researchers have exploited different methods to study the impact of urban events. Liang et al. [10] use the LBSN sign-in data to model the size and duration of crowd gathering in urban events. However, some of the social network sign-in data is delayed, and there is an error in the analysis of the duration. Other work is to detect city events from a large number of text streams in Twitter posts, such as Sakaki et al. [7] detecting typhoon and earthquakes from Twitter posts, Li et al. [8] analyzing the similarity and frequency of Twitter to find out the city events mentioned in Twitter. Agarwal et al. [9] applied a graph clustering algorithm for event detection by discovering some dense subgraphs in a graph. Zhang et al. [12] propose a method that can not only discover the happening time and venue of events from abnormal social activities, but also measure the scale of events through changes in such activities. Among the various types of impacts that occur in urban events, the most prominent is the movement of people under the influence of events. Regional dynamic reflects the traffic conditions around the region, which is the key to the normal operation of the urban. Well-known changes in crowd dynamic for urban management department is particularly important. After knowing the possible impact area of the event, corresponding measures can be deployed in advance of the area to prevent all accidents from happening. Therefore, not only to accurately detect urban events, but also grasp the impact of urban events on the dynamics of urban areas. The above methods for detecting urban events are very accurate, but they only detect events in urban areas. They do not analyze in depth the impact of urban events on crowd dynamic and do not find out the reasons for these impacts. In this paper, we extract regional daily traffic flow characteristics, then use anomaly detection algorithm to detect abnormal traffic flow and analyze urban events and analyze the impact of urban events in depth, and then classify urban events. These analysis results can be used as a reference for the deployment of traffic management department and the general public travel program reference.

3 Dataset Description

In this section, we mainly present the two datasets used in this paper, viz., the Xiamen Taxi GPS Dataset and the Xiamen Social Media Dataset.

3.1 Taxi GPS Dataset

This paper used taxi GPS trajectory data in Xiamen City in September 2016. Vehicles with GPS can dynamically sense the urban road information: road network traffic conditions, congestion conditions, node traffic conditions, taxi trajectory space-time characteristic. As of the end of 2016, the number of taxis in Xiamen City is about 6,000. This car GPS will upload the vehicle information to the Xiamen Satellite information center every 15 s. In September, about 377 million records were generated, with an average of 68,666 records per taxi. The raw data was stored in an Oracle database in which the main fields of the taxi's GPS data include:

- VEHICLE_ID: the unique ID of each taxi;
- LONGITUDE: the current longitude;
- LATITUDE: the current latitude;
- STATE: indicates whether the taxi meter is running; In other words, whether or not the taxi has a passenger: 1 if the taxi is occupied and 0 if it is vacant;
- VELOCITY: the current taxi speed in km/h;
- TIME: the sample timestamp "YYY-MM-DD HH:MM:SS".

3.2 Social Media Dataset

The social media is widely used, record information about the immediate surroundings of urban life, traffic conditions, the effects of events, and upcoming or emerging disasters and threats to the general public. Moreover, social media data are usually authentic and timely. For the occurrence of events in the city, there are detailed records, including the place and time of the events, the government promulgated policies to reduce the impact of events and so on. Therefore, the social media data can serve as the basis for our in-depth analysis of the impact of urban events.

4 Framework Overview

The main purpose of this paper is to analyze the dynamics of crowd movement in different urban structures in a low-cost and automatic manner. As shown in Fig. 1, our framework consists of two phases, i.e., urban structure portrait phase, and crowd dynamic characterization phase. In the *urban structure portrait* phase, first, the urban area is meshed by the features of urban geographical area, then the taxi GPS trajectory data is mapped into the corresponding area grid. Secondly, we extract the spatial-temporal characteristics of the daily flow in a single grid, the distance-constrained clustering algorithm (DCCA) [11] is used to cluster the regional grids according to their structural similarity and then the flow characteristics of the clustered areas are extracted. In the *crowd dynamic characterization* phase, we detect anomaly traffic flow for each cluster area, interpret urban events for each abnormal flow point, and correlate urban events to the interpretation results. Finally, we determine the scope of urban events and use visualization techniques to demonstrate the impact of different types of urban events.

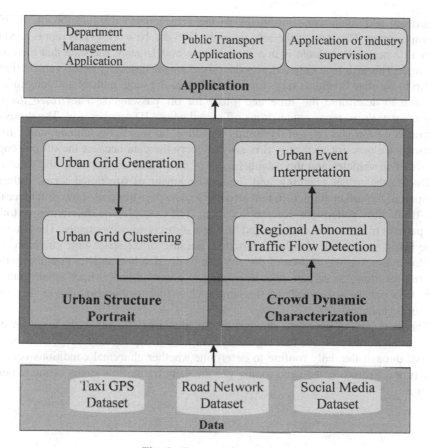

Fig. 1. Framework overview

5 Urban Structure Portrait

In this chapter, our objective is to structurally divide the urban area. Each of the divided area consists of small areas of similar structure. In order to achieve this result, we firstly extract the passenger trajectory data from GPS of taxi and grid the urban area according to the characteristics of urban geographic structure. The vehicle data is mapped to the corresponding grid according to its latitude and longitude to form a three-dimensional traffic flow matrix. Then the clustering algorithm with limited distance is used to gather grids with similar traffic flow characteristics according to the daily traffic flow characteristics extracted from each grid. Finally, the urban area has the structural characteristics.

In this paper, we mainly use the GPS data of taxi, each record in the database contains its timestamp, latitude and longitude coordinates. Since vehicles travel on urban roads, these longitude and latitude records are scattered over the urban geographic area and can be mapped one by one. If the density of the trajectory data is large enough, the road network in the urban area can be displayed. As shown in Fig. 2, map the latitude and longitude coordinates of the vehicle onto the location of the city geographic map. The state of the vehicle in the GPS database STATE is represented by 0 and 1, 0 means the vehicle is in a no-load state and 1 means the state that there is a passenger. This article mainly extracts the off-passenger trajectory of the vehicle that is 1 jump to the state 0 vehicle trajectory, according to the time, latitude and longitude at this time to determine the time and place for off passengers. The $trajec\_drop = (id, lon, lat, t\_drop)$ represents a drops off record whose ID number is id. The drops off location is the lon and lat, lon represents longitude, lat represents latitude, and t_drop represents the time for alighting. It is easy to query the data because the off passenger records of all vehicles and the establishment of time index are sorted in the database.

Due to the urban planning and the establishment of functional areas, different locations of the urban show different structural characteristics and show differences in daily traffic flow patterns in different locations. For example, the differences of traffic flow patterns in the school districts and business districts are vary greatly. The structure of the urban needs to be clearly constructed. Because of the space-time characteristics of vehicle data, the whole area of the city can be divided into smaller granularity than the area according to certain rules according to a certain order. Each small granular traffic flow pattern is studied and the similar clustering together to analyze the urban Structural characteristics.

The daily state of the urban shows a certain regularity. However, the occurrence of an urban event often breaks this regularity. We need to tap the city's regularity of this routine, through the daily routine to determine whether abnormal conditions occur at other times. Therefore, we need to analyze the daily traffic flow status of the regional structure.

Fig. 2. Map the latitude and longitude coordinates of the vehicle onto the city geographic map

5.1   Urban Grid Generation

In order to divide urban areas into smaller ones, urban areas need to be gridded [17, 18]. According to the city's geographical area and geological characteristics of the city is divided into $M \times N$ matrix, which is the formation of a consistent size and arrangement of small rectangular. Then, according to the latitude and longitude range of the small rectangle, latitude is between lat_min and lat_max, longitude is between lon_min and lon_max. All the drops off information in this range are all mapped in a small rectangle, sorted according to the timestamp of each record, and the number of records within a certain time t_drop range is counted $g_{i,t}$.

Criterion1: the number of drop off vehicles in the area grid is counted

$$g_{i,t} = \{count(trajec\_drop)/t = t\_drop, lon\_min \leq lon \leq lon\_max, lat\_min \leq lat \leq lat\_max\} \tag{1}$$

Where each grid area has a time series $G_i = \{g_{i,1}, g_{i,2}, g_{i,3} \cdots g_{i,n}\}$ representation, which is the drops off traffic feature of the area where the matrix is located, and this traffic feature has the property of space-time. We form a three-dimensional matrix of $M * N * T$ by analyzing its time series for each grid. As show in Fig. 3.

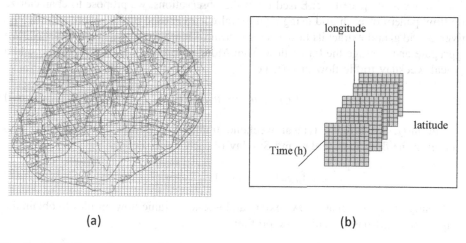

(a)	(b)

Fig. 3. (a) Grid to city area; (b) Analyze its time series for each grid, forming a three-dimensional matrix of $M * N * T$

5.2   Urban Grid Clustering

Through the division of regional grid, we observe that the traffic pattern of a grid is highly dynamic under different temporal contexts.

As shown in the Fig. 4. We can see clear weekday-weekend patterns as a result of regular patterns, the traffic flow patterns of similar distances are similar, and the traffic flow patterns of the distant grids have a large difference by comparing the traffic characteristics of different grids. It shows that the grid's daily flow patterns have the

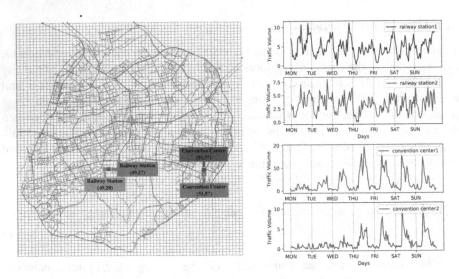

Fig. 4. This picture shows the flow patterns of the four regional grids. Convention Center area taken two grids, the railway station area took two grids.

characteristics of space-time. Based upon the observations, we propose to characterize the flow pattern of each grid using a temporal-context-based profile. More specifically, given a grid g_i and extract its flow vector measured in hours from its time series G_i. We aggregate and average the traffic flow from Monday to Friday in each week to build a typical weekday traffic flow profile, i.e.

$$f_w(g_i) = [u_1, u_2, u_3, \ldots, u_{120}] \tag{2}$$

Similarly, we build a typical weekend traffic flow profile by aggregating and averaging the flow in Saturday and Sunday of each week, i.e.

$$f_n(g_i) = [v_1, v_2, v_3, \ldots, v_{24}] \tag{3}$$

Finally, we concatenate the weekday and weekend traffic flow profiles to obtain the temporal-context-based traffic flow profile:

$$f(g_i) = [f_w(g_i), f_n(g_i)] \tag{4}$$

Therefore, similar grids with similar distances need to be clustered together. Firstly, we construct a weighted graph model to represent the relationship of grids. We exploit graph links to express the grid distance constraints, weighted graph $G = (V, E)$, where $V = \{g_1, g_2, \ldots, g_N\}$ denotes the set of N grids, and E denotes the set of links between two grids.

Secondly, we define the adjacency matrix A of graph G, which is a $N \times N$ symmetric matrix with entries $a_{i,j} = 1$ when there is a link between grid g_i and grid g_j, and

$a_{i,j} = 0$ otherwise. We use the geographic distance of two grids to determine whether they are adjacent or not. More specifically, for grid g_i and grid g_j, we define:

$$a_{i,j} = \begin{cases} 1, & \text{if } dist(g_i, g_j) \leq \tau \\ 0, & \text{otherwise} \end{cases} \tag{5}$$

where $dist(g_i, g_j)$ is the geographic distance between the two grids, and τ is a neighborhood threshold controlling the geographic distance of neighboring grids.

Given two neighboring grids, we use their similarity measurement to determine their link weight, i.e.

$$w(g_i, g_j) = (SIMDIST\{f(g_i), f(g_j)\}) * a_{i,j} \tag{6}$$

We note that $w(g_i, g_j) = 0$ when there is no link between g_i and g_j ($a_{i,j} = 0$).

Given graph $G = (V, E)$, we first define a set of clusters $R = \{C_1, \cdots, C_K\}$, where $\cup_{\forall C_k \in R} = V$, $\cap_{\forall C_k \in R} = 0$. Then, given a grid v, we define the connectivity of v to a cluster C as the sum of link weights between v and the grids in the cluster C: $con(v, C) = \sum_{v' \in C} w_{v,v'}$. Finally, we define the adjacent clusters $C(v)$ of v as $C(v) = \{C | con(v, C) > 0, C \in R\}$.

With the above definition, our objective is to find an optimal set of clusters R, so that the internal connectivity within a cluster is higher than the inter-cluster connectivity, i.e.

$$\forall_v \in C_k, con(v, C_k) \geq \max\{con(v, C_l), C_l \in R\} \tag{7}$$

We also need to bound the distance span of a cluster within the neighborhood threshold, i.e.

$$\forall_{v,v'} \in C_k, dist(v, v') \leq \tau \tag{8}$$

Based on [15, 16], we use a Distance Constrained Clustering Algorithm (DCCA) to cluster area grids. The basic idea of DCCA is iteratively assigning grids to the adjacent clusters, where the gain of assigning grid v to cluster C is iteratively evaluated by a value function as follows:

$$value(v, C) = con(v, C) \times \log\left(\frac{\tau}{\max(dist(v, v'))}\right) \tag{9}$$

The DCCA greedily assigns the grids to the adjacent cluster with highest value until none of the grids are moves among clusters. As the convergence of such a greedy approach is difficult to prove, (1) the user specified maximum iteration number max_iter is reached, or (2) none of the grids are moved among clusters.

6   Crowd Dynamic Characterization

In this section, the main objective is to explain the impact of the corresponding urban events on crowd dynamic by examining the anomalous traffic flow in the urban structured areas. The occurrence of urban events can affect the daily traffic characteristics in urban areas. The impact on crowd dynamic is different because of the different types of events in the city. For example, when concerts are held in one place, the number of people in the vicinity of this area increases before the concert starts, resulting in increased activity in the area. In addition, if there is a natural disaster in the city, such as a typhoon, it will lead to stagnation of urban production, paralysis of traffic and reduction of people's travel. In many parts of the city, there is less activity when the disaster happened.

We need to analyze the most influential urban events because some urban events have a great influence on urban life. Through the analysis of urban events, it is found that its impact on the crowd dynamic can helps the urban management department to arrange the management and deploy the works of the city. The analysis is mainly divided into two steps. Firstly, we use anomaly detection algorithms to detect abnormal traffic in the daily traffic characteristics of each region. Secondly, we analyze every abnormal traffic point. We find abnormal traffic-related urban events based on the social media news, and then analyze the impact of urban events on the dynamics of urban areas.

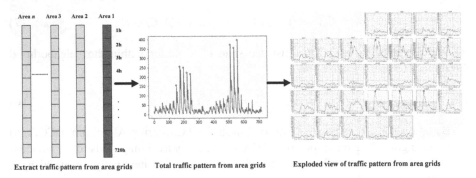

Fig. 5. This picture shows the flow pattern of the extracted area grid, draw a 30-day general flow characteristic chart of the grid in one area, and a daily flow chart. Showing abnormal flow points.

6.1   Regional Abnormal Traffic Flow Detection

Through the structural analysis of the urban area in the previous chapter, we can extract the daily historical traffic flow characteristics of each regional. If the urban event occurs in any day and the event is more influential, so the flow characteristics of the area will change greatly. As show in Fig. 5. By observing the 30-day total flow chart of a region and decomposing it into daily flow chart, we can find red anomalous traffic points. Therefore, we detect irregularities from all basic patterns at once using the ARIMA Outlier Detection [14] method.

More specifically, we first extract the daily historical traffic flow characteristics $A(r,n)$ of each area r,n represents the area r consists of n area grids, i.e.

$$A(r,n) = \{a_1, a_2, \cdots, a_k\} \tag{10}$$

Where a_i $(1 \leq i \leq k)$ represents the sum of the flow in the ith hour of the n regional grids. The traffic flow characteristics vector $A(r,n)$ is divided into two parts, namely test set and training set. Then we use ARIMA to train the flow basic patterns model $Model(r,j)$ when there is no event.

$$Train = \{a_1, a_2, \cdots, a_j\} \tag{11}$$

$$Test = \{a_{j+1}, a_2, \cdots, a_k\} \tag{12}$$

$$Model(r,j) = ARIMA(Train) \tag{13}$$

In this way, we use the model $Model(r,j)$ to predict traffic $a'_{j+1}, a'_{j+2}, \cdots, a'_{j+t}$ for the next time t, detect the abnormal traffic flow with the anomaly detection algorithm [19], this method requires individual threshold δ_m to control the irregularity significance for each basic pattern of each region.

$$a'_m \in \left(a'_{j+1}, a'_{j+2}, \cdots, a'_{j+t}\right), a_m \in \left(a_{j+1}, a_{j+2}, \cdots, a_{j+t}\right) \in Test, \left|a_m - a'_m\right| \geq \delta_m \tag{14}$$

If $\left|a_m - a'_m\right| \geq \delta_m$, it represents abnormal traffic flow point appeared in the m moment. So, we can find out the abnormal traffic flow in all areas at any time by this method.

6.2 Urban Event Interpretation

We searched social media for urban events that coincided with the time and place of anomalous flow peaks, interpreted urban events for each abnormal flow point, and correlated urban events to the interpretation results. This paper analyzes the impacts of such urban events on crowd dynamic, including the scope of the affected area and the length of time. By utilizing the visual display technology, we show the impact areas of urban events, which helps urban management departments and the general public to understand more deeply about the impact of urban events. When a similar event occurs in the future, it can be used as a reference.

7 Evaluation

In this section, we use Xiamen taxi data and social media data to verify the correctness of this method. Firstly, we need to introduce the experiment settings. Secondly, we present the evaluation results. Finally, we use a visual interface to display our analysis results.

7.1 Experiment Settings

Evaluation Plan
Firstly, we map the grids to the coverage areas of Siming and Huli districts in Xiamen and aggregate the traffic data to the corresponding area grids. We select the data between 09/01/2016 and 09/30/2016 to generate traffic profile of the area grids, and use the DCCA algorithm to cluster area grids to urban structure areas. Then, we extract the daily traffic characteristics of each cluster area, and use anomaly detection algorithm to detect abnormal traffic points. Finally, we interpret the urban events of each abnormal traffic point, and display the distribution of the impact of each urban on the dynamic of urban areas visually.

Evaluation Metrics
We evaluate the anomaly detection algorithm that we used by computing the accuracy of the anomaly traffic detection results. We find that some abnormal points can correspond to the urban events because some abnormal traffic points are caused by urban events. We calculate the *precision* and *recall* by comparing the anomaly traffic detection results with the abnormal traffic points caused by all urban events in the real world. If a detected traffic abnormal point has a temporal overlapping with the real world urban event, we mark the detection as a hit. Otherwise, ignore it. Based on this idea in this paper, the precision and recall are calculated as follows:

$$precision = \frac{A}{B} \tag{15}$$

$$recall = \frac{A}{C} \tag{16}$$

A denotes the detected abnormal points that corresponding to urban events
B denotes all detected abnormal points
C denotes the abnormal points caused by all urban events.

In addition, we calculate the $F1$-*Score* as

$$F1\text{-}Score = \frac{2 \times precision \times recall}{precision + recall} \tag{17}$$

to assess the performance of the anomaly detection algorithm that we used.

Baseline Method
We compare our abnormal traffic detection algorithm with the upper and lower quartile threshold method. This method use the upper and lower quartiles to establish a threshold range, and then traffic values above the threshold range as outliers. However, this method simply sets the threshold for the traffic characteristics and does not establish the model. The model predicts the flow value in the next time range and compares it with the real value, and then sets the threshold to detect the abnormal value.

7.2   Evaluation Results

Urban Structure Portrait Results

The geographical area of Xiamen is partitioned into 77 * 68 grids with the grid size is about 200 * 200 square meters. In each grid, a normalized traffic intensity is recorded on an hourly basis. It is meaningless to analyze all the traffic grids and it will waste computing resources because that the number of grids is too large and the small traffic volume in some grids which cannot affect the traffic greatly. Therefore, we check the traffic volume of all grids and select 1655 grids. As show in Fig. 6.

The traffic characteristic curve of a single grid is particularly unstable, which is not conducive to analyze the regional structure. Therefore, we choose the maximum number of iterations max_*iter* = 20 and distance threshold $\tau = 1$ km, we clustered 1665 regional grids into 147 regions using the DCCA clustering algorithm. The results of the clustering areas are shown in the Fig. 7, grids of the same color at a closer distance represent a regional structure.

Fig. 6. Selected 1655 grids distribution   **Fig. 7.** The area grids are clustered into 147 areas

We find that the flow characteristic curve of a single area after clustering is relatively stable and has spatiotemporal characteristics. There are differences in the characteristics of traffic at different locations, and the trend of changes in traffic at different times in the same location is regular. As show in Fig. 8.

Anomalous Traffic Flow Detection Results

The performance comparison of two abnormal traffic flow detection algorithms is shown in the Fig. 9. ARIMA anomaly detection algorithm is consistently better than the baselines over all metrics, in particular the F1-Score achieves 80.7%. So this paper mainly detects abnormal traffic flow through ARIMA anomaly detection algorithm.

Firstly, we use the ARIMA to train the model in terms of traffic characteristics from 2016/09/01 to 2016/09/07 in each clustered area, and then predict the next six hours traffic by using the training model. Secondly, the real traffic flow of six hours are added

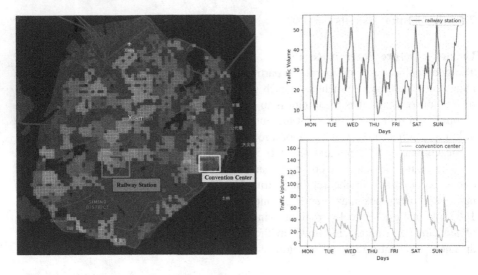

Fig. 8. This picture shows the traffic pattern curve of the railway station and the convention center formed by the regional grid clustering

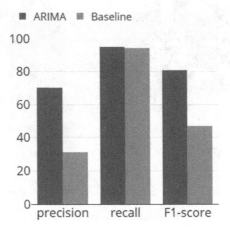

Fig. 9. Anomaly traffic detection accuracy

to the original training data set to establish a new model. Finally, we set the anomaly detection threshold to detect abnormal traffic points by comparing the original traffic data with the traffic data which is predicted by the model. We use social media data to interpreted urban events which have these anomalies and then find the urban events associated with each outlier. As shown in the Fig. 10. This is the result of anomaly detection on the traffic characteristics of Xiamen Convention Center. We can see that September 8, 2016 to November 11, 2016, and September 23, 2016 to November 25, 2016 have abnormal points. We find that Xiamen Convention Center hold large-scale commercial exhibition at the time when the anomalous points appeared by interpreting city events through each outlier.

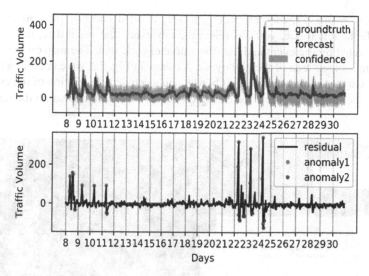

Fig. 10. Convention and Exhibition Center anomaly detection map

7.3 Visualization Display of Crowd Dynamic

The occurrence of urban events often leads to the change of crowd dynamic, which is mainly reflected the abnormal traffic flow in some areas. Firstly, we find their corresponding urban events and then count the impact areas of each urban event by analyzing the abnormal traffic flow points in each area. In order to demonstrate the impact of urban events on city dynamics more clearly, we have created a visual display interface to show the impact areas of each urban event and show the traffic flow changes in each area.

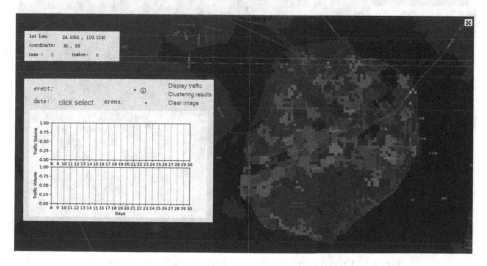

Fig. 11. Crowd dynamic display interface: overview.

As show in Fig. 11, this interface is mainly divided into two parts to display the analysis. In the left frame, you can select the city event, the event date, the number of the area affected by the event and the flow curve of each area. The map on the right can show the influenced area of the city event. As show in Fig. 12, we can arbitrarily select an event from the event list of the left border and a detailed description of the event will appear. At the same time, the affected region of this event will appear on the right map.

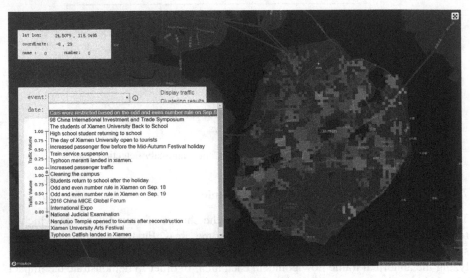

Fig. 12. Crowd dynamic display interface: a typical day.

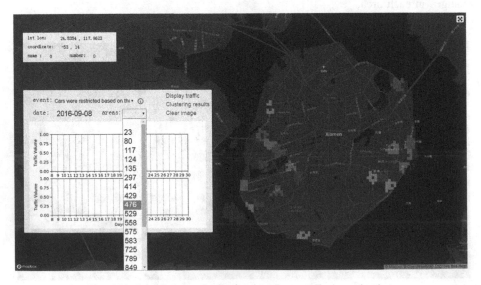

Fig. 13. Crowd dynamic display interface: traffic investigation.

As show in Figs. 13 and 14, we can select a region from the list of areas impacted by event and show the changes of the traffic within 22 days and the distribution of abnormal traffic points in this area.

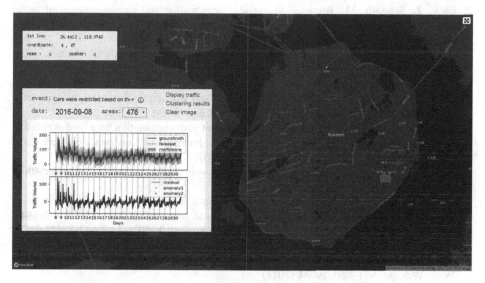

Fig. 14. Crowd dynamic display interface: traffic investigation.

8 Conclusion

The occurrence of urban events affects all aspects of urban life, including the traffic conditions of the cities and the convenience of citizens' normal travel. Therefore, it's necessary to capture the areas that affected by urban events and analyze the reasons of these impacts. In this paper, we characterize the regular crowd movement patterns in different areas, detect the peaks of abnormal crowd movement flow, and then interpret the influences of different types of urban events. Firstly, we divide the geographical regions of the city into fine-grained regions and cluster them according to the similarity of crowd movement characteristics. Secondly, we detect anomalous traffic flow for each cluster area, interpret each abnormal flow point using urban events, and count all interpretation results. Finally, we determine the scope and impact of urban events and visualize them. In the future, we plan to build a real-time monitoring system for the city dynamic, which system will bring convenience to urban authorities to monitor the state of the city's dynamics.

Acknowledgment. We would like to thank the reviewers for their constructive suggestions. This research was supported by Fujian Collaborative Innovation Center for Big Data Applications in Governments, and the China Fundamental Research Funds for the Central Universities No. 0630/ZK1074, and NSF of China No. U1605254, No. 61371144, No. 61300232.

References

1. Yuan, N.J., Zheng, Y., Xie, X.: Segmentation of urban areas using road networks. MSR-TR-2012–65, Technical report (2012)
2. Esch, T., Schmidt, M., Breunig, M., et al.: Identification and characterization of urban structures using VHR SAR data. In: 2011 IEEE International Geoscience and Remote Sensing Symposium (IGARSS), pp. 1413–1416 (2011)
3. Chen, S., Wu, H., Tu, L., et al.: Identifying hot lines of urban spatial structure using cellphone call detail record data, Ubiquitous Intelligence and Computing. In: 2014 IEEE 11th International Conference on Autonomic and Trusted Computing, and IEEE 14th International Conference on Scalable Computing and Communications and its Associated Workshops (UTC-ATC-ScalCom), pp. 299–304. IEEE (2014)
4. Gonzalez, H., Han, J., Li, X., et al.: Adaptive fastest path computation on a road network: a traffic mining approach. In: Proceedings of the 33rd International Conference on Very Large Data Bases. VLDB Endowment, pp. 794–805 (2007)
5. Krumm, J., Horvitz, E.: Predestination: where do you want to go today. Computer **40**(4), 105–107 (2007)
6. Powell, J.W., Huang, Y., Bastani, F., Ji, M.: Towards reducing taxicab cruising time using spatio-temporal profitability maps. In: Pfoser, D., et al. (eds.) SSTD 2011. LNCS, vol. 6849, pp. 242–260. Springer, Heidelberg (2011). https://doi.org/10.1007/978-3-642-22922-0_15
7. Sakaki, T., Okazaki, M., Matsuo, Y.: Earthquake shakes Twitter users: real-time event detection by social sensors. In: Proceedings of the 19th International Conference on World Wide Web, pp. 851–860. ACM (2011)
8. Li, C., Sun, A., Datta, A.: Twevent: segment-based event detection from tweets. In: Proceedings of the 21st ACM International Conference on Information and Knowledge Management, pp. 155–164. ACM (2012)
9. Agarwal, M.K., Ramamritham, K., Bhide, M.: Real time discovery of dense clusters in highly dynamic graphs: identifying real world events in highly dynamic environments. In: Proceedings of the VLDB Endowment, pp. 980–991 (2012)
10. Liang, Y., Caverlee, J., Cheng, Z., et al.: How big is the crowd?: event and location based population modeling in social media. In: Proceedings of the 24th ACM Conference on Hypertext and Social Media, pp. 99–108. ACM (2013)
11. Chen, L., Pan, G., Jakubowicz, J., et al.: Complementary base station clustering for cost-effective and energy-efficient cloud-RAN. In: 14th IEEE International Conference on Ubiquitous Intelligence and Computing (UIC 2017) (2017)
12. Zhang, W., Qi, G., Pan, G., et al.: City-scale social event detection and evaluation with taxi traces. ACM Trans. Intell. Syst. Technol. (TIST) **6**(3), 40 (2015)
13. Li, H., Ji, H., Zhao, L.: Social event extraction: task, challenges and techniques. In: 2015 IEEE/ACM International Conference on Advances in Social Networks Analysis and Mining (ASONAM), pp. 526–532. IEEE (2015)
14. Ali, S.M.: Time series analysis of Baghdad rainfall using ARIMA method. Iraqi J. Sci. **54**, 1136–1142 (2013)
15. Chen, L., Zhang, D., Wang, L., et al.: Dynamic cluster-based over-demand prediction in bike sharing systems. In: Proceedings of the 2016 ACM International Joint Conference on Pervasive and Ubiquitous Computing, pp. 841–852. ACM (2016)
16. Raghavan, U.N., Albert, R., Kumara, S.: Near linear time algorithm to detect community structures in large-scale networks. Phys. Rev. E **76**(3), 036106 (2007)

17. Veloso, M., Phithakkitnukoon, S., Bento, C.: Urban mobility study using taxi traces. In: Proceedings of the 2011 International Workshop on Trajectory Data Mining and Analysis, pp. 23–30. ACM (2011)
18. Tostes, A.I.J., de LP Duarte-Figueiredo, F., Assunção, R., et al.: From data to knowledge: city-wide traffic flows analysis and prediction using bing maps. In: Proceedings of the 2nd ACM SIGKDD International Workshop on Urban Computing, p. 12. ACM (2013)
19. Li, H., Wu, Q., Dou, A.: Abnormal traffic events detection based on short-time constant velocity model and spatio-temporal trajectory analysis. J. Inf. Comput. Sci. **10**, 5233–5241 (2013)
20. Wang, L., Zhang, D., Wang, Y., Chen, C., Han, X., M'hamed, A.: Sparse mobile crowdsensing: challenges and opportunities. IEEE Commun. Mag. **54**(7), 161–167 (2016)
21. Yang, D., Zhang, D., Zheng, V.W., Yu, Z.: Modeling user activity preference by leveraging user spatial temporal characteristics in LBSNs. IEEE Trans. on Syst. Man Cybern. Syst. (TSMC) **45**(1), 129–142 (2015)

A Multiple Factor Bike Usage Prediction Model in Bike-Sharing System

Zengwei Zheng[1], Yanzhen Zhou[1,2], and Lin Sun[1(✉)]

[1] Hangzhou Key Laboratory for IoT Technology and Application,
Zhejiang University City College, Hangzhou, China
{zhengzw, sunl}@zucc.edu.cn
[2] College of Computer Science and Technology, Zhejiang University,
Hangzhou, China
zhouyanzhen@zju.edu.cn

Abstract. Bike-sharing is becoming popular in the world, providing a convenient service for citizens. The system has to redistribute bikes among different stations frequently to solve the imbalance of spatial distribution. Real-time monitoring doesn't solve this problem well, since it takes too much time to redistribute the bike and affects the user experience. In this paper, we first analyze the influence of factors such as time, weather, the location of stations. Then we cluster neighboring stations with similar usage pattern, and propose a lagged variable to simulate the effect of weather conditions in usage number. Finally, a multiple factor regression model with ARMA error (MFR-ARMA) is proposed to predict the check-out/in number of bikes in each cluster in a period of time. Evaluation dataset is from New York Bike Sharing System. The prediction results of the model are compared with four baseline methods. The experiments show a lower RMSLE and ER for check-out/in number prediction in our model.

Keywords: Bike sharing system · Regression prediction model ·
ARMA error · Cluster-level

1 Introduction

In recent years, bike-sharing has made rapid development, originating from Europe, and spreading to other areas, such as America, Asia and Australia [1]. To satisfy customers, system manager needs to ensure high availability of bikes as far as possible [2, 3]. Hence, bike sharing systems face challenges in reconfiguration of bikes between stations [4]. Essentially, the bike usage pattern is varying with changes of time and location. Therefore, there may be some stations full, resulting in not enough slots to return the bike sometimes. Similarly, stations would be empty which can't meet the users' need for bikes. Monitoring the number of bicycles in real time can't settle the problem completely, since it may be too late to rebalance bikes after imbalanced distribution [5, 6].

To solve this problem, we predict the bikes demand of individual station in the next period of time. The prediction of bikes demand helps managers to control the system effectively, improves resource utilization and provides a better service to users. However, achieving this goal is very difficult because traffic is affected by many complex factors, such as weather conditions, impacts between neighboring stations, different days of the week (like weekday or weekend) and different times of the day. For example, weekdays show usage peaks in the morning and late afternoon, while weekends show usage peaks in early afternoon [7]. Besides, users who choose to ride bikes usually randomly select a station near their origins or destinations without following regular pattern.

In this paper, we propose a multiple factors regression model with ARMA error to predict check-out/in number in bike-sharing system. First, we extract the hourly check-out/in data of each station, and separate data into weekday and weekend/holiday. Then, we explore the impact of time factor, weather condition and temperature. Finally, we cluster the neighboring stations with similar usage pattern into groups, propose a lagged variable to simulate the effect of weather conditions, and then provide a prediction model of the check-out/in number for the next period of time.

In particular, the main contributions of this paper are mainly as follows:

1. To the best of our knowledge, it is the first time to use a lagged variable to simulate the effect of weather condition in demand of bike usage.
2. We propose a multiple regression factor model with ARMA error (MFR-ARMA) which considers weather condition, temperature, temporal factor. Prediction result of the model shows higher prediction accuracy and lower RMSLE and ER in bikes system compared with other base methods.
3. We evaluate the clustering result by different weight setting and the prediction result by different parameters of lagged variable. As the experiment result shows, neighboring stations with similar usage pattern can be clustered together effectively by weighted k-means clustering.

2 Related Work

Bike-sharing system has received extensive researchers' attention in recent years. Earlier studies focused on predicting the number of available bikes and locks or the usage pattern at each station. For example, Kaltenbrunner et al. [8] showed an Autoregressive Moving Average Model (ARMA) to predict the available number of bikes. Yoon et al. [9] showed an Autoregressive Integrated Moving Average Model (ARIMA) that combines the correlation between stations to predict the number of available bikes for each station in a short and long term. Yang et al. [10] first proposed a spatio-temporal mobility model based on historical data and then predicted check-out/in number for each station based on the model. But usage pattern of stations is affected by many factors in an area [11, 12], the prediction based on station-level is not very accurate.

To tackle this problem, researches have been mainly focused on cluster-level prediction. Researchers clustered the similar stations into clusters, and then predict the usage number for each cluster. For example, Li et al. [5] presented a bipartite clustering algorithm to cluster stations into groups, and forecasted the usage number of each cluster by a multi-similarity-based inference model. Chen et al. [11] estimated the base bike demand of a cluster by using the average value in historical data, and multiplied the base bike demand by the overall inflation rate to calculate the final prediction for each cluster. Singhvi et al. [12] aggregated bike stations within neighborhoods to predict pairwise bike demand for New York Citi Bike system.

3   Preliminary and Framework

The section presents the overview of MFR-ARMA model framework, and defines terms and the notations used in this paper in Table 1.

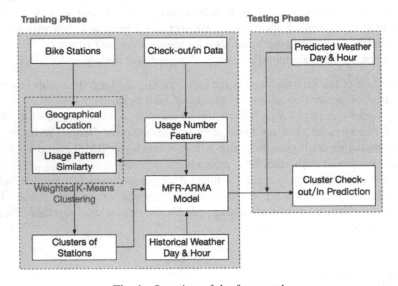

Fig. 1. Overview of the framework.

The overview of framework is shown in Fig. 1. The training phase is divided into two stages. In first stage, to tackle the problem that irregular usage pattern happens at each station, we first extract usage number features from the check-out/in data at each station. Then, we use the weighted k-means clustering algorithm to group similar stations into clusters. As a result, stations in a cluster are closed to each other and have similar usage pattern. In second stage, we propose MFR-ARMA model by integrating the meteorological data, time of day, day of week, and temperature, then use the historical check-out/in data to fit model. In testing phase, we add predicted meteorological data and use model to predict cluster check-out/in number in a period of time.

Definition 1: Bike Usage Number, represents the sum of absolute values of the number of check-out bikes and check-in bikes for a station or a cluster.

Definition 2: Bike Usage Pattern, indicates the time distribution of the user's bike usage.

Table 1. The top-10 most cared aspects of different categories.

Notation	Description
$\{x_{ij}\}$	Dataset of usage number and geographic location
U^{T_j}	Usage number in time span T_j
C_l	The lth cluster
m_l	The center point of lth cluster
Y_t	Check-out/in number in period $[t,\ t + 1)$
$Y_t^{(C_l)}$	Check-out/in number in cluster C_l
μ_v	Parameters of lagged variable by weather conditions
G_t'	Total effect of weather conditions in bike usage
n	Number of stations

4 Problem Analysis

4.1 Temporal Factor

The usage pattern of the stations is similar every day, every week [10, 11]. Time of day, day of week are two important time factors. Figure 2 shows hourly bike usage number in all stations during Aug. 1st–Sep. 30th, 2014. In Fig. 2, the color depth of the heatmap represents the usage number value, the darker color indicates the larger usage number. In Fig. 2(a), the bike usage number is concentrated in 7:00–10:00 and 16:00–20:00 on weekday, while it is well-distributed in 9:00–19:00 on weekend in Fig. 2(b). From the two figures, we observe different bike usage patterns between weekday and weekend, and the usage pattern is regular in different days on weekdays despite the existence of some traffic noise.

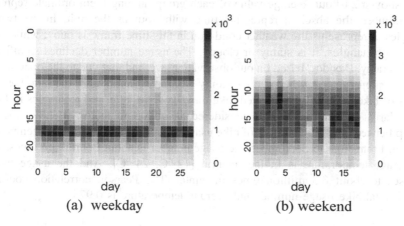

(a) weekday   (b) weekend

Fig. 2. The bike usage number of all station in two months on weekday and weekend.

4.2   Meteorological Factor

More people would rather ride a bike on sunny day than on a rainy day. Similarly, more people may tend to check out/in bike on a warm day than on a cold day. Compared to some meteorological factor like temperature and humidity, weather conditions like rainy and snowy have a greater impact on the bike usage pattern.

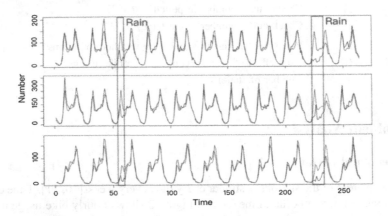

Fig. 3. The bike usage number series of three groups on weekdays.

As presented in previous studies [10, 11, 13], bike usage pattern is significantly influenced by weather conditions and may vary remarkably under different weather conditions. We aggregate neighboring stations into groups, figure out the bike usage number on weekday in each group, and show three groups data series of them. As shown in Fig. 3, black curves show the ground truth usage number over time, blue curves show a 24-hour average value of each group, among them ordinate represents usage number, the abscissa represents time with hour as the unit. In the two red rectangles, it represents the weather condition in this time frame is rainy, while out of these two rectangles, it is sunny or cloudy. The usage number declines significantly when it's rainy. Besides, based on our observation, we find that due to the open-air bike station or slippery road, when it turns sunny within two hours after the rain, the usage number of bikes will still be affected, like the data series after red rectangles in Fig. 3.

Similarly, temperature is also considered an important factor. It shows the relationship between the total amount of bike usage per month and the average temperature in Fig. 4. From June to September, the usage number is about 0.9 million times, when the average temperature is in the interval [12 °C, 24 °C]. And the usage number decreases to about 0.3 million times in winter. The Pearson correlation coefficient between total bike usage number and average temperature is 0.975.

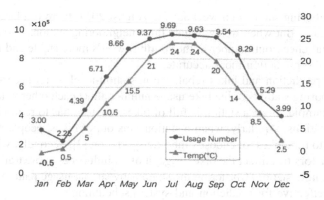

Fig. 4. The strong correlation between total bike usage number and average temperature in each month.

5 Modeling

5.1 Meteorological Factor

Based on researches [8, 10, 11], the usage number of a single station is not regular with relatively large fluctuations, but the regularity of total usage number of stations in a certain area is quite obvious. As shown in Fig. 5, there are three check-out number

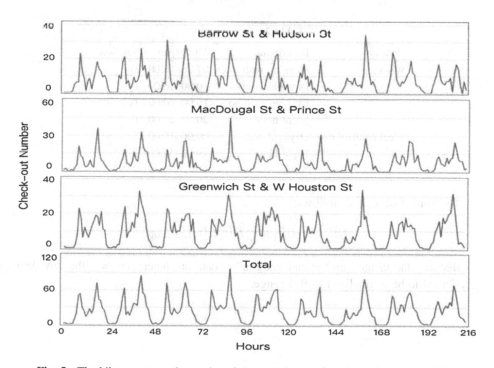

Fig. 5. The bike usage number series of three stations and total number on weekday

series of neighboring stations on weekdays in August 2014, there is a large fluctuation in a single station. But when we aggregate three neighboring stations into a group, and present the total check-out number in the fourth row, it's more stable and more regular which makes the prediction more accurate.

In reality, prediction and bikes rebalance in station-level is not necessary. First of all, the neighboring stations of the bike usage number affect each other when it's full or empty. For example, when a station is full of bikes, the user has to return bike at other neighboring stations. Similarly, when a station runs out of bikes, people need to go to other stations to rent bikes, or choose other transportation [2]. Secondly, when there are some social factors that affect the bike usage, it often influences entire traffic of an area, rather than a single station. Clustering the stations by similar usage pattern and location will be more effective for prediction and system scheduling.

We use a weighted k-means clustering algorithm based on usage pattern and geographical location to cluster similar stations. Based on such observations in Fig. 2, we divide the time span into six groups according to difference usage pattern as shown in Table 2. In order to take into account the similarity of usage pattern of stations in clustering phase, weight values for usage pattern and geographical location are added into model. We first counted usage number of each station in each time span, and integrate it with station geographical location, latitude and longitude as $\{x_{ij}\}, i = 1 \ldots n, j = 1 \ldots 8$. When $j \leq 6$, x_{ij} is a usage of ith station during the jth time span, when $j \geq 7$ it means the latitude and longitude of ith station. Then, the constraints of weight values are as follows: $w_1 = w_j (j \leq 6), w_7 = w_8$.

Table 2. Time groups

Data type	Description	Time span	Natation
Weekdays	Morning rush hour	07:00–10:00	T_1
	Daylight hour	10:00–16:00	T_2
	Evening rush hour	16:00–20:00	T_3
	Night hour	20:00–24:00	T_4
Weekend/Holiday	Daylight hour	09:00–19:00	T_5
	Night hour	19:00–54:00	T_6

The format of x_i is as follows:

$$\langle U^{T_1}, U^{T_2}, U^{T_3}, U^{T_4}, U^{T_5}, U^{T_6}, \text{Lat}, \text{Lon} \rangle$$

Because the usage number and geographic data are heterogeneous, the raw data $\{x_{ij}\}$ need to be normalized to 0–1 range.

The weighted k-means algorithm that minimizes the objective function (1), using Eqs. (2)(3) as follows:

$$E = \sum_{l=1}^{K} \sum_{x_i \in C_l} \sum_{j=1}^{J} w_j \left(x_{ij} - m_{lj} \right)^2 \tag{1}$$

$$m_l^{(r+1)} = \frac{1}{\left| C_l^{(r)} \right|} \sum_{x_i \in C_l^{(r)}} x_i \tag{2}$$

$$C_l^{(r)} = \left\{ x_i : \sum_{j=1}^{d} w_j \left(x_{ij} - m_{lj} \right)^2 \le \sum_{j=1}^{d} w_j \left(x_{ij} - m_{kj} \right)^2, \forall k, 1 \le k \le K \right\} \tag{3}$$

where E is the within-cluster sum of squares, $C_l^{(r)}$ is a cluster during r^{th} iteration, $m_l^{(r)}$ is the center point of $C_l^{(r)}$ during rth iteration, K is the total number of clusters, w_j is a weight of jth dimension of the data. Each station is assigned to exactly one cluster $C_l^{(r)}$ using Eq. (2).

5.2   Lagged Variable of Weather Condition

As shown in Fig. 6, we provide a quantitative analysis of the relationship between the hourly bike usage number of all stations and the weather conditions in the 2014 hourly weather forecast data. We extracted the following five weather conditions: clear, overcast, cloudy, light rain, and rain/snow, and then we counted the average hourly bike usage number of all stations under different weather conditions. We found that the average usage number dropped sharply during the raining days and snowing days.

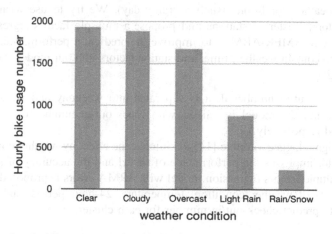

Fig. 6. The hourly bike usage number of all stations under different weather conditions.

Since there is no explicit interaction between bike usage and weather conditions, based on the previous analysis and influence of the weather condition, we initially convert the weather conditions to integer values in Table 3 (each 200 decrease in the average usage number, there is a 1 point increase in the G value), and then determine lagged variable G'_t to calculate the total effect of weather conditions in bike usage.

$$G'_t = \sum_{v=0}^{2} \mu_v G_{t-v} \tag{4}$$

where G_{t-v} is weather condition value in period $[t - v, t - v + 1)$, and μ_v represents the contribution of G_{t-v} in the total effect. The larger value of G the greater influence on usage number. Since the effect of the previous weather condition on the bike demand decreases over time, we add the constraint: $\mu_0 \geq \mu_1 \geq \mu_2$.

Table 3. Weather condition conversion

Weather condition	G value
Clear/Cloudy	0
Overcast	1
Light rain	5
Rain/Snow	8

5.3   MFR-ARMA Model

Recently, many researches [8, 9] used Auto-regressive Moving Average (ARMA) model to predict the available number of bikes or usage number. However, the ARMA model is applied to the stationary model and does not consider the situation of sudden changes in weather conditions (such as rainy day). We try to use weather factors, temporal factors, cluster of stations and propose a multiple factors regression model with ARMA error (MFR-ARMA) to improve the prediction performance.

In order to contain weather, time, and station factors into our model, we have done the following:

(1) The check-out/in number data is separated into weekday and weekend at pre-process, and the model is trained with check-out/in number on weekday and weekend respectively.

(2) We design a lagged variable [14] to simulate the effect by weather conditions and this factor improves the performance of model and the accuracy of predictions.

(3) The multiple factors regression model with ARMA errors is proposed to combine the effects of weather condition, temperature, 24-hour period, and we use the model to predict check-out/in number for each cluster.

MFR-ARMA model is described by:

$$Y_t = y_t + n_t$$

$$y_t = \sum_{d=1}^{D} \alpha_d \sin\left(\frac{2\pi dt}{L}\right) + \beta_d \cos\left(\frac{2\pi dt}{L}\right) + \eta G'_t + \delta T_t \qquad (5)$$

where y_t is the multiple regression model, it contains a Fourier series model with $D = 10$ and two parameters about the effect of weather conditions G'_t and temperature T_t. L is a 24-hours period in our model.

ARMA (p, q) error is described by:

$$n_t = \sum_{i=1}^{p} \theta_i n_{t-i} + \theta_0 + \sum_{j=1}^{q} \gamma_j \varepsilon_{t-j} \qquad (7)$$

where n_t is the ARMA error process. p, q are the orders of AR, MA respectively, θ_i, γ_j are coefficients of AR, MA respectively.

The coefficients α_d, β_d, η, δ, θ_i, γ_j depend on the constituent columns of Y_t, and the value can be estimated by ordinary least squares during data training phase [15].

6 Experiments

6.1 Datasets

We collect trip data from the website of Citi Bike system [16]. The trip dataset format is: (trip duration, start station name, end station name, start time, end time, user type) in New York, which covers one year from Jan. 1st 2014 to Dec. 31st 2016. We extract the check-out/in number of bikes in stations per hour from trip data. To compare with method [5], training set is selected from the data from Jun. 1st to Sep. 10th in 2014 and testing set is selected from the data from Sep. 10th to Sep. 30th in 2014. In the meantime, to validate the performance of our model, we also experimented with data sets in 2015 and 2016, compared with the other two baselines.

We use crawler to collect the hourly weather data from the website of Weather Underground Inc. [17], and divide data into weather condition and temperature.

6.2 Baselines and Metric

We provide four cluster-level prediction methods. Two of them are simple baseline methods, named Historic Mean (HM) and ARMA based on weighted k-means clustering. The remaining two prediction methods are called HP-KNN, HP-MSI based on bipartite station clustering (BC) [5]. The usage number of bikes Y_t for each cluster will be represented at following formulas.

Historic Mean (HM): The model adopts the average of check-out/in number of bikes in each cluster over the past k days at $[t, t + \Delta t)$ period. For example, if we want to predict the check-out number in 8:00am–9:00am, we calculate the average value of the data in the training set in 8:00am–9:00am.

ARMA: The method models the usage number in each cluster as a time series, and uses an auto-regressive moving average (ARMA) model to predict the number in the future.

HP-KNN(BC): The method uses hierarchical prediction method based on KNN. They first clustered each station into groups based bipartite station clustering [5], predicted the entire traffic of the city and then allocate the traffic to each cluster.

HP-MSI(BC): The method integrates multiple similarities between expected and historical periods, they first clustered each station into groups based bipartite station clustering, and predicted the check-out/in of each station cluster [5].

Metric: The metric we adopt is the Root Mean Squared Logarithmic Error (RMSLE), ER in check-out/in number of bikes for performance evaluation.

$$RMSLE = \sqrt{\frac{1}{K}\sum_{l=1}^{K}\left(\log\left(\hat{Y}_t^{C_l}+1\right) - \log\left(Y_t^{(C_l)}+1\right)\right)^2}$$

$$ER = \frac{\sum_{l=1}^{K}\left|\hat{Y}_t^{C_l} - Y_t^{(C_l)}\right|}{\sum_{l=1}^{K}Y_t^{(C_l)}}$$

where $\hat{Y}_t^{C_l}$ is the ground truth of the check-out/in of cluster C_l during t while $Y_t^{(C_l)}$ is the prediction value.

6.3   Clustering Study

Based on our observations, if we set the number of clusters K to n, each station will be a cluster, the usage pattern is not obvious. If the number of clusters is 1, all stations will be considered as a whole. Even if the regularity of usage pattern is obvious, it's not helpful to scheduling which only predict the bike usage number of whole city. In our experiments, K is set to 23.

On the choice of weight, there are three different weight schemes.

(1) Case A that all weights are equal. $w_j = 1 (j \leq 8)$
(2) Case B that the weights of latitude and longitude are larger. $w_j = 1 (j \leq 6)$ *and* $w_j = 5 (6 < j \leq 8)$
(3) Case C that only considers latitude and longitude. $w_j = 0 (j \leq 6)$ and $w_j = 1 (6 < j \leq 8)$.

In Fig. 7, it shows the clustering result in different weight schemes. In Case A, stations in a cluster are scattered and not suitable for the development of scheduling schemes. In Case B and Case C, stations in a cluster are more concentrated, but clustering in Case B, we also consider similar usage pattern which is helpful for prediction model training.

In each case, we calculate the average hourly usage number of each cluster from the training data as the value of usage pattern in each cluster, and then we calculate RMSLE between the training data and the usage pattern shown in Table 4. In the clustering results, if the usage pattern of cluster is more regular and more stable, it will get a relatively smaller RMSLE. From Table 4, we can see that Case B which considers the similar usage pattern and geographic location, achieves the relatively smaller RMSLE. We set the weight according to Case B for the subsequent experiments.

| (a) Case A | (b) Case B | (c) Case C |

Fig. 7. Clustering results of three cases.

Table 4. RMSLE between training data and usage pattern

Case	Check-out number		Check-in number	
	Weekday	Weekend	Weekday	Weekend
CaseA	0.306	0.359	0.305	0.351
CaseB	**0.289**	**0.350**	**0.287**	**0.345**
CaseC	0.314	0.362	0.310	0.357

6.4   Weather Condition Study

As the demand of bike usage is affected by the weather condition, we present a lagged variable to simulate the impact of weather conditions. However, due to the relationship between weather condition and demand is not obvious, the choice of parameters μ_0, μ_1, μ_2

is a big issue. In our experiment, we choose three types of parameter selections. (A) $(\mu_0, \mu_1, \mu_2) = (1, 0, 0)$, in this case, $G'_t = G_t$ means that the usage number is only correlated to the weather condition at the time. (B) $(\mu_0, \mu_1, \mu_2) = (\frac{1}{3}, \frac{1}{3}, \frac{1}{3})$, assumes weather conditions in the last two hours have the same effect on usage number. (C) $(\mu_0, \mu_1, \mu_2) = (\frac{6}{10}, \frac{3}{10}, \frac{1}{10})$, assumes effect of weather conditions in last two hours decreases over time.

Table 5. Result of different parameters

μ_0, μ_1, μ_2	Check-out number		Check-in number	
	RMSLE	ER	RMSLE	ER
$(1, 0, 0)$	0.351	0.311	0.354	0.312
$(\frac{1}{3}, \frac{1}{3}, \frac{1}{3})$	0.349	0.306	0.345	0.301
$(\frac{6}{10}, \frac{3}{10}, \frac{1}{10})$	**0.332**	**0.277**	**0.334**	**0.280**

We use the MRF-ARMA model with three types of parameter selections to predict check-out/in number in 2014 and calculate the RMSLE and ER of the result. As shown in Table 5, when $(\mu_0, \mu_1, \mu_2) = (\frac{6}{10}, \frac{3}{10}, \frac{1}{10})$, the RMSLE and ER of result outperform other two selections.

6.5   Evaluation Result

In this section, we set $(\mu_0, \mu_1, \mu_2) = (\frac{6}{10}, \frac{3}{10}, \frac{1}{10})$, and use MRF-ARMA based on clustering with Case B. Then, we compare the result of our model with four cluster-level prediction methods.

We present the average RMSLE and ER of all the prediction hours in 2014 in Table 6. We compare the prediction result difference between HM, ARMA, HP-KNN, HP-MSI and our model. The prediction result by using HM, ARMA method are very close, because the usage pattern of each cluster itself is regular and the usage number series fluctuations in the historical average number after clustering similar stations into clusters. HP-KNN and HP-MSI has a lower ER than HM and ARMA, because they take into account the effects of weather condition, temperature and wind speed. From this comparison point of view, considering the effects of weather factors really help to improve the performance of prediction. In the result of check-out prediction, our MFR-ARMA model achieves 0.332 RMSLE and 0.277 ER, outperforming all the baseline methods. As for the result of check-in prediction, we can see that our model achieves 0.334 RMSLE and 0.280 ER, also outperforms the other four methods.

Besides, we show the comparison of the RMSLE and ER under the three prediction methods in 2014, 2015, 2016. In Fig. 8, we can see that MFR-ARMA achieves both lower RMSLE and error rate in Check-out/in prediction in three years. It is more obvious in the prediction in 2014, 2016. Firstly, the weather conditions in test set in 2015 are mostly sunny and cloudy, our model doesn't show much improvement in prediction accuracy, only 2% more than HM and ARMA on the error rate. The prediction error rate in 2014 and 2016 improve about 8%. Secondly, the average error rate of our model remains at around 0.28, and the average value of RMSLE is about 0.33.

Table 6. Prediction error of cluster check-out/in number in 2014

Method	Check-out number		Check-in number	
	RMSLE	ER	RMSLE	ER
HM	0.358	0.355	0.354	0.353
ARMA	0.359	0.357	0.356	0.354
HP-KNN(BC)	0.358	0.299	0.306	0.295
HP-MSI(BC)	0.349	0.282	0.350	0.290
MFR-ARMA	**0.332**	**0.277**	**0.334**	**0.280**

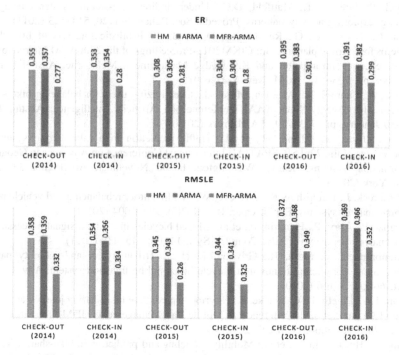

Fig. 8. The comparison of the RMSLE and ER in 2014–2016.

7 Conclusion

In this paper, we first analyze the effects of some factors on the bike usage including meteorological and temporal factors. We discuss the availability of cluster-level prediction, and cluster all neighboring stations into clusters with similar usage pattern.

Then, we propose a lagged variable to simulate the effect of weather conditions in usage number, and a multiple factor regression model with ARMA error (MFR-ARMA). Finally, we use the model to predict the bike check-out/in number. Our model outperforms other models which has a lower average RMSLE and ER in check-out/in number prediction. In the future, we will consider the use of data sets in different cities for learning and prediction. And we will consider social factors and traffic factors to improve the model, such as traffic events or large activities.

Acknowledgment. This work is supported by Zhejiang Provincial Natural Science Foundation of China (No. LY17F020008).

References

1. Shaheen, S., Guzman, S., Zhang, H.: Bikesharing in Europe, the Americas, and Asia: past, present, and future. Transp. Res. Rec. J. Transp. Res. Board **2143**, 159–167 (2010)
2. Vogel, P., Greiser, T., Mattfeld, D.C.: Understanding bike-sharing systems using data mining: exploring activity patterns. Procedia Soc. Behav. Sci. **20**, 514–523 (2011)
3. Gast, N., Massonnet, G., Reijsbergen, D., et al.: Probabilistic forecasts of bike-sharing systems for journey planning. In: CIKM2015-Proceedings of the 24th ACM International on Conference on Information and Knowledge Management, Melbourne, Australia, 18–23, October, pp. 703–712. ACM, New York (2015)
4. Singla, A., Santoni, M., Meenen, M., et al.: Incentivizing users for balancing bike sharing systems. In: Twenty-Ninth AAAI Conference on Artificial Intelligence, Austin, Texas, 25–30 January, pp. 723–729. AAAI Press (2015)
5. Li, Y., Zheng, Y., Zhang, H., et al.: Traffic prediction in a bike-sharing system. In: Proceedings of the 23rd SIGSPATIAL International Conference on Advances in Geographic Information Systems, Seattle, Washington, 03–06 November, Article no. 33. ACM, New York (2015)
6. Schuijbroek, J., Hampshire, R.C., Hoeve, W.J.V.: Inventory rebalancing and vehicle routing in bike sharing systems. Eur. J. Oper. Res. **257**(3), 992–1004 (2017)
7. Borgnat, P., Abry, P., Flandrin, P., et al.: Shared bicycles in a city: a signal processing and data analysis perspective. Adv. Complex Syst. **14**(03), 415–438 (2011)
8. Kaltenbrunner, A., Meza, R., Grivolla, J., et al.: Urban cycles and mobility patterns: exploring and predicting trends in a bicycle-based public transport system. Adv. Complex Syst. **6**(4), 455–466 (2010)
9. Yoon, J.W., Pinelli, F., Calabrese, F.: Cityride: a predictive bike sharing journey advisor. In: IEEE 13th International Conference on Mobile Data Management (MDM), Karnataka, India, 23–26 July, pp. 306–311. IEEE (2012)
10. Yang, Z., Hu, J., Shu, Y., et al.: Mobility modeling and prediction in bike-sharing systems. In: Proceedings of the 14th Annual International Conference on Mobile Systems, Applications, and Services, Singapore, 26–30 June, pp. 165–178. ACM, New York (2016)
11. Chen, L., Zhang, D., Wang, L., et al.: Dynamic cluster-based over-demand prediction in bike sharing systems. In: Proceedings of the 2016 ACM International Joint Conference on Pervasive and Ubiquitous Computing, Heidelberg, Germany, 12–16 September, pp. 841–852. ACM, New York (2016)
12. Singhvi, D., Singhvi, S., Frazier, P.I., et al.: Predicting bike usage for New York City's bike sharing system. In: Workshops at the Twenty-Ninth AAAI Conference on Artificial Intelligence: Computational Sustainability (2015)

13. Gebhart, K., Noland, R.B.: The impact of weather conditions on bikeshare trips in Washington D.C. Transportation **41**(6), 1205–1225 (2014)
14. Reed, W.R.: On the practice of lagging variables to avoid simultaneity. Oxford Bull. Econ. Stat. **77**(6), 897–905 (2015)
15. Box, G.E.P., Jenkins, G.M., Reinsel, G.C., et al.: Time Series Analysis: Forecasting and Control. Wiley, Hoboken (2015)
16. Citi Bike: Citi Bike System Data (2014). http://www.citibikenyc.com/system-data
17. Weather Underground Inc.: Weather history (2014). https://www.wunderground.com/history/

Data Mining and Knowledge Mining

Talents Recommendation with Multi-Aspect Preference Learning

Fei Yi[✉], Zhiwen Yu, Huang Xu, and Bin Guo

Northwestern Polytechnical University,
Xi'an 710072, Shaanxi, People's Republic of China
yifeinwpu@gmail.com, xuhuang601@gmail.com,
{zhiwenyu, guob}@nwpu.edu.cn

Abstract. Discovering talents has always been a crucial mission in recruitment and applicant selection program. Traditionally, hunting and identifying the best candidate for a particular job is executed by specialists in human resources department, which requires complex manual data collection and analysis. In this paper, we propose to seek talents for companies by leveraging a variety of data from not only online professional networks (e.g., LinkedIn), but also other popular social networks (e.g., Foursquare and Last.fm). Specifically, we extract three distinct features, namely global, user and job preference to understand the patterns of talent recruitment, and then a Multi-Aspect Preference Learning (MAPL) model for applicant recommendation is proposed. Experimental results based on a real-world dataset validate the effectiveness and usability of our proposed method, which can achieve nearly 75% accuracy at best in recommending candidates for job positions.

Keywords: Talent recommendation · Multi-Aspects Preference Learning

1 Introduction

In this era of talent-hunger economy, how to efficiently discover and recruit the right talent for the right job has become a critical problem. Companies usually conduct several procedures to hire the most desired and suitable person, including resume checking, job knowledge testing, cognitive testing and even personality testing. However, all these works are triggered only after some applicants submit their resumes. This passive manner for companies in recruitment is called "post and pray", which describes a situation that companies can only idly hope good candidates swim up to the bait after they advertised the job opening. Although there are many solid recruiting methods [7–9] developed by specialists in human resources management, the technological advances can now streamline almost every business operation in talent recruitment.

Apart from some traditional methods [16, 18] for personnel selection, the booming of Internet and social networking services are changing the mind of human resources managers in talent hunting. Many job-related social network platforms like Indeed [28], Glassdoor [29], Monster [30] and LinkedIn have provided a massive scale of open data of employee's profile for companies, based on which, companies are now becoming

© Springer Nature Switzerland AG 2019
S. Li (Ed.): GPC 2018, LNCS 11204, pp. 409–423, 2019.
https://doi.org/10.1007/978-3-030-15093-8_29

more proactive in discovering their desired talents. However, it is reported [25] that recruiting online has a sort of paradox of choice, and the major problem is how to find the top-notch employees who are the most suitable persons for the target position. Despite the substantial manpower and resources allocated to assess the quality of the discovered potential employees, the rapid growth of job seekers online has made it arduous for companies to hunt the most desired talents. Therefore, an intelligent system that can automatically find out the most valuable talents becomes critical.

In this paper, we aim to develop such a system that helps companies to hunt for their most wanted talents. Specifically, we first collect employees' professional trajectories from LinkedIn and personal profiles from Last.fm and FourSquare. After that, we extract a number of preference features both for potential employees and job positions. Finally, we fuse all these features into a Multi-Aspect Preference Learning model for talents recommendation. Our work has made the following unique contributions:

1. We define three different features for both applicants and job positions that may reflect the results in personal selection, namely global, user, and job preference.
2. We develop a Multi-Aspect Preference Learning (MAPL) model that fuses the three defined preference features instead of separately taking one of them into consideration. And different learning algorithms are proposed to extract patterns for different features.
3. We conduct extensive experiments on 10,274 employees' career trajectories from popular professional networks. Our results not only prove the effectiveness of extracted features, but also validate the usability and effectiveness of our proposed recommendation method.

The remainder of this paper is organized as follows. In Sect. 2, we review the related work. The problem statement and system framework are illustrated in Sect. 3. Section 4 explains how we link/integrate different sources of data. In Sect. 5, we present models of feature extraction and then discuss the recommendation model in Sect. 6. Details of our extensive experiments are presented in Sect. 7. Finally, we conclude our work and discuss possible future works in Sect. 8.

2 Related Works

2.1 Job Recommendation

In recent years, the study of methods for recommending jobs to applicants as well as systems that provide job searching services have been an extensive topic of research in the area of labor economics. Malinowski et al. [24] proposed an approach applying two distinct recommendation systems to the field in order to improve the matching between people and jobs. Paparrizos et al. [23] addressed the problem of recommending suitable jobs to people who are seeking a new job. And Irving et al. [2] studied the problem of how to assign students to suitable hospitals with right job positions. Apart from these, as social networks become popular, researchers have paid more attention to learn job-related patterns from online information. Cheng et al. [22] analyzed the job information from popular social networks like Twitter and LinkedIn. Al-Otaibi et al. [21] gave us an overview of e-recruiting process and related issues for building personalized recommendation systems of candidates/job matching. Besides, Zhang et al. [3] proposed to offer applicants a personalized online service that can help them find ideal jobs quickly and conveniently.

2.2 Personnel Selection

Matching jobs and applicants are bilateral problem, which means apart from recommending jobs to applicants, personnel selection is also a critical problem. Xu et al. [20] aggregated the work experiences and check-in records of individuals to model the job change motivations to target down the applicant set who are willing to change jobs. They further proposed to discover talent circles based on job transition records in a recent work [19]. Besides, Chien et al. [18] proposed to fill the gap by developing a data mining framework based on decision tree and association rules to generate useful rules for personnel selection. Mehrabad et al. [17] developed an expert system for effective selection and appointment of the job applicants, and Hooper et al. [16] also paid their attention on using expert system in personnel selection process.

The main difference of our work with previous works is that we employ information from multi-aspects of heterogeneous data sources, including professional and social networks (e.g., LinkedIn, Foursquare and Last.fm), to understand multiple patterns in job recruitment and talent recommendation. Specifically, we defined three aspects of features that capture both company and applicant preferences, and a Multi-Aspect Preference Learning (MAPL) model is proposed to fuse all these features for talent recommendation.

3 Problem Statement and System Overview

3.1 Problem Statement

In this paper, we aim to recommend the most suitable talents for given jobs. And the problem is formulated as follows. In professional networks, there are two domain sets, namely a job set $\mathcal{J} = \{j_1, j_2, \ldots, j_n\}$ where each $j \in \mathcal{J}$ represents a specific job profile, and a user set $\mathcal{U} = \{u_1, u_2, \ldots, u_m\}$ where each $u \in \mathcal{U}$ represents the talent candidates. Thus, given a job profile $j_i \in \mathcal{J}$, the problem in this paper is to find a group of user $u_{best} \in \mathcal{U}$ which contains the most suitable users for job j_i.

Specifically, for each $j \in \mathcal{J}$ and $u \in \mathcal{U}$, there is a fitness probability $Fit(u|j)$ that captures the degree how well user u fits the job j. Therefore, we can investigate this probability for every user $u \in \mathcal{U}$ to find out the most suitable users as $\mathcal{U}_{best}$. In details, we considered three main factors that contribute to the fitness probability $Fit(u|j)$. (1) Global Preference. This factor (GP) captures the overall job transition patterns for all applicants and jobs. For example, most applicants usually prefer to search for jobs that provided by big companies. (2) User Preference. This factor measures the individual job preference for different applicants, which can be extracted from the historical professional trajectories of a specific user and represented as User Preference (UP).

(3) Job Preference. Different job position usually have different requirements for applicants, hence we extract Job Preference (*JP*) based on the historical recruitment results for every jobs to capture such pattern.

According to the three factors we proposed above, the fitness probability $Fit(u|j)$ can be extended to $Fit(u|j, GP, UP_u, JP_j)$ where GP, UP_u and JP_j are the global, user and job preference respectively. And each of the factors can be deduced through their parameters such as $GP \propto f(\Omega)$, $UP \propto f(\Phi)$ and $JP \propto f(\Theta)$. Therefore, the final version of $Fit(u|j)$ is transformed into $Fit(u|j, \Omega, \Phi, \Theta)$. And the ultimate goal of our work is to learn this function using the massive scale of professional trajectories as well as different sources of related datasets. The learning method for this fitness function will be illustrated more comprehensively in the following sections.

3.2 System Overview

In this section, we propose a bottom-up architecture of framework as shown in Fig. 1, which contains four main components: *data collection*, *feature extraction*, *parameter learning*, and *prediction and recommendation*.

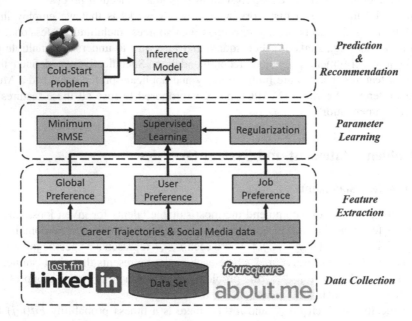

Fig. 1. System overview for talents recommendation.

Data Collection

A branch of job-searching or professional networks are launched, especially LinkedIn, the world's most popular professional network. To achieve the goal of talents recommendation, we mainly collected the career trajectory data from LinkedIn for each crawled user. Furthermore, we enrich this dataset with user's profile information through their homepage on aboutme.com, which includes their check-ins from Foursquare and play lists from Last.fm.

Feature Extraction
This component mainly focuses on how to extract related features from raw dataset. Specifically, we propose three distinct features: global, user and job preference. They describe the explicit or implicit factors that may influence the process in talents recruitment. The global preference is extracted using Non-negative Matrix Factorization (NNMF) [4] based on a matrix that is built with applicants and jobs, the other two features are modeled from the historical trajectories of each individual applicants and the previous personnel recruitment records of every jobs using a linear regression algorithm. Detailed information and method for extracting features will be described in Sect. 5.

Parameter Learning
In this component, we aim at training our proposed talents hunting model using labeled dataset. Specifically, since our datasets are all labeled, we train our proposed model in a supervised learning manner. Besides, in order to prevent overfitting, we introduce regularization in our learning algorithm.

Prediction and Recommendation
After determining the parameters in our model, it is able for us to apply this function on every applicant to find out the most suitable ones when given a specific job profile. However, to build a more robust model, we also take "cold-start" [1] problem into consideration. Finally, our recommendation model can reach up to more than 70% accuracy in selecting the best applicants.

4 Data Collection

One of the major issues in our work is how to link/integrate various social media data at the individual level. Despite the several models that cover the problem of entity linking [26, 27], there are many personal web hosting service platforms like "about.me", which offer registered users a homepage that can allow them to link their multiple online identities. Hence, we directly crawl user's multi-aspects information from those homepages on about.me. Specifically, the homepage of each person on about.me contains information from different aspects of social networks, including Facebook, LinkedIn and Twitter. And all these homepages are publicly available that can be crawled. Finally, we build our dataset for each user with multi-aspect information.

5 Preference Feature Extraction

5.1 Preliminaries

The large amount of job positions and job seekers make it troublesome to manually understand the requirements of job recruitment and individual preferences. However, every job and applicant can be classified into different attribute sets w.r.t different criterion. For example, company size can be classified into large, middle and small according to the total number of employees. Then, we can narrow down the data space for better discussion and learning. The following criterion are proposed for job and applicant.

Job Features

Company Size (Com). According to the number of employees (*NoE*) in corresponding companies, we measure the company size as follows:

$$Com_i = \frac{lg^{NoE_i}}{\sum_{i=1}^{n} lg^{NoE_i}} \tag{1}$$

where n is the total number of companies, thus a larger NoE_i means a bigger Com_i for ith company.

Job Level (JL). Similar to the definition of *Com* above, we have the following equation to generate a scalar value of this feature:

$$JL_i = \exp\left(-\frac{lg^{NoJ_i}}{\sum_{i=1}^{n} lg^{NoJ_i}}\right) \tag{2}$$

where n is the number of job categories and NoJ_i represents the number of jobs that have similar category with ith job. Therefore, if many jobs' description are similar to ith job, we will have a low value of JL_i. For example, the number of staffs is always larger than the number of managers, but obviously, the job level for staff is always lower than that of manager.

Contract Year (CY). The number of years that a job requires the applicant to sign with the company is an influential attribute that can affect recruitment result, which is defined as: $CY = x$ where x is the contract year that an applicant need to sign with.

Job Mobility Rate (Jmr). This attribute measures the extent whether a specific job position changes its employees frequently or not as defined:

$$Jmr = \frac{1}{12} \sum_{i=1}^{12} \frac{M_j(i)}{M(i)} \tag{3}$$

where $M_j(i)$ represents the number of transitions of job j in month i, including the number of resignations and recruitments. $M(i)$ is the total transitions on all jobs in month i.

Applicant Features

Education Background (Edu). Job recruitment is sometimes a process of education-bias [6]. In order to quantify this factor, we define the education background as follows:

$$Edu_i = \exp\left(-\frac{lg^{NoD_i}}{\sum_{i=1}^{n} lg^{NoD_i}}\right) \tag{4}$$

where n is the total category of education background, including Ph.D, master and bachelor degree, and NoD_i represents the number of degrees that is equal with ith applicant. Obviously, higher degree usually has lower numbers, thus will result in higher *Edu*.

Gender (Gen). It has been discussed [15] that gender can be an influential factor in recruitment. Thus, we give a binary representation *(Gen)* to describe an applicant's gender. Specifically, we set *Gen* = 1 when the applicant is male, otherwise *Gen* = 0.

Work Experience (WE). Work experience is one important factor in recruitment, and can be quantified according to the specific applicants' historical professional trajectories, which is defined as:

$$WE_i = \sum_{j=1}^{n} Com_{(i,j)} * JL_{(i,j)} * CY_{(i,j)} \tag{5}$$

where n is the number of previous jobs that ith applicant has experienced, and $Com_{(i,j)}$, $JL_{(i,j)}$ and $CY_{(i,j)}$ are the corresponding company size, job level and contract year during his/her jth job.

Mobility Pattern (MP). Xu et al. [20] have discussed that a user's mobility pattern can influence whether a employee is going to change his/her current job. Thus, we take mobility pattern into consideration as follows:

$$MP_i = \sum_{j=1}^{n} \frac{lgR_j}{lgN_j} \tag{6}$$

where n is the number of recent months of a user's check-in records. And N_j represents the check-in rate at last jth month, R_j is the check-in radius at last jth month which is further calculated as:

$$R = \sqrt{\frac{1}{N} \sum_{t=1}^{N} (r_t - \tilde{r})^2} \tag{7}$$

where N represents the number of check-ins, r_t is the location of every check-ins and $\tilde{r}$ is the center location of all available users' according to their check-ins.

Finally, according to all these pre-defined features, we can easily represent the attribute of a job or an applicant as $Job_i = \langle Com_i, JL_i, CY_i, Jmr_i \rangle$ and $User_j = \langle Edu_j, Gen_j, WE_j, MP_j \rangle$ respectively, which can be further used in extracting user and job preference.

5.2 Global Preference

The massive scale of job transition and recruitment records contain useful knowledge for us to understand the relationship between jobs and talents. Hence, in this paper, we first build a matrix consists of relationship between applicants and jobs to extract the global pattern of talents recruitment. Specifically, we cluster all the collected jobs into n clusters according to their features as defined above, and then if we have totally m individuals on professional network, we can construct a $m \times n$ matrix M named *"User-Job Matrix"*. And each element $val_{i,j}$ is a binary value 0 or 1 in this matrix, which indicates that if user i has experienced a job that belongs to jth job cluster, $val_{i,j}$ will be 1, otherwise it will be 0. After that, we apply Non-negative Matrix Factorization (NNMF) to decompose this matrix M such that $M \approx P \times Q^T$, notice that P is a $m \times k$ and Q is a

$n \times k$ matrix respectively, where k is the number of latent factors. Then, the global preference between ith user and jth job cluster can be calculated as:

$$GP_{i,j} = p_i q_j^T \tag{8}$$

and according to the choice of k, we would have different results of $GP_{i,j}$. Despite other parameters that could influence the factorization result, k is one of the most significant factors, thus we have $GP \propto f(k)$.

5.3   User Preference

In this part, we extract each user preference based on their career trajectories respectively. Specifically, there is a historical professional trajectory $Traj = \langle Job_1, Job_2, \ldots, Job_n \rangle$ that records all the jobs a user has experienced. Then, according to the proposed four features of job, we can represent each Job_i as $\langle Com_i, JL_i, CY_i, Jmr_i \rangle$, hence, the ith user preference to jth job is defined as the linear combination of these features:

$$UP_{i,j} = \alpha_i^1 Com_j + \alpha_i^2 JL_j + \alpha_i^3 CY_j + \alpha_i^4 Jmr_j \tag{9}$$

where $\alpha_i^n, n \in \{1,2,3,4\}$ are the parameters that control the weights of each job features in influencing the user's choice when deciding to accept a job offer. Thus, we have $UP_i \propto f(\alpha_i^n, n \in \{1,2,3,4\})$ for each user i.

5.4   Job Preference

Similar with user preference, there is a set of applicants $Appl_j = \langle User_1, User_2, \ldots, User_n \rangle$ containing totally n users that the corresponding job has hired. Then, we can represent every user i in this set as $\langle Edu_i, Gen_i, WE_i, MP_i \rangle$, therefore, the jth job preference to ith user is defined as:

$$JP_{j,i} = \beta_j^1 Edu_i + \beta_j^2 Gen_i + \beta_j^3 WE_i + \beta_j^4 MP_i \tag{10}$$

where $\beta_j^n, n \in \{1,2,3,4\}$ are the parameters, and we can deduce that $JP_i \propto f(\beta_j^n, n \in \{1,2,3,4\})$ for each job j.

6   Inference Model

6.1   General Model

As we have discussed in Sect. 3.1, the goal of our work is to learn the fitness probability $Fit(u|j, GP, UP_u, JP_j)$ that generates the estimated fitness value $\widehat{fit}_{i,j}$ for u_i and job_j, where GP, UP_u and JP_j are the proposed global, user and job preference respectively, and they are controlled by several parameters as discussed in Sect. 4. The loss function for learning our fitness probability is defined as follows:

$$min \sum\nolimits_{i,j \in R} \left(fit_{i,j} - \widehat{fit}_{i,j}\right)^2 + \lambda\left(||k||^2 + \alpha^2 + \beta^2\right) \qquad (11)$$

and $\widehat{fit}_{i,j}$ can be inferred by:

$$\widehat{fit}_{i,j} = GP_{i,j}|k + UP_{i,j}|\alpha + JP_{i,j}|\beta \qquad (12)$$

where $\widehat{fit}_{i,j}$ is the linear combination on the proposed three features. We also apply regularization terms to prevent overfitting when learning parameters in Eq. 11. Where λ controls the degree of regularization, which is pre-defined as 0.01 according to [14]. In order to judge whether a u_i is suitable for job_j, we empirically set the threshold as 0.5 for $\widehat{fit}_{i,j}$, which means that when $\widehat{fit}_{i,j} > 0.5$, u_i is fit for job_j, otherwise not.

6.2 Cold-Start Problem

"Cold-Start" [1] is one classic problems in recommendation system, and we apply collaborative filtering [5] to deal with such problem. Specifically, suppose there is a new job $job_{new} \notin \mathcal{J}$ where $\mathcal{J}$ is the existing job set we have. We first extract the four features mentioned above of this new job, then we calculate the similarity between this new job and all existing jobs based on their features as follows:

$$Sim(job_{new}, job_i) = \frac{job_{new} \cdot job_i}{||job_{new}||||job_i||} \qquad (13)$$

After that, we select the "Top-K" similar jobs as the foundation for this new job, in which the number K is calculated using $K = [\lg N]$ where N is the total number of current jobs. Therefore, for each potential user u_i and the target new job position job_{new}, we utilize the parameters among this K similar jobs in Eq. 12 to infer K fitness values, and the final fitness value $fit_{i,new}$ is defined as the mean value among all these K fitness values.

Table 1. Data illustration.

Description	Number of records
User	10,274
Job	55,439
Ratio of gender	1.63:1 (male: female)
Check-ins	690,823

7 Experiments

7.1 Data Illustration

We totally collected 10,274 users' professional trajectories in LinkedIn and in other social networks like Foursqaure and Last.fm. The detailed information of this dataset is illustrated in Table 1. Figure 2 shows the distribution of company size. It indicates that companies with less than 500 employees are the majority across all companies, which takes nearly 80%. In Fig. 3, we illustrate the changes in job mobility rate for all types of jobs. Specifically, we observe that nearly 6% employees would change their work when they have worked for one year on their current job position, and this rate decreases as the number of months increases. Besides, since employees usually sign offers with contract years as the integer multiples of one year, we can also see that there are always significant increases of job mobility rate every one year. Figure 4 shows the gender distribution over different type of company size and job level. We observe that larger companies have a more balanced gender distribution than smaller companies.

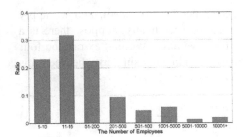

Fig. 2. Company size distribution.

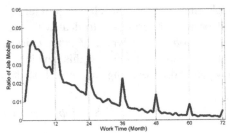

Fig. 3. Job mobility rate distribution.

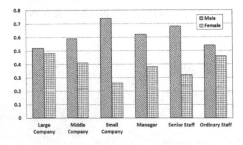

Fig. 4. Gender distribution.

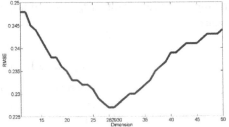

Fig. 5. Parameter learning in NNMF.

7.2 Parameter Learning

To obtain the best NNMF results on *"User-Job Matrix"*, we need to determine the most suitable value of dimension parameter k for matrix P and Q. Here, we apply Root Mean Square Error (RMSE) as object function to learn parameter k and the object function of RMSE is defined as:

$$RMSE = \sqrt{\frac{1}{n}\sum_{i,j}\left|fit_{i,j} - \widehat{fit}_{i,j}\right|^2}\qquad(14)$$

where $fit_{i,j}$ represents the real fitness value and $\widehat{fit}_{i,j}$ is the estimated fitness value after NNMF, n is the number of all elements in *"User-Job Matrix"*. We chose different value of k to minimize this object function, the results are shown in Fig. 5. As shown in Fig. 5, we discover that there is a turning point when $k = 28$. Before that, the RMSE shows a downward trend as k increases, and it shows a upward trend as k increases after that. Thus, we adopted 28 as the value of k, in other words, the dimension to decompose the original matrix, which reaches the lowest RMSE as 0.227. After that, we applied gradient descent to optimize the object function in Eq. 14.

In order to balance the computation complexity with the optimized result, we need to find an appropriate iteration time in optimizing that object function. As shown in Fig. 6, with the number of iteration time increases, the value of the object function decreases. And when we iterate 12 times, the result of object function reaches 0.002, which is a relatively small value. After that, when iteration time exceeds 49, the RMSE no longer changes and remains stabled. Therefore, we chose 49 as the number of iteration time in optimizing that object function.

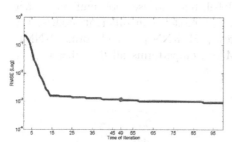

Fig. 6. Iteration performance.

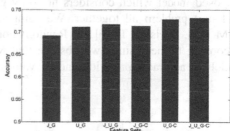

Fig. 7. Recommendation results based on different feature sets.

7.3 Feature Evaluation

When the model parameters are all determined, we turn to see what feature can be effective in recommendation, and whether the consideration of cold-start problem can elevate the performance of our proposed model. As shown in Fig. 7, we compare different combinations across different features with/without considering cold-start problem. Specifically, the combinations that are used in recommendation are illustrated in Table 2. In Fig. 7, we discover that when taking cold-start problem into account, the overall recommendation performance is better than those do not consider cold-start problem. For example, feature set J_U_G-C achieves over 73% accuracy while J_U_G is nearly 71%. Meanwhile, it can be observed that considering user preference usually get higher accuracy than only taking job preference into account, such that U_G-C out-preforms J_G-C and U_G also gains higher accuracy than J_G, which indicates that user preference is much effective in recommendation. In conclusion, taking all preference features as well as the cold-start problem into consideration can achieve the best performance among all feature combinations.

Table 2. Feature sets.

Features	Description
J_G	Job and global preferences
U_G	User and global preferences
J_U_G	All three preferences
J_G-C	Job, global preferences and cold-start problem
U_G-C	User, global preferences and cold-start problem
J_U_G-C	All three preferences and cold-start problem

7.4 Recommendation Results

After we determined all the parameters for fitness probability, we then apply it in recommending applicants to specific jobs. And we evaluated the performance across several models using 10-fold cross-validation. As shown in Fig. 8, MAPL is our proposed model which considers not only global, but also user, job preferences and cold-start problem all together. We compare our model with different models, i.e., SVM [11], Decision Tree [10], Random Forest [12], KNN [13] and normal NNMF. According to the results, we observe that MAPL outperforms all the other popular models in accuracy, precision and F1 values.

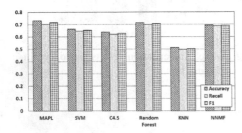

Fig. 8. Recommendation results on different models.

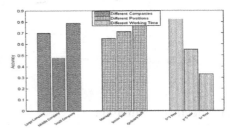

Fig. 9. Recommendation under different conditions.

Specifically, only random forest and MAPL exceed 70% accuracy, while MAPL preforms better than random forest (73.1% vs. 71.5%). Among all the models, KNN is the worst method. Besides, comparing with normal NNMF, MAPL preforms better due to it considers not only latent factors after matrix decomposition, but also takes user and job preferences into account. However, other models like random forest and decision tree only learn patterns from user and job preferences, they do not consider latent factors which make them preforms worse than our proposed model. Furthermore, we apply our proposed MAPL method on different conditions as shown in Fig. 9. We can observe that the recommendation performance fluctuates across different conditions, for example, recommending talents for small company can achieve nearly 80% accuracy while that is lesser than 50% for middle company, and similar situation appears in other conditions. Data imbalance could be one of the significant reasons for this performance fluctuation, specifically, we have more than 70% company profiles that belong to small companies, which may lead to a better model than others. Besides, there are other factors influencing the performance, such that large companies usually maintain similar and stable recruitment strategies, while others like middle companies have variety of recruitment methods, which makes it much harder to learn an accurate model for middle companies than larger ones. Therefore, our model performs worse in middle company than large company.

8 Conclusion and Future Works

In this paper, we tackled the problem of talents recommendation for a specific job. Instead of applying traditional talent recruitment methods, we proposed to solve this problem by using large amount of data that were collected from popular online professional networks like LinkedIn as well as some other user-related data from social networks like Foursquare and Last.fm. Specifically, we proposed a Multi-Aspect Preference Learning (MAPL) model, which takes global, user and job preferences all together as we defined in this paper. Based on the relationship between these features and job recruitment, we learned a fitness probability $Fit(u|j)$ that can determine whether a given applicant is suitable for the target job position. We further consider cold-start problem in our recommendation algorithm to build a more robust method. Experiments on the real-world data clearly validated that our model achieves effective and accurate results.

In future works, since content-based information like user/job descriptions are also important to model user/job preferences, we attempt to collect more contents to enrich the meaning of our defined features to learn a more effective model for talents recommendation.

References

1. Lam, X.N., et al.: Addressing cold-start problem in recommendation systems. In: Proceedings of the 2nd International Conference on Ubiquitous Information Management and Communication. ACM (2008)
2. Irving, R.W.: Matching medical students to pairs of hospitals: a new variation on a well-known theme. In: Bilardi, G., Italiano, G.F., Pietracaprina, A., Pucci, G. (eds.) ESA 1998. LNCS, vol. 1461, pp. 381–392. Springer, Heidelberg (1998). https://doi.org/10.1007/3-540-68530-8_32
3. Zhang, Y., Yang, C., Niu, Z.: A research of job recommendation system based on collaborative filtering. In: 2014 Seventh International Symposium on Computational Intelligence and Design (ISCID). vol. 1. IEEE (2014)
4. Sra, S., Dhillon, I.S.: Generalized nonnegative matrix approximations with Bregman divergences. Advances in neural information processing systems (2006)
5. Sarwar, B., et al.: Item-based collaborative filtering recommendation algorithms. In: Proceedings of the 10th International Conference on World Wide Web. ACM (2001)
6. Wegener, B.: Job mobility and social ties: social resources, prior job, and status attainment. Am. Sociol. Rev. **56**, 60–71 (1991)
7. Parry, E., Tyson, S.: An analysis of the use and success of online recruitment methods in the UK. Hum. Resour. Manag. J. **18**(3), 257–274 (2008)
8. Carroll, M., et al.: Recruitment in small firms: processes, methods and problems. Empl. Relat. **21**(3), 236–250 (1999)
9. Marsden, P.V.: The hiring process: recruitment methods. Am. Behav. Sci. **37**(7), 979–991 (1994)
10. Quinlan, J.R.: Simplifying decision trees. Int. J. Man-Mach. Stud. **27**(3), 221–234 (1987)
11. Cortes, C., Vapnik, V.: Support-vector networks. Mach. Learn. **20**(3), 273–297 (1995)
12. Ho, T.K.: The random subspace method for constructing decision forests. IEEE Trans. Pattern Anal. Mach. Intell. **20**(8), 832–844 (1998)
13. Altman, N.S.: An introduction to kernel and nearest-neighbor nonparametric regression. Am. Stat. **46**(3), 175–185 (1992)
14. Du, R., et al.: Predicting activity attendance in event-based social networks: content, context and social influence. In: Proceedings of the 2014 ACM International Joint Conference on Pervasive and Ubiquitous Computing. ACM (2014)
15. Correll, S.J.: Gender and the career choice process: the role of biased self-assessments. Am. J. Sociol. **106**(6), 1691–1730 (2001)
16. Hooper, R.S., et al.: Use of an expert system in a personnel selection process. Expert. Syst. Appl. **14**(4), 425–432 (1998)
17. Mehrabad, M.S., Brojeny, M.F.: The development of an expert system for effective selection and appointment of the jobs applicants in human resource management. Comput. Ind. Eng. **53**(2), 306–312 (2007)
18. Chien, C.-F., Chen, L.-F.: Data mining to improve personnel selection and enhance human capital: a case study in high-technology industry. Expert Syst. Appl. **34**(1), 280–290 (2008)
19. Xu, H., et al.: Talent circle detection in job transition networks. In: Proceedings of the 22nd ACM SIGKDD International Conference on Knowledge Discovery and Data Mining. ACM (2016)
20. Xu, H., et al.: Learning career mobility and human activity patterns for job change analysis. In: 2015 IEEE International Conference on Data Mining (ICDM). IEEE (2015)
21. Al-Otaibi, S.T., Ykhlef, M.: Job recommendation systems for enhancing e-recruitment process, Las Vegas Nevada, pp. 433–440 (2013)

22. Cheng, Y., et al.: Jobminer: a real-time system for mining job-related patterns from social media. In: Proceedings of the 19th ACM SIGKDD International Conference on Knowledge Discovery and Data Mining. ACM (2013)
23. Paparrizos, I., Cambazoglu, B.B., Gionis, A.: Machine learned job recommendation. In: Proceedings of the Fifth ACM Conference on Recommender Systems. ACM (2011)
24. Malinowski, J., et al.: Matching people and jobs: a bilateral recommendation approach. In: Proceedings of the 39th Annual Hawaii International Conference on System Sciences, HICSS 2006, vol. 6. IEEE (2006)
25. https://justworks.com/blog/online-recruiting-tips-finding-candidates-quickly-successfully. Accessed 19 Jan 2017
26. Yang, Y., et al.: Entity matching across heterogeneous sources. In: Proceedings of the 21th ACM SIGKDD International Conference on Knowledge Discovery and Data Mining. ACM (2015)
27. Sun, Y., Han, J.: Mining heterogeneous information networks: a structural analysis approach. ACM SIGKDD Explor. Newsl. **14**(2), 20–28 (2013)
28. https://www.indeed.com
29. https://www.glassdoor.com/index.htm
30. https://www.monster.com

A Temporal Learning Framework: From Experience of Artificial Cultivation to Knowledge

Lin Sun[1(✉)], Zengwei Zheng[1], Jianzhong Wu[1], and JianFeng Zhu[2]

[1] Intelligent Plant Factory of Zhejiang Province Engineering Lab,
Zhejiang University City College, Hangzhou, China
{sunl, zhengzw, wujianzhong}@zucc.edu.cn
[2] College of Computer Science and Technology, Zhejiang University,
Hangzhou, China

Abstract. This paper presents a novel learning framework to generate fine-grained temporal cultivation knowledge from large climatic sensor data. Compared with human-experience based control, the machine-learned cultivation knowledge can provide precise climatic descriptions in temporal domain during the growth of plants. In the paper, the temporal characteristics of the sensor data are analyzed with heat maps in different temporal aspects. A merging algorithm on temporal segments, which are initialized with respect to the regularity of the heat maps, is designed to create climatic labels. Then the training samples consisting of temporal attributes and climatic labels are constructed for knowledge learning, which is represented as a collection of tree-structured classifiers. The experiments are carried out on the cultivation of a valuable Chinese herbal medicine. A cultivation knowledge cube in month, day and hour dimensions is illustrated. The results show that about 80% climatic conditions in the past successful cases can be duplicated to guide the future artificial cultivation by our method. The framework can also be applied to learn the knowledge of cultivation practices for other plants.

1 Introduction

Temporal data are of increasing importance in a variety of fields, such as financial forecasting, social network, geographic information system, healthcare and environmental science. Temporal data mining deals with the discovery of useful information from a large amount of temporal data. The patterns are required to be new, useful, and understand able to humans. Over the last decade many interesting techniques of temporal data mining were proposed and shown to be useful in many applications [1–5].

Precision agriculture is a modern farming practice to improve product quality and labor efficiency [6–8]. In recent years, data-driven precision agriculture, which is strongly connected with data mining techniques [9–12], has a great impact on traditional farming [13]. The task of data mining deals with the harvesting of useful information from sensor data for yield predication, soil classification, food grading, etc. [14].

The advent of Wireless Sensor Networks (WSNs) spurred a new direction of research in agricultural and farming domain. WSNs are used for environmental

© Springer Nature Switzerland AG 2019
S. Li (Ed.): GPC 2018, LNCS 11204, pp. 424–439, 2019.
https://doi.org/10.1007/978-3-030-15093-8_30

monitoring and widely applied in artificial cultivation such as greenhouse and plant factory [15–17]. The optimal ambient temperature and relative humidity to benefit plant growth are usually provided by the experts. This kind of knowledge has two drawbacks:

1. It was coarse-grained in temporal dimension. The smallest temporal granularity of ambient requirements was usually at growth stage level, e.g. seeding, vegetable or fruiting [18]. It was impossible to obtain ambient requirements at granularity level of day or hour from farming experience.
2. It required a lot of cultivation practices over a long period of time to witness many successes and failures to summarize complete expert knowledge.

The goal of our work is to learn fine-grained climate knowledge from temporal sensor data of the past successful cultivations. Our main contributions are summarized as follows:

1. The framework has investigated the characteristics of heat maps at different temporal granularity. The clustering of sensor data proceeds on the temporal segments, which are initialized by referring to the regularity of temporal patterns in the heat maps. After clustering, time series data can be transformed to a new temporal label set for fine-grained temporal knowledge learning.
2. Tree-structured models consisting of temporal attributes are built by the temporal label set. A fine-grained temporal cultivation knowledge cube in month, day and hour dimensions is presented. Farmers without cultivation expertise of Dendrobium officinale can easily implement optimal climate control with the generated knowledge cube.
3. The experiments are carried out on the cultivation of Dendrobium officinale, a very valuable Chinese herbal medicine, which quality is sensitive to the environment at the different growth stages. The experimental results show the validity (81% past artificial cultivation experience can be leaned), novelty (involving a new framework of temporal knowledge learning), usefulness (guidance for cultivation of Dendrobium officinale in future), and understandability (the knowledge cube is understandable to humans) of the method.

2 Related Work

The tasks of temporal data mining are clustering [19], classification [20, 21], pattern analysis [22, 23], association rules [24, 25], prediction [26, 27], outlier detection [5], etc. Nowadays, time-series analysis covers a wide range of real-life problems in various fields of research, such as economic forecasting [28], intrusion detection [29], healthcare [30] and hydrology [31]. Batal et al. [30] proposed a temporal pattern mining approach for classifying complex electronic health record data. Relying on temporal abstraction and temporal pattern, the authors extracted classification task relevant features and demonstrated the usefulness of the framework on patients prediction. Ouyang et al. [31] used K-mean clustering to segment the annual process of the daily discharge and discovered the relationship with temperature and precipitation. Zhong et al. [29] investigated multiple centroid-based unsupervised clustering algorithms for intrusion detection and proposed an effective self-labeling heuristic for detecting attacks.

A new trend of precision agriculture is to emphasize temporal data analysis [32–35]. Franke et al. [32] carried out analysis of the temporal occurrence of wheat diseases based on multi-temporal remote sensing. Mahlein et al. [33] summarized recent advances in precision crop protection and pointed out that precision disease control required temporal information regarding the status of crop growth-relevant parameter. Ibrahim et al. [34] characterized temporal patterns of soil water regarding seasonal precipitation to determine crop yield in dryland. Diacono et al. [35] assessed the temporal variability of attributes related to the yield and quality of durum wheat production, and estimated the spatial dependence as well as their temporal stability. The reason why temporal aspect is concerned in greenhouse control is that the environmental requirement in different stages of the growth cycle of a plant is always temporal variability.

WSNs are used for continuously monitoring some physical phenomenon like temperature, humidity etc. This raw data, if efficiently analyzed and transformed to usable information through data mining, can facilitate automated or human-induced strategic decision. Tripathy et al. [36] developed data mining techniques, such as Naive Bayes classification, rapid association rule and multivariate regression, to turn the data into useful knowledge to understand the relation of crop-weather-environment-disease. Traditional data mining techniques are not directly applicable to WSNs due to the nature of sensor data, their special characteristics [37]. Therefore, new mining paradigms should be designed to achieve good performance for applications.

In recent artificial cultivation system, the knowledge of environmental requirements is always represented as single logical expressions, e.g. temperature should be lower or higher than a threshold. Erazo et al. [38] defined high and low critical range for sensors to avoid disease, e.g., rose bud not developed ($T > 25$ °C), plant tissue death ($T < 4$ °C) or plague putrefaction ($RH > 75\%$). Park et al. [39] proposed the system to prevent dew on the surface of leave of crop. When dew temperature was lower than temperature of leaves, the relay was off. In the work of [40], a warning threshold for temperature or humidity was predefinded. If any abnormal event occurred, warning signal and mail will be sent to remote client. Shamshiri et al. [18] presented the ideal climatic conditions of tomato in five growth stages. Temperature and VPD (Vapor Pressure Deficit) ranges were defined with respect to weather condition (Sunny, Cloudy or Night) and growth stage (Seeding, Vegetable or Fruiting). The weakness of the above knowledge expressions is that it lacks or only has coarse-grained temporal variability. To solve this problem, a temporal knowledge learning framework is proposed in the following section.

3 Temporal Knowledge Learning Framework

The framework for temporal knowledge learning on time series data of WSNs is shown in Fig. 1. The sensor data in greenhouse or plant factory, e.g., temperature or humidity, is collected and in the format of

$$[\#Hourse, DataTime, Temperature|Humidity]. \tag{1}$$

The overall process of temporal data mining involves the following steps:

(1) **Temporal characteristics Analysis** We use 2D heat map, which is the histogram of the sensor value in the different temporal aspects, to illustrated the temporal characteristics of the sensor data. Temporal dimension in heat maps can be hour, day or month. The purpose of this stage is to find out the regularity of heat maps in different temporal aspects, and choose appropriate temporal dimensions to create the initial merging segments. The importance of temporal characteristic analysis is to ensure that the following clustering step will create more effective climatic clusters.

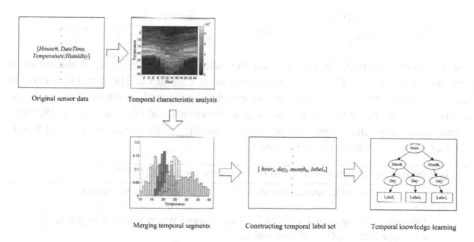

Fig. 1. The proposed temporal knowledge learning framework

(2) **Merging temporal segments** According to the regularity of heat maps in different temporal aspects, time-series data are divided to a collection of temporal segments, e.g. $\Omega = \{C_{<hour_i,month_k>}\}$ if it is regular in the heat maps of hour and month dimensions. Each temporal segment $C_{<>}$ is a set of the values in a specific hour, day or month, e.g. $C_{<hour_i,month_k>} = \{T_{<hour_i,day_j,month_k>}\}$, where $T_{<hour_i,day_j,month_k>}$ is a temperature value at the ith hour of the jth day in the kth month. A merging algorithm on temporal segments (MTS), shown in Sect. 3.1, is proposed to cluster the temporal segments of Ω.

(3) **Constructing temporal label set** After merging temporal segments, the sensor
values are separated into several intervals by Bayesian decision rule, see Sect. 3.2.
The time series data will be transformed into temporal label samples in the form of

$$[hour_i, day_j, month_k, l_n],\tag{2}$$

where l_n is the label of a clustering interval and $T_{<hour_i, day_j, month_k>}$ should be
within the interval of cluster l_n as temperature example.

(4) **Temporal knowledge learning** The learning procedure is performed on the tree
structured model. The nodes of the decision tree are temporal attributes, and the
leaves are climate classification labels, see DT learning block in Fig. 1. The
detailed description is in Sect. 3.3.

3.1   MTS Algorithm

A merging algorithm on temporal segments (MTS) shown in Algorithm 1, is designed
to merge temporal segments of Ω into a few clusters. In MTS algorithm, the merging
criteria $D_{p,q}$ between temporal segments C_p and C_q, where $C_p, C_q \in \Omega$, is defined as,

$$D_{p,q} = \frac{count(C_p)}{count(C_p \cup C_q)}\left|\mu_{C_p} - \mu_{C_p \cup C_q}\right| + \frac{count(C_q)}{count(C_p \cup C_q)}\left|\mu_{C_q} - \mu_{C_p \cup C_q}\right|\tag{3}$$

where function $count(A)$ is to calculate the number of the set A, μ_A is the mean of
elements in set A, $C_p \cup C_q$ denotes the union set of C_p and C_q. The merging criteria in
Eq. (3) is a weighted metric, where the weights correspond to the proportion of the
number of temporal segments C_p and C_q. If the criteria between two sets is the minimum, then merge these two sets. It ensures that the center of the merged clusters is
most close to the centers of two clusters.

Algorithm 1. A merging algorithm on temporal segments (MTS).

---

    input : A collection of temporal segments $\Omega = \{C_k\}$, the number of clusters N.

1 **repeat**

2   Compute merging criteria $D_{p,q}$ of Eq. (3) between any two temporal segments C_p
    and C_q, where $C_p, C_q \in \Omega$ and $p \neq q$;

3   Find minimum criteria $D_{s,t}$ in $\{D_{p,q}\}$;

4   Remove C_s, C_t from Ω and put $C_s \cup C_t$ into Ω, i.e.,
    $\Omega = (\Omega - \{C_s, C_t\}) \cup \{C_s \cup C_t\}$;

5 **until** *The number of the elements in Ω equals to N*;

3.2 Computing Intervals Using Bayesian Decision Rule

The result of MTS algorithm is overlapping clusters, shown in Figs. 5 and 6. Bayesian decision rule [41] is applied to decide which of clusters the sensor value belongs to. Probability of the value s belongs to label l_i, $P(l_i|s)$, is provided by Bayes' Theorem,

$$P(l_i|s) = \frac{P(s|l_i)P(l_i)}{P(s)} \tag{4}$$

$P(s)$ is the evidence of histogram shown in Fig. 2, $P(l_i)$ is the prior probability, the proportion of the cluster l_i to the whole data. $P(s|l_i)$ is the likelihood, the percentage of the value s in the cluster l_i, e.g. the normalized distribution of the temperature value in the same color histogram in Fig. 5(b). The decision rule is defined as,

$$\begin{cases} s \in l_i & \text{if } P(l_i|s) > P(l_j|s) \\ s \in l_j & \text{otherwise} \end{cases} \tag{5}$$

3.3 Decision Tree Learning

The performance of tree ensemble models, e.g. bagging, boosting and random forests (RF), is always better than a single decision tree because it can decrease the variance of the model without increasing the bias [42, 43]. Random forests are more robust with respect to noise [44]. A random forest is defined as a classifier consisting of a collection of tree-structured classifiers $\{ h(X, \Theta_k), k = 1, \ldots, M \}$, where M is the number of trees, $\{\Theta_k\}$ are independent identically distributed random variables and X is input data. Each tree is trained on a bootstrap sample of the training data X. The best variable among Θ_k is picked to split at each node when growing a tree. Finally, the prediction of random forests for classification is voting the predictions of all trees. Random forests do not overfit when more trees are added and produce a limiting value of the generalization error [44].

4   An Artificial Cultivation Experience Learning Example

In the past decade, Dendrobium officinale, a valuable Chinese herbal medicine, has been cultivated in the greenhouse for medicine production. The high quality of Dendrobium officinale always needs daily care of the experts. The challenge of the cultivation of Dendrobium officinale is that the quality of Dendrobium officinale is sensitive to climate effects at the different growth stages [45]. Therefore, the coarse-grained cultivation rules in WSNs monitoring system can not guarantee the quality of Dendrobium officinale stable.

The growth of Dendrobium officinale requires 11-12 months. Dendrobium officinale grows in tissue culture bottle in the first four months and in the following 7–8 months it grows in the greenhouse. In our experiment, it took seven months to collect sensor data in ten greenhouses of Ningbo Feng Kang Biological Technology Co., Ltd in 2015. These seven months were the important growth period of Dendrobium officinale in greenhouse. The quality of Dendrobium officinale in these ten greenhouses was high and close to wild ones. Figure 2 shows the histograms of temperature and humidity on the collected data. There is a strong peak in the distribution of humidity because the relative humidity in the greenhouse always needs to be kept above 80%, since Dendrobium officinale loves moist, humid conditions.

In the following sections, we will discuss how to use the proposed temporal learning framework to construct fine-grained cultivation knowledge.

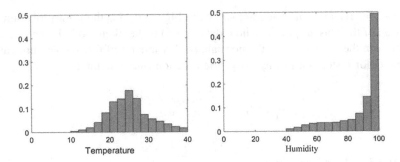

Fig. 2. Histograms of temperature and humidity on the collected data

4.1   Temporal Characteristic Analysis

We investigate temporal characteristics of temperature and humidity in heatmap representation, shown in Figs. 3 and 4, where the vertical axis is temperature or humidity value and horizontal ones are hour, day and month respectively. The colormap represents the percentage of the sensor value in the whole dataset.

The heat maps of temperature in Fig. 3 are regular in hour and month dimensions because temperature at noon is higher than other hours, and in month 7, 8 is higher than other months. In comparison to hour and month dimensions, there is no regularity in day dimension. Therefore, the set of temporal segments, Ω_T, is defined as,

$$\Omega_T = \left\{ C_{<hour_i,month_k>} \right\} \tag{6}$$

$$C_{<hour_i,month_k>} = \left\{ T_{<hour_i,day_j,month_k>}, \forall day_j \right\}. \tag{7}$$

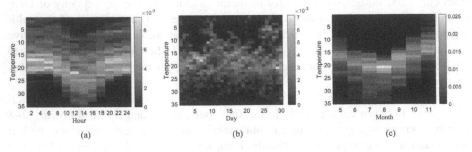

Fig. 3. Temporal heat maps of temperature data

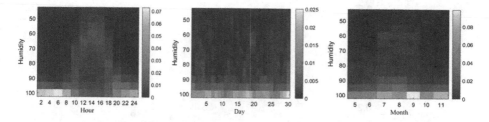

Fig. 4. Temporal heat maps of humidity data

There is no obvious temporal regularity for humidity in day and month dimensions, shown in Fig. 3. However, in hour dimension, humidity is usually high at midnight and in the early morning, drops rapidly, after the sunrises, until it is lowest just after midday. This physical phenomenon occurs because the air humidity usually decreases when the temperature increases. So the set of temporal segments, Ω_H, is defined as,

$$\Omega_H = \{C_{<hour_i>}\} \tag{8}$$

$$C_{<hour_i>} = \{H_{<hour_i, day_j, month_k>}, \forall day_j, \forall month_k\} \tag{9}$$

4.2 Results of MTS Algorithm

There are 168 (24 h × 7 months) temperature temporal segments in Ω initially. The termination condition N for the iteration in Algorithm 1 is set 5 or 4. Figure 5 shows the results of MTS algorithm on temperature data. A color histogram stands for a normalized distribution of a cluster generated by Algorithm 1. Figure 5(a) is the result of $N = 5$ and Fig. 5(b) is that of $N = 4$. Two nearest clusters, yellow and green ones in (a), are merged from five clusters to four clusters. The merging result of four clusters is better than that of five clusters. There are 24 (24 h) temporal segments initially with respect to humidity. Figure 6 shows the result of MTS algorithm on humidity data

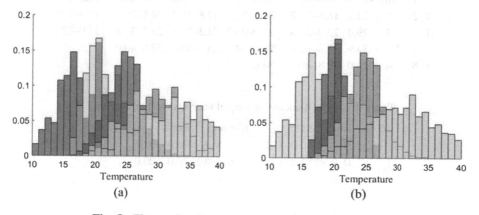

Fig. 5. The results of MTS algorithm on temperature data.

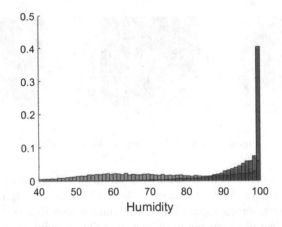

Fig. 6. The result of MTS algorithm on humidity data.

while $N = 2$. Then we use Eq. (5) in Sect. 3.2 to identify which label the temperature or humidity values belongs to. The sensor values are divided into several intervals, shown in the MTS columns of Tables 1 and 2 respectively.

Table 1 shows spatial merging on temporal segments (MTS), K-means and equal intervals on temperature data. MTS_5 intervals are the result of 5 merging clusters using Algorithm 1 and Bayesian decision rules, and MTS_4 intervals are that of 4 clusters. We also cluster the collected data by a classical clustering algorithm, K-means [46]. K-means_5 and K-means_4 intervals in Table 1 show the temperature clustering results of K-means with 5 and 4 clusters respectively. In the temperature histogram of Fig. 2, the temperature value varies from 10 °C to 40 °C. So equal intervals are divided into five or four intervals in [10, 40], shown in Table 1.

Table 1. MTS, K-means and equal intervals on temperature (°C)

$label_T$	MTS_5	K-means_5	5 equal	MTS_4	K-means_4	4 equal
T_1	10.0–17.8	10.0–18.6	10.0–16.0	10.0–17.8	10.0–20.8	10.0–17.5
T_2	17.8–21.2	18.6–23.2	16.0–22.0	17.8–21.2	20.8–25.3	17.5–25.0
T_3	21.2–29.4	23.2–27.8	22.0–28.0	21.2–28.8	25.3–30.8	25.0–32.5
T_4	29.4–34.9	27.8–32.8	28.0–34.0	28.8–40.0	30.8–40.0	32.5–40.0
T_5	34.9–40.0	32.8–40.0	34.0–40.0			

Table 2. MTS, K-means and equal intervals on humidity (%)

$label_H$	MTS	K-means	Equal
H_1	40.0–81.2	40.0–81.6	40.0–70.0
H_2	81.2–100.0	81.6–100.0	70.0–100.0

Table 2 shows MTS, K-means and equal intervals on humidity data, where MTS intervals are the result of 2 merging clusters in Fig. 6 and K-means intervals are the results of K-means clustering. In the humidity histogram of Fig. 2, the humidity value varies from 40% to 100%. So equal intervals are divided into five or two intervals in [40, 100], shown in Table 2.

Finally, the climate classification labels are defined as $<label_T, label_H>$. For example, there are 8 climatic labels if 4 temperature clusters $\times$ 2 humidity clusters. In the next section, we will evaluate the performance of tree-structured classifiers on the different clustering results, which are shown in Tables 1 and 2. The performance comparisons in Sect. 4.3 will testify which the most optimal climatic labels are.

4.3 K-Fold Cross Validation Results

The goal of our work is to duplicate the vast majority of climatic knowledge from the historical successful cultivation cases and reproduce it in the future farming practices. Therefore, how much correct information is learned from the historical cases is an important metric. In the experiments, k-fold cross validation is performed. Ten greenhouses are divided into five groups. Four groups are randomly selected for training and the rest one for testing. The form of samples is shown in Eq. (2).

To evaluate the performance of classifiers, we define the following metrics, true positive rate (TPR) and false positive rate (FPR) [47],

$$TPR = \frac{|TP|}{|TP| + |FN|} \quad FPR = \frac{|FP|}{|FP| + |TN|}, \quad (10)$$

where TP is true positive, FN is false negative, FP is false positive and TN is true negative. TPR measures the proportion of positives that are correctly identified. It means how many climate conditions will reappear in other greenhouses. FPR represents the proportion of positives that are incorrectly identified. It means how many climate conditions will not reappear in other greenhouses. We also compute the AUC (Area Under ROC Curve) values as another metric to evaluate the performance.

Table 3. Performance of MTS in 5-fold cross validation

	TPR	FPR	AUC
10 climatic classes			
C4.5	73.7%	13.6%	0.924
Bagging trees	74.0%	14.4%	0.931
Random forecasts	78.5%	11.1%	0.959
8 climatic classes			
C4.5	76.6%	12.5%	0.926
Bagging trees	76.8%	12.5%	0.934
Random forecasts	**81.2%**	**9.9%**	**0.963**

In decision tree training, the minimum number of instances per leaf is set 200 in prepruning and confidence factor is set 0.25 in pessimistic post-pruning. In random forests, the number of trees is set 100 and one variable is randomly selected from three variables, which are hour, day and month. Table 3 shows the comparisons of C4.5, bagging trees and random forests on MTS intervals by 5-fold cross validation. The performances of 8 climatic classes (4 temperature classes × 2 humidity classes) are better than those of 10 climatic classes (5 temperature classes × 2 humidity classes) in decision tree and tree ensemble models. The performance of random forests outperforms that of C4.5 and bagging trees in both 8 and 10 climatic classes while the performance of bagging trees is close to that of C4.5.

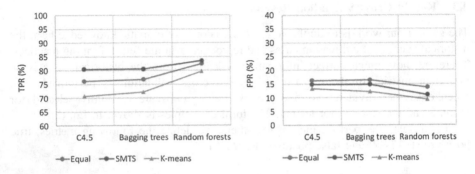

Fig. 7. Comparisons of MTS, K-means and equal intervals on temperature.

In order to investigate the performances of MTS, K-means and equal intervals in detail, experiments have been done on temperature and humidity independently. Figure 7 shows the comparisons of TPR and FPR on 4 temperature intervals of MTS, K-means and equal. Figure 8 shows the comparisons of TPR and FPR on 2 humidity intervals of MTS, K-means and equal. From Fig. 7, we can see that the predication result of MTS intervals is better than those of K-means and equal ones. The performances of MTS in C4.5 and bagging trees are increased approximately 4% on TPR metric and also better on FPR metric in comparison with equal. Although FPR metric of MTS is averagely 1.4% higher than K-means, TPR metric of MTS increased approximately 9% in C4.5 and bagging trees. Meanwhile, the percentage of outliers in MTS 4 is 2.7%, which is less than 3.6% in K-means 4. In Fig. 8, the TPR result of MTS intervals is slightly lower than those of equal ones, but the FPR result of MTS intervals dramatically decreases averagely 14% in comparison with equal ones. The intervals of MTS and K-means on humidity are almost the same, shown in Table 2, so TPR and FPR metrics of MTS and K-means are almost equal.

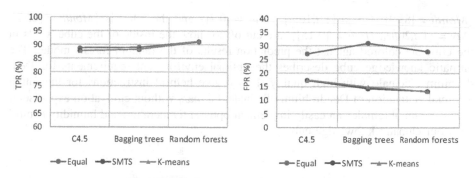

Fig. 8. Comparisons of MTS, K-means and equal intervals on humidity.

4.4 Out-of-Bag Estimate of Random Forests

Given a specific training set X, predictors $\{h_t(X)\}$ are constructed using boost strap samples X_k from training set X. Out-of-bag (OOB) estimate, also called OOB error, is the error rate of the aggregation classifiers, which not contain $x_i \in X$ in their boost strap samples. Out-of-bag estimate can be given for the generalization error of bagged predictors [44]. Figure 9 shows OOB error of random forests in M trees and one random variable. It figures out the approximate number of trees $M = 100$, obtaining the optimal prediction. This explains the reason for the parameter setting of the number of trees and one random variable in Sect. 4.3. OOB error, which is approximately 7.3%, also shows good generalization performance of our method.

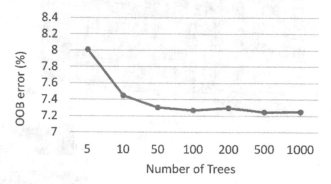

Fig. 9. OOB error of random forests.

4.5 Climatic Knowledge Cube for Dendrobium Officinale Cultivation

At the end of this paper, we present a climatic knowledge cube for Dendrobium officinale cultivation. The cube has three dimensions, which are hour, day and month. The color of each cell represents one class label $< label_T, label_H >$, where $label_T$ is defined in Table 1 and $label_H$ in Table 2. There are 8 climatic classes (4 temperature classes × 2 humidity classes) shown in the colormap of Fig. 10. Two blocks, one is

{*Month* = [8, 9], *Day* = [1*st*,10*th*], *Hour* = [0,7]} and the other is {*Month* = [6, 7], *Day* = [1*st*,20*th*], *Hour* = [0,15]}, are cut off to show the inside of the cube. Cells in the cube are colored against the prediction results of the random forests model. The climatic knowledge cube describes the optimal environment requirements for Dendrobium officinale with high quality. It will give helpful instructions for the future artificial cultivation of Dendrobium officinale. Farmers without cultivation expertise of Dendrobium officinale can easily implement optimal temperature and humidity control with our climatic knowledge cube.

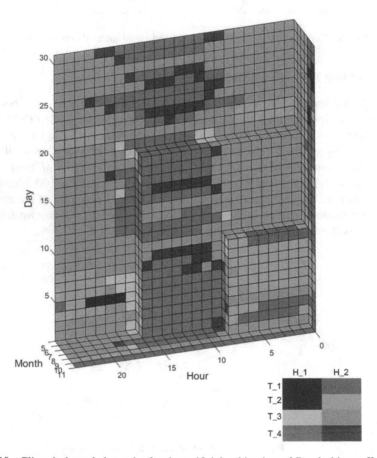

Fig. 10. Climatic knowledge cube for the artificial cultivation of Dendrobium officinale

5 Conclusion

This paper presents a novel data mining framework to generate fine-grained temporal knowledge for Dendrobium officinale cultivation from large environmental sensor data. Compared with human-experience based decision rules, the advantage of our method is that the fine-grained rules can provide precise climate description in temporal domain during the growth of plants. A spatial merging algorithm on temporal segments, which are initialized regarding the temporal patterns in heat maps, is proposed to cluster sensor data. The tree-structured model consisting of temporal attributes is built. The predication results with spatial merging intervals are all better than K-means and equal intervals, increasing TPR approximately 9% and 4% respectively on temperature and decreasing FPR of equal intervals averagely 14% on humidity. C4.5 algorithm, bagging tree and random forests are tested in the experiments and random forests performs with 4 merging intervals performs the best. The results show that about 80% climatic conditions in the past successful cultivation can be used to guide the future artificial cultivation by our fine-grained temporal knowledge. Farmers without cultivation expertise of Dendrobium officinale can implement optimal ambient control with the climatic knowledge cube. Our method can also be extended to construct environmental knowledge for artificial cultivation of other plants.

Acknowledgment. The authors would like to thank Prof. Yong He of Zhejiang University for providing the sensor data. This work is supported by Zhejiang Provincial Natural Science Foundation of China (No. LY17F020008), Hangzhou Science and Technology Development Plan Project (No. 20150432B17, 20162012A06, 20170432B30).

References

1. Mörchen, F.: Time Series Knowledge Mining. Görich & Weiershäuser, Marburg (2006)
2. Mitsa, T.: Temporal Data Mining. Chapman and Hall/CRC, New York (2010)
3. Fu, T.C.: A review on time series data mining. Eng. Appl. Artif. Intell. **24**(1), 164–181 (2011)
4. Esling, P., Agon, C.: Time-series data mining. ACM Comput. Surv. (CSUR) **45**(1), 12 (2012)
5. Gupta, M., Gao, J., Aggarwal, C.C., Han, J.: Outlier detection for temporal data: a survey. IEEE Trans. Knowl. Data Eng. **26**(9), 2250–2267 (2014)
6. McBratney, A., Whelan, B., Ancev, T., Bouma, J.: Future directions of precision agriculture. Precision Agric. **6**(1), 7–23 (2005)
7. Mulla, D.J.: Twenty five years of remote sensing in precision agriculture: key advances and remaining knowledge gaps. Biosyst. Eng. **114**(4), 358–371 (2013)
8. Ojha, T., Misra, S., Raghuwanshi, N.S.: Wireless sensor networks for agriculture: The state of-the-art in practice and future challenges. Comput. Electron. Agric. **118**, 66–84 (2015)
9. Ruß, G., Brenning, A.: Data mining in precision agriculture: management of spatial information. In: Hüllermeier, E., Kruse, R., Hoffmann, F. (eds.) IPMU 2010. LNCS (LNAI), vol. 6178, pp. 350–359. Springer, Heidelberg (2010). https://doi.org/10.1007/978-3-642-14049-5_36
10. Tripathy, A., et al.: Data mining and wireless sensor network for agriculture pest/disease predictions. In: 2011 World Congress on Information and Communication Technologies (WICT), pp. 1229–1234. IEEE (2011)
11. Ruß, G., Kruse, R.: Exploratory hierarchical clustering for management zone delineation in precision agriculture. In: Perner, P. (ed.) ICDM 2011. LNCS (LNAI), vol. 6870, pp. 161–173. Springer, Heidelberg (2011). https://doi.org/10.1007/978-3-642-23184-1_13
12. Ramesh, D., Vardhan, B.V.: Data mining techniques and applications to agricultural yield data. Int. J. Adv. Res. Comput. Commun. Eng. **2**(9), 3477–3480 (2013)

13. Guo, W., Cui, S., Torrion, J., Rajan, N.: Data-driven precision agriculture: opportunities and challenges. In: Soil-Specific Farming, pp. 353–372 (2015)
14. Patel, H., Patel, D.: A brief survey of data mining techniques applied to agricultural data. International Journal of Computer Applications 95(9) (2014)
15. Mohanty, N.R., Patil, C.: Wireless sensor network design for greenhouse automation. Int. J. Eng. Innovative Technol. 3(2), 257–262 (2012)
16. Chaudhary, D., Nayse, S., Waghmare, L.: Application of wireless sensor networks for greenhouse parameter control in precision agriculture. Int. J. Wireless Mobile Netw. (IJWMN) 3(1), 140–149 (2011)
17. Ferentinos, K.P., Katsoulas, N., Tzounis, A., Bartzanas, T., Kittas, C.: Wireless sensor networks for greenhouse climate and plant condition assessment. Biosys. Eng. 153, 70–81 (2017)
18. Shamshiri, R., Ismail, W.I.W.: A review of greenhouse climate control and automation systems in tropical regions. J. Agric. Sci. Appl. 2(3), 176–183 (2013)
19. Rani, S., Sikka, G.: Recent techniques of clustering of time series data: a survey. International Journal of Computer Applications 52(15) (2012)
20. Zhang, X., Wu, J., Yang, X., Ou, H., Lv, T.: A novel pattern extraction method for time series classification. Optim. Eng. 10(2), 253–271 (2009)
21. Pree, H., Herwig, B., Gruber, T., Sick, B., David, K., Lukowicz, P.: On general purpose time series similarity measures and their use as kernel functions in support vector machines. Inf. Sci. 281, 478–495 (2014)
22. Batal, I., Fradkin, D., Harrison, J., Moerchen, F., Hauskrecht, M.: Mining recent temporal patterns for event detection in multivariate time series data. In: Proceedings of the 18th ACM SIGKDD International Conference on Knowledge Discovery and Data Mining, pp. 280–288. ACM (2012)
23. Wang, F., Lee, N., Hu, J., Sun, J., Ebadollahi, S., Laine, A.F.: A framework for mining signatures from event sequences and its applications in healthcare data. IEEE Trans. Pattern Anal. Mach. Intell. 35(2), 272–285 (2013)
24. Liu, X., Zhai, K., Pedrycz, W.: An improved association rules mining method. Expert Syst. Appl. 39(1), 1362–1374 (2012)
25. Nguyen, L.T., Vo, B., Hong, T.P., Thanh, H.C.: Car-miner: an efficient algorithm for mining class-association rules. Expert Syst. Appl. 40(6), 2305–2311 (2013)
26. Noulas, A., Scellato, S., Lathia, N., Mascolo, C.: Mining user mobility features for next place prediction in location-based services. In: 2012 IEEE 12th International Conference on Datamining (ICDM), pp. 1038–1043. IEEE (2012)
27. Ying, J.J.C., Lee, W.C., Tseng, V.S.: Mining geographic-temporal-semantic patterns in trajectories for location prediction. ACM Trans. Intell. Syst. Technol. (TIST) 5(1), 2 (2013)
28. Song, H., Li, G.: Tourism demand modelling and forecasting a review of recent research. Tour. Manag. 29(2), 203–220 (2008)
29. Zhong, S., Khoshgoftaar, T.M., Seliya, N.: Clustering-based network intrusion detection. Int. J. Reliab. Qual. Saf. Eng. 14(02), 169–187 (2007)
30. Batal, I., Valizadegan, H., Cooper, G.F., Hauskrecht, M.: A temporal pattern mining approach for classifying electronic health record data. ACM Trans. Intell. Syst. Technol. (TIST) 4(4), 63 (2013)
31. Ouyang, R., Ren, L., Cheng, W., Zhou, C.: Similarity search and pattern discovery in hydrological time series data mining. Hydrol. Process. 24(9), 1198–1210 (2010)
32. Franke, J., Menz, G.: Multi-temporal wheat disease detection by multi-spectral remote sensing. Precision Agric. 8(3), 161–172 (2007)
33. Mahlein, A.K., Oerke, E.C., Steiner, U., Dehne, H.W.: Recent advances in sensing plant diseases for precision crop protection. Eur. J. Plant Pathol. 133(1), 197–209 (2012)

34. Ibrahim, H.M., Huggins, D.R.: Spatio-temporal patterns of soil water storage under dryland agriculture at the watershed scale. J. Hydrol. **404**(3), 186–197 (2011)
35. Diacono, M., Castrignanò, A., Troccoli, A., DeBenedetto, D., Basso, B., Rubino, P.: Spatial and temporal variability of wheat grain yield and quality in a mediterranean environment: a multivariate geostatistical approach. Field Crops Res. **131**, 49–62 (2012)
36. Tripathy, A., et al.: Knowledge discovery and leaf spot dynamics of groundnut crop through wireless sensor network and data mining techniques. Comput. Electron. Agric. **107**, 104–114 (2014)
37. Mahmood, A., Shi, K., Khatoon, S., Xiao, M.: Data mining techniques for wireless sensor networks: a survey. Int. J. Distrib. Sens. Netw. **9**(7), 406316 (2013)
38. Erazo, M., et al.: Design and implementation of a wireless sensor network for rose greenhouses monitoring. In: 2015 6th International Conference on Automation, Robotics and Applications (ICARA), pp. 256–261. IEEE (2015)
39. Park, D., Cho, S., Park, J.: The realization of greenhouse monitoring and auto control system using wireless sensor network for fungus propagation prevention in leaf of crop. In: Ślęzak, D., Kim, T.-H., Stoica, A., Kang, B.-H. (eds.) CA 2009. CCIS, vol. 65, pp. 28–34. Springer, Heidelberg (2009). https://doi.org/10.1007/978-3-642-10741-2_4
40. Cao, W., Xu, G., Yaprak, E., Lockhart, R., Yang, T., Gao, Y.: Using wireless sensor networking (wsn) to manage micro-climate in greenhouse. In: IEEE/ASME International Conference on Mechatronic and Embedded Systems and Applications. MESA 2008, pp. 636–641. IEEE (2008)
41. Bernardo, J.M., Smith, A.F.: Bayesian theory (2001)
42. Breiman, L.: Bagging predictors. Mach. Learn. **24**(2), 123–140 (1996)
43. Freund, Y., Schapire, Robert E.: A desicion-theoretic generalization of on-line learning and an application to boosting. In: Vitányi, P. (ed.) EuroCOLT 1995. LNCS, vol. 904, pp. 23–37. Springer, Heidelberg (1995). https://doi.org/10.1007/3-540-59119-2_166
44. Breiman, L.: Random forests. Mach. Learn. **45**(1), 5–32 (2001)
45. Juan, A., Ning, Y., Hong, II., ShuYun, L., et al.: Effects of temperature on the growth and physiological characteristics of dendrobium officinale (orchidaceae). Acta Botanica Yunnanica **32**(5), 420–426 (2010)
46. Jain, A.K.: Data clustering: 50 years beyond k-means. Pattern Recogn. Lett. **31**(8), 651–666 (2010)
47. Sokolova, M., Lapalme, G.: A systematic analysis of performance measures for classification tasks. Inf. Process. Manage. **45**(4), 427–437 (2009)

A Recency Effect Hidden Markov Model for Repeat Consumption Behavior Prediction

Zengwei Zheng[1], Yanzhen Zhou[1,2], and Lin Sun[1(✉)]

[1] Hangzhou Key Laboratory for IoT Technology and Application, Zhejiang
University City College, Hangzhou, China
{zhengzw, sunl}@zucc.edu.cn
[2] College of Computer Science and Technology, Zhejiang University,
Hangzhou, China
zhouyanzhen@zju.edu.cn

Abstract. With the rapid development of mobile payment technology in China, people can use smartphone with some mobile payment apps (such as Alipay, WeChat pay and Apple pay etc.) to pay bills instead of paying cash. Some commercial platforms accumulated large transaction date from users' smartphones. In the repeat consumption activities, the final few (or recency) consumption has a great impact on current consumption than long-ago consumption. But traditional HMM can't deal with this recency effect in our repeat consumption case. This paper proposes a modified HMM method based on recency effect to predict the users' repeat consumption behavior. We introduce a factor to represent the different recency effect of different time distance. An empirical study on real-world data sets shows encouraging results on our approach, especially on the consumer group which has the most uncertain consumption behavior.

Keywords: Hidden Markov Model · Recency effect · Repeat consumption

1 Introduction

Nowadays, with the rapid development of mobile payment technology, people can make payment by smartphones apps (such as Alipay, WeChat pay and Apple pay etc.) instead of by cash. Therefore, how to use previous consumption record and model user's repeat consumption behavior and predict which store the user likely to go in future time is very important. The study of consumption behavior is to know the way an individual spends his resources in the process of consuming items. This is an approach that comprises of studies of the items that they buy and the reason for buying and the timing. It is also about where they make the purchase and how frequently. In this paper, we concerned on the repeat consumption, because repeat consumption accounts for a major portion of people's daily consumption behavior, such as shopping at a same fruit shop, eating regularly at a same restaurant.

In the repeat consumption activities, the final few (or recency) consumption has a great impact on current consumption than long-ago consumption. This feature called the recency effect. Hidden Markov Model (HMM) always used on time series problem,

S. Li (Ed.): GPC 2018, LNCS 11204, pp. 440–450, 2019.
https://doi.org/10.1007/978-3-030-15093-8_31

but it treats every training data equally. So, the traditional HMM is not suitable for repeat consumption problem. In this paper, we proposed a modified HMM method based on recency effect to predict the users' repeat consumption behavior. This approach contains a recency effect factor can deal with this recency effect. An empirical study on real-world data sets shows encouraging results on our approach. In the most unpredictable consumer group, the best prediction accuracy of re-HMM is 62.17%, improved 4.22% than traditional HMM.

2 Related Work

Consumption behavior is an approach to know the way an individual spends his resources in the process of consuming items [1]. Consumption behavior analysis is critically extending the domain of behavior analysis and behavioral economics into marketing theory [2]. Early ways of predicting the consumer market's behavior involved time series analysis and other methods based on historical data. Yilmaz et al. [3] develop a prediction model to determine purchase behavior of consumer for remanufactured products. However, more recent research has found that data mining is an increasingly more effective method. Zheng et al. [4] proved that neural networks are consistently accurate with a high probability in predicting customer restaurant preference. Štencl et al. [5] compared several artificial neural network models of accuracy level of consuming behavior forecasted. Li et al. [6] used feature engineering to predict user purchase behavior and won ranks 4th in Ali Mobile Recommendation Competition held in 2015. In the same competition, Yi et al. [7] employed a weight sampling method and achieved the top 10 results. Another data mining approach is the decision tree which have been used by Fokin [8] to predict the user behavior in the online consumer market.

But the study of repeat consumption behavior is a bit different of consumption behavior, it focusses on predicting whether or not the user will repeat consume items which he has consumed in previous time. Many previously work studied the problem of people's repeat consumption behavior. Ashton and Ravi et al. [9] proposed a hybrid model combine Quality model and Recency model to model people's repeat consumption behavior. Christina and Lars [10] developed the multinomial SVM (Support Vector Machine) item recommender system MN-SVM-IR to calculate personalized item recommendation for a repeat-buying scenario. Chen et al. [11] proposed method factorizes the temporal user-item interaction via learning the mappings from the behavioral features in observable space to the preference features in latent space, then combines users' static and dynamic preferences together in recommendation. Besides, there are many other methods can be used in this filed. However, those methods still have some shortcoming, we proposed a method based on time weight Hidden Markov Model to predict the consumer behavior.

HMM is a powerful statistical tool for modeling generative sequences that can be characterized by an underlying process generating an observable sequence. It's one of the most basic and extensively used statistical tools for modeling the discrete time series. Thus, HMM often used in time series data model and prediction. HMM were introduced in the beginning of the 1970's as a tool in speech recognition and become increasingly popular in the recent years because of its strong mathematical structure and theoretical. Garcia et al. [12] proposed a trip destination prediction method based in past GPS log using a HMM. Gupta and Dhingra [13] presented HMM approach for forecasting stock price compared with ANN and Arima. Raghavan et al. [14] proposed a coupled HMM for user activity in social network and validate the model using a significant corpus of user activity traces on Twitter. Si et al. [15] introduced a technology of mobility prediction base on HMM to improve communication performance in cellular. Ridi et al. [16] used HMM for human activity recognition with an accuracy rate of 84.6%. Mathew et al. [17] presented a hybrid method for predicting human mobility on the basis of HMM. Nowadays, the HMM is still successfully be used in many other research domains such as natural language processing, DNA sequence analysis, handwritten characters recognition etc.

3   Methodology

In this section, we propose a prediction model based on recency effect Hidden Markov Model to predict the repeat consumption behavior that which store the customer most likely to repeat consume in the future time.

HMM is the simplest dynamic Bayesian network, and it's a finite state machine which has some fixed number of state [18]. This model is a well-known directed probabilistic graphical model provides a probabilistic framework for modelling a time series of multivariate observations. It provides a simple Markov process is shown in Fig. 1, the variables in a HMM can be divided into two groups. $\{s_1, s_2, ..., s_n\}$ is the state variable, where s_i represent the state at time i. It is often assumed that state variable is hidden and cannot be observed, so state variable is also called hidden variable. $\{o_1, o_2, ..., o_n\}$ is the observed variable, where o_i represent the observed value at time i.

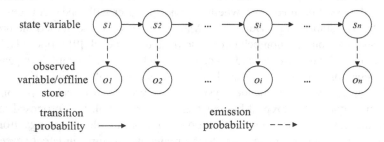

Fig. 1.   The graph structure of HMM

For the practical prediction question in this article, the observed variable is the different store that the consumer had consumed. So, the observed sequence is the store sequence that user consumed at the previous time.

The arrows in Fig. 1 indicate the dependencies between variables, the solid line arrow represents the transition probability and the dotted line arrow represent the emission probability. At any moment, the value of the observed variable x_t only determined by the state variable y_t, independent of other state variables and observations. Meanwhile, y_t, the state at time t, depends only on y_{t-1} which is the state at time $t - 1$, and independent of the other $n - 2$ state. This dependency relationship called the Markov chain that the state of next time is determined only by the current state and independent of any previous state. Based on the Markov chain, the joint probability distribution for all variables is shown as

$$P(x_1, y_1, \ldots, x_n, y_n) = P(y_1)P(x_1|y_1)\prod_{i=2}^{n} P(y_i|y_{i-1})P(x_i|y_i) \qquad (1)$$

In the real repeat consumption activity, we consider the people's memory feature that people may forget which restaurant they have been visited a year ago, but we can easily remember which they have been visited last weekend. So, we think the recency consumption activities have the bigger impact on the current consumption than the early consumption activities. This phenomenon is called recency effect. For this reason, we proposed the recency effect Hidden Markov Model(re-HMM) which include a recency effect factor $\omega(\Delta t)$. This recency effect factor $\omega(\Delta t)$ represent the impact of the previous consumption Δt time ago on the current consumption. According to the recency effect, factor $\omega(\Delta t)$ must a monotonically decreasing function. After introduced the recency effect factor $\omega(\Delta t)$, the forward-backward algorithm which used in parameters training process and prediction process can be extended, shown in Fig. 2. And the traditional HMM is a special case of re-HMM, if the recency time effect factor $\omega(\Delta t)$ are equal.

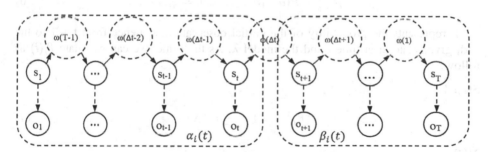

Fig. 2. The forward-backward algorithm in re-HMM

In the new forward-backward algorithm, the forward variable defined as

$$\alpha_t(i) = P(o_1 o_2 \ldots o_t, i_t = q_i|\lambda) \qquad (2)$$

represent the probability of the partial observation sequence $o_1 o_2 \ldots o_t$ (until time t) and state q_i at time t, given the model λ. We can calculate $\alpha_t(i)$ as follow:

Step 1.

$$\alpha_1(i) = \pi_1 b_j(o_1), (1 \leq j \leq N) \qquad (3)$$

where π_i is initial state probability represent the probability of being in state i at the beginning of the experiment, and $b_j(o_k). = P(x_t = o_k \mid y_t = s_j)$ $(1 \le j \le N, 1 \le k \le M)$ is emission probability represent the probability of observing o_k at state s_j.

Step 2.

$$\alpha_{t+1}(j) = \left[\sum_{i=1}^{N} \alpha_{t+1}(i) a_{ij} \omega(T-t)\right] b_j(o_{t+1}),$$

$$(1 \le t \le T - 1, 1 \le j \le N).$$

$$(4)$$

where $a_{ij} = P(y_{t+1} = s_j \mid y_t = s_i)$ is the transition probabilities represent probability of transition to state s_j at time $t + 1$ from state s_i at time t.

Step 3.

$$P(O|\lambda) = \sum_{i=1}^{N} \alpha_T(i) \tag{5}$$

Step 1 initiates the forward probabilities with the joint probability of state si and initial observation $o1$. Step 2 inductively calculate $\alpha t + 1(j)$ by $\alpha(i)$, and shows how the states sj is reached at time $t + 1$ from the N possible states si. Finally, Step 3 calculate $P(O|\lambda)$ as the sum of the terminal forward variables $\alpha T(i)$.

Then, the backward variable defined as

$$\beta_t(i) = P(o_{t+1}o_{t+2}\ldots o_T|i_t = q_i, \lambda) \tag{6}$$

It represents the probability of the partial observation sequence from t + 1 to the end, given state qi at time t and the model λ. As the same, we can calculate $\beta_t(i)$ as follow:

Step 1.

$$\beta_T(i) = 1, (1 \le i \le N) \tag{7}$$

Step 2.

$$\beta_t(i) = \sum_{j=1}^{N} a_{ij} \omega(T - t - 1) b_j(o_{t+1}) \beta_{t+1}(j),$$

$$(t = T - 1, T - 2, \ldots, 1, 1 \le i \le N)$$

$$(8)$$

Step 3.

$$P(O|\lambda) = \sum_{i=1}^{N} \pi_i b_i(o_1) \beta_1(i) \tag{9}$$

Step 1 arbitrarily defines $\beta T(i)$ to be 1 for all i. Step 2 inductively calculate $\beta t(i)$ by $\beta t + 1(j)$, and shows that in order to have been in state si at time t, and to account for the observation sequence from time t + 1 on, we have to consider all possible states sj at time t + 1. Step3 calculates $(O|\lambda)$ as the sum of the terminal forward variables $\beta 1(i)$.

With the forward variables $\alpha_t(i)$ and backward variables $\beta_t(i)$, the probability of observation sequence $O\{o_1 o_2 \ldots o_T\}$ and given the model $\lambda = (A, B, \pi)$ can be calculated

$$P(O|\lambda) = \sum_{i=1}^{N} \alpha_t(i)\beta_t(i) = \sum_{i=1}^{N} \sum_{j=1}^{N} \alpha_t(i)a_{ij}\omega(T-t-1)b_j(o_{t+1})\beta_{t+1}(j)$$

$$(10)$$

4   Experiment

4.1   Dataset

The dataset we used in this study is a part of the data source that used in Tianchi big data contest [20]. It's the consumption record of consumer to use Alipay at offline store. This dataset includes 2000 shops in different city over the country. The dataset time covers from July 1st 2015 to October 31th 2016. We select 1057 consumer who consumed more than 120 times and consumed more than 3 different stores.

Due to different consuming habits of different consumers, we calculated the information entropy of user's consumption sequence as shown as

$$H(x) = E[I(x_i)] = E[\log_2(1/p(x_i))] = -\sum p(x_i) \log_2(1/p(x_i)),$$
$$(i = 1, 2, \ldots, n)$$

$$(11)$$

where $p(x_i)$ represent the probability of random variables event x_i.

The information entropy can be used to measure the uncertainty of random variables events. If the consumer always consumption in a same store, the consumption sequence will be certainty absolutely, so the information entropy will be very close to 0. If the user visit 2 stores equally, the information entropy will be 1; if the user visit 4 stores equally, the information entropy will be 2. The distribution of consumption sequence information entropy is shown in Fig. 3. Then we divide the all consumer to 3 groups by the information entropy: Group1 include 448 consumers, information

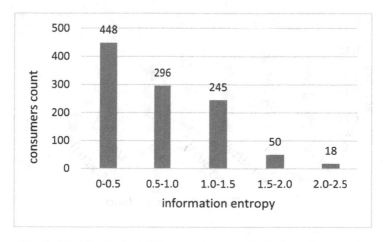

Fig. 3. The distribution of consumption sequence information entropy

entropy between 0 to 0.5, the consumer in this Group almost visit a same store, so their consumption behavior is highly certain; Group2 include 296 consumers, their information entropy between 0.5 to 1.0, so their consumption behavior is middle certain; and Group3 include 313 consumers, information entropy greater than 1.0, their consumption behavior is most uncertain.

4.2 Recency Effect Factor Parameters Choosing

We selected several different monotonically decreasing functions as the time weight parameters in order to find the best function can fit our data. We implemented the experiment in group3 with training length 100, and the result is shown in Table 1 and Fig. 4. With the comparison of those function, we found $exp(-\Delta t/10)$ got the best performance, then we think this function can best suited our data. Therefore, we chose $exp(-\Delta t/10)$ as the final recency time factor function to implement the experiment in the next sub-section.

Table 1. The accuracy of re-HMM in 3 groups with different effect factor functions

$\omega(\Delta t)$	Group1	Group2	Group3	Average
$exp(-\Delta t)$	0.9587	0.8147	0.5993	0.8119
$exp(-\Delta t/5)$	0.9618	0.8153	0.6093	0.8164
$exp(-\Delta t/10)$	**0.9635**	**0.8172**	**0.6168**	**0.8199**
$exp(-\Delta t/15)$	0.9605	0.8143	0.6125	0.8165
$1/\Delta t$	0.9608	0.8123	0.6098	0.8153
$0.1/\Delta t$	0.9580	0.8071	0.4094	0.7533
$1/\log(\Delta t+1)$	0.9572	0.8063	0.3289	0.7289
$T-\Delta t$	0.9513	0.8059	0.3051	0.7192

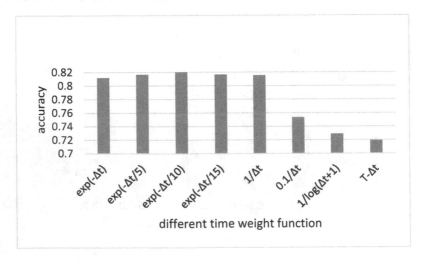

Fig. 4. Performance of re-HMM in 3 groups with different recency effect functions

4.3   Result Analysis

We compare the performance of the proposed method with some baseline. These methods are:

MF: We considerate the consumption frequency is a particular aspect of people's consumption behavior, so we choose the most frequent as the baseline of our data experiment.

HMM: The traditional HMM method to predict the repeat consumption behavior.

re-HMM: The recency effect HMM method we proposed in this paper.

The prediction result of above mentioned methods is shown in Table 2 and Fig. 5.

Table 2. The prediction accuracy of three different methods

	Group	MF	HMM	re-HMM
Train-length=80	1	**0.9529**	0.9526	**0.9529**
	2	**0.8199**	0.8171	0.8186
	3	0.5747	0.5921	**0.6120**
Average		0.8037	0.8079	**0.8143**
Train-length=60	1	**0.9558**	0.9552	0.9557
	2	0.8252	0.8264	**0.8299**
	3	0.5751	0.5942	**0.6119**
Average		0.8065	0.8122	**0.8187**
Train-length=40	1	0.9576	**0.9579**	0.9574
	2	0.8354	0.8364	**0.8372**
	3	0.5868	0.5903	**0.6174**
Average		0.8136	0.8150	**0.8231**
Train-length=20	1	**0.9597**	0.9592	0.9593
	2	**0.8444**	0.8400	0.8424
	3	0.5909	0.5965	**0.6217**
Average		0.8182	0.8184	**0.8266**

As the Table 2 and Fig. 5 shows, the re-HMM gain the best prediction accuracy in the data experiment. The Group1 gain the highest prediction accuracy, and the Group3 gain the lowest accuracy, which means the data in Group1 is easy to prediction and Group3 is hard, because Group1 has the highest information entropy and Group3 has the lowest. For the entire dataset, the re-HMM got better performance than the traditional HMM and won the best performance among the three methods. But due to the consumption behavior of consumers in Group1 and Group2 are very certain, the performances of the three methods is not much difference. And because of consumption behavior in Group3 consumer is most uncertain, the three different methods got the largest performance difference in this group.

The training length will affect the performance of prediction method. As the Fig. 6 shows, we select 4 different training length (20, 40, 60, 80) to estimate the data in Group3. The re-HMM gain the highest accuracy when the training length is 20.

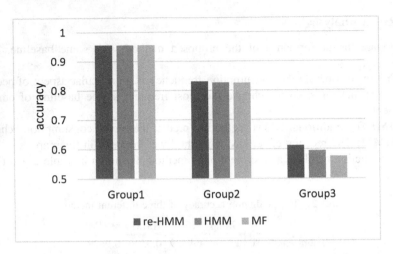

Fig. 5. Comparison of three methods in 3 groups when training length is 60

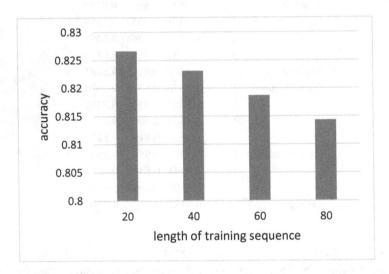

Fig. 6. The prediction accuracy of RE-HMM on different training length (Group3)

Besides, according to Table 2, we can find that all three methods get the best performance when the training length is 20, and prediction accuracy almost decrease when the training length increase, because the longer training sequence can introduce more noise data.

5   Conclusion

In this paper, we propose a prediction framework that based on re-HMM to predict the customer's repeat consumption behavior. This method contains a recency time effect factor $\omega(\Delta t)$ to match the time dependency of repeat consumption problem. The data experiment implement on the dataset of 2017 Ali Tianchi big data contest. We compare the method with traditional HMM and most frequent(MF). The result shows that our re-HMM get better performance than the other two methods. In the most unpredictable group, the re-HMM got best accuracy 62.17% when training length is 20, improved 4.22% than traditional HMM and 5.21% than MF. And we find the length of training data can will affect the performance of our prediction method. Shorter training data can achieve higher prediction accuracy. Therefore, to a certain extent, our research has practical significance for predicting the repeat consumption behavior.

Acknowledgment. This work is supported by Zhejiang Provincial Natural Science Foundation of China (No. LY17F020008).

References

1. Schiffman, L., O'Cass, A., Paladino, A., et al.: Consumer Behaviour. Pearson Higher Education, AU (2013)
2. Foxall, G.R.: Intentionality, symbol, and situation in the interpretation of consumer choice. Mark. Theory **13**(1), 105–127 (2013)
3. Yilmaz, K.G., Belbag, S.: Prediction of consumer behavior regarding purchasing reman-ufactured products: a logistics regression model. Int. J. Bus. Soc. Res. **6**(2), 01–10 (2016)
4. Zheng, B., Thompson, K., Lam, S.S., et al.: Customers' behavior prediction using artificial neural network. In: Proceedings of the IIE Annual Conference on Institute of Industrial and Systems Engineers (IISE), p. 700 (2013)
5. Štencl, M., Popelka, O., Šťastný, J.: Forecast of consumer behaviour based on neural networks models comparison. Acta Universitatis Agriculturae et Silviculturae Mendelianae Brunensis **60**, 437–442 (2012)
6. Li, D., Zhao, G., Wang, Z., et al.: A method of purchase prediction based on user behavior log. In: IEEE International Conference on Data Mining Workshop, pp. 1031–1039. IEEE Computer Society (2015)
7. Yi, Z., Wang, D., Hu, K., et al.: Purchase behavior prediction in m-commerce with an optimized sampling methods. In: IEEE International Conference on Data Mining Workshop, pp. 1085–1092. IEEE Computer Society (2015)
8. Fokin, D., Hagrot, J.: Constructing decision trees for user behavior prediction in the online consumer market (2016)
9. Benson, A.R., Kumar, R., Tomkins, A.: Modeling user consumption sequences. In: International Conference on World Wide Web International World Wide Web Conferences Steering Committee, pp. 519–529 (2016)
10. Lichtenthäler, C., Schmidt-Thieme, L.: Multinomial SVM item recommender for repeat-buying scenarios. In: Spiliopoulou, M., Schmidt-Thieme, L., Janning, R. (eds.) Data Analysis, Machine Learning and Knowledge Discovery. SCDAKO, pp. 189–197. Springer, Cham (2014). https://doi.org/10.1007/978-3-319-01595-8_21

11. Chen, J., et al.: Recommendation for repeat consumption from user implicit feedback. IEEE Trans. Knowl. Data Eng. **28**(11), 3083–3097 (2016)
12. Alvarez-Garcia, J.A., Ortega, J.A., Gonzalez-Abril, L., et al.: Trip destination prediction based on past GPS log using a Hidden Markov model. Expert Syst. Appl. **37**(12), 8166–8171 (2016)
13. Gupta, A., Dhingra, B.: Stock market prediction using Hidden Markov Models. In: Engineering and Systems, pp. 1–4. IEEE (2012)
14. Raghavan, V., Ver Steeg, G., Galstyan, A., et al.: Coupled hidden markov models for user activity in social networks. In: Multimedia and Expo Workshops (ICMEW), pp. 1–6. IEEE (2013)
15. Si, H., Wang, Y., Yuan, J., et al.: Mobility prediction in cellular network using hidden markov model. In: 7th Consumer Communications and Networking Conference (CCNC), pp. 1–5. IEEE (2010)
16. Ridi, A., Zarkadis, N., Gisler, C., et al.: Duration models for activity recognition and prediction in buildings using Hidden Markov Models. In: IEEE International Conference on Data Science and Advanced Analytics, pp. 1–10. IEEE (2015)
17. Mathew, W., Raposo, R., Martins, B.: Predicting future locations with hidden Markov models. In: ACM Conference on Ubiquitous Computing, pp. 911–918. ACM (2012)
18. Rabiner, L., Juang, B.: An introduction to hidden Markov models. IEEE ASSP Magazine **3**(1), 4–16 (1986)
19. Rabiner, L.R.: A tutorial on hidden Markov models and selected applications in speech recognition. Proc. IEEE **77**(2), 257–286 (1989)
20. Tianchi big data contest. https://tianchi.aliyun.com/competition/information.htm?spm=5176. 100067.5678.2.8cfDIU&raceId=231591

Forecast of Port Container Throughput Based on TEI@I Methodology

Qingfei Liu, Laisheng Xiang, and Xiyu Liu[(✉)]

College of Management Science and Engineering, Shandong Normal University,
East of Wenhua Road No. 88, Jinan 250014, Shandong, China
liuqingfei1201@163.com, xls3366@163.com,
sdxyliu@163.com

Abstract. Forecasting container throughput accurately is crucial to the success of any port operation policy. At present, prediction of container throughput is mainly based on traditional time series analysis or single artificial neural network technology. Recent study shows that the combined forecast model enjoys more precise forecast result than monomial forecast approach. In this study, a TEI@I hybrid forecasting model is proposed, which is based on ARIMA (autoregressive integrated moving average model) and BP neural network. Under the proposed framework, ARIMA model can be first used to predict linear component, then using BP neural network to predict the error of ARIMA model which is the nonlinear component. The new method is applied to forecasting the container throughput of Qingdao Port, one of the most important ports of China. The empirical results show that this prediction method has higher prediction accuracy than the single prediction method.

Keywords: TEI@I methodology · BP neural network · ARIMA model · Container throughput · Forecasting

1 Introduction

Forecasting container throughput has an enormously influence on decision-making such as port layout, berth reconstruction, operational and development strategies. Many efforts have been made to the development of new models which are able to analyze and predict container throughput [1–5]. There are many prediction time series methods. Such as regression analysis, exponential smoothing and time series analysis [6, 7]. However, each of these prediction methods has their own characters which all lead to low precision of container throughput forecasting. ARIMA widely used as an effective linear model [8], BPNN famous as a powerful nonlinear model [9]. Constructed on the foundation of BP neural network, the combination forecast model performs well in time series forecasting and solves the problem excellently.

This paper proposes a new integrated forecasting framework to predict port container throughput based on TEI@I methodology, which can effectively solve many problems in complex systems such as house price forecasting [10], inflation forecasting [11], grain Production forecasting [12] and crude oil price forecasting [13] etc. In order to explain the methodology and to verify the effectiveness of the integrated forecasting

© Springer Nature Switzerland AG 2019
S. Li (Ed.): GPC 2018, LNCS 11204, pp. 451–461, 2019.
https://doi.org/10.1007/978-3-030-15093-8_32

framework, a case study of empirical analysis to Qingdao port's container throughput is presented.

The rest of this paper is organized as follows. Section 2 presents a general description of TEI@I methodology and forecasting framework. Main methods used in the framework are introduced in Sect. 3. Section 4 presents a case study of Qingdao port's container throughput forecasting using this framework. Finally, the conclusions and future research directions are discussed in Sect. 5.

2 TEI@I Based Hybrid Forecasting Model

As a new methodology for studying complex systems, TEI@I is used by Qian [14]. And further developed by Wang [15]. It can be applied to analyze the complicated systems which are emergent, unstable, nonlinear and uncertain. The name TEI@I is based on the integration (@ integration) of text mining + econometrics + intelligence (intelligent algorithms). Using "@" to replace "+" is to emphasize the central role of integration. In this methodology, econometric models are used to model linear components and artificial intelligence technologies are used to model nonlinear components. So it can combine the advantages of both models. In addition, the effects of infrequent and irregular events are explored by text mining technique and expert system technique.

Figure 1 shows the basic forecasting framework within TEI@I methodology. Given the fluctuation characteristic of container throughput, this paper uses ARIMA model to capture the main trends and computes the simulated values. Then the prediction error is adjusted by BP neural networks model to get the more accurate forecast result. The process of analyzing the effects of irregular events is basically the same as Wang et al. [11], so we do not give the details here. At last, the final prediction result is obtained by integrating above forecast results as shown in Fig. 1.

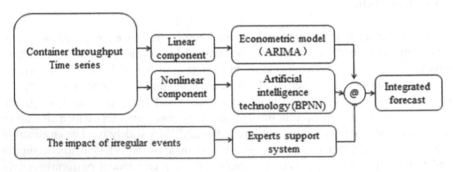

Fig. 1. Container throughput forecasting framework based on TEI@I

3 Model Introduction

3.1 ARIMA Model

The ARIMA model is introduced by Box-Jenkins, which is the autoregressive integrated moving average model. ARIMA model is a high precision linear time series forecasting methods. An ARIMA model typically consists of three parts i.e. auto regression AR (order p), moving average MA (order q) and differencing in order to strip off the integration of the series (order d) and then form ARIMA(p, d, q). The general expression of ARIMA model is

$$\phi(B)(1 - B)^d y_t = \theta(B)\varepsilon_t \qquad (1)$$

Where y_t is the non-stationary series

$$\phi(B) = 1 - \phi_1 B - \phi_2 B^2 - \ldots - \phi_P B^P$$

$$\theta(B) = 1 - \theta_1 B - \theta_2 B^2 - \ldots - \theta_q B^q$$

d is the differential order; ε_t is the white noise sequence. The modeling process of ARIMA time series forecast model can be expressed as follows:

(1) Checked for Stationarity. If they are not stationary, a differencing operation is performed. If the data are still nonstationary, differencing is again performed until the data are finally made stationary. If the differencing is performed d times, the integration order of the ARIMA method is said to be d.
(2) Identifying the model order or identifying p and q; Identifying the orders p, q is done using correlation analysis. Usually there are 4 kinds of methods for fixing order of ARIMA(p, d, q) model, method of sample auto correlation function (ACF) and partial correlation function (PACF), method of final prediction error (FTP), method of minimum Akaike Information Criterion (AIC), and method of minimum Schwarz Criterion. Here mainly use the method of sample ACF and PACF, AIC and SC to determine the order p and q.
(3) Model Testing: The model parameters are estimated. If the model error is the white noise, the model passed the test, otherwise the order and parameter estimation should be refreshed.
(4) Prediction: Forecasting by the selected model with appropriate parameters.

3.2 BP Neural Networks Model

Figure 2 shows the structure of a typical BP neural networks model, which includes three layers: input layer, hide layer and output layer. Each layer includes several nodes, which represent a neuron, upper node and lower node was connected by the weight, nodes between layers were connected with the form of whole internet, and there is no connection between nodes within each layer as shown in Fig. 2.

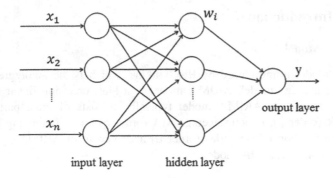

Fig. 2. The structure of BP neural networks.

Artificial Neural Network (ANN) is an important branch of intelligent control, which provide a new thought for modeling and controlling uncertainty nonlinear systems because of its favorable nonlinear mapping capability and self-adaptive learning capability [16]. BP neural networks is a multilayer feedforward network, which can map any complicated nonlinear relationship, more than this the learning rule is simple. The input value is $x_i(i = 1, 2, ..., n)$, $w_i(i = 1, 2,... n)$ is the weight coefficient. θ is the threshold. The relationship between output and input of the neuron is expressed as follows:

$$net = \sum_{i=1}^{n} w_i x_i - \theta \tag{2}$$

$$y = f(net) \tag{3}$$

Where y is the neuron output, net is called net activation. f is called activation function, there are many kinds of activation function, such as sigmoid, tanh, relu, maxout, and some variants of relu such as leaky relu, elu. Activation function commonly used the parameter of S type, such as

$$f(x) = 1/(1 + e^{-ax}) \tag{4}$$

where a is the S type hyperparameter of adjusting the form of activation function.

4   Case Study

The Port of Qingdao is located on the west of the Yellow Sea and at the entrance to Jiaozhou Bay. Qingdao Port is an important international trade port and maritime transport hub on the west coast of the Pacific Ocean. New Qingdao port contains Dagang, Huangdao, Qianwan and Dongjiakou. Among them, Qianwan specialises in container and bulk shipping, with the capacity to handle a large vessel up to 18,000 TEUs 3E container ship and took 80% of the throughput of Qingdao Port. From 2007, Qingdao started to rank among the top ten container ports in the world. In 2016, the port reached a total throughput of 17.47 million TEUs in container shipping.

4.1 Data Description and Evaluation Criteria

Figure 3 shows the monthly Qingdao port's monthly container throughput data used in this study. From the Fig. 3 we can find that Qingdao Port's container throughput is not stable in the level.

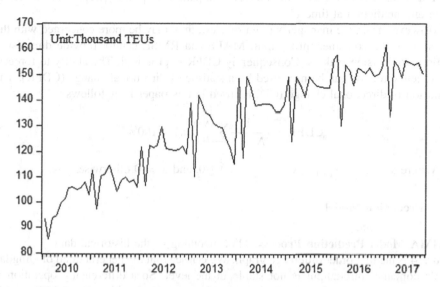

Fig. 3. Qingdao Port's container throughput during Jan. 2010–Oct. 2017.

The monthly data of Qingdao Port's container throughput used in this study are from CEIC macroeconomic database (http://www.ceicdata.com), covering the period from January 2010 to October 2012, with a total of 94 observations, as illustrated in Fig. 3. The data from January 2010 to December 2016 are used as the training data set for training purpose and the remainder from January 2017 to October 2017 are used as the test data set to evaluate the forecasting performance of prediction based on some evaluation criteria.

4.2 Performance Measures

For comparison, three evaluation criteria are used in this section, mean absolute percentage error (MAPE), root mean squared error (RMSE) and the correct direction Forecast rate (CDFR). The following performance measures are used:

Mean Absolute Percentage Error (MAPE)

$$\text{MAPE} = \frac{1}{N} \sum\nolimits_{i=1}^{N} |\frac{y_i - y_i'}{y_i}| \tag{5}$$

Root Mean Squared Error (RMSE)

$$\text{RMSE} = \sqrt{\frac{1}{N}\sum_{i=1}^{N}(y_i - y_i')^2} \tag{6}$$

The symbol N is the total number of data patterns. y_i and y_i' represent the actual value and prediction at time i.

However, from the manager's point of view, they may be more concerned with the direction of the container throughput. MAPE and RMSE cannot provide direct suggestions to decision makers. Consequently CDFR is proposed. The ability to forecast movement direction can be measured by a statistic of directional change (CDFR). The definition of directional change (CDFR) used in this paper is as follows

$$\text{CDFR} = \left(\frac{1}{N-1}\sum_{i=1}^{N-1} a_t\right) \times 100\% \tag{7}$$

Where $a_t = 1$, if $(y_{i+1} - y_i)(y_{i+1}' - y_i) \geq 0$, and $a_t = 0$ otherwise.

4.3   Prediction Model

ARIMA Model Prediction Process. (1) Smoothing of the historical data.
Figure 4 is the first order difference graph. We can see that the general trend of Qingdao Port's container throughput is not stable in the level. So a differencing operation is performed as shown in Fig. 4 which shows the original variables are basically stable after he first order difference, so set the parameters of ARIMA model d = 1.

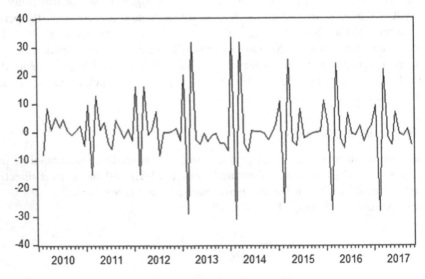

Fig. 4. The graph of first order difference

(2) Determination of p and q for ARIMA model.

Table 1 shows the AIC value and SC value of different ARIMA model. The best-suited ARIMA model has the minimum value of the AIC and SC.

Table 1. The comparison of ARIMA(p, 1, q) models

Number	Model	AIC value	SC value
1	ARIMA(1, 1, 1)	6.946755	7.055684
2	ARIMA(1, 1, 2)	6.965321	7.101482
3	ARIMA(2, 1, 1)	6.965141	7.101302
4	ARIMA(2, 1, 2)	6.986268	7.149662
5	ARIMA(3, 1, 1)	6.955640	7.079033
6	ARIMA(3, 1, 2)	7.035077	7.198470

Several models have been analyzed, and try to find the optimal model. Generally, the relative optimal model has the smallest value of AIC and SC by comparing the AIC and SC criterion. By comparison, the ARIMA(1, 1, 1) model is the optimal model. And the values of AIC and SC are shown in Table 1.

(3) Application of ARIMA(1, 1, 1) forecast model for Qingdao Port's container throughput prediction.

Using ARIMA(1, 1, 1) model to forecast the Qingdao Port's container throughput, statistical data is between January 2010 to December 2017. The fitted result is shown in Fig. 5. As shown in this picture, ARIMA model can well fit the linear trend of the time series. The prediction error of ARIMA model can be seen as nonlinear component, which can be simulated by BP neural network.

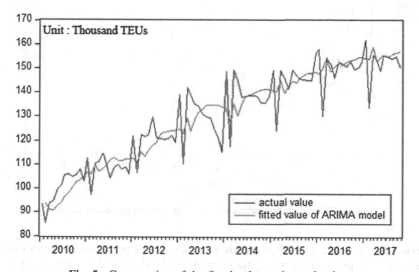

Fig. 5. Comparation of the fitted value and actual value

The Process of Forecasting the Error of ARIMA Model by BP Neural Network.
In this paper, BP neural network is used to forecast the error sequence predicted by
ARIMA model. Due to the prediction error of ARIMA model is Feb. 2010 to Oct.
2017, so the capacity of the total sample of BP neural network is 93. The forecast error
data of February to April in 2010, March to September in 2010, …, July to September
in 2017 as the input of network, data of May in 2010, June in 2003, …, October in
2017 as the ideal output, So the input neurons of BP network is 3, the output neurons is
1, and the neuron number of middle layer is determined by comparing the optimization.
So 93 data can be divided into 90 sets data. The data of first 80 groups is as training
sample output, the data of other 10 groups is used to verify. The training model uses
3-layer neural network, and hidden layer node is determined to 3 by trial method,
namely the network structure is 3-3-1. And parameters are set up such as learning rate
is 0.1, the maximum training step is 2500, training error goal is 0.0001, so as to
establish BP neural network.

Figure 6 shows the training results of BP neural network on the prediction error of
ARIMA model from Jan. 2017 to Oct. 2017, which shows that BPNN have a good
non-linear fitting ability.

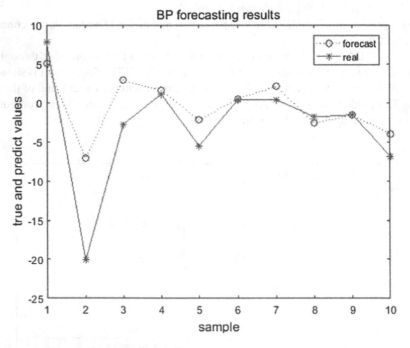

Fig. 6. The forcasting result of irregular component using BPNN

Integrate Result with TEI@I Methodology. Up to this point, we have got two predicted values about container throughput, which are ARIMA model simulation, BP error
simulation. In order to integrate the predicted results and tests the validity of the model.

The single BP neural Networks is compared. The integrate result obtained by the combined model of ARIMA and BP neural Networks and single BP neural Networks is showed as Table 2. We can draw the conclusion that TEI@I based integrated model indeed enhances forecasting performance.

Table 2. The integrate results of container throughput

Time	Actual value	ARIMA simulation	ARIMA error	BP error simulation	TEI@I integrate	BPNN simulation
2017-01	161.88	154.22	7.78	5.03	159.25	152.39
2017-02	133.78	154.08	−20.08	−7.1	146.98	155.82
2017-03	155.58	158.79	−2.79	2.86	161.66	166.83
2017-04	153.58	152.91	1.09	1.6	154.51	153.93
2017-05	148.9	154.53	−5.53	−2.22	152.31	143.18
2017-06	155.76	155.68	0.32	0.48	156.16	152.39
2017-07	155.11	154.63	0.37	2.04	156.67	155.6
2017-08	153.82	155.83	−1.83	−2.63	153.21	147.56
2017-09	154.98	156.59	−1.59	−1.61	154.98	153.38
2017-10	150.27	156.84	−6.84	−4.03	152.81	153.77

Table 3 compares their performance of ARIMA, BPNN and TEI@I methodology in terms of MAPE, RMSE and CDFR. In light of the evaluation results shown in Table 3, the TEI@I-based integrated framework significantly outperforms other single models in all of the three evaluation criteria. Which clearly shows that the TEI@I-based integrated framework are capable to improve the forecasting performance.

Table 3. Performance comparison of the 3 models.

Model	MAPE	RMSE	CDFR (%)
ARIMA	0.033	7.505	66.7
BPNN	0.043	8.95	77.8
TEI@I	0.021	4.898	88.9

5 Conclusion

The container throughput forecast determines the entire development of the port. In this paper, the port container throughput forecasting problem is discussed. Considering the complexity of the problem, TEI@I methodology is used to forecast port's container throughput via decomposing port's container throughput series to three components: linear component, nonlinear component and the effects of irregular and infrequent events' impact, then forecast each component separately and integrate them eventually. In the first step, ARIMA model is used to fit linear component. In the second step, BP neural network is used to fit nonlinear component. This method combines time series analysis ARIMA model and BP neural network, and the advantages of the two methods are complementary, avoiding the bigger prediction error of time series and the local minimum of BP neural network.

To verify the effectiveness of the TEI@I-based integrated forecasting framework, a case study on Qingdao port's container throughput is presented. Through the comparison and analysis of the Qingdao container throughput data in the experiment, the TEI@I forecasting framework have high forecast accuracy. The comparison results indicate that TEI@I methodology can significantly improve the performance and the TEI@I model performs best in terms of all the evaluation criteria. For further research.

Acknowledgements. This work is supported by the Natural Science Foundation of China (nos. 61472231, 61502283, 61640201), Ministry of Education of Humanities and Social Science Research Project, China (12YJA630152), Social Science Fund Project of Shandong Province, China (16BGLJ06, 11CGLJ22).

References

1. Huang, A., Lai, K., Yinhua, L.I., et al.: Forecasting container throughput of Qingdao port with a hybrid model. J. Syst. Sci. Complex. **28**(1), 105–121 (2015)
2. Xie, G., Wang, S., Zhao, Y., et al.: Hybrid approaches based on LSSVR model for container throughput forecasting: a comparative study. Appl. Soft Comput. J. **13**(5), 2232–2241 (2013)
3. Peng, W.Y., Chu, C.W.: A comparison of univariate methods for forecasting container throughput volumes. Math. Comput. Model. **50**(7), 1045–1057 (2009)
4. Dragan, D., Kramberger, T., Intihar, M.: A comparison of methods for forecasting the container throughput in north Adriatic ports. In: IAME 2014 Conference, Norfolk, VA, USA (2014)
5. Huang, W.C., Wu, S.C., Cheng, P.L., Yu, Z.H.: Application of grey theory to the transport demand forecast from the viewpoint of life cycle-example for the container port in Taiwan. J. Marit. Sci. **12**, 171–185 (2003)
6. Zhang, C., Huang, L., Zhao, Z.: Research on combination forecast of port cargo throughput based on time series and causality analysis. J. Ind. Eng. Manag. **6**(1), 124–134 (2013)
7. Chen, S.H., Chen, J.N.: Forecasting container throughputs at ports using genetic programming. Expert Syst. Appl. **37**(3), 2054–2058 (2010)
8. Mombeni, H.A., Rezaei, S., Nadarajah, S., et al.: Estimation of water demand in Iran based on SARIMA models. Environ. Model. Assess. **18**(5), 559–565 (2013)
9. Kumru, M., Kumru, P.Y.: Using artificial neural networks to forecast operation times in metal industry. Int. J. Comput. Integr. Manuf. **27**(1), 48–59 (2014)
10. Yan, Y., Xu, W., Bu, H., et al.: Housing price forecasting method based on TEI@I methodology. J. Syst. Eng. Theory Pract. **27**(7), 1–9 (2007)
11. Zhang, J.W., Suo, L.N., Qi, X.N., et al.: Inflation forecasting method based on TEI@I methodology. Syst. Eng. Theory Pract. **30**(12), 2157–2164 (2010)
12. Chen, Q., Zhang, C.: Grey prediction of China grain production with TEI@I methodology. In: IEEE International Conference on Grey Systems and Intelligent Services, pp. 253–260. IEEE (2015)

13. Wang, S.: Crude oil price forecasting with TEI@I methodology. J. Syst. Sci. Complex. **18**(2), 145–166 (2005)
14. Qian, X.S.: Creating Systematology. Shanxi Science and Technology Publishing House, Taiyuan (2001)
15. Wang, S.Y.: TEI@I: a new methodology for studying complex systems. In: R. Workshop on Complexity Science, Tsukuba, pp. 22–23 (2004)
16. Zhong, S.: Stability Theory of Neural Network. First edn. Science Press, Beijing

Posters

Named Entity Recognition Based on BiRHN and CRF

DongYang Zhao[✉]

National University of Defense Technology, Changsha 410000, Hunan, China
Zhaodyl988@gmail.com

Abstract. Named entity recognition is one of the basic work in the field of natural language processing. By utilizing bidirectional LSTM, Lample achieved the best results in the field of named entity recognition in 2016. In this paper, we propose a new neural network structure based on Recurrent Highway Networks (BiRHN for short) and Conditional Random Field (CRF for short). RHN is a good solution to the problem caused by gradients, which extends the LSTM architecture to allow step-to-step transition depths larger than one. Experiments on several datasets show that our model achieves better results (F1 values) than Lample.

Keywords: BiRHN · CRF · NER

1 Introduction

Named entity recognition refers to the identification of texts with specific meanings, including names, place names, organization names, proper nouns and so on, which play an important role in the process of natural language processing. Named entity recognition has undergone three stages of development. The first stage is based on rules and manual annotation methods. The second stage is a statistical based method represented by conditional random fields and support vector machines. Statistical-based methods have higher requirements for feature selection and greater reliance on corpora, but there are few large-scale common corpus used to evaluate named entity recognition systems. The third stage is a machine learning method represented by recurrent neural network. It is also the mainstream method of research on named entity recognition. Recurrent neural networks have been proposed for nearly 30 years. With more and more fields of application, researchers have proposed various variations of it. One of the most popular results is LSTM [5]. Although LSTMs perform well on gradient problems and become the mainstream research directions for recurrent neural networks, it still has some problems. Another more famous variant is GRU [2]. Its structure is simpler than that of LSTM, but only slightly improved on the gradient problem. Greff and Britz Studied various variants of LSTM, and proved the equivalence of the effects of each variant [1, 4]. Therefore, new ideas and methods are needed to solve this problem. Zilly et al. Recently proposed a recurrent highway network (RHN) [11] and provided substantial theoretical support for the improved gradient flow of the architecture. Experiments show that RHN performs better than LSTM in some aspects. In this paper, we use a RHN-based approach to name entity recognition. Testing in multiple corpora shows

© Springer Nature Switzerland AG 2019
S. Li (Ed.): GPC 2018, LNCS 11204, pp. 465–473, 2019.
https://doi.org/10.1007/978-3-030-15093-8_33

that our experimental results provide comparable results to those proposed by Lample [1]. In the second part, we introduce the structure of the experiment model. In the third part, we introduce how we get the vector representation of the word. The fourth part introduces the experiment, including the experimental parameters, data sets, training methods, Experimental results. Finally, in the fifth part, we summarize this article.

2   Model

2.1   RHN

Highway layers [10] enable easy training of very deep feedforward networks through the use of adaptive computation, The Highway layer computation is defined as:

$$y = h \cdot t + x \cdot c$$

Among them:

$$h = H(x, W_H); t = T(x, W_T); c = C(x, W_C)$$

Where H/T/C is the activation function, $W_H/W_T/W_C$ is the weight matrix. Recurrent Highway Network (RHN) is an extension of the LSTM framework. The transition depth between LSTM steps is 1, while RHN allows this depth to be higher than 1, further learning the complex relationship between signals. We use one forward RHN and one backward RHN to learn word representations with left and right contextual information, respectively, as Graves did in LSTM. Different parameters are used for forward and backward RHN. The input to the forward RHN is the forward sequence of words in the sentence. The backward RHN input is the reverse sequence of words in the sentence. The output of each RHN is a word with the sentence's left and right context information, and we concatenate the two output vector representations to get the final word representation with contextual information.

For a standard RHN, the formula for its recurring unit is as Fig. 1:

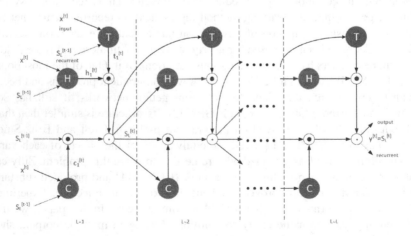

Fig. 1. Schematic showing computation within an RHN layer inside the recurrent loop.

The recurring unit of a recurrent highway network is a structure containing L highway layers. We can easily find that when L is 1, RHN is equivalent to a variant of the standard LSTM. When L is greater than 1, the RHN recurring unit structure as shown in Fig. 1. Then an RHN layer with a recurrence depth of L is described by:

$$s_l^{[t]} = h_l^{[t]} \cdot t_l^{[t]} + s_{l-1}^{[t]} \cdot c_l^{[t]}$$

Where

$$h_l^{[t]} = \tanh(W_H x^{[t]} \prod_{\{l=1\}} + R_{H_l} s_{l-1}^{[t]} + b_{H_l})$$
$$t_l^{[t]} = \sigma(W_T x^{[t]} \prod_{\{l=1\}} + R_{T_l} s_{l-1}^{[t]} + b_{T_l})$$
$$c_l^{[t]} = \sigma(W_C x^{[t]} \prod_{\{l=1\}} + R_{C_l} s_{l-1}^{[t]} + b_{C_l})$$

$\prod_{\{\}}$ is the indicator function.

2.2 CRF

The words in the sequence are grammatical relevance, so it is unwise to decide independently which label each word is labeled. Like Lafferty et al. [7], we use a CRF layer above the RHN layer to calculate the probability of a sentence being marked as a certain sequence. The specific calculation is as follows, For an input sentence:

$$X = (x_1, x_2, \ldots\ldots, x_n)$$

Where x_1 is the i-th word in the input sentence, and we enter it into RHN by expressing it as a word vector. The output of RHN is a k-dimensional vector, where k is the number of different tags and each element of the vector represents the fraction of the corresponding tags for that word. So for a single input sentence, our RHN can get an n * k fractional matrix. For a tag sequence:

$$y = (y_1, y_2, \ldots\ldots, y_n)$$

We define the score of this sequence as:

$$s(X, y) = \sum_{i=0}^{n} A_{y_i, y_{i+1}} + \sum_{i=1}^{n} P_{i, y_i}$$

Where A represents the fractional matrix transferred from one to another, we set a start mark and an end mark, BOS and EOS, so matrix A is a (k + 2) dimensional square matrix. We use a softmax layer to calculate the probability of each token sequence of an input statement:

$$p(y|X) = \frac{e^{s(X,y)}}{\sum\limits_{\tilde{y} \in Y_X} e^{s(X,\tilde{y})}}$$

During the training phase, we maximize the log-likelihood of correctly labeled sequences:

$$\log(p(y|X)) = s(X, y) - \log\left(\sum_{\tilde{y} \in Y_X} e^{s(X,\tilde{y})}\right)$$

Y_X represents all possible mark sequences in an input statement. We predict the output sequence that obtained the maximum score given by:

$$y^* = \arg\max_{\tilde{y} \in Y_X} s(X, \tilde{y})$$

The above formula is binary, so we can use dynamic programming to solve it.

2.3   Structure

$P_{i,j}$ is defined as the dot product of the outputs of bidirectional LSTMs, which added to the corresponding transfer score, yields the final score of a certain marking sequence of the sentence. The whole model is structured as Fig. 2:

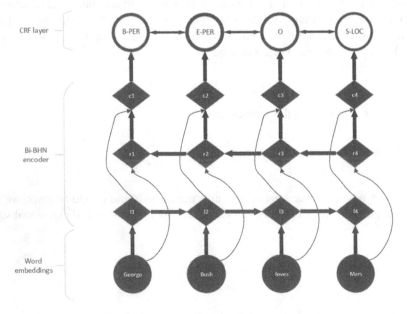

Fig. 2. Main architecture of the network.

From the figure we can see that the parameters required for our model are: bigram compatibility scores A; the parameters of the bidirectional RHN; the linear feature weights; the word embeddings. From Fig. 1, we can see that our word with contextual information is denoted as vector Ci, and then we use this vector as input to add a hidden layer and an output layer, where the number of units in the output layer is the same as the number of different labels. Add hidden layer can get better results. Instead of using softmax to separately score each label at the output level, we use a CRF to jointly decode it to produce a final prediction for each word. The parameters are trained to maximize $\log(p(y|X))$ of observed sequences of NER tags in an annotated corpus, given the observed words.

2.4 Tagging Schemes

Each named entity may consist of more than one word. Each word in a sentence may be a single named entity, a part of a named entity, or not a named entity. We mark every word in a sentence using a tagging scheme of IOB. Although there are other tagging schemes like IOBES that have their own advantages, they do not show significant performance gains, so we chose the IOB tagging Program. During testing, words that do not have an embedding in the lookup table are mapped to a UNK embedding. To train the UNK embedding, we replace singletons with the UNK embedding with a probability 0.5.

3 Word Embeddings

To capture character-level features and morphological features of words, we use BiLSTM to model character-based word embedding. We can benefit from this representation for morphologically rich languages. Ling proves experimentally [9] that word-based word embedding can effectively improve model performance.

3.1 The Overall Structure of Word Embedding

The overall structure of word embedding we generated is shown in Fig. 3.

First, we collect all the characters, initialize each character randomly into a vector, and these character-vector pairs form a look-up table. For a word, we input the vectors corresponding to the characters that comprise the word into the graph in order and reverse order. The character-based word embedding we get is a concatenation of the outputs of the forward and reverse LSTMs. In addition, we use a word embedding in the word look-up table to get a vector representation of words. The final representation of words is the concatenation of character-based word embedding and query-based word embedding. During training, if a word does not have a corresponding word vector (probably because the frequency is too low, resulting in being filtered out), then a UNK term is used to embed the representation. To train the UNK embedding, we replace singletons with the UNK embedding with a probability 0.5 [8].

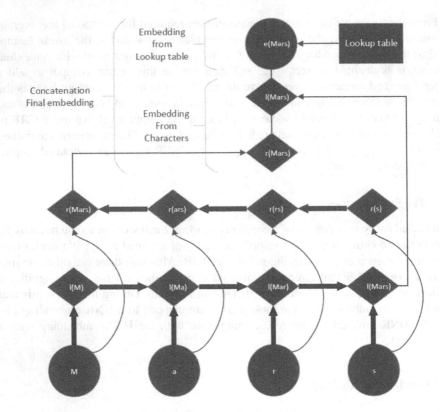

Fig. 3. How do we get the final embedding.

3.2 Pretrained Embeddings

In the previous section, we concatenate character-based word embedding with query-based word embedding. As in Collobert et al. [3], we use pretrained word embeddings to initialize our word lookup table. Embeddings are pretrained using skip-n-gram [9]. These embeddings are fine-tuned during training. The training corpus we use is the CoNLL-2002. We use an embedding dimension of 100 for English.

3.3 Dropout Training

We use Dropout training in two places, the first one before the input of Fig. 1 and the second one is the input of the CRF layer. The purpose of the former is to allow the model to balance the use of two kinds of word embedded representation, to avoid relying heavily on one of them; the latter is to prevent over-fitting.

4 Experiment

This section describes the experiment-related content, including the experimental method, the experimental parameter configuration, training data, compared with other models.

4.1 Training

We used a stochastic gradient descent algorithm with a learning rate of 0.01 to update the weights in each round of training. Although our model optimizes the gradient problem, there is still the possibility of a gradient explosion. So we use a gradient clipping of 5.0. Our BiRHN-CRF model uses one forward and one backward RHN, each of which has an RHN layer with a depth of 10 highway layers. The model's input is a 100-dimensional vector. Here we set a dropout layer, dropout rate is still set to 0.5, because too high or too low will have a negative impact.

4.2 Training

We use CoNLL-2002 and CoNLL-2003 datasets, which contain four languages. We chose one of the Spanish and English tests. Both datasets contain four types of entities: person name, place name, organization name, others.

4.3 Results

Table 1 presents our comparisons with other models for named entity recognition in English. From the table we can see that the best performance is achieved by Luo et al., they get this result by use external labeled data. Our model performance is higher than the other listed models, including passos et al. 'model with external labeled data. Tables 2 presents our comparisons with other models for named entity recognition in Spanish. We can see that our model is slightly better than lample's model and has some advantages over other models that use external labeled resources. Figure 4 compares the F1 value of our model and the BiLSTM-CRF model with the length of the sentence. It can be seen from the figure that as the length of the sentence increases, the F1 value

Table 1. English NER results. * indicates modes trained with the use of external labeled data

Model	F1
Passos et al. (2014) [12]	90.05
Passos et al. (2014)* [12]	90.90
Huang et al. (2015)* [13]	90.10
Luo et al. (2015)* + gaz + linking [14]	91.20
Chiu and Nichols (2015)* [15]	90.77
LSTM-CRF (2016) [16]	90.94
Our model: BiRHN-CRF	90.96

Table 2. Spanish NER results

Model	F1
Gillick et al. (2015) [17]	81.83
Santos and Guimaraes (2015) [18]	82.21
Gillick et al. (2015)* [17]	82.95
LSTM-CRF (2016) [16]	85.75
Our model: BiRHN-CRF	85.83

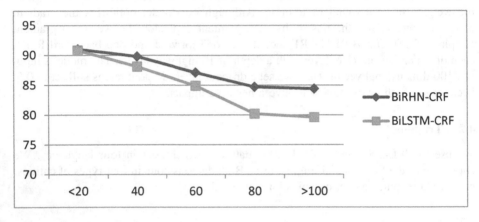

Fig. 4. F1 against sentence length.

of our model decreases more slowly, while the F1 value of the BiLSTM-CRF model decreases faster, indicating that our model can handle the dependence of longer sequences and Remember the information for a longer time step.

5 Conclusion

In this paper we propose a new model that is similar in structure to BiLSTM-CRF. Experiments on multiple datasets show that our model achieved better results. The key to our model's success is our flexible use of RHN. At the same time we use the dropout layer in two places also play a role in improving the performance of the model. In addition, by using the CRF to decipher label sequences rather than treating them as unrelated, independent tags, our model has been shown to be ideal.

References

1. Britz, D., Goldie, A., Luong, T., Le, Q.: Massive exploration of neural machine translation architectures. arXiv preprint arXiv:1703.03906 (2017)
2. Chung, J., Gulcehre, C., Cho, K., Bengio, Y.: Empirical evaluation of gated recurrent neural networks on sequence modeling. arXiv preprint arXiv:1412.3555 (2014)

3. Collobert, R., Weston, J., Bottou, L., Karlen, M., Kavukcuoglu, K., Kuksa, P.: Natural language processing (almost) from scratch. J. Mach. Learn. Res. **12**, 2493–2537 (2011)
4. Greff, K., Srivastava, R.K., Koutnik, J., Steunebrink, B.R., Schmidhuber, J.: LSTM: a search space odyssey. IEEE Trans. Neural Netw. Learn. Syst. **28**(10), 2222–2223 (2017)
5. Hochreiter, S., Schmidhuber, J.: Long short-term memory. Neural Comput. **9**(8), 1735–1780 (1997)
6. Huang, Z., Xu, W., Yu, K.: Bidirectional LSTM-CRF models for sequence tagging. arXiv preprint arXiv:1508.01991 (2015)
7. Lafferty, J.D., Mccallum, A., Pereira, F.C.N.: Conditional random fields: probabilistic models for segmenting and labeling sequence data. In: Eighteenth International Conference on Machine Learning, pp. 282–289 (2001)
8. Lample, G., Ballesteros, M., Subramanian, S., Kawakami, K., Dyer, C.: Neural architectures for named entity recognition. arXiv preprint arXiv:1603.01360 (2016)
9. Ling, W., et al.: Not all contexts are created equal: better word representations with variable attention. In: Conference on Empirical Methods in Natural Language Processing, pp. 1367–1372 (2015)
10. Srivastava, R.K., Greff, K., Schmidhuber, J.: Training very deep networks. In: Advances in Neural Information Processing Systems, pp. 2377–2385 (2015)
11. Zilly, J.G., Srivastava, R.K., Koutnik, J., Schmidhuber, J.: Recurrent highway networks. arXiv preprint arXiv:1607.03474 (2016)
12. Passos, A., Kumar, V., Mccallum, A.: Lexicon infused phrase embeddings for named entity resolution. Computer Science (2014)
13. Huang, Z., Xu, W., Yu, K.: Bidirectional LSTM-CRF models for sequence tagging. Computer Science (2015)
14. Luo, G., Huang, X., Lin, C.Y., et al.: Joint entity recognition and disambiguation. In: Conference on Empirical Methods in Natural Language Processing, pp. 879–888 (2016)
15. Chiu, J.P.C., Nichols, E.: Named entity recognition with bidirectional LSTM-CNNs. Computer Science (2016)
16. Lample, G., Ballesteros, M., Subramanian, S., et al.: Neural architectures for named entity recognition, pp. 260–270 (2016)
17. Gillick, D., Brunk, C., Vinyals, O., et al.: Multilingual language processing from bytes. Computer Science (2016)
18. Santos, C.N.D., Guimarães, V.: Boosting named entity recognition with neural character embeddings. Computer Science (2015)

Quadratic Permutation Polynomials-Based Sliding Window Network Coding in MANETs

Chao Gui[1], Baolin Sun[1(✉)], Xiong Liu[1], Ruifan Zhang[1],
and Chengli Huang[2(✉)]

[1] School of Information and Engineering, Hubei University of Economics,
Wuhan 430205, China
gui_chao@126.com, blsun@163.com, 1215522520@qq.com,
1535833618@qq.com
[2] Computer Engineering Department, Guangdong Youth Vocational College,
Guangzhou 510507, China
chengli_huang@163.com

Abstract. Quadratic permutation polynomials provide very good coding performance, and they also support a particular specific conflict-free parallel access. Network coding (NC) is a technique where relay nodes mix packets using mathematical operations, which can increase the network throughput and data persistence in Mobile Ad hoc NETwork (MANET). In this paper, we propose a Quadratic Permutation Polynomials-based Sliding Window Network Coding in MANETs (QPP-SWNC). QPP-SWNC enables to control the decoding complexity of each sliding-window independently from the packets received and recover the original data. The performance of the QPP-SWNC is studied using NS2 and evaluated in terms of the encoding overhead, decoding delay and throughput when a packet is transmitted. The simulations result shows that the QPP-SWNC with our proposition can significantly improve the network throughput and encoding efficiency.

Keywords: MANET · Network coding · Quadratic permutation polynomials · Sliding window

1 Introduction

Mobile Ad hoc network (MANETs) is composed of a large number of mobile nodes, which are randomly deployed within a certain range. Mobile nodes have mobility, computing power and communication capability. The MANET is a multi-hop self-organizing network system. When a node moves out of or into the transmission range of another node, the wireless link between the two will break or connect [1–5].

Network coding (NC) is one of the recent breakthroughs in communications research. It has first been proposed in [1] and potentially impacts all areas of communications and networking. NC was in fact devised based on algebraic operations

This work is supported by The National Natural Science Foundation of China (No. 61572012), The Natural Science Foundation of Hubei Province of China (No. 2017CFB773, 2018CFB661).

S. Li (Ed.): GPC 2018, LNCS 11204, pp. 474–481, 2019.
https://doi.org/10.1007/978-3-030-15093-8_34

over the finite fields $GF(2^q)$, which yields to very complex multiplications from the computational and thus energetic perspective. However, the advantages of network coding come at the expense of additional computational complexity, mainly due to the processing power of packet encoding and decoding. Ahlswede et al. [1] considered network coding for solving energy consumption in MANETs. However, their study did not give a specific network coding implementation method. Network coding is nothing new. Its theoretical basis has been applied in different systems, which is emphasized by Yang [2] when discussing the historical perspective of network coding. In [3] reviews the recent work in NC for multimedia applications and research on the techniques the gap between NC theory and practical applications. It outlines the benefits of NC and presents the open challenges in this area.

In our previous study, we proposed a network coding multipath routing algorithm in WSN (NC-WSN) [4], which is typically proposed in order to increase the reliability of data transmission or to provide load balancing. In the literature [5], we present an efficient approach to construct adaptive length sliding window and network coding coefficients in a pseudo-random manner on each user. We provide a thorough description of adaptive length sliding window, network coding, and energy efficient in MANETs (ALSW-NCEE), a novel class of network codes.

The rest of the paper is organized as follows. Section 2 discusses the some related work. Section 3 provides a model of quadratic permutation polynomials. Section 4 describes models of sliding encoding window methods in MANET. Some simulating results are provided in Sect. 5. Finally, the paper concludes in Sect. 6.

2 Related Works

A large number of studies have been conducted on the real-time data flow of MANET network, some of which are focused on minimizing delays [3–5], while others studying focus on the quadratic permutation polynomial of butterfly networks [6–8]. There has also been work on optimizing the overlay structure or a multi-radio multi-channel network [11–14]. These literature studies some new wireless real-time data flow transmission mechanisms with collaborative mechanism and NC coding data, as well as its joint optimization scheme.

Ayatollahi et al. [6] discussed the caching mechanism of wireless networks to improve the performance of the network and the transmission quality of data streaming. Bayat et al. [7] proposed a peer-to-peer (P2P) framework for the deployment of live video streaming applications over P2P overlay networks. Zhang et al. [8] introduce a decoding algorithm based on maximum ratio combination to realize the encoding gain and diversity gain. Nieminen [9] presents a quadratic replacement polynomial with butterfly network, as an interconnection network between the decoder unit and memory, and supports the parallel access method of various flexible and conflict-free turbo codes. Guan et al. [10] propose a different irregular Block-LDPC code based on quadratic permutation polynomial (QPP) instead of identity matrices. This scheme can provide better error performance and reduce complexity.

In the literature [11], the author considered a comprehensive data recovery scenario to achieve full recovery of data, but has not considered the joint problem of packet pulling and sharing. Zhang *et al.* [12] study a novel live free viewpoint video streaming network where each user pulls a subset of anchors from the server via a primary channel. In order to reduce energy consumption, Kim *et al.* [13] proposed a self-adaptive code propagation algorithm based on link quality (ACODI), which can improve energy efficiency better. Ostovari *et al.* [14] purposed a lightweight triangular inter-layer NC instead of the general form of inter-layer NC, to reduce the time complexity of the optimization.

3 Quadratic Permutation Polynomials

The set of integers from 0 to $k - 1$ is denoted by integer set Z_k for an integer $k \geq 1$. We say a function from Z_N to Z_N is an interleaver of a length N if it is one-to-one on Z_N. Such functions are also called permutations polynomial. A polynomial p, defined by $p(x) = a_k x^k + \ldots + a_1 x + a_0$ (mod N) for all x in Z_N with a_i are nonnegative integer coefficients, is said to be a permutation polynomial over Z_N when $p(x)$ permutes $\{0, 1, 2, \ldots, N - 1\}$.

Theorem 1. If and only if the following two conditions are satisfied, $p(x) = a_k x^k + \ldots + a_1 x + a_0$ (mod N) is the permutation polynomial on Z_N.

(1) The prime factor the N, and all the prime numbers are $p(x) = a_k x^k + \ldots + a_1 x + a_0$ (mod N) are the permutation polynomials.
(2) For the maximum of the upper n, satisfy $p^n | N$.

Theorem 2. Let $p(x) = a_k x^k + \ldots + a_1 x + a_0$ be a polynomial with integer coefficients. $p(x)$ is a permutation polynomial over the integer ring $N = 2^n$ if and only if (1) a_0 is odd, (2) $a_1 + a_3 + a_5 + \ldots$ is even, (3) $a_2 + a_4 + a_6 + \ldots$ is even.

Theorem 3. To prime numbers set $P = \{2, 3, 5, 7, \ldots\}$, the factor of any positive integer N can be expressed as.

$$N = \prod_{i=1}^{np_i} p^{np_i} \tag{1}$$

where $p_i s$ are distinct prime numbers, $p(x)$ is a permutation polynomial modulo N if and only if $p(x)$ is also a permutation polynomial modulo $p_i^{np_i}$, $\forall$ i. For example, $N = 112 = 2^4 7^1$.

Theorem 4. The following is the QPPs of an infinite sequence that generate maximum-spread interleavers:

$$p(x) = 2^{k+1} x^2 + (2^k - 1)x \qquad k = 1, 2, 3, \ldots \tag{2}$$

Strictly, we have QPP interleavers only when $k > 3$. The first observation is that for $k = 1$ and $k = 2$, the corresponding. The maximum spread QPP interval is shown in Table 1.

Table 1. For example maximum spread QPP interleaver

k	N	$p(x)$	Sqr(2N)
2	8	$p(x) = 8x^2 + 3x$	4
3	32	$p(x) = 16x^2 + 7x$	8
4	128	$p(x) = 32x^2 + 15x$	16

Figure 1 illustrates how the external values are stored by the maximal uncontended parallel access method when $N = 32$, the degree of parallel access method is 4, and the length of sub windows is 8. The number in each cell stands for the index of the extrinsic value e_i for $i = 0, 1, 2, \ldots, 31$.

0	1	2	...	6	7

8	9	10	...	14	15

16	17	18	...	22	23

24	25	26	...	30	31

Fig. 1. The length of sub windows is 8 and extrinsic values for $N = 32$.

4 Sliding Encoding Window Model

4.1 Sliding Window Mechanism

For each packet P to transmit, the source selects the blocks $(x_0, x_1, \ldots, x_{M-1})$ and coding vector $(c_0, c_1, \ldots, c_{M-1})$ to combine with in a sliding encoding window of size $1 \leq SW \leq M$, The elements of the encoding vector that do not belong to the encoding window are equal to zero. The size of the sliding window $SW = e - f + 1$, for N elements, there are $M\text{-}SW + 1$ possible sliding windows of size SW. A sliding encoding window of size W is a sequence of blocks $(x_f, \ldots, x_l)$ where $0 \leq f, l \leq M - 1$ and $f \leq l$ and $l - f + 1 = SW$. Once f has been drawn, the trailing edge of the sliding encoding window is calculated as $l = f + SW - 1$. We define f_i and l_i the leading edge and the trailing edge of the i-th sliding encoding window $SW_i = e_i - f_i + 1$. Figure 2 shows the encoding vector for a generation of size $M = 8$.

Window W_i

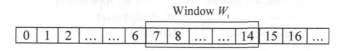

Fig. 2. Coding window M = 8.

4.2 Network Coding Mechanism

In this paper, we use a linear network coding scheme. The linear network coding scheme is an encoding method such that coding vector $c_i = (c_{i0}, c_{i1}, \ldots, c_{iM-1})$ is given, and input packet $X = (x_0, x_1, \ldots, x_{M-1})$ is converted into output packet P_i by the following expression.

$$P_i = \sum_{j=0}^{M-1} c_{ij} x_j \tag{3}$$

Then, the elements c_i of the encoding vector g are set to one with probability $p = 1/2$ for $i \in [f_i, l_i]$, with probability $p = 0$ otherwise. The destination node can decode input packets because the coding vector $c_i = (c_{i0}, c_{i1}, \ldots, c_{iM-1})$ and output packet data $P = (P_0, P_1, \ldots, P_{M-1})$ are obtained from the received packets, and an inverse matrix exists in G.

5 Simulation Experiments

5.1 Simulation Scenario

In this section, we present various simulation results for the proposed a Quadratic Permutation Polynomials-based Sliding Window Network Coding in MANETs (QPP-SWNC) algorithm. We evaluate QPP-SWNC in a data streaming scenario where one source distributes a data sequence to multiple cooperating receiver nodes. Nodes are randomly and uniformly located over a 1000 m × 1000 m area, with a node transmission range of 250 m. Node mobility follows the random waypoint model. Link loss rate for anchor/NC packets is independent and identically distributed with mean 2%. The generation distributed by the source is called seeding position within the data stream (Table 2).

Table 2. Simulation parameters

Number of nodes	100
Network area	1000 m × 1000 m
Transmission range	250 m
Simulation time	600 s
Beacon period	100 ms
Communication model	Constant Bit Rate (CBR)
Message size (b_{msg})	512 bytes/packet
Beacon period	100 ms

5.2   Simulation Results

We analyze the performance of QPP-SWNC from the point of view of the encoding efficiency. We use the NS-2 simulator [15] to evaluate the QPP-SWNC. The encoding efficiency of QPP-SWNC depends on the generation size N and on the sliding encoding window size M, M range from 8, 16, and 24.

Firstly, we test the QPP-SWNC performance in encoding overhead and network size. We plot the encoding overhead versus network size as shown in Fig. 3. Figure 3 shows that the encoding overhead of QPP-SWNC algorithm depends on sliding window M and is of great relevance. When the value of the sliding window $M = 16$, the encoding overhead is the smallest, which showed the best performance.

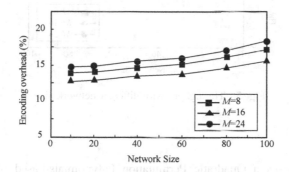

Fig. 3. Encoding overhead with different network size.

Then, we test and analysis the QPP-SWNC algorithm performance in decoding delays parameters. Figure 4 shows the three different sizes of sliding window under the comparison of decoding delay performance. When the network node number increases, the sliding window $M = 8$, showing the smallest decoding delay. The Fig. 4 shows that QPP-SWNC enables to change by adjusting sliding encoding window size M. Lower is better.

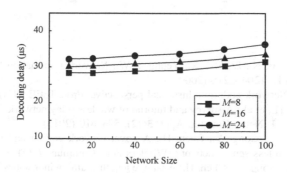

Fig. 4. Decoding delay with different network size.

We analyze the performance of QPP-SWNC from the point of view of the throughput. Figure 5 shows a comparison the sliding window network coding algorithm in terms of throughput as a function of network size load by adjusting sliding encoding window size W. In the Fig. 5, when the sliding window $M = 16$, network throughput is the highest. Higher is better.

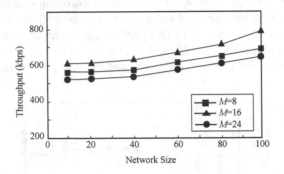

Fig. 5. Throughput with different network size.

6 Conclusion

This paper proposes a Quadratic Permutation Polynomials-based Sliding Window Network Coding in MANETs (QPP-SWNC). First, we provide a thorough description of quadratic permutation polynomials and sliding-window network coding in MANET. Second, we design an analytical model of QPP-SWNC decoding complexity is derived allowing to matching the decoding computational cost to the capacity of the mobile node. Third, the performance of the QPP-SWNC is studied using NS2 and experimentation to assess the encoding efficiency, the decoding complexity of QPP-SWNC enabled mobile node. The simulation result shows that QPP-SWNC produce encoding overhead, decoding delay, packet loss probability and throughput. This technique can guarantee the same reliability while consume the least energy.

References

1. Ahlswede, R., Cai, N., Li, S.-Y.R., Yeung, R.W.: Network information flow. IEEE Trans. Inf. Theory **46**(4), 1204–1216 (2000)
2. Yeung, R.W.: Network coding: a historical perspective. Proc. IEEE **99**(3), 366–371 (2011)
3. Mohammed, A.H., et al.: A survey and tutorial of wireless relay network protocols based on network coding. J. Netw. Comput. Appl. **36**(2), 593–610 (2013)
4. Sun, B.L., Gui, C., Song, Y., Chen, H.: A novel network coding and multi-path routing approach for wireless sensor network. Wireless Pers. Commun. **77**(1), 87–99 (2014)
5. Sun, B., Gui, C., Song, Y., Chen, H.: Adaptive length sliding window-based network coding for energy efficient algorithm in MANETs. In: Shi, X., An, H., Wang, C., Kandemir, M., Jin, H. (eds.) NPC 2017. LNCS, vol. 10578, pp. 13–23. Springer, Cham (2017). https://doi.org/10.1007/978-3-319-68210-5_2

6. Ayatollahi, H., Khansari, M., Rabiee, H.R.: A push-pull network coding protocol for live peer-to-peer streaming. Comput. Netw. **130**, 145–155 (2018)
7. Bayat, N., Lutfiyya, H.: Network coding for coping with flash crowd in P2P multi-channel live video streaming. In: 11th International Conference on the Design of Reliable Communication Networks (DRCN), Kansas, KS, USA, 24–27 March 2015, pp. 243–246 (2015)
8. Zhang, S.W., Song, R.F., Hong, T.: Network-coding-based two-way relay cooperation with energy harvesting. Int. J. Distrib. Sens. Netw. **13**(4) (2017). https://doi.org/10.1177/1550147717706437
9. Nieminen, E.: On quadratic permutation polynomials, turbo codes, and butterfly networks. IEEE Trans. Inf. Theory **63**(9), 5793–5801 (2017)
10. Guan, W., Liang, L.: Construction of block-LDPC codes based on quadratic permutation polynomials. J. Commun. Netw. **17**(2), 157–161 (2015)
11. Aboutorab, N., Sadeghi, P., Sorour, S.: Enabling a tradeoff between completion time and decoding delay in instantly decodable network coded systems. IEEE Trans. Commun. **62**(4), 1296–1309 (2014)
12. Zhang, B., Liu, Z., Gary Chan, S.-H., Cheung, G.: Collaborative wireless freeview video streaming with network coding. IEEE Trans. Multimed. **18**(3), 521–536 (2016)
13. Kim, D., Nam, H., Kim, D.: Adaptive code dissemination based on link quality in wireless sensor networks. IEEE Internet Things J. **4**(3), 685–695 (2017)
14. Ostovari, P., Wu, J., Khreishah, A., Shroff, N.B.: Scalable video streaming with helper nodes using random linear network coding. IEEE/ACM Trans. Netw. **24**(3), 1574–1587 (2016)
15. The Network Simulator - NS-2. http://www.isi.edu/nsnam/ns/

Image Retrieval Using Inception Structure with Hash Layer for Intelligent Monitoring Platform

BaoHua Qiang[1], Xina Shi[1], Yufeng Wang[2], Zhi Xu[1], Wu Xie[1(✉)],
Xianjun Chen[1], Xingchao Zhao[1], and Xukang Zhou[1]

[1] Guangxi Key Laboratory of Trusted Software,
Guilin University of Electronic Technology, Guilin 541004, China
xiewu588@126.com
[2] The 54th Research Institute of China Electronics
Technology Group Corporation, Shijiazhuang 050081, China

Abstract. In view of the problem of low efficiency and accuracy in traditional image retrievals, a method using inception structure with hash layers of image retrieval is presented for intelligent monitoring platform. The main idea of our work is to add hash layers into the inception structure of deep neural network, which can be used to transform the global average pooling features into low dimensional binary hash codes. Our method is utilized to not only ensure the sparseness of the neural network, but also avoid the overfitting phenomenon. Experimental results via the MNIST and CIFAR-10 datasets show that the retrieval efficiency and accuracy can be higher using our methods than before.

Keywords: Image retrieval · Inception structure · Intelligent platform · Deep learning

1 Introduction

Image retrieval has been widely researched and applied in such fields as computer vision since the 1990s. For instance, with the development of big data technology, the number of online pictures has been rapidly increased, and image retrieval is often used to solve the problem of search the users' pictures from massive images. To retrieve accurately, new ways were adopted with the development from CBIR (Content Based Image Retrieval) [1] to deep learning. In the renowned image retrieval competition ILSVRC (Large Scale Visual Recognition Challenge), deep learning models have some excellent results. Hinton using the deep learning model AlexNet [2] reduced the error rate to 15.315%, and the GoogLeNet [3] model has the error rate of 6.657%. Kaiming He et al. [4] proposed a deep residual learning for image recognition tasks, which was substantially deeper than those used previous deep learning models. Chandrasekhar et al. [5] put forward a method of deep compression to improve the efficiency of image retrieval. Nikkam et al. [6] and Chathurani et al. [7] researched the key point selection feature fusion method in CBIR. However, these methods are often used to retrieve images slowly. How to improve the retrieval speed has become a challenge especially

© Springer Nature Switzerland AG 2019
S. Li (Ed.): GPC 2018, LNCS 11204, pp. 482–488, 2019.
https://doi.org/10.1007/978-3-030-15093-8_35

for massive image retrievals. Xia et al. [8] combined hash functions with CNN (Convolutional Neural Network) [9] to improve the retrieval efficiency. These fruits above have attracted much academic attention from diverse fields of image retrieval theories and applications.

However, the accuracy that is obtained by hash layer is not good enough. It is still needed to research to improve the retrieval efficiency and accuracy. In this paper, we added a hash layer to change the inception structure. Using inception structure, the computational complexity of the neural network is reduced to avoid the overfitting phenomenon in the fully-connected structure, and the retrieval efficiency will be enhanced via this hash layer without losing the retrieval accuracy.

2 Inception Structure with Hash Layer for Image Retrieval

2.1 Proposed Framework of Deep Learning

Deep neural network is a new machine learning way replacing of CBIR, which is different from automata learning [10]. To retrieve image from massive data sets, our scheme using deep neural network includes three main parts as in Fig. 1. The first is the supervised pre-training via the large-scale ImageNet dataset. The second is fine-tuning the network with hash layer from the target domain dataset. The third retrieves similar images by calculating the Hamming distance.

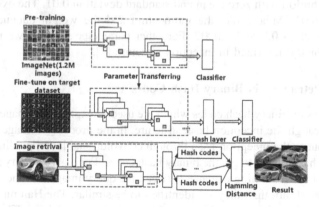

Fig. 1. The deep learning scheme of inception structure with hash layer for image retrieval

The pre-training component is mainly used to supervise the ImageNet dataset which contains more than 1.2 million images categorized into 1000 object classes. Fine-turning component is adopted to obtain deep semantic representations of images by feature extraction in MNIST and CIFAR-10 datasets. The neural network structure is added with a hash layer between the classifier layer of the inception structure and the Fc layer, the features processed by global average pooling are then converted into low-dimensional Hash features by Hash layers. Through binarization processing, the image features become binary hash codes. Retrieving similar images is done by calculating the Hamming distance between binary hash codes.

2.2 Inception Structure with Hash Layer

In generally, the most direct way to enhance the CNN is to increase the depth and width of the network, which means that it will generate a large number of parameters and increase the amount of computation. The inception structure can be used to reduce the parameters of model by adding a 1×1 convolution kernel in front of the 3×3 and 5×5 convolution kernels. The main idea of the inception structure [3] is maintaining the sparsity of the network and improving the computational efficiency by reducing the number of feature filters in each layer.

To improve the image retrieval efficiency and reduce the storage space, here a new method inception-hash structure is introduced by adding a Hash layer to inception structure. The Hash layer is located between inception structure and the classification layer. Compared with the deep hashing [11], SH [12], CNNH [8], DLBHC [13], our method is adopted to not only avoid the over fitting problem in the full-connectivity structure, but also improve the retrieval accuracy.

After constructing the inception-hash network, the Caffe (Convolutional Architecture for Fast Feature Embedding) tool [14] is used to train it. With these parameters, our network will be the optimal state. The parameters are initialized by drawing from a Gaussian distribution with zero mean and standard deviation 0.01. The overall learning rate is set to 0.01. Meanwhile, the momentum and the weight attenuation penalty coefficient are set to 0.9 and 0.0002. For other parameter settings, we use the same values as originally described in inception structure [3].

2.3 Image Retrieval via Binary Hash Codes

Different lengths of binary hash codes which are used to express for image retrieval can be obtained through the inception-hash structure. In the process of image retrieval, the images are input to the inception-hash through convolutional layer, pooling layer, fully connected and hash layers to get the binary hash codes. For a given query image and its binary codes, if the Hamming distance between the query image and the training set is lower than a threshold, then they are identified to be similar. The Hamming distance is the number of bits which are differing from two binary strings. The smaller the Hamming distance, the higher the similarity is. Here, the Hamming distance formula is:

$$d_h = (a, b) \sum_{i=1}^{d} xor(a_i, b_i) \tag{1}$$

Here d is the length of a and b. xor is exclusive OR between a_i and b_i.

The proposed method is evaluated by means of MAP (Mean Average Precision). MAP is a sort-based measure of accuracy. The formula is:

$$\text{MAP} = \sum_{i}^{k} Rel(i)/l_r \tag{2}$$

$Rel(i)$ is denoted with the true values of the query image i and the i-th ranked image.

3 Experiment Results

In our work, the MNIST and CIFAR-10 datasets are used to verify the benefits of inception-hash structure. The depth of the network is 22 layers with inception-hash structure. By setting different dimensions, we can get different length of binary codes. The hardware environment is a personal computer with NVIDIA Quadro K4200, and the operating system is a 64-bit Linux. The software environment is Python 2.7. Two indicators are used to test whether the training model can be used for image retrieval or not. Firstly, we deal with accuracy of training to test the classification effect of the model. Then, we adopt the hash coding with different length to conduct experimental analysis on the MAP standard.

(1) Results on MNIST and CIFAR-10 Datasets. MNIST [15] is the most widely used dataset in the field of image recognition, which contains binary handwritten numeric characters. CIFAR-10 [16] consists of four types of transport vehicles such as aircraft and automobiles, and six types of animals such as birds and deer with color picture of 32×32 pixel sizes. There are 50000 training samples and 10000 test samples.

In the experiment, the batchsize is set to 32 and the number of iterations is 50000 to get a stable accuracy value. The experimental results on different CNN are shown as Tables 1 and 2.

Table 1. Experimental results of MNIST datasets on different networks.

Algorithm	MAP
Inception-hash	0.966
Caffenet	0.959

Table 2. Experimental results of CIFAR-10 dataset on different networks.

Algorithm	MAP
Inception-hash	0.899
Caffenet	0.863

The impact of the current mainstream hash method on MAP values over binary hash codes of different lengths is shown as Figs. 2 and 3.

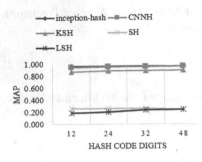

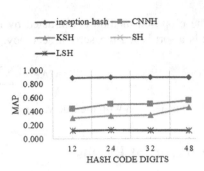

Fig. 2. Result on the MNIST   **Fig. 3.** Result on the CIFAR-10

Figures 2 and 3 shows that the results via the MNIST and CIFAR-10 datasets using our method is better than that of other hash methods by comparing the experiments with 12, 24, 32, and 48 different length hash codes. Using the inception-hash method, our work has achieved higher MAP value than other networks.

(2) Image retrieval efficiency. To verify the retrieval efficiency, the Hamming distance and Euclidean distance are adopted in our experiments using the same parameters and environments. By comparing the Euclidean distance and the Hamming distance from 10000 iterations to millions of iterations, the method using Hamming distance proved to be more effective than that using the other. The result is shown as Fig. 4.

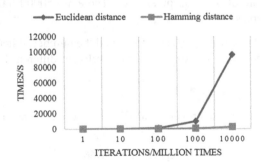

Fig. 4. The calculation time using Euclidean distance and Hamming distance

From Fig. 4, it can be seen that the way with Hamming distance is more effective in compare with the approach of the Euclidean distance via calculations. With the increasing numbers of the similarity alignments, the computational advantage in Hamming space becomes more and more obvious. The deep learning method of image retrieval is far less in space and time complexity than the depth-based image retrieval method in both retrieval accuracy and retrieval efficiency. Due to its binary nature, bit-wise storage can be used to save space. 1 MB of space can be stored by more than 20 thousand 48-bit binary hash code and up to one-million-level hamming distance calculating costs no more than 1 s, improving the search efficiency.

4 Conclusion

In our work, a new large-scale image retrieval method via the hash layer and inception structure has been proposed for such applications as intelligent monitoring platforms. Our method can be used to solve the problem of slow speed of retrieval in deep learning. By using the global average pooling and multi-size filtering in inception structure, a hash layer is added before the classifier, the extracted features are compressed into low-dimensional binary hash values, and the similarity values are computed by the hamming distance of the features of different pictures. The experimental results based on public datasets shows that our method can well improve the retrieval efficiency while ensuring the accuracy. In future, this method will be applied in such fields as intelligent monitoring platform, video surveillance, etc.

Acknowledgments. This work is supported by the National Marine Technology Program for Public Welfare (No. 201505002), Guangxi Science and Technology Development Project (No. 1598018-6), the National Natural Science Foundation of China under Grant No. 61462020, No. 61762025 and No. 61662014, Guangxi Natural Science Foundation under Grant No. 2017GXNSFAA198226, Guangxi Key Research and Development Program (No. AB17195053), Guangxi Key Laboratory of Trusted Software (kx201510, kx201413), Guangxi Colleges and Universities Key Laboratory of Cloud Computing and Complex System (Nos. 14106, 15204), the Innovation Project of GUET Graduate Education (Nos. 2017YJCX52, 2018YJCX42), Guangxi Cooperative Innovation Center of Cloud Computing and Big Data (Nos. YD16E01, YD16E04, YD1703, YD1712, YD1713, YD1714). Guangxi Colleges and Universities Key Laboratory of Intelligent Processing of Computer Image and Graphics (No. GIIP201603).

References

1. Blanco, G., Bedo, M.V., Cazzolato, M.T., et al.: A label-scaled similarity measure for content-based image retrieval. In: IEEE International Symposium on Multimedia, pp. 20–25. IEEE (2016)
2. Krizhevsky, A., Sutskever, I., Hinton, G.E.: ImageNet classification with deep convolutional neural networks. In: International Conference on Neural Information Processing Systems, pp. 1097–1105 (2012)
3. Szegedy, C., Liu, W., Jia, Y., et al.: Going deeper with convolutions. In: Proceedings of the IEEE Conference on Computer Vision and Pattern Recognition, pp. 1–9. IEEE (2015)
4. He, K., Zhang, X., Ren, S., et al.: Deep residual learning for image recognition. In: Proceedings of the IEEE Conference on Computer Vision and Pattern Recognition, pp. 770–778. IEEE (2016)
5. Chandrasekhar, V., Lin, J., Liao, Q., et al.: Compression of deep neural networks for image instance retrieval. In: Data Compression Conference, pp. 300–309. IEEE (2017)
6. Nikkam, P.S., Reddy, E.B.: A key point selection shape technique for content based image retrieval system. In: International Journal of Computer Vision and Image Processing, pp. 54–70 (2016). IJCVIP
7. Chathurani, N., Geva, S., Chandran, V., et al.: Image retrieval based on multi-feature fusion for heterogeneous image databases. Int. J. Comput. Inf. Eng. **10**(10), 1797–1802 (2016)

8. Xia, R., Pan, Y., Lai, H., Liu, C., Yan, S.: Supervised hashing for image retrieval via image representation learning. In: Twenty-Eighth AAAI Conference on Artificial Intelligence AAAI (2014)
9. Hershey, S., Chaudhuri, S., Ellis, D.P.W., et al.: CNN architectures for large-scale audio classification. In: IEEE International Conference on Acoustics, Speech and Signal Processing, pp. 131–135. IEEE (2017)
10. Zhang, H., Feng, L., Wu, N., Li, Z.: Integration of learning-based testing and supervisory control for requirements conformance of black-box reactive systems. IEEE Trans. Autom. Sci. Eng. 15(1), 2–15 (2018)
11. Liong, V.E., Lu, J., Wang, G., et al.: Deep hashing for compact binary codes learning. In: Computer Vision and Pattern Recognition, pp. 2475–2483. IEEE (2015)
12. Salakhutdinov, R., Hinton, G.E.: Semantic hashing. Int. J. Approx. Reason. 50(7), 969–978 (2009)
13. Lin, K., Yang, H.F., Hsiao, J.H., et al.: Deep learning of binary hash codes for fast image retrieval. In: Computer Vision and Pattern Recognition Workshops, pp. 27–35. IEEE (2015)
14. Jia, Y., Shelhamer, E., et al.: Caffe: convolutional architecture for fast feature embedding. In: Proceedings of the 22nd ACM International Conference on Multimedia, pp. 675–678 (2014)
15. Lecun, Y., Cortes, C.: The MNIST database of handwritten digits (2010)
16. Krizhevsky, A.: Learning multiple layers of features from tiny images (2009)

Retail Consumer Traffic Multiple Factors Analysis and Forecasting Model Based on Sparse Regression

Zengwei Zheng[1], Junjie Du[1,2], Yanzhen Zhou[1,2], Lin Sun[1(✉)],
Meimei Huo[1], and Jianzhong Wu[1]

[1] Intelligent Plant Factory of Zhejiang Province Engineering Lab,
Zhejiang University City College, Hangzhou, China
{zhengzw,sunl,huomm}@zucc.edu.cn
[2] College of Computer Science and Technology,
Zhejiang University, Hangzhou, China
{junjiedu,zhouyanzhen}@zju.edu.cn

Abstract. The rapid development of O2O business has increased the competition among offline shops in China. Accurate prediction of the shop's customer traffic can help the stores to change the strategy of sales timely and improve their competitiveness. Customer traffic forecast is more than a problem of time series. In fact, customer traffic for the next period is related to some external factors except for historical traffic. In this paper, the external factors affecting the customer traffic are analyzed using sparse coding, and we propose a sparse regression forecasting model with these external factors. The obtained results show that these external factors have varying degrees of impact on consumer traffic, and the prediction accuracy is significantly improved after considering these factors.

Keywords: Consumer traffic forecasting · Multiple factor · Sparse regression

1 Introduction

With the rapid development of mobile Internet and O2O (online to offline) business model, people can easily use their mobile phone to obtain the detail information of nearby shops, which make the retail store in a more and more increasingly competition. Therefore, it is particularly important for the retail store to make the right marketing strategy according to the accurate consumer traffic forecast.

Retail business consumer traffic forecast is short term forecast of time series, it relies on the historical data to forecast the future consumer traffic. But there are many factors affect the consumer traffic, including holiday, weather, temperature, etc. These factors coupled together, which result in the consumer traffic prediction becomes a complex and highly uncertain problem that is difficult to establish an efficient related mathematical model.

© Springer Nature Switzerland AG 2019
S. Li (Ed.): GPC 2018, LNCS 11204, pp. 489–494, 2019.
https://doi.org/10.1007/978-3-030-15093-8_36

In this paper, we propose a sparse regression forecasting model combined with holiday, weather and temperature factors in 2000 retail shops. The impacts of these factors on the prediction are analyzed. The experiment results show that these additional factors can improve the forecast accuracy.

2   Methods

2.1   Forecasting Framework Overview

The forecasting framework is shown in Fig. 1. Initially, we need to input the historical data y (time from t_1 to t_k) into preprocess module to filter the noise data and get result y'. After that, we use the Discrete Cosine Transform [1], Kronecker Delta [2] Dictionary and the external factors (include holiday, temperature and weather) to build an overcomplete dictionary D. Then, we figure out α according to $y' = D_{t_1 \ldots t_k} \alpha$. In this process, the most important is to make the α as sparse as possible. Finally, the α is used to forecast consumers traffic for the next 14 days (time from t_{k+1} to t_{k+14}) combined with $D_{t_{k+1} \ldots t_{k+14}}$.

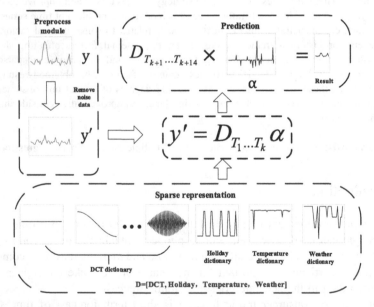

Fig. 1. Multi-factors sparse regression prediction framework

2.2   Multi-factor Dictionary

In this work, we build an overcomplete dictionary D for sparse coding based on Discrete Cosine Transform dictionary combine with holiday dictionary, temperature dictionary and weather dictionary.

Overcomplete DCT Dictionary. We use orthogonal DCT-II dictionary [1] shown in Eq. (1) and Kronecker Delta Dictionary shown in Eq. (2) to construct an overcomplete DCT dictionary.

$$\phi_j(n) = \begin{cases} \frac{1}{\sqrt{N}} & j = 0 \\ \sqrt{\frac{2}{N}}\cos(\frac{\pi}{N}(n+\frac{1}{2})j) & j = 1, 2, \ldots, N-1 \end{cases} \tag{1}$$

$$K_j(n) = \delta(n-j) \quad j = N, N+1, \ldots, 2N-1 \tag{2}$$

Holiday Dictionary. In this work, we construct a holiday dictionary, and use **1** represents holidays and **0** for working days. And we have manually adjusted the dictionary according to Legal Holidays in China in order to meet the actual situation better in China.

Temperature Dictionary. Considering the relatively stable weather conditions in China, the extreme temperature is rare. Therefore, the temperature dictionary is defined as follows in this paper.

$$T = \frac{1}{1 + e^{(-\frac{high+low}{2}+10)}}, \tag{3}$$

where *high* is the highest temperature of the day, and *low* is the lowest temperature of the day, *T* is the result.

Weather Dictionary. We classify the weather conditions into several different situations based on the common weather conditions and severity, and assign different labels, as shown in Table 1:

Table 1. Different weather conditions and their labels.

Weather condition	Label
Sunny	0
Little rain	−0.5
Heavy rain	−1
Little snow	−1.5
Heavy snow	−2

2.3   Multi-factors Sparse Coding

The goal of sparse representation [3] is to define a given data vector *y* as a weighted linear combination of a small number of basis atoms, it can be expressed as Eq. (4).

In this paper, in order to predict consumer traffic more accurately. we construct a new dictionary D using external factors, as shown in (5),

$$y = \alpha D \tag{4}$$

$$D = [DCT, H, T, W] \tag{5}$$

$$\alpha = [\alpha_{DCT}, \alpha_H, \alpha_T, \alpha_W], \tag{6}$$

where DCT is the Discrete Cosine Transform dictionary, H is the holiday factor, T is the temperature factor, W is the weather factor. α_{DCT}, α_H, α_T, α_W are the weights of DCT dictionary, holiday, temperature and weather dictionary, respectively.

After constructing the dictionary D, our purpose is to represent y sparsely in the domain defined by dictionary D. the above problem can be strictly defined as an optimization problem for solving α, as shown in (7) and we use the method proposed by Koh [4] to solve α.

$$minimize \ \|y - \alpha D\|^2 + \delta \|\alpha\|_1 \tag{7}$$

3   Experiment and Result

3.1   Setting

Dataset. In this paper, we use the dataset from the Tianchi Big Data contest hosted by Alibaba (IJCAI-17 Consumers Traffic Forecast [5]). Dataset time range is from July 1th, 2015 to October 31th, 2016 (except for December 12th, 2015).

Metric. We use the average forecast error of 2000 shops as the error to measure the performance of the forecast methods. We use the Eq. (8) as the calculation method.

$$error = \frac{1}{nT} \sum_{i}^{n} \sum_{t}^{T} \left| \frac{c_{it} - c_{it}^g}{c_{it} + c_{it}^g} \right|, \tag{8}$$

where n is the number of shops. T is the number of days predicted, C_{it} is the predictive value for day i, and C_{it}^g is the actual value of day i.

3.2   Impact of Different Factors on Accuracy

In this work, we construct sparse regression model with no factor, one factor, all factors respectively, and calculate the average prediction error separately. We show the error below to compare the influence of factors on the performance of the model, where W/O means no factor, W, T and H are weather, temperature and holiday factor respectively. Obviously, the prediction accuracy is significantly improved after adding all factors (Table 2).

Table 2. The prediction error of different factors.

Different factors	Error
W/O	0.1181
W	0.1184
T	0.1181
H	0.1044
H+T+W	**0.0938**

Figure 2 shows the weight distribution of three factors in 2000 shops which corresponding to α_H, α_T, α_W in formula (6). We can find that holiday factor has the greater influence than temperature and weather factor, which corresponding to the above experimental conclusion.

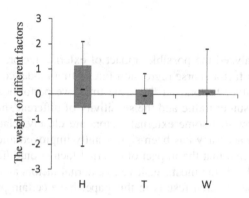

Fig. 2. The distribution of different factors weight

3.3 The Prediction Result in Different Shop Types

The sensitivities of different types retails to factors are not the same because of the nature difference of the retail stores. In 2,000 shops, there are **579** supermarkets, **639** restaurants and **782** snack bars. We experiment in three types of retails and analyze the impact of external factors on these retail stores.

As shown in Fig. 3. We can find that the forecasting model in the supermarket perform best, and the worst in snack bar. Compared with supermarket and restaurant, the size of snack bar is much smaller, so the daily consumer traffic is relatively less stable, which result in the worst forecast accuracy in snack bar. After all external factors were added, the forecast accuracy of supermarket improved by **1.97%** and the prediction accuracy of restaurant increased by **2.44%**. It is noteworthy that the forecast accuracy of snack bar improved by **2.73%**, which means snack bar is more sensitive to external factors than supermarket and restaurant. The reason for the result is the essence difference in retail stores. Compare to restaurant and snack bar. People generally buy necessities in supermarket that is rigid demand, which led to the relatively small affected by external factors. In contrast, the consumption in restaurants and snack bars is elastic demand which is more vulnerable to the environment.

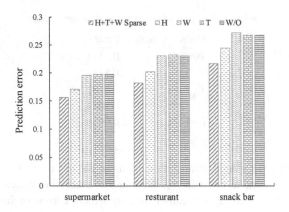

Fig. 3. The prediction error of different shops

4 Conclusion

In this paper, we analyzed the possible impact of external factors on consumer traffic and proposed a multi factor sparse regression forecasting model combine with holiday, temperature and weather factors. At the same time, we performed the impact of different factors on consumer traffic and the sensitivity of different shops to these factors. The experiments show that some external factors are closely related to the consumer traffic and prediction accuracy has been significantly improved after considering these external factors. We find that the impact of external factors on different types of shops are not same, snack bar is the most sensitive to external factors and holiday is the most influential factor. The research results in this paper have certain practical significance for improving the forecasting accuracy of consumer traffic.

Acknowledgment. This work is supported by Zhejiang Provincial Natural Science Foundation of China (No. LY17F020008).

References

1. Zhang, H., Yang, Y.J., Han, H.Y.: Novel binary-grouping watermarking algorithm based on DCT. Appl. Res. Comput. **11**, 3473–3476 (2014)
2. Shi, Y., Gao, Y., Yang, Y., et al.: Multimodal sparse representation-based classification for lung needle biopsy images. IEEE Trans. Biomed. Eng. **60**(10), 2675–2685 (2013)
3. Zhang, Q., et al.: Sparse representation based multi-sensor image fusion for multi-focus and multi-modality images: a review. Inf. Fusion **40**, 57–75 (2018)
4. Koh, K., Kim, S.-J., Boyd, S.: An interior-point method for large-scale l 1-regularized logistic regression. IEEE J. Sel. Top. Signal Process. **1**(4), 606–617 (2008)
5. https://tianchi.aliyun.com/competition/introduction.htm?spm=5176.11409391.333.4. 58ee49feRuZHFB&raceId=231591

Consulting and Forecasting Model of Tourist Dispute Based on LSTM Neural Network

Yiren Du, Jun Liu[(✉)], and Shuoping Wang

Zhejiang University City College, Hangzhou, China
Liuj@zucc.edu.com

Abstract. To fill the vacancy of tourist dispute in legal consultation resources, the consulting model of tourist dispute is proposed. The legal consultation model studied in this paper is based on the Long Short-Term Memory (LSTM) network. In terms of natural language processing, the Chinese word segmentation tool jieba popular in Python is adopted, to realize dialogue through the sequential translation model seq2seq and solve the long input sequence being covered or diluted with the help of Attention model. Finally, Google's second generation of artificial intelligence learning system TensorFlow based on DistBelief is adopted to train and optimize the model, so as to realize and train the forecasting model in this research.

Keywords: Sequential translation · LSTM · Tourist dispute

1 Overview

As people's living standards develop, tourism has become an essential component in the national life. The gap between the tourism-related service and the tourists' demand always results in dispute. The corpus adopted by the simulated training in this paper needs to be classified in detail, the information filtering-involved text classification is required for label processing.

Word segmentation processing is one of the most important issues in the study of natural language processing. Due to the different scenes of dialogue application, in Chinese. For instance, the reverse matching algorithm [1] being adopted to improve the matching efficiency and classification accuracy, can be applied to realize the classification system with high performance. On the other hand, the jieba plug-in unit has realized the Chinese word segmentation on the basis of reverse matching algorithm, so as to cut the Chinese text into the format suitable for text analysis.

In the following text, the functions of seq2seq translation model for tourist dispute forecast and the Attention model will be the first to introduce. At last, the experiments carried out on the basis of data in existing tourist dispute corpus will be researched and the forecasting results of the model will be verified.

S. Li (Ed.): GPC 2018, LNCS 11204, pp. 495–500, 2019.
https://doi.org/10.1007/978-3-030-15093-8_37

2   Forecasting Model

The Encoder-Decoder [3] structure as in Fig. 1 of seq2seq model has broken through the fixed input size of the past Chinese, i.e. the idea of realizing the sequence-sequence mapping through the neutral networks. The major idea is to map the input sequence into the output sequence and realize it through the coding input and decoding output, the encoder and decoder are composed of the recurrent neural networks as shown in Fig. 1. The input sequence ABC is input into the neurons by frame in the order of A, B and C; <EOS> (end of sentence) represents the end of input, it will be mapped into the W, X, Y Z sequence and output as WXYZ <EOF>. Obviously, this model has solved the problems in word-for-word translation and fixed sequences of input and output.

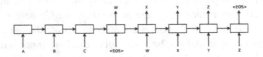

Fig. 1.   [4] Mapping of input and output in the seq2seq model

Return to the Encoder-Decoder as shown in Fig. 2, the first word X_1 to the last word X_T will be input into the Encoder structure by the timestep; most of the structural units in Encoder will be taken as the neurons in RNN (i.e. RNN, GRU, LSTM etc.), these words will be compressed into a specific semantic vector c, which can be understand as the semantics of a paragraph after taking each word into consideration.

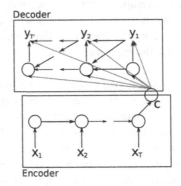

Fig. 2.   [3] Encoder-Decoder structure developing in timestep

As the traditional RNN has the phenomenon of increased depth yet disappeared gradient, at the time of different timestep inputs, the different timestep inputs on time axis equal to the multilayer depth, the gradient disappearance is unavoidable; therefore, the RNN applied in this model is a RNN variant, the Long Short-Term Memory [5] (LSTM) has solved the gradient disappearance generated by the RNN network under different timestep inputs (Fig. 3).

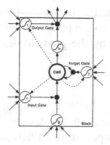

Fig. 3. [6] Long short-term memory block

Although LSTM is adopted as the structural unit, there is still a problem in the Encoder-Decoder model, namely the encoder and decoder are connected only by the semantic vector, as the timestep increases, the earlier the information are input, the easier they will be diluted. This is because when the sentence is relatively longer, the semantics can only be represented by the semantic vectors; if each word is of the same weight, the information input earlier will be diluted; the Attention model [7] is introduced to overcome the problems of semantic loss and information dilution. The major idea of Attention model is to input the different weights corresponding to the different timesteps in accordance with the information on the hidden layer and distribute the different weights of different words through the probability distribution of the attention (Fig. 4).

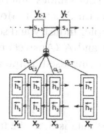

Fig. 4. [7] Structural diagram of Attention model

3 Building the Forecasting Model

3.1 Model Training

First of all, the placeholders of weights shall be initialized, the other parameters shall be set to obtain the loss, the smaller loss can be obtained through optimization; the gradient descent method shall be cyclically used for optimization. In the gradient descent class of TensorFlow, the learning rate shall be firstly set for the initialization; after the learning rate is defined, the parameters by gradient calculation shall be set, in the gradient calculation, the most important parameter is loss, the value of loss error shall be provided by seq2seq; finally, the parameter updating at the time of gradient descent will be updated according to the returned value of the gradient calculation. The loss shall be gradually weakened by the gradient descent of the cyclic iteration.

3.2 Model Execution

After making the loss reach the ideal state by training the mode, the model needs to be stored, so as to carry out the effective forecasting. Input at the time of forecasting mustn't be taken as the Decoder input, so the different parameters will be configured to make sure the same configuration is not adopted during the training and forecasting; in which the most important is that former participation is required for the input during the forecasting. It will be realized by the feed previous parameter in the seq2seq model.

4 Experiment

4.1 Experimental Data

Upon the above research and discussion, valid data shall be selected from the corpus to be used and divided into the Question part and Answer part for respective storage; the text will be classified into the training set and test set through the random assortment. The corpus sources are mainly focused on the CCL corpus retrieval system according to the vertical crawler in the tourist dispute field and a large number of China's established laws and regulations such as Management Rules for Travel Agency, Regulations on Administration of Tour Guides, Interim Measures for the Management of Scenic Spots, Interim Provisions on Retention Money of Travel Agency, Measures for the Control of Security in the Hotel Industry and son on. The language materials are divided into the Q (Question) part and A (Answer) part for the model realization and training. A list of 1000 sentences being used as training samples in experiment.

4.2 Experimental Parameters

The training ordinal length of input data is 5, as the training sample can be longer than the test sample in length, a parameter 0 shall be set for filling, the output homing sequence shall be marked while the setting of other parameters can be presented through Table 1. The feed_previous factor is set as false at training and true at forecasting to realize the mapping.

Table 1.

Model parameter	Value
encoder_inputs	5
decoder_inputs	5
num_encoder_symbols	20
num_decoder_symbols	25
embedding_size	8

During initializing the training, the set parameters will be a huge influence on the model training; under the interface given by TensorFlow, the common learning rate will be preset as 0.1. After the neurons being initialized through the initialization method given by TensorFlow, the cyclic iteration function will be compiled for gradient descent. Due to the huge number of corpus, during the training, there might be problems i.e. largely taken memory and prolonged training iteration time; therefore, solutions such as batch gradient descent can be adopted, namely, batch loading is available, so that the iteration can be carried out upon several sample being obtained. The parameters are presented through Table 2.

Table 2.

Model parameter	Value
learning_rate	0.1
grad_lose	none
global_loss	none

4.3   Experimental Result

As shown in Fig. 5, the value of loss reaches an acceptable level after 10,000 times of Itertions. Meanwhile the conditions of different tensors and hidden layer can be observed clearly during the process of training with the help of visualization tools.

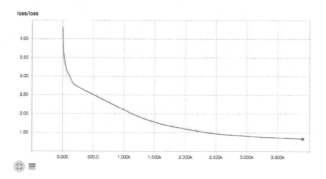

Fig. 5. Loss rate scalars

5   Conclusion

A consulting and forecasting model based on the LSTM neural networks under the seq2seq model is studied in this paper. What's more, the corpus classified into the training set and test set is adopted to carry out training and test on the model, so as to obtain a forecasting model that can be used to provide the legal dispute consultation; in terms of contract dispute and infringement dispute, after offering the related description of the problems, advices including solutions i.e. settlement by agreement, complaint from relevant departments, agreement arbitration and initiation of proceedings will be provided.

References

1. Xin, L., Liu, D.: Technology for Chinese text categorization based on reverse matching algorithm. Comput. Appl. **28**(4), 945–947 (2008)
2. Howard, M.W., Kahana, M.J.: A distributed representation of temporal context ☆. J. Math. Psychol. **46**(3), 269–299 (2002)
3. Cho, K., Merrienboer, B.V., Gulcehre, C., et al.: Learning phrase representations using RNN encoder-decoder for statistical machine translation. Comput. Sci. (2014)
4. Sutskever, I., Vinyals, O., Le, Q.V.: Sequence to sequence learning with neural networks **4**, 3104–3112 (2014)
5. Hochreiter, S., Schmidhuber, J.: Long Short-term memory. Neural Comput. **9**(8), 1735–1780 (1997)
6. Graves, A.: Supervised Sequence Labelling with Recurrent Neural Networks. Studies in Computational Intelligence. Springer, Heidelberg (2012). https://doi.org/10.1007/978-3-642-24797-2
7. Bahdanau, D., Cho, K., Bengio, Y.: Neural machine translation by jointly learning to align and translate. Comput. Sci. (2014)

Research and Design of Cloud Storage Platform for Field Observation Data in Alpine Area

Jiuyuan Huo[1,2(✉)]

[1] School of Electronic and Information Engineering,
Lanzhou Jiaotong University, Lanzhou 730070, People's Republic of China
huojy@mail.lzjtu.cn
[2] CERNET Co., Ltd., Beijing 100084, People's Republic of China

Abstract. With the rapid increase of field observation data volume in alpine regions, there are many problems exist in the data storage for geoscience researchers, such as lack of sufficient hardware storage devices, high maintenance costs, and incomplete storage environment. Nowadays, Cloud Storage technology which based on the open source Cloud Computing platform can effectively solve these problems. Therefore, this paper constructs and designs the field observation data Cloud storage platform in the alpine region based on the Apache Hadoop Cloud platform to realize the functions of creating, uploading and browsing of field observation data files in the Cloud Storage, so as to meet the needs of researchers to store observation data, share information and backup and so on. The system also can enable the efficient management of server resources, and provide large-scale data processing capabilities.

Keywords: Cloud storage · Field observation data · Hadoop · HDFS

1 Introduction

Field observation station is an important part of science and technology infrastructure, and it has a very important and irreplaceable role for the development of disciplines and national construction [1, 2]. In the background of rapid development of massive observation data, the current decentralized and intensive data storage mode can not provide good scalability and has the bottlenecks of data access. Cloud Computing is a new type of Internet applications and integrated services [3], and it can integrate a large number of computing resources, storage resources and software resources together and can provide efficient computing and massive data processing, at the same time can significantly reduce the cost of management [4].

Because Cloud Computing platform adopts the replication storage technology, when one of the copies of the data file is lost, it can quickly recover from the other copies [5]. Dynamic migration technology and dynamic scheduling technology of Cloud Computing platform can adjust physical resources according to the operation of the entire system which can efficiently improve utilization of physical resources and reduce energy consumption [6].

S. Li (Ed.): GPC 2018, LNCS 11204, pp. 501–505, 2019.
https://doi.org/10.1007/978-3-030-15093-8_38

Therefore, it is necessary to research and design a Cloud Storage management system to meet the requirements of scalability, high performance and high security for field observation data.

2 Related Works

Since Google put forward the concept of Cloud Computing in 2006, it has set off an upsurge of researches in the world. Cloud Storage is a solution based on Cloud Computing technology to solve massive data storage [7]. For example, Amazon Company launched the Amazon S3 (Simple Storage Service) [8] to provide online storage service. Users can easily save data files to server through the Web service interface. Google Cloud storage system GFS (Google File System) [9] is a large distributed file system that can be deployed on inexpensive common hardware with high fault tolerance for large, distributed mass data storage service. And the HDFS [10] distributed file system in Apache Hadoop is an open source implementation of Google's GFS file system. It has the advantages of highly fault tolerant, reliable, and stable to meet the needs of enterprises and scientific researches. Thus, in this paper, we design a Cloud storage platform running under the Apache Hadoop environment, manage the files by the HDFS distributed file system and provide the file storage related services to the users via the Internet through the WEB manner.

3 Cloud Storage Platform Design

3.1 System Architecture Design

As shown in Fig. 1, according to the actual field observation environment and the demand analysis of scientific research information in field stations in alpine areas, this paper logically divides the Cloud platform into the following four layers.

(1) Storage Layer
   The Storage layer is the underlying infrastructure of the system, and it is the physical storage device for data file. Through the Hadoop cluster software, the physical disks in normal PCs were integrated into a unified pool of resources to achieve the unified mass storage of data, as well as the centralized management of storage devices, status monitoring and dynamic expansion of capacity.
(2) Data Management Layer
   Through the calling of HDFS API interface to Hadoop cluster, this layer is to realize the functions of data management, task allocation, data read and write, delete, backup and share. The underlying storage seamlessly links with the upper applications through the unified user management, security management, and replication management and so on.
(3) Data Aggregation Layer
   The main task of Data Aggregation Layer is to receive the multi-sources, heterogeneous, multi-space-time, multi-scale field station data, and store them in the field observation data Cloud platform.

(4) Cloud Platform Application Layer

Application Layer is the system's service layer. Based on the field observation data files which unified stored in the Data Aggregation Layer, researchers are provided data services such as viewing, downloading, uploading and visualizing of observation data through the graphical user interface.

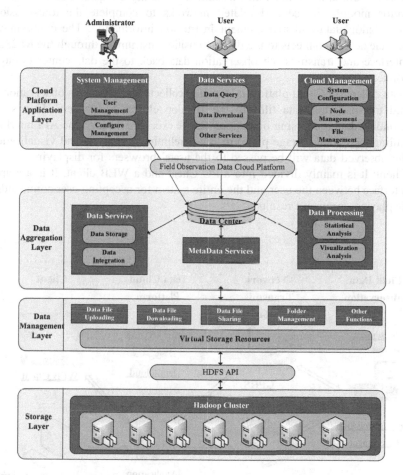

Fig. 1. System architecture of field observation data cloud platform in alpine area

3.2 Overall Structure Design

The overall structure of the prototype of the field observation data Cloud platform in the alpine region is shown in Fig. 2, which is mainly composed of four parts: field data observation, network communication transmission, data collection and analysis service, and client. The system is designed to collect the geosciences data in real time, and store the collected data in the field observation data Cloud platform, and then the data can be accessed, analyzed and visualized.

(1) Field data observation: it is mainly composed of observation sensors and data acquisition instruments to complete the recording and storage of observation data of observation elements. After sensors completing the observation, the electrical signals will be transmitted to the digital acquisition instrument, which digitizes the signals in data and stores them in the storage device.

(2) Network communication transmission: It is mainly composed of the communication module device and related networks to complete the transmission of observation data and management instruction information. The communication module device connects to the data acquisition instrument through the RS232/485 interface and transmits the observation data back to the data center through its connected network.

(3) Data collection Cloud platform: It mainly collects and processes observation data, and transmits the data files to the Hadoop cluster file system. User's corresponding data processing operations can be executed through the API interface of calling the Cloud storage platform. The preliminary analysis and visualization of the observed data will be passed to the users' browsers for displaying.

(4) Client: It is mainly divided into a PC Client and a WEB client. It is a graphical interface between the user and the entire system for accepting user commands and displaying command results.

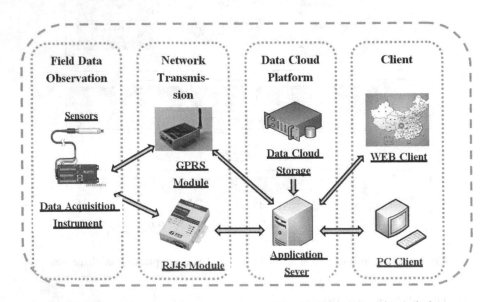

Fig. 2. General structure of prototype for the field observation data cloud platform

4 Conclusions

Given the rapid development of mass observational data, the traditional data storage model cannot keep up with the trends of data growth and can not meet the requirements of current data storage. Therefore, a field observation data Cloud storage platform for the alpine area was constructed and designed based on the Apache Hadoop platform to provide massive storage space and large-scale data processing capabilities. It will help researchers for observation data storage, information sharing and backup.

Acknowledgement. This work is supported by the CERNET Innovation Project (No. NGII20160111) and the Gansu Science and Technology Support Program (Grant number: 1606RJZA004).

References

1. Fu, B.J., Niu, D., Yu, G.R.: Ecosystem observation and research network in earth system science. Prog. Geogr. **26**(1), 1–16 (2007)
2. Ding, Y.J.: China Cold and Arid Regions Environmental and Engineering Science for 50 years. Scientific Press, Beijing (2009)
3. Peng, L.: Cloud Computing. Electronic Industry Press, Beijing (2010)
4. Wu, Z.H.: Core Technology Analysis of Cloud Computing. People's Posts and Telecommunications Press, Beijing (2011)
5. Armbrust, M., Fox, A., Griffith, R., et al.: A view of Cloud Computing. Commun. ACM **53**(4), 50–58 (2010)
6. Zhang, Q., Cheng, L., Boutaba, R.: Cloud Computing: state-of-the-art and research challenges. J. Internet Serv. Appl. **1**(1), 7–18 (2010)
7. Yang, N.: Cloud storage technology and its applications. Electron. Technol. Softw. Eng. **21**, 220 (2014)
8. Palankar, M.R., Iamnitchi, A., Ripeanu, M., et al.: Amazon S3 for science grids: a viable solution? Proceedings of the 2008 International Workshop on Data-Aware Distributed Computing, pp. 55–64. ACM (2008)
9. Ghemawat, S., Gobioff, H., Leung, S.T.: The Google file system. In: Proceedings of the Nineteenth ACM Symposium on Operating Systems Principles, SOSP 2003, pp. 29–43. ACM, New York (2003)
10. Yu, Q., Ling, J.: Research of Cloud storage security technology based on HDFS. Comput. Eng. Des. **34**(8), 2700–2705 (2013)

A Novel PSO Algorithm for Traveling Salesman Problem Based on Dynamic Membrane System

Yanmeng Wei and Xiyu Liu[✉]

School of Management Science and Engineering,
Shandong Normal University, Jinan, China
15154122301@163.com, sdxyliu@163.com

Abstract. Membrane computing is a class of distributed parallel computing model. In this paper, we propose a novel evolutionary computation method based on dynamic active membrane system. First, an improved particle swarm optimization based on neighborhood searching of every particle that called NPSO is proposed. That is, instead of learning from Pbest and Gbest during the whole evolution, the proposed NPSO learns from Pbest and NPbest (the NPbest is selected by the Neighborhood Searching Based Learning Strategy) in the early stage to preserve swarm diversity. After the predefined number of iterations, the NPSO switches into the conventional global version PSO to accelerate convergence speed. Second, in order to avoid suffering from premature convergence in the early stage, NPSO is partitioned into two stages that in the first stage is to preserve swarm diversity and in the second stage is to enhance the convergence speed towards global optimum. The classic Traveling Salesman Problem (TSP) is one of the most significant stochastic routing problems so we use the proposed NPSO to solve it. In fact, the NPSO can achieve better balance between exploration and exploitation as well. Experimental results show that the proposed NPSO algorithm is more superior or competitive.

Keywords: Membrane computing · Particle swarm optimization algorithm · TSP

1 Introduction

1.1 A Subsection Sample

Membrane computing, initiated by Gh. Păun was also called P system, as a new branch of natural computing, aims to abstract computing models from the structure and functioning of living cells [1, 2, 4]. Generally, they are characterized by three elements: (i) membrane structure, (ii) multisets and (iii) evolution rules (Fig. 1). The multisets of objects are placed in compartments surrounded by membranes, and evolved by some given rules. The membrane system has great parallelism. P systems have several interesting features: non-determinism, programmability, extensibility, readability, they are easy to communicate, etc., and many variants have been proposed. Most P systems

© Springer Nature Switzerland AG 2019
S. Li (Ed.): GPC 2018, LNCS 11204, pp. 506–515, 2019.
https://doi.org/10.1007/978-3-030-15093-8_39

variants have proved to be powerful (in the sense of Turing completeness) and effective (since they have successfully solved a large number of NP-hard problems in a linear or polynomial time) [3, 4]. In recent years, the potentiality and characteristics of membrane computing have attracted much attention in relation to real-life applications, such as membrane algorithms for solving optimization problems, and fuzzy spiking neural P systems for dealing with knowledge representation and fault diagnosis. It could reduce the computational time complexity and is suitable for solving combinatorial problems like the clustering problem [5, 6]. A variety of inherent mechanisms and characteristics that P systems possess are able to provide a new incentive for the research of clustering analysis.

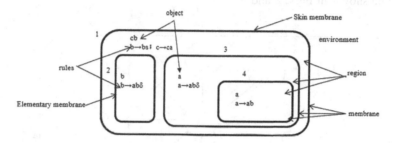

Fig. 1. The structure of membrane system.

The Traveling Salesman Problem (TSP) has plenty of applications in operations research. It also has a nice structure and is one of the most investigated NP-complete problems of combinatorial optimization. The TSP is one of the most widely studied NP-hard combinatorial optimization problem which cannot be solved exactly in polynomial time. It is also an important research topic. This problem is, for a given set of n cities with travel costs (or distances) c_{ij} between each pair of cities i, j $\in$ [1: n], to determine a minimum cost (or distance) circuit (Hamiltonian circuit or cycle) passing through each vertex once and only once. Every such tour together with a start city can be characterized by the permutation of all cities as they are visited along the tour. TSP has important applications to real world problems, such as vehicle routing problem, mixed chinese postman problems, and printed circuit board punching sequence problems.

The rest of the paper is organized as follows. Section 2 introduces the theoretical background. In Sect. 3, the proposed NPSO algorithm process is discussed in detail. Section 4 provides the experiment setup, other algorithms compared, experimental results and analysis. Finally, Sect. 5 draws the conclusions.

2 Theoretical Background

2.1 Conceptions of Star Active Membrane System

Membrane systems with active membranes have been introduced in recent years to overcome the restriction of usual membrane systems. Different from usual membrane systems, active membranes system possesses not only the evolution-communication mechanism of objects but also membrane evolution mechanism, such as membranes dissolution and membranes creation. Therefore, membrane systems with active membranes have a dynamical membrane structure during evolution. The dynamic active membrane system we designed as computing framework is illustrated in details in this section and shown in Figs. 2 and 3.

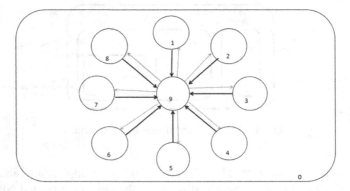

Fig. 2. The dynamic active membrane system

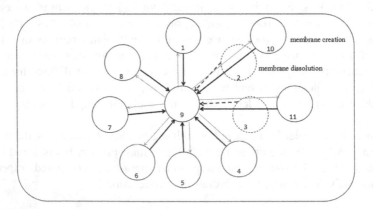

Fig. 3. Example of membrane dissolution and creation of the system

In the dynamic active membrane system, there are nine elementary membranes from number 1 to 9 and a skin membrane number 0. Membrane 0 is called skin membrane and is used for storing the final optimal value after every computation. The elementary membranes from 1 to 8 is used for computation/evolution and the number 9 is used for storing the global optimum found so far from the other elementary membranes in every iteration. And, the other elementary membranes from 1 to 8 will communicate with membrane 9 at the same time. The optimal value from any other elementary membranes will be replaced in membrane 9 as long as there is a better value in it in every iteration. And meanwhile, if the optimal value in any other elementary membranes still not change for $t(5 < t < 10)$ iterations, the value will be replaced by the optimal value stored in membrane 0. After one computation for the whole system, all the membranes will be dissolved and recreated new ones to continue the next computation for searching better optimal value shown in Fig. 3. The whole star active membrane system still keeps eight elementary membranes during each computation to achieve more evolution at the limited time.

2.2 Conceptions of Particle Swarm Optimization Algorithm

PSO algorithm is an evolutionary computation technique which inspired by social behavior of bird flocking, it uses the physical movements of individual or birds in the swarm. Each bird adjusts to their flight path according to the flying experience of its own and global individual in every iteration. Through this natural phenomenon, Dr. Eberhart and Dr. Kennedy proposed PSO algorithm in 1995 [4]. Compared with other intelligence algorithms, the characteristics of PSO algorithm are high efficiency, quick convergence, good robustness, easy implementation and there are few parameters to adjust.

In the PSO algorithm, the population size of particle swarm is S, D is the dimension of search space, f is the fitness function, T_{max} max is the maximum iteration number, each particle has a position vector $x_i = \{x_{i1}, x_{i2}, \cdots x_{iD}\}$, and a velocity vector $v_i = \{v_{i1}, v_{i2}, \cdots v_{iD}\}$, $p_i = \{p_{i1}, p_{i2}, \cdots p_{iD}\}$ is the best position of particle i in history, $p_g = \{p_{g1}, p_{g2}, \cdots p_{gD}\}$ is the best particle in the swarm also called global particle. At iteration t, the new position and velocity of each particle i is determined by the following equations:

$$v_{ij}(t+1) = w * v_{ij}(t) + c_1 * r_1 * \left(p_{ij}(t) - x_{ij}(t)\right) + c_2 * r_2 * \left(p_{gj}(t) - x_{ij}(t)\right); \tag{1}$$

$$x_{ij}(t+1) = x_{ij}(t) + v_{ij}(t+1); \tag{2}$$

w: inertia weight. c_1, c_2: acceleration coefficients which control the maximum step size. t: iteration counter. r_1, r_2: two independently uniformly that distributed random variables. The termination criterion is determined to whether the max generation or the defined values is reached.

3 Methodology

To preserve swarm diversity in the early stage without slowing down convergence significantly, a two-stage NPSO is proposed. Instead of learning from Pbest and Gbest during the whole evolution, the NPSO learns from Pbest and NPbest (NPbest is selected by the RSN learning strategy) in the early stage to preserve swarm diversity. After the predefined number of iterations, the NPSO switches into the global version to accelerate convergence speed.

3.1 Neighborhood Searching Based Learning Strategy

$$v_{ij}(t+1) = w * v_{ij}(t) + c_1 * r_1 * \left(p_{ij}(t) - x_{ij}(t)\right) + c_2 * r_2 * \left(npbest_{ij}(t) - x_{ij}(t)\right); \tag{3}$$

In Eq. (3), NPbest$_{ij}$ denotes the best particle with the lowest fitness value among the randomly selected neighboring particles. In the initialization, each particle chooses a few neighboring particles randomly from the whole population, except its own Pbest. Among these neighboring particles, the particle with the lowest fitness value is selected as NPbest. Unlike the conventional PSO, in which a group of particles share the same Localbest, in the proposed NPSO each particle has its own NPbest. There is an important parameter in NPSO algorithm, the neighborhood size which indicates the number of randomly selected neighboring particles of every particle. To illustrate the effects of neighborhood size of NPSO, a simulated test is carried out, with the increase of neighborhood size, the high-ranking particles have higher chances while the low-ranking particles have lower chances to be selected. It is turned out that large neighborhood size can speed up the convergence rate at the cost of swarm diversity loss. With small neighborhood size, the low-ranking particles have higher chances to be learned by other particles to enhance swarm diversity. Finally, as a result, neighborhood size = 5 can achieve good balance between exploration and exploitation.

4 Algorithm Test: The NPSO for Clustering Problem

In membrane system, elementary membranes are designed to solve cluster problems. Therefore, the objects in elementary membranes are considered to be the sets of cluster centers, each object in the elementary membranes can be described by initial particles that each particle contains k data points stored in a vector and represents a solution. Besides, the clustering metric is defined as the sum of the Euclidean distances of the data points to their corresponding cluster centers, where x_j (j = 1, 2, …, m) are data points, C_i (i = 1, 2, ..., k) are k clusters and z_i (i = 1, 2, ..., k) are the corresponding cluster centers. The goal is to search for the optimal cluster centers that have the smallest value.

In the proposed Neighborhood Based Particle Swarm Optimization, we divided it into two stages. In the first stage, we use Eq. (3) to preserve the swarm diversity and avoid trapping into local optimum because lacking of diversity. The iterations we decided is 200. And after 200 iterations, the algorithm switches into our conventional PSO to use Gbest to achieve global optimum.

4.1   Datasets and Parameters' Setting

To test the proposed NPSO, we used a dataset which contains 300 two-dimensional datas. Besides, the parameters are needed to be illustrated in this part. K is the clustering number of the problem. The inertia weight w is constant 1; The acceleration coefficients C_1 and C_2 are two positive constants empirically advised to be 2.0 (Kennedy and Eberhart 1995); R_1 and R_2 are two random functions ranging from 0 to 1; Iteration1 and Iteration2 denote the iterations number of the first stage and second stage respectively; N represents the total number of initial particles; d denotes the dimension of the dataset. The parameters are listed in Table 1.

Table 1. Parameters of the NPSO algorithm.

Parameter	Description	Value
K	Cluster number	5
Iteration1	Iterations of first stage	200
Iteration2	Iterations of second stage	200
N	Number of particles	50
C_1	Cognitive parameter	2
C_2	Social parameter	2
R_1	Random value	[0, 1]
R_2	Random value	[0, 1]
w	Inertia weight	1

4.2   Experimental Results

Figure 4 illustrated the comparative results clearly. It is evident that the convergent rate of our NPSO is faster than PSO and K-means in almost every computation, because the proposed NPSO can effectively avoid the Best Value trapping into local optimum. What's more, as we can see the Best Value of the NPSO is smaller than PSO in every iteration. And the convergent rate of our NPSO is much faster than PSO and K-means in almost every computation, because the proposed NPSO can effectively avoid the

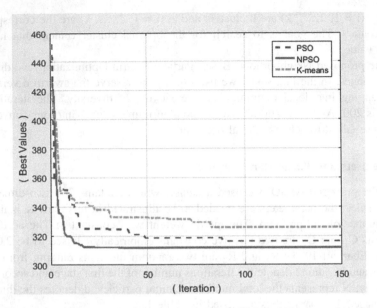

Fig. 4. Comparation of convergence of NPSO, PSO and K-means algorithm

Best Value trapping into local optimum. As we can see the Best Value of the NPSO is smaller than PSO and K-Means in almost every iteration and the Best Value is found fastest.

5 NPSO Algorithm for Traveling Salesman Problem

We have testify the proposed NPSO algorithm in Sect. 4 and the NPSO really out-performs many other algorithm which not only can preserve the diversity of particle swarms avoid trapping premature convergence but also better balance the exploitation and exploration. So in this section, we use the proposed NPSO to solve the classic travel salesman problem (Figs. 5, 6 and 7).

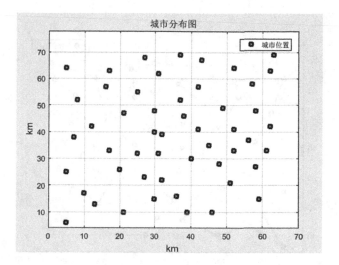

Fig. 5. The distribution of cities

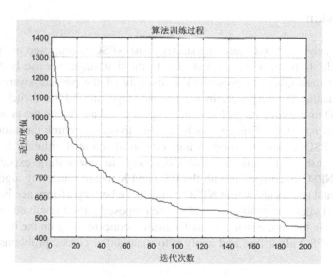

Fig. 6. The convergence of the algorithm

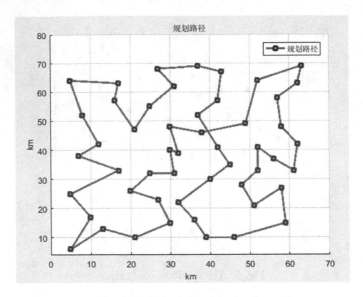

Fig. 7. Result of the path planning

6   Conclusion

Based on the neighborhood searching strategy, the NPSO is proposed in this study. In fact, we divide its process into two stages. In the first stage, we use both Pbest and NPbest rather than Gbest to preserve the diversity of the particle swarm and avoid trapping premature convergence. In the second stage, the NPSO switches into conventional PSO to improve its global search capacity and accelerate the convergent rate towards global optimum. This approach can effectively overcome the shortcoming of conventional PSO which is prone to trap into local minimal solution in the former iterations because of lacking of swarm diversity. In our testing experiments, our approach of NPSO is compared with the PSO and K-means clustering algorithms, and the experimental results show that NPSO can achieve better performance in most cases. Last, the travel salesman problem based on the proposed NPSO is solved and get better optimal value. In the future work, we will continue concentrating on the issue of how to enhance the NPSO algorithm and apply it to many other classic problems.

References

1. Păun, Gh.: Computing with membranes. J. Comput. Syst. Sci. **61**(1), 108–143 (2000)
2. Martin-Vide, C., Pazos, J., Păun, Gh.: A Rodriguez-Paton. Tissue P systems. Theoret. Comput. Sci. **296**(2), 295–326 (2003)
3. Ionescu, M., Păun, Gh., Yokomori, T.: Spiking neural P systems. Fundamenta Informaticae **71**(2–3), 279–308 (2006)

4. Păun, Gh., Rozenberg, G., Salomaa, A.: Membrane Computing. Oxford University Press, New York (2010)
5. Freund, R., Păun, Gh., Pérez-Jiménez, M.J.: Tissue-like P systems with channel-states. Theoret. Comput. Sci. **330**(1), 101–116 (2005)
6. Păun, A., Păun, Gh.: The power of communication: P systems with symport/antiport. New Gen. Comput. **20**, 295–305 (2002)

Breaking Though the Limitation of Test Components Using in Authentication Test

Meng-meng Yao[✉] and Hai-ping Xia

Jiangnan Computing Technology Research Institute, Wuxi 214083, China
wellstudy@163.com

Abstract. In order to break through the limitation that test components in authentication test cannot be encrypted, researchers have conducted plenty of extension study into strand space and made some achievements. Firstly, we analysis the new definitions and improved theorems raised by those researchers and point out their restriction and inaccuracy by way of strand theory and examples in this paper. Secondly, we propose in this paper a new definition named minimum encryption term, effectively limiting the number of forms in which components appear in strand space, lessening the redundancy of authentication test and simplifying the analysis process of nested term encryption. And, based on minimum encryption term, we provide improved authentication test theorems: NE outgoing test, NE incoming test and NE unsolicited test, which help to testify symmetric protocol and discover its flaws, that is, the protocol is an easy target of Man-In-The-Middle attack. These improved theorems increase the accuracy of authentication tests and extend its scope of use to both symmetric and asymmetric cryptosystem.

Keywords: Strand space · Authentication test · Protocol

1 Introduction

Security protocol analysis and verification are extremely important to ensure the operation of the security protocol correctly. The analysis of security protocols is the key to ensuring protocol security. At present, formal analysis method is considered an effective method of guaranteeing security protocols.

Because of the simplicity and intuitive of authentication test, researchers have been studying this field in recent years. Reference [1] develops authentication tests based on Distributed Temporal Protocol Logic and propose a generic strategy to find out Man-In-The-Middle attack on authentication protocols. Guttman extends the strand pace model and makes it possible to analysis fair exchange protocol [2]. Reference [3] builds a model-theoretic viewpoint on security goals model-theoretic viewpoint on security goals based on strand space. Reference [4] proposes four new authentication test for encryption, signature, identity-making and hash. Consistence of strand parameters, for protocol principals in authentication test theory, is studied in reference [5]. Reference [6] proposes a model and algorithm of automatic verification of security protocols based on authentication test.

© Springer Nature Switzerland AG 2019
S. Li (Ed.): GPC 2018, LNCS 11204, pp. 516–524, 2019.
https://doi.org/10.1007/978-3-030-15093-8_40

But, the researchers above do not cover that the authentication test does not apply to the test component nested encryption [7]. The strong authentication tests theorems given in reference [8] can be applied to analyze the nested encryption of authentication tests and reduce redundancy of result of authentication test. But strong authentication tests theorems can not be used in symmetric cryptography, and four complicated definitions are proposed such that the difficulty of protocol analysis is increased. And the strong unsolicited test in reference [8] is incorrect.

Based on above discussion, we embody these ideas in improved authentication tests theorems (NE incoming test, NE outgoing test and NE unsolicited test). These theorems allow us to analyze the nested encryption of authentication tests in symmetric cryptography and asymmetric encryption. Only one simple definition is proposed, and we do not increase the complexity of proof. The results show that NE authentication tests theorems are intuitive and effective.

In Sect. 2, we introduce the basic concepts in this paper. we introduce the limitation of that nested encryption components are not suitable for authentication test in Sect. 3. We point out the weakness of the theorem in [8] in Sect. 4. New concepts and improved theorems by which we prove the Woo-Lam protocol are proposed in Sect. 5.

2 Basic Notions

2.1 Symbols

The meaning of the symbols used in this paper are as follows:

Italic characters indicate variables. A, B, S, Z represent the principal part; Kas represents the shared key between A and S; Ka represents the key of A; N_a represents random number generated by principal A; $Ksym$ represents the set of symmetric key set; $\{M\}_K$ indicates that the message M is encrypted with the key K; P indicates all the set of messages obtained by penetrators (e.g., keys, messages, etc.).

The inverse of the key is inv: $K \rightarrow K$, denoted $\text{inv}(K) = K^{-1}$. If $K \in K_{sym}$, then $k = K^{-1}$ [4]. It is pointed out that K^{-1} does not specifically refer to private keys. If K is the encryption key, K^{-1} is the decryption key; similarly K is the decryption key, and K^{-1} is the encryption key.

2.2 Notions

The basic concepts and propositions used in this paper are as follows [7, 8], and the proof of the propositions is in reference [7].

Definition 1. (1) The atomic term t is its own subterm, $t \subset t$; (2) If and only if $t' \subset t$, then $t' \subset \{t\}_K$; (3) If and only if $t \subset t_1$ or $t \subset t_2$, then $t \subset t_1 t_2$. If $t' \subset t$ but $t' \neq t$, then t' is a proper subterm of t.

Definition 2. If t' is not a concatenated term (like $t_1 \| t_2$) with $t' \subset t$; $t' \neq t$; and $t' \subset t''$ $\subset t$, then t'' is concatenated term and t' is called a component of t.

Definition 3. Let $n = <s, i>$ and $t \subset term(n)$. If $t \not\subset term(n')$ for every $j < i$, then t is called a new component of n.

Definition 4. Let $a \subset t$; $t = \{h\}_K$; the regular node $n \in \sum$; and $t \subset term(n)$. If there is any regular node $n' \in \sum$ with $t \subset t' \subset term(n')$, then $t' = t$. t is not a proper subterm of the components of other regular nodes. t is called a test component for a in n and a is called test element.

Definition 5. In edge $n \Rightarrow^+ n'$, node n is positive and node n' is negative. Let $a \subset term$ (n); $a \subset t'$; $t' \subset term(n')$; and t' is a new component of n'. Then edge $n \Rightarrow^+ n$ is called transformed edge for a.

Definition 6. In edge $n \Rightarrow^+ n'$, node n is negative and node n' is positive. Let $a \subset term$ (n); $a \subset t'$; $t' \subset term(n')$; and t' is a new component of n'. Then edge $n \Rightarrow^+ n$ is called transforming edge for a.

Definition 7. If regular positive node $n = \langle s, i \rangle$; $t \subset term(n)$; and $t \nsubseteq term(\langle s, j \rangle)$ for every $j < i$, t is originating at the node n. If t only originates at n in strand space $\sum$, then t is uniquely originating at n.

Definition 8. Let $n \Rightarrow^+ n'$ is transformed edge for a and a is uniquely originating at n, then $n \Rightarrow^+ n'$ is a test for a.

Definition 9. Let $n \Rightarrow^+ n'$ is a test for a and let $t = \{h\}_K$ is a test component for a in node n in which $K^{-1} \notin P$, then $n \Rightarrow^+ n'$ is an outgoing test for a in t.

Definition 10. Let $n \Rightarrow^+ n'$ is a test for a and let $t = \{h\}_K$ is a test component for a in node n in which $K \notin P$, then $n \Rightarrow^+ n'$ is an incoming test for a in t.

Definition 11. Let $t = \{h\}_K$ is a test component for a in node n in which $K \notin P$, then $n \Rightarrow^+ n'$ is an unsolicited test for a in t.

Proposition 1 (Outgoing test theorem). Let C be a bundle with n, $n' \in C$ and let $n \Rightarrow^+ n'$ is an outgoing test for a in $t = \{h\}_K$. Then:

(1) There must be regular nodes m, $m' \in C$ such that $t \subset term(m)$ and edge $m \Rightarrow^+ m'$ is the transforming edge for a.
(2) Suppose that a occurs only in the component $t0 = \{h0\}_{K0}$ of m', that $t0$ is not a proper subterm of any other regular component, and that $K0^{-1} \notin P$. Then there is a regular negative node m'' and $t0$ is the component at m''.

Proposition 2 (Incoming test theorem). Let C be a bundle with n, $n' \in C$ and let $n \Rightarrow^+ n'$ is an incoming test for a in $t = \{h\}_K$. Then there must be regular nodes m, m' $\in C$ such that $t \subset term(m')$ and edge $m \Rightarrow^+ m'$ is the transforming edge for a.

Proposition 3 (Unsolicited test theorem). Let C be a bundle with $n \in C$ and let $n \Rightarrow^+ n'$ is an unsolicited test for a in $t = \{h\}_K$. Then there is a regular positive node m with $t \subset term(m)$.

3 Limitations of Nested Encryption Test Component

By Definition 4, test component are not a proper subterm of other regular nodes. So if test component exists on other regular nodes, its form can not be re-encrypted. Nested encryption test components bringing about uncertainty and redundancy are not suitable for authentication test.

3.1 Uncertainty

On incoming test in Fig. 1, a attacker can make the transition from $\{\{N_b\}K^{-1}\}K1^{-1}$ to $\{N_b\}K^{-1}$ with $K^{-1} \notin P$. Similarly, a attacker can make the transition from $\{\{N_b\}_K\}_{K1}$ to $\{N_b\}_K$ with $K \notin P$ *and* $K1 \notin P$ on outgoing test. So, we can know that authentication tests do not apply to that the test component is nested encryption. If replacing $\{N_b\}K^{-1}$ with $\{AN_b\}K^{-1}$, attacker can do nothing. Therefore, authentication tests apply to that the test component is nested encryption by limiting the form of test component.

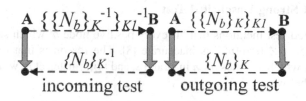

incoming test   outgoing test

Fig. 1. Nested encryption

3.2 Redundancy

More complicated protocol may lead to more serious redundancy. As shown in Fig. 2, choosing $\{N\}_{K1}$ as the test component do not satisfy use conditions of incoming test. This is due to $\{N\}_{K1} \subset term(<S, 2>)$. If choosing $\{\{N\}_{K1}\}_{K2}$ or $\{\{\{N\}_{K1}\}_{K2}\}_{K3}$ as the test component, Fig. 2 satisfies use conditions of incoming test and no matter which test component to choose gets the identical results. In fact, choosing different test components may lead to the different situations that whether or not to satisfy use conditions of incoming test or outgoing test. Redundancy will be reduced by limiting the form and choice of test component.

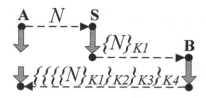

Fig. 2. Redundancy of incoming test

4 Limitations of Strong Authentication Test Theorems

Strong authentication test theorems in reference [8] still have limitations.

4.1 Scope of the Definitions Use

Reference [8] proposes security encryption item, security package, class elements and equivalence class to limit the form of test component in strand space. These definitions extend the knowledge set of the authentication tests theorem, it can solve the problem of redundancy caused by the different test components, while it increases the complexity of protocol analysis.

The definitions proposed in reference [8] requires $K_{i-1} \in P$, which limits the scope of application. In the PKI system, the public key is publicly released (i.e., public key $\in P$); but the private key is uncompromised (i.e., private key $\notin P$). In symmetric encrypted security protocols, $K, K^{-1} \notin P$.

Obviously, the concept of equivalence class and related theorems does not apply to the cryptographic protocol of symmetric key system, such as Woo-Lam protocol.

4.2 Incorrect Strong Unsolicited Test

Strong unsolicited test mentions that "If every positive node n' which satisfies $n' <_c n$ in the strand s, then $t \not\subset term(n')$" is inaccurate [8]. The reason is that there must be a positive node m which satisfies $n' <_c n$ in strand s, and t originates at m with $t \subset term(m)$. Proof is given below.

Proof. Since $a \subset t \subset t'$, let $t = t'$. Let the nodes which include t as components in bundle C construct a set denoted as Ψ, and every node n'' in Ψ satisfies $t \subset term(n'')$. Since $n \in \Psi$, Ψ is not empty such that Ψ contains the minimal element denoted as m, and the symbol is positive with $t \subset term(m)$. Since the node m generates a new component, m can not be in the penetrator strand: M-strands, K-strands, C-strands, S-strands, T-strands, and F-strands. Since $K \notin P$, m can not be in the E-strand (Definition 18 in the reference [4]). Suppose that m is in the D-strands, since m is positive, then m is the third positive node of the D-strands and m' is the second negative node of the D-strands. Let $\{h'\}_{K'} \subset term(m')$ and since $m' \Rightarrow m$; $\{h\}_K \subset h'$; $h' \subset \{h'\}_{K'}$, then $t \subset \{h'\}_{K'}$ (i.e., $t \subset term(m)$). By above discussion, we can get $m' \in \Psi$ and $m' <_c m$ which contradicts that m is the minimal element of Ψ. This implies that m is not in the D-strand. Therefore, m can only be a regular node and $m <_c n$ and $t \subset term (m)$ conflict with strong unsolicited test theorem.

5 Improvement

5.1 Improved Definition

Definition 12. Let a is a atomic term, $\{\{\{\ldots a \ldots\}_{K0} \ldots\}_{K1} \ldots\}_{Kn}$ is called nested encryption term for a and $\{\ldots a \ldots\}_{K0}$ is called the minimum encryption term for a, Where $n \geq 0$.

Proposition 4. Let $a, t \in \sum$ and a is a atomic term. Supposing minimum encryption term $\{\ldots a \ldots\}_K \not\subset t$, then the nested encryption term $\{\{\{\ldots a \ldots\}_{K0} \ldots\}_{K1} \ldots\}_{Kn}$ $\not\subset t$, where $n \geq 0$.

Proof. If $\{\ldots a \ldots\}_{K0} \not\subset t$. We assume that $\{\{\{\ldots a \ldots\}_{K0} \ldots\}_{K1} \ldots\}_{Kn} \subset t$. We can get $\{\ldots a \ldots\}_{K0} \subset t$ because of $\{\ldots a \ldots\}_{K0} \subset \{\{\{\ldots a \ldots\}_{K0} \ldots\}_{K1} \ldots\}_{Kn}$. There is a contradiction with $\{\ldots a \ldots\}_{K0} \not\subset t$. So Proposition 4 is completely true.

Security protocol in the design process as far as possible to follow the principle of simple and efficient security [9]. Too many layers of nested encryption generally does not appear, so ideas in reference [8] make the design of protocol complicated. By 3.2 and Proposition 4, we can know that choosing every term as test component in nested encryption term can get the same result. So, we follow a simple principle of choosing the minimum encryption term as test component for reducing complexity and redundancy. The definition in this section is simple and effective, and will be used in improved theorems.

5.2 Improved Theorem

NE incoming test: Let C be a bundle with edge $n \Rightarrow^+ n' \in C$. n is a positive node and n' is a negative node. The atomic term a is uniquely originating at node n. $t = \{h\}_K$ is the minimum encryption term for a and is the test component of negative node n' in which $K \notin P$. All positive nodes denoted as n'' in strand S of n satisfy $n'' \varsubsetneq n'$ with $t \not\subset term(n'')$, then there must be regular nodes $m, m' \in C$ such that $t \subset term(m')$ and edge $m \Rightarrow^+ m'$ is the transforming edge for a.

NE outgoing test: Let C be a bundle with edge $n \Rightarrow^+ n' \in C$. n is a positive node and n' is a negative node. The atomic term a is uniquely originating at node n. $t = \{h\}_K$ is the minimum encryption term for a with $a \subset term(n')$ and $t \not\subset term(n')$ in which $K^{-1} \notin P$. All positive nodes denoted as n'' in strand S of n satisfy $\varsubsetneq n'$, and every subterm $t0 \subset term(n'')$. If $a \subset t0$, then $t \subset t0$, then:

(1) There are regular nodes $m, m' \in C$ such that $t \subset term(m)$ and edge $m \Rightarrow^+ m'$ is the transforming edge for a.
(2) Suppose that a occurs only in the component $t0 = \{h0\}_{K0}$ of m', that $t0$ is not a proper subterm of any other regular component, and that $K0^{-1} \notin P$. Then there is a regular negative node m'' and $t0$ is the component of m''.

NE unsolicited test: Let C be a bundle with negative node $n \in C$. a is a atomic term. $a \subset t = \{h\}_K \subset term(n)$ and $K \notin P$. Then there must be a regular positive node

m in C with $t \subset term(m)$ and t is originating at node m. If t is not uniquely originating at m, there is regular positive nodes m' with $t \subset term(m')$ and $m' \in C'$.

The proof of the relevant NE theorems is similar to strong authentication tests theorems in this section [7, 8], so this paper does not expound the process of proof.

5.3   Using NE Theorems

Woo-Lam protocol is an one-way authentication protocol designed by Simon S. Lam and Thomas Woo. Responder authenticates the identity of the initiator in Woo-Lam with key distribution center. However, the protocols suffer from various security flaws, so researchers are constantly improving the protocol. We will prove improved Woo-Lam protocol.

We fix a string space $\sum$ of improved Woo-Lam protocol in Fig. 3. From the form, the protocol of Fig. 3 involves four types of regular strands:

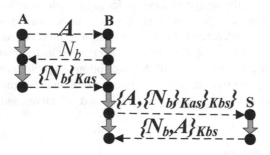

Fig. 3. Bundle of improved Woo-Lam protocol

(1) Initiator strands: $init[A, B, S, N_b]$, trace of message $< +A, -N_b, + \{N_b\}_{Kas} >$;
(2) Responder strands: $resp[A, B, S, N_b, H = \{A, \{\{N_b\}_{Kas}\}_{Kbs}]$, trace of message $< -A, +N_b, -\{N_b\}_{Kas}, +H, -\{N_b, A\}_{Kbs} >$;
(3) Server strands: $resp[A, B, S, N_b]$, trace of message $< -\{A, \{\{N_b\}_{Kas}\}_{Kbs}, + \{N_b, A\}_{Kbs} >$;
(4) Penetrator strands P.

H in strands indicates that the principal B can not know encrypted content.

Supposing that C is a bundle of $\sum$ space, if strand $Sr \in resp[A, B, S, N_b]$ is in bundle C with C-height 5. N_b is uniquely originating in $\sum$, and K_{as}, $K_{bs} \notin P$. Then there is a server strand $S_s \in serv[A, B, S, N_b]$ with C-height 2 and there is a response strand $S_i \in init[A, B, S, N_b]$ with C-height 2 at least.

Proof. N_b is uniquely originating at node $<S_r, 2>$. Node $<S_r, 5>$ receives the message term $\{N_b, A\}_{Kbs}$, $K_{bs} \notin P$. It follows that edge $<S_r, 2> \Rightarrow^+ <S_r, 5>$ is a incoming test for N_b and $\{N_b, A\}_{Kbs}$ is the test component. By the NE incoming test, there must be regular nodes m, m' with $m \Rightarrow^+ m'$ which is a transforming edge for N_b to $\{N_b, A\}_{Kbs}$, and N_b is the component of m'. It can be seen from Fig. 3 that the negative node $<S_r, 5>$ receives the term $\{N_b, A\}_{Kbs}$, and only the positive node $<Ss, 2>$ of the server strand sends $\{A, N_b\}_{Kbs}$, Therefore, there must be server strand $S_s \in serv[A, B, S, N_b]$ in bundle C with C-height 2.

The negative node $<S_s, 2>$ in Fig. 3 with $N_b \subset \{N_b\}_{Kas} \subset \{\{N_b\}_{Kas}\}_{Kbs}$ form an unsolicited test for $\{N_b\}_{Kas}$. By NE unsolicited test, there is a regular positive node m with $\{N_b\}_{Kas} \subset term(m)$ in C and $\{N_b\}_{Kas}$ is originating at node m. In Fig. 3, N_b is not uniquely originating at m, so $\{N_b\}_{Kas}$ is not uniquely originating at m. $\{N_b\}_{Kas}$ may be a component of the node in other bundle, as shown in the following Fig. 4.

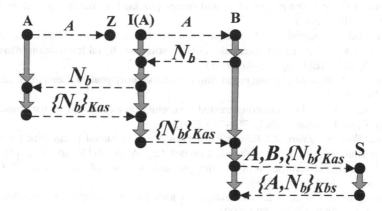

Fig. 4. Attack message of Woo-Lam

When A communicates with Z, the penetrator intercepts message from A to Z and impersonates A to send it to B. After the penetrator received message N_b form B, the penetrator impersonates Z to send N_b to A. A sends $\{N_b\}_{Kas}$ to Z after A received message N_b. The penetrator intercepts message $\{N_b\}_{Kas}$ from A to Z and impersonates A to send $\{N_b\}_{Kas}$ to B. At this point, normal communication with A and Z is blocked, and the penetrator successfully made B certified its identity. The penetrator successes. So, the protocol is attacked by a man-in-the-middle. By NE unsolicited test, we find out the flaw of protocol.

Incoming test and outgoing test are based on challenge and responsive mode, yet the unsolicited test is based on the responsive mode [6]. So the unsolicited test needs to identify the received message. In fact, in the design of the security protocol, it is still not enough to emphasize that n is uniquely originating at m, and the negative node n can recognize the uniquely originating of n in the same strand space, such as the identity of both parts.

6 Conclusion

In this paper, we propose NE theorems by which we find out a flaw of improved Woo-Lam. The NE theorems are simple and effective, and break through limitations of nested encryption test component using in authentication tests.

Based on the strand space model, improving or developing automatic analysis protocol tools used for the analysis of the security protocol has been a hot research field (Reference [10]). We will develop automation tools based on the improved theorems in this paper.

References

1. Guttman, J.D.: State and progress in strand spaces: proving fair exchange. J. Autom. Reason. **48**(2), 159–195 (2012)
2. Muhammad, S.: Applying authentication tests to discover Man-In-The-Middle attack in security protocols. In: Eighth International Conference on Digital Information Management (ICDIM), 2013. IEEE, pp. 35–40 (2013)
3. Guttman, J.D.: Establishing and preserving protocol security goals. J. Comput. Secur. **22**(2), 203–267 (2014)
4. Feng, W., Feng, D.-G.: Analyzing trusted computing protocol based on the strand spaces model. Chin. J. Comput. **38**(4), 701–716 (2015)
5. Yu, L., Wei, S., Zhuo, Z.: Research on consistence of strand parameters for protocol principals in authentication test theory. Comput. Eng. Appl. **51**(13), 86–91 (2015)
6. Liu, J.: Automatic verification of security protocols with strand space theory. J. Comput. Appl. **35**(7), 1870–1876 (2015)
7. Guttman, J.D., Thayer, F.J.: Authentication tests and the structure of bundles. Theor. Comput. Sci. **283**(2), 333–380 (2002)
8. Song, W.-T., Hu, B.: One strong authentication test suitable for analysis of nested encryption protocols. Comput. Sci. **42**(1), 149–169 (2015)
9. Yuan, B.-A., Liu, J., Zhou, H.-G.: Study and development on cryptographic protocol. J. Mil. Commun. Technol. **38**(1), 90–94 (2017)
10. Zhang, H.-G., Wu, F.-S., Wang, H.-Z., Wang, Z.-Y.: A survey: security verification analysis of cryptographic protocols implementations on real code. Chin. J. Comput. **38**(1), 90–94 (2017)

Author Index